Interdisciplinary Biotechnological Advances

This series is an authoritative resource that explains recent advances and emerging directions in biotechnology, reflecting the forefront of research clearly and reliably, without excessive hype. Each volume is written by authors with excellent reputations and acknowledged expertise in the topic under discussion. The volumes span the entire field from an interdisciplinary perspective, covering everything from biotechnology principles and methods to applications in areas including genetic engineering, transgenic plants and animals, environmental problems, genomics, proteomics, diagnosis of disease, gene therapy, and biomedicine. The significance of these applications for the achievement of UN Sustainable Development Goals is highlighted. The series will be highly relevant for Master's and PhD students in Biotechnology, Nanochemistry, Biochemical Engineering, and Microbiology, medical students, academic and industrial researchers, agricultural scientists, farmers, clinicians, industry personnel, and entrepreneurs.

Komal Agrawal • Pradeep Verma
Editors

Biotechnological Advances in Biorefinery

Springer

Editors
Komal Agrawal
Department of Microbiology, School of Bio Engineering and Biosciences
Lovely Professional University
Phagwara, Punjab, India

Pradeep Verma
Bioprocess and Bioenergy Laboratory (BPEL) Department of Microbiology
Central University of Rajasthan
Kishangarh, Ajmer, Rajasthan, India

ISSN 2730-7069 ISSN 2730-7077 (electronic)
Interdisciplinary Biotechnological Advances
ISBN 978-981-97-5543-1 ISBN 978-981-97-5544-8 (eBook)
https://doi.org/10.1007/978-981-97-5544-8

This Springer imprint is published by the registered company Springer Nature Singapore Pte Ltd.
The registered company address is: 152 Beach Road, #21-01/04 Gateway East, Singapore 189721, Singapore

Preface

With the depletion of conventional and nonrenewable resources, there has been a global shift toward utilizing renewable resources for bioenergy production and value-added commodities. However, implementation of this process in real time is still a major challenge, which hinders its effective process conversion and commercialization. With advancements in technology and science, global research has witnessed a paradigmatic shift. This book aims to provide insight into the current global scenarios in various dimensions of biorefinery. The chapters in this book explore the significance of the transformation of renewable resources into biorefinery products through sustainable approaches. Further, it also discusses the challenges, environmental-economic feasibility, recent advances, and various technological advancements that can facilitate its scale-up and implementation.

Lignocellulosic biomass (LCB) is widely acknowledged as a renewable carbon sink that holds significant promise for the sustainable production of biofuels and value-added products. To fully recognize its potential within biorefineries, various biological systems are currently being explored. Thus, the discussion is focused on the remarkable potential of various microorganisms, such as bacteria and fungi, to produce extracellular enzymes that can facilitate the breakdown of lignocellulosic biomass into useful products for biorefinery. The enzymatic breakdown of lignocellulosic biomass is a crucial step in the production of biofuels, and the ability of microorganisms to produce such enzymes could lead to more efficient and sustainable biorefinery processes.

The global emergence of zero-waste algal biorefinery concepts has significantly amplified and enhanced the current research model. This, in turn, has led to an in-depth recognition of the diverse potential of algae in the production of biofuels and value-added products. To provide a comprehensive insight into the intricacies of algae and its potential, the importance of natural resources from the perspective of sustainable habits and academic research is highlighted.

To widen the scope of current insight beyond the recognized system, it is imperative to summarize the significant advancements in both biotechnology and nanotechnology. These innovations have constructed transformative opportunities across various industries. Furthermore, the extensive implementation of biorefinery systems requires the development and design of suitable bioreactor systems. This has been categorized and summarized in terms of the various types of bioreactor systems, as well as their respective designs and development processes.

A global vision of a sustainable future has increased the world's demand for sustainable materials, thus making it clear that biorefinery industry will play a pivotal role in the global bioeconomy. Thus, to meet the same various approaches are being designed such as *in silico* studies which are paving the way for future research. Further, tools like techno-economic evaluation (TEA), life cycle assessment (LCA), and value-sensitive design (VSD) also need to be investigated. The TEA aids in cost evaluation, while LCA measures environmental impact. Further, VSD integrates the ethical, social, and technical aspects into technology system design, thereby presenting a promising strategy to address intricate ethical and societal issues related to various aspects of biorefinery. It also involves stakeholders (local communities, environmental advocates, and industry experts), thereby enabling a comprehensive understanding of diverse aspects that are crucial to the design process in biorefinery, which aids the decision-makers to navigate the ethical and societal dimensions of their projects, striving for more inclusive, responsible, and sustainable outcomes.

Phagwara, Punjab, India — Komal Agrawal
Kishangarh, Rajasthan, India — Pradeep Verma

Acknowledgments

We express our sincere gratitude to Springer Nature for accepting our proposal to act as the editor of this book. We are deeply indebted to the researchers and academicians who have contributed to this book. Their prompt responses to our invitations and their valuable assistance during the revision stage have been invaluable, and we are grateful for their contributions. We would also like to extend our appreciation to the Central University of Rajasthan (CURAJ), Ajmer, and Lovely Professional University, India, for providing the necessary infrastructure and a conducive research environment. Our heartfelt thanks are extended to the authors whose contributions have played a pivotal role in shaping the content of this book and making it a success. We also acknowledge the efforts of the reviewers whose constructive feedback has improved the quality of this book. Finally, we dedicate this work to our families, thanking them for their support.

Contents

Editors and Contributors

About the Editors

Komal Agrawal completed her BSc in Microbiology at University of North Bengal, India, and her MSc in Microbiology from Sikkim University (Central University in India). She later joined as Research Scholar at Department of Microbiology, Central University of Rajasthan, India, and was awarded her PhD in 2021. The doctoral journey was about addressing the critical gaps of laccases from non-model fungal cultures. After which she joined as Assistant Professor in the Department of Microbiology, Lovely Professional University, Punjab India.

The scientific contribution and hard work have resulted in acknowledgment of her work and she was awarded with Best Paper Award at the first International Conference on Biotechnology and Biological Sciences—BIOSPECTRUM 2017. Later, in 2023, she was awarded with the Aditi Pant Women Scientist Award at International Conference on Microbial Bioprospecting Towards Sustainable Development Goals (ICBMSDG) 2023. In the same year, her work was acknowledged and she was included in the prestigious list of Top 2% Scientist as per Stanford University 2023.

Since the start of her scientific career, she has contributed to 80+ publications consisting of research articles, review papers, and book chapters. Since 2018 she has an h-index of 20+, over 1500 citations, and an i10 index of 40+.

Pradeep Verma completed his PhD at Sardar Patel University in Gujarat, India, in 2002. In the same year, he joined the Czech Academy of Sciences in Prague, Czech Republic, as UNESCO fellow. He later moved to Charles University, Prague, to work as a Post-Doctoral Fellow. In 2004, he joined as a visiting scientist at UFZ Centre for Environmental Research, Halle, Germany. He was awarded a DFG fellowship which provided him another opportunity to work as a Post-Doctoral Fellow at Gottingen University, Germany. He moved to India in 2007 and joined Reliance Life Sciences, Mumbai, and worked on biobutanol production which attributed to a few patents. Later, he was awarded the JSPS Post-Doctoral Fellowship Programme and joined the Laboratory of Biomass Conversion, Research Institute for Sustainable Humanosphere (RISH), Kyoto University, Japan. He has been awarded the prestigious Fellow Award from the Mycological Society of India (2020), P.C. Jain Memorial Award (MSI), and Biotech Research Society of India (2021) for his contributions. He was also awarded Fellow of the Biotechnology Research Society of India (BRSI) (2021) and Fellow of the Academy of Sciences of AMI India (FAMSc) (2021). He is also a recipient of various prestigious awards such as the Ron Cockcroft Award by the Swedish Society, and UNESCO Fellow ASCR Prague. Furthermore, he has also been awarded the JSPS Bridge Fellow Award (2022). He was included in the Top 2% of Scientists as per Stanford University (2023) and named Top Scholar by Scholar GPS a California-based company in Bioprocess Engineering, Lifetime #67, and prior 5 years #15, and in Microbiology, #249.

He began his independent academic career in 2009 as a Reader and Founder Head at the Department of Microbiology at Assam University. In 2011, he moved to the Department of Biotechnology at Guru Ghasidas Vishwavidyalaya (A Central University), Bilaspur, and served as an Associate Professor till 2013. He works as a Professor at the Department of Microbiology, CURAJ. He has more than 95 research/review articles in peer-reviewed international journals and contributed to 62+ book chapters in different edited books. He has an h-index of 45, over 7318 citations, and an i10 index of 108. He has also edited 24+ books in international publication houses and is also a Guest Editor for several journals of high repute. Prof. Verma has presented his work at numerous international conferences, where his findings led to the wood waste biorefinery Circular Economy Concept and Zero Waste Approach, which have received widespread acclaim. Notably, he completed major research grants on Biomass worth 150 Million INR.

Contributors

Mansi Achhoda Bioprocess Laboratory, Department of Biotechnology, Thapar Institute of Engineering and Technology, Patiala, Punjab, India

Arindham Adhikary Department of Medical Microbiology, University School of Allied Health Sciences, Rayat Bahra University, Mohali, Punjab, India

Komal Agrawal Department of Microbiology, School of Bio Engineering and Biosciences, Lovely Professional University, Phagwara, Punjab, India

Anamika Department of Biotechnology, Dr. B. R. Ambedkar National Institute of Technology, Jalandhar, Punjab, India

M. Reshma Anjum Sri Padmavathi Mahila Visvavidyalayam (Women's University), Andhra Pradesh, Tirupati, India

Anindita Banerjee UG, PG & Research Department of Microbiology, St. Xavier's College (Autonomous), Kolkata, WB, India

Bunushree Behera Bioprocess Laboratory, Department of Biotechnology, Thapar Institute of Engineering and Technology, Patiala, Punjab, India

Esmil Beliya School of Studies in Biotechnology, Pt. Ravishankar Shukla University, Raipur, Chhattisgarh, India

Department of Botany, Government College Bichhua, Chhindwara, Madhya Pradesh, India

Pavan Kumar Bijalwan Department of Physics, R.C.U. Gov. P. G. College, Uttarkashi, Uttarakhand, India

R. Bobita Biofuel Research Laboratory, Department of Microbiology, School of Life Sciences, Central University of Tamil Nadu, Thiruvarur, Tamil Nadu, India

Chandrama Chanda Department of Life Science & Bioinformatics, Har Gobind Khurana School of Life Sciences Assam University, Silchar, Assam, India

Arju Chauhan Department of Biotechnology, Thapar Institute of Engineering and Technology, Patiala, Punjab, India

Vivek Chintada Department of Zoology, Sri Venkateswara University, Tirupati, Andhra Pradesh, India

Sayak Das Department of Life Science & Bioinformatics, Har Gobind Khurana School of Life Sciences Assam University, Silchar, Assam, India

Swagat Kumar Das Department of Biotechnology, Odisha University of Technology and Research, Bhubaneswar, Odisha, India

Ashwini Kumar Dash Department of Textile Engineering, Odisha University of Technology and Research, Bhubaneswar, Odisha, India

Deepshikha Department of Microbiology, School of Bio Engineering and Biosciences, Lovely Professional University, Phagwara, Punjab, India

G. P. Dhanvanti Biofuel Research Laboratory, Department of Microbiology, School of Life Sciences, Central University of Tamil Nadu, Thiruvarur, Tamil Nadu, India

Narasimha Golla Department of Virology, Sri Venkateswara University, Tirupati, Andhra Pradesh, India

Payal Guleria Department of Biotechnology, Dr. B. R. Ambedkar National Institute of Technology, Jalandhar, Punjab, India

Nisha Gupta School of Studies in Biotechnology, Pt. Ravishankar Shukla University, Raipur, Chhattisgarh, India

Nirmalya Haldar Bioprocess Laboratory, Department of Biotechnology, Thapar Institute of Engineering and Technology, Patiala, Punjab, India

S. K. Jadhav School of Studies in Biotechnology, Pt. Ravishankar Shukla University, Raipur, Chhattisgarh, India

K. R. Jeya Department of Biotechnology, Government Arts and Science College (Women) Veppur, Perambalur, Tamil Nadu, India

Sundeep Kaur Department of Biotechnology, Dr. B. R. Ambedkar National Institute of Technology, Jalandhar, Punjab, India

Santosh Kumar Department of Microbiology, School of Life Sciences, Sikkim University, Gangtok, Sikkim, India

Mrigank Mayank Department of Biotechnology, Dr. B. R. Ambedkar National Institute of Technology, Jalandhar, Punjab, India

Sumer Singh Meena Department of Biotechnology, Dr. B. R. Ambedkar National Institute of Technology, Jalandhar, Punjab, India

Sankari Mohan Sri Padmavathi Mahila Visvavidyalayam (Women's University), Tirupati, Andhra Pradesh, India

Anee Mohanty Department of Biotechnology, Dr. B. R. Ambedkar National Institute of Technology, Jalandhar, Punjab, India

Nirza Moktan Department of Microbiology, Dhruba Chand Halder College, Dakshin Barasat, WB, India

UG, PG & Research Department of Microbiology, St. Xavier's College (Autonomous), Kolkata, WB, India

Amisha Patel P. G. Department of Biosciences, Sardar Patel University, Satellite Campus, Anand, Gujarat, India

Ravi Patel Faculty of Science, Department of Botany, The Maharaja Sayajirao University of Baroda, Vadodara, Gujarat, India

Jai Shankar Paul School of Studies in Biotechnology, Pt. Ravishankar Shukla University, Raipur, India

Devi Sri Rajendran Integrated Bioprocessing Laboratory, Department of Biotechnology, School of Bioengineering, Faculty of Engineering and Technology, SRM Institute of Science and Technology (SRM IST), Chengalpattu, Tamil Nadu, India

S. Rakesh Biofuel Research Laboratory, Department of Microbiology, School of Life Sciences, Central University of Tamil Nadu, Thiruvarur, Tamil Nadu, India

G. Sai Sri Rekha Sri Padmavathi Mahila Visvavidyalayam (Women's University), Tirupati, Andhra Pradesh, India

Prangya Ranjan Rout Department of Biotechnology, Dr. B. R. Ambedkar National Institute of Technology, Jalandhar, Punjab, India

Chirantan Saha Department of Microbiology, School of Life Sciences, Sikkim University, Gangtok, Sikkim, India

A. Sankaranarayanan Department of Life Sciences, Sri Sathya Sai University for Human Excellence Navanihal, Kalaburagi, Karnataka, India

Mansi Sharma Bioprocess Laboratory, Department of Biotechnology, Thapar Institute of Engineering and Technology, Patiala, Punjab, India

Shreya Sharma Department of Microbiology, School of Bio Engineering and Biosciences, Lovely Professional University, Phagwara, Punjab, India

Mingma Thundu Sherpa Department of Microbiology, Nar Bahadur Bhandari Government College, Tadong, Sikkim, India

Subir Kumar Shill Department of Life Science & Bioinformatics, Har Gobind Khurana School of Life Sciences Assam University, Silchar, Assam, India

Mohan Bir Subba Department of Mathematics, K. K. Das College, Kolkata, WB, India

Kolli Venkata Supraja Waste Treatment Laboratory, Department of Biochemical Engineering and Biotechnology, Indian Institute of Technology Delhi, Hauz Khas, New Delhi, India

Nagendra Thakur Department of Microbiology, School of Life Sciences, Sikkim University, Gangtok, Sikkim, India

Nandini Thakur Bioprocess Laboratory, Department of Biotechnology, Thapar Institute of Engineering and Technology, Patiala, Punjab, India

J. Tharunkumar Biofuel Research Laboratory, Department of Microbiology, School of Life Sciences, Central University of Tamil Nadu, Thiruvarur, Tamil Nadu, India

Sunil S. Thorat Institute of Bioresources and Sustainable Development (IBSD)—Meghalaya, Upper Shillong, Meghalaya, India

Shivani Uniyal Department of Botany, H.N.B. Garhwal University, BGR Campus, Pauri, India

Vinoth Kumar Vaidyanathan Integrated Bioprocessing Laboratory, Department of Biotechnology, School of Bioengineering, Faculty of Engineering and Technology, SRM Institute of Science and Technology (SRM IST), Chengalpattu, Tamil Nadu, India

K. Veeraiah Department of Zoology, Acharya Nagarjuna University, Guntur, Andhra Pradesh, India

M. Veerapagu Department of Biotechnology, Microbiology and Bioinformatics, National College (Autonomous), Tiruchirappalli, Tamil Nadu, India

Swethaa Venkatraman Integrated Bioprocessing Laboratory, Department of Biotechnology, School of Bioengineering, Faculty of Engineering and Technology, SRM Institute of Science and Technology (SRM IST), Chengalpattu, Tamil Nadu, India

Pradeep Verma Bioprocess and Bioenergy Laboratory (BPEL), Department of Microbiology, Central University of Rajasthan, Kishangarh, Ajmer, Rajasthan, India

Lignocellulosic Biomass as Replenishable Source for the Derivation of Biofuel and Value-Added Products of Global Significance

Sundeep Kaur, Payal Guleria, Arju Chauhan, Prangya Ranjan Rout, Anee Mohanty, Sumer Singh Meena, and Bunushree Behera

1 Introduction

Burgeoning global fuel and energy demands driven by rapid economic expansion like industrialization and urbanization in recent decades have extensively exploited conventional non-renewable energy reserves. These reserves include coal, crude oil, and natural gas, which constitute about 86% of total global energy requirements (Shahbeik et al. 2022; Shafizadeh et al. 2023). Extensive dependency on depleting fossil reserves has not only driven up fuel costs but has also poised significant environmental risks, including an increase in noxious gaseous emissions, contributing to higher levels of global warming, which in turn has caused profound impacts over the climate, causing disturbance to ecological biodiversity (Chen et al. 2023). To accommodate the demands of the continually growing population, it is crucial to investigate renewable resources to meet the future demands, simultaneously reducing the reliance over conventional fossil fuels. Biomass-based alternatives are natural sustainable resources that appear to be long-lasting reservoirs to guarantee a steady carbon feedstock availability for bioenergy and the day-to-day essential commodities via green synthesis (Singh et al. 2021).

Sundeep Kaur and Payal Guleria contributed equally with all other contributors.

S. Kaur · P. Guleria · P. R. Rout (✉) · A. Mohanty · S. S. Meena
Department of Biotechnology, Dr. B. R. Ambedkar National Institute of Technology, Jalandhar, Punjab, India
e-mail: routpr@nitj.ac.in

A. Chauhan · B. Behera (✉)
Department of Biotechnology, Thapar Institute of Engineering and Technology, Patiala, Punjab, India
e-mail: bunushree.behera@thapar.edu

K. Agrawal, P. Verma (eds.), *Biotechnological Advances in Biorefinery*, Interdisciplinary Biotechnological Advances,
https://doi.org/10.1007/978-981-97-5544-8_1

Biomass falls into three broad categories: starchy food crops, lignin-containing or lignocellulosic biomass (LCB) (Agrawal et al. 2023), and third-generation algal biomass (Goswami et al. 2022). LCB acts as the most promising feedstock that has garnered worldwide attention with high potential in biorefineries owing to its abundant availability, renewability, and comparatively cheap price (Sidana et al. 2023). Although LCB is inevitable, the major sources are agricultural and forest residues, energy crops, wastes generated by municipal communities, organic municipal solid waste, and by-products mostly comprising of waste pulp, wood, or papers (Roy et al. 2021). Despite an annual production of approximately 181.5 billion tons of LCB, constituting about 90% of total biomass, only 3% is effectively utilized (Dahmen et al. 2019). Cellulose, hemicellulose, and lignin contents of LCB primarily vary in the range of 40–50%, 25–35%, and 16–30%, respectively, which forms a highly recalcitrant network. Additionally, it consists of different fractions of ash, moisture, lipids, proteins, minerals, etc. In the integrated biorefinery approach, each of these elements serves as the primary precursor of several commercial bio-based products (Akbarian et al. 2022; Bianchini et al. 2022).

LCB biorefinery requires a pretreatment process for obtaining value-added products. The pretreatment step is essential to loosen the LCB structure that enables its constituents accessible to different hydrolytic enzymes to liberate fermentable sugars. Further, these sugars can be hydrolyzed or fermented to produce biofuels, biochemicals, and biomaterials (Usmani et al. 2021). Significantly, the pretreatment process largely determines the overall success or failure of biorefinery. Several physical and chemical techniques, and their combination thereof, have been in use to generate more than 200 value-added products from LCB (Isikgor and Becer 2015). Sugar-based degradative by-products are often produced during the application of heat, pressure, and exposure to chemicals (acid/alkali) during the conventional pretreatment methods. These by-products often inhibit the yield and performance of subsequent fermentative reactions. To avert shortcomings of physicochemical strategies for lignin degradation, several advanced methods of ultrasonication, hydrodynamic cavitations, electron beam irradiation, pulse electric field, plasma treatment, etc., have recently been explored (Sidana et al. 2022).

This book chapter discusses the conversion techniques of recalcitrant LCB into biofuels, bioenergy, and platform chemicals. Moreover, the need for enviro-economic assessment and recent advancements in LCB-based biorefinery is also elucidated.

2 Recalcitrance of LCB: Need and Assessment of Pretreatment Strategies

LCB from forest residues, agricultural waste, and grasses is extensively exploited for conversion to biofuels and chemicals and is considered as a viable technical scheme. For the effective utilization of LCB, the biomass recalcitrance, an inherent property of plant cell wall that protects the biomass from biological deconstruction, needs to be reduced. Biomass recalcitrance is mostly attributed to the complicated structure of LCB containing an intricate network of cellulosic and hemi-cellulosic fibers with lignin, along with other cell wall components comprising of lipids and proteins (Xu et al. 2019; Banu et al. 2021). Chemical constituents in biomass aid in facilitating protection from enzymatic or microbial degradation, thereby making the bioconversion process difficult and inefficient. Recalcitrance intensity differs depending on the biochemical constituents present in LCB, which is linked with the genotype of the plant and ecological growth conditions along with human activities such as crop and plant part management strategies (Surendra et al. 2018). Zoghlami and Paës (2019) reported that the recalcitrance in LCB is governed by factors concerning the lignin barrier around cellulose and hemicellulose fibers. The degree of crystallinity and polymerization of cellulose also determines its recalcitrance. All these factors can be reduced or altered via pretreatment to increase the accessibility of the enzymes to cellulosic and hemicellulosic portion of biomass (Moradian et al. 2021). A maximum of 20% reducing sugar yield (theoretical) is believed to be achieved from biomass using enzymatic hydrolysis, without any pretreatment. The 80% loss in residual sugars is due to the presence of recalcitrant lignin, which hampers the process of enzyme action (Akhtar et al. 2016). Hence, to achieve efficient enzymatic hydrolysis and optimize sugar recovery from biomass, it is essential to pretreat LCB. This process opens up the complex structure by breaking down the underlying interlinkages (Kumar et al. 2009). Pretreating process consumes a huge amount of energy and is regarded as an expensive step that determines the fate of biomass conversion processes (Kumar et al. 2020). Negligence in this step can result in considerable loss in sugar recovery and surge in the overall process cost (Prakesh et al. 2021). In addition to sugar recovery, a promising pretreatment technique is defined by its mild operational conditions, low catalyst cost, ability to process large amount of biomass, biomass conversion efficiency with least production of inhibitory compounds, and frugal investment (Kaur et al. 2022; Mosier et al. 2005). Pretreatment techniques differ significantly from each other with regard to their function and LCB, for instance, the acidic pretreatment often degrades the hemicellulose constituent biomass, converting it to monosaccharides, whereas alkali pretreatment causes swelling of the biomass and removes the lignin and partly some of the hemicellulose. These treatments also aid in reducing the crystalline nature of cellulose for effective enzymatic hydrolysis, while others help in the solubilization of lignin (Mirmohamadsadeghi et al. 2021). The hydrothermal technique solubilizes the hemicellulosic fraction without complete hydrolysis of LCB (Jönsson and Martín 2016). Thus, different pretreatment methods are employed depending on the feedstock categories and its chemical as well as structural composition (Fig. 1).

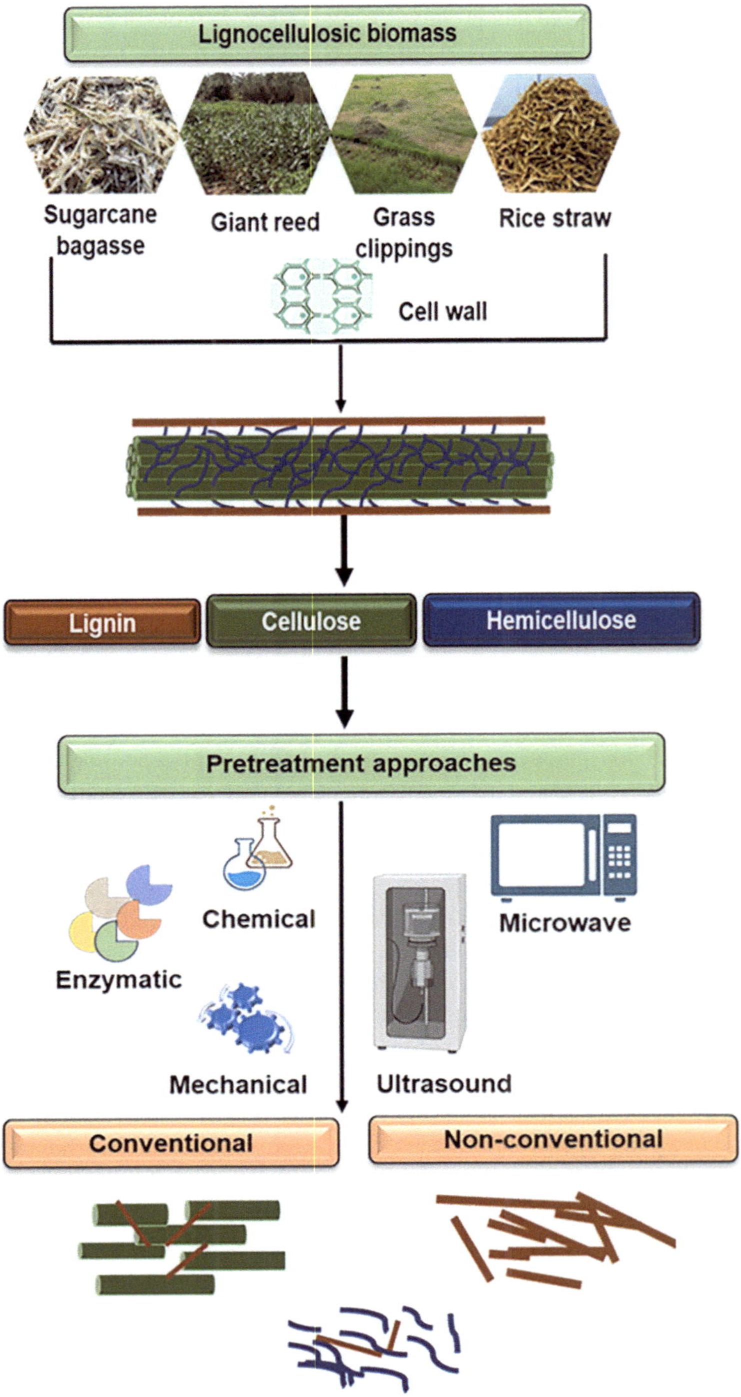

Fig. 1 Structural composition and pretreatment technologies for LCB

3 Conversion Techniques of LCB to Bioenergy

LCB can be transformed into biofuels and bioenergy via two primary processes, i.e., thermochemical and biochemical techniques (Fig. 2). In thermochemical conversion, LCB undergoes treatment with heat and chemicals resulting in different energy/fuel variants like biochar, bio-oil, as well as syngas. The conversion of LCB into bioenergy via thermochemical method takes place by various processes, viz., torrefaction, pyrolysis, gasification, and combustion. In contrast, biochemical conversion relies on enzymes or microorganisms and includes processes, such as enzyme pretreatment, saccharification, fermentation, anaerobic digestion, transesterification, etc. aimed to produce bioethanol, biobutanol, and bio-hydrogen (Devaraja et al. 2022; Manikandan et al. 2023). Different thermochemical and biochemical methods for the production of sustainable bioenergy options are detailed in Table 1. The production pathways for biofuel pose considerable challenges, but the sequential and streamlined technologies like pretreatment, enzymatic saccharification, microbial fermentation, and purification contribute to advancements in process efficiency. These processes are widely adopted for increasing the efficacy of fermentation technique in terms of better yield of products with reduced waste generation (Shahbeik et al. 2022; Elgarahy et al. 2023).

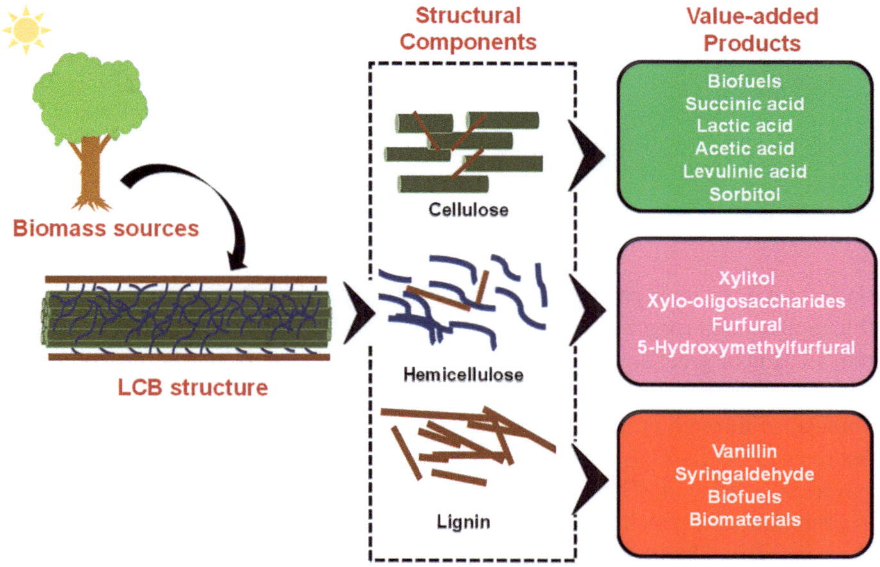

Fig. 2 Conversion of LCB components to biofuels and value-added chemicals

Table 1 Different thermochemical and biochemical conversion strategies for the production of bioenergy from LCB

Conversion process	Biomass	Operational temperature (°C)	Bioenergy products	References
Thermochemical				
Torrefaction	Sugarcane bagasse, cotton stalk, rice straw	250–300	Biochar	Amer et al. (2019)
Pyrolysis	Pigeon pea stalk, bamboo	400–600	Biochar	Sahoo et al. (2021)
	Miscanthus	350–550	Bio-oil	Wang and Lee (2019)
	Wheat straw, sugarcane bagasse, rice husk, coconut shell, cotton stalk, palm kernel shell	550	Syngas	Akubo et al. (2019)
	Switchgrass, *Miscanthus*	500	Biochar Bio-oil Syngas	Zaimes et al. (2015)
Liquefaction	Corn stover	250–375	Hydrochar Bio-oil	Mathanker et al. (2020)
Gasification	Willow wood chips	600–850	Syngas	Hai et al. (2019)
Biochemical				
Anaerobic digestion	Olive mill residues	35–37	Biomethane	Messineo et al. (2020)
Fermentation	Giant reed	26–30	Bioethanol	Sidana et al. (2022)
	Sugarcane bagasse	37	Biobutanol	Pratto et al. (2020)

3.1 Bioethanol

Bioethanol is the most promising green fuel globally and is being explored as a possible replacement for gasoline in transportation sector. The production of bioethanol from LCB is of great significance because of its clean combustion properties, high octane number, low carbon emission, and volatile nature (Sharma et al. 2023). It is synthesized through either a thermochemical or biological process, wherein complex recalcitrant lignin is degraded to form simpler by-products, subsequently being converted to ethanol. The most widely used thermochemical process is Fischer-Tropsch catalysis, which produces syngas, and then converted into ethanol, butanol, green diesel, and hydrocarbon fuels (Limayem and Ricke 2012). Alternatively, syngas fermentation can also be done via hydrolysis involving enzymes, resulting in bioethanol. Enzymes like cellulase and hemicellulase aid in the breakdown of cellulose/hemicellulose following lignin degradation to form monosaccharide units, which are then subjected to microbial fermentation to yield ethanol. Bioethanol, when blended with existing conventional fuel, has the capacity to decline noxious gaseous emissions caused by petroleum-based sources. Although bioethanol production from LCB is in the early stages, several renowned industries, including Abengoa, GranBio, DuPont, Poet-DSM, and Raízen, have made investments to establish commercial plants (Ashokkumar et al. 2022).

3.2 *Biodiesel*

Biodiesel, an emerging renewable biofuel, often a mixture of fatty acid alkyl ester molecules, is obtained from hydrodeoxygenation of edible or nonedible waste oils. Commonly recognized as green diesel, this fuel stands out for its renewable, biodegradable, and sustainable characteristics, which could be used as a standalone fuel or in blended forms (Veza et al. 2022). In contrast to conventional diesel, it has lesser associated environmental impacts. These advantages are linked to its 10–11% oxygen content, minimal nitrogen content, and lack of sulfur (Silitonga et al. 2013; Hazrat et al. 2019). LCB is processed to obtain lipids, and, subsequently, esterification or transesterification methods are used to convert these precursors into biodiesel (Ashokkumar et al. 2022). Pyrolysis and microemulsion-based methods could also be employed for biodiesel production. Due to environmental concerns, several countries are encouraging this alternative fuel by implementing volumetric standards and offering incentives (Singh et al. 2023). However, advanced research should prioritize a comprehensive life cycle assessment, considering factors like energy consumption, economic viability, and environmental friendliness of the biofuel.

3.3 *Jet Fuels*

Jet fuel is a special type of aviation fuel (kerosene) specifically designed for use in commercial and combat aircrafts. It usually consists of several organic chemicals, including paraffins, olefins, aromatics, and naphthenes (Ashokkumar et al. 2022). As the aviation industry is associated with major greenhouse emissions, there is growing interest in transition to sustainable aviation fuel from non-fossil fuel carbon feedstock and LCB (waste organics, agricultural residues) to produce bio-kerosene. Bio-kerosene can be synthesized through thermochemical and biochemical routes by transforming various hydro-treated vegetable oil and biogenic wastes (Eswaran et al. 2021). There are two approaches to produce bio-kerosene: first, the feedstock undergoes a combination of thermal and chemical treatment to generate intermediate products, which are thereby upgraded or transformed into final product. The second approach involves the production of biokerosene via hydrolysis of carbon-based sugar compounds. However, the primary challenge in commercializing biokerosene lies in its high costs that hinders its process scaleup and commercialization (Thilo 2018; Cheng and Brewer 2017). Addressing these challenges may involve the implementation of effective global policies to reduce taxes on biofuel supply. Additionally, developing alternative pathways coupled with the manufacture of bio-based commodities could reduce the commercial cost, resulting in practical utility.

3.4 Biogas

Biogas contains a blend of gases, particularly methane and carbon dioxide, along with trace amounts of water vapor, hydrogen sulfides, and siloxanes produced during the decomposition of organic matter via anaerobic digestion or bio-methanation (Singh et al. 2023). The bio-methanation process is a complex microbial approach that unfolds in anaerobic conditions and is categorized into four stages: (1) hydrolysis, (2) acidogenesis, (3) acetogenesis, and (4) methanogenesis (Bazaluk et al. 2020). Thermophilic conditions are maintained during hydrolysis and acidogenesis, whereas methanogenesis is favorable in mesophilic conditions. Productivity and yield of biogas is dependent over various parameters like organic loading and carbon to nitrogen ratio, along with operational temperature, pH, and hydraulic retention time (Chen et al. 2023). For biogas production, various organic matters such as municipal, agricultural, and food waste have been used for decomposition. The utilization of LCB is limited due to its recalcitrant structure, leading to suboptimal digestion and biodegradation. Strategies involving pretreating the biomass, along with codigestion to optimize the carbon-nitrogen ratio, are currently being focused to improvise the underlying microbe-based metabolism in biogas plants.

4 Conversion Techniques of LCB to High-Value Commodities

Apart from bioenergy alternatives, LCB has the potential for the production of high-value platform chemicals and bio-based materials (Devi et al. 2022). LCB-based cellulosic polymers can be valorized as per the reports of the Department of Energy (DOE). The agency revealed the presence of 12 chemicals having the capacity to act as building blocks for platform chemicals, ten of which can be obtained from the biomass-derived sugars (Werpy and Petersen 2004). The primary chemicals produced from LCB are hexoses and pentoses, which are further convertible to high-value platform chemicals such as xylitol, sorbitol, acid/aldehyde, furfural, lactic, succinic, and acetic acids using the process of fermentation. Among others, 5-hydroxymethyl furfural, 2,5-furandicarboxylic acid, and 3-hydroxypropionic are also produced commonly. Lignin component of LCB is not involved in fermentation step but can be processed into high-value chemicals under biorefinery concept, making it economically feasible (Garlapati et al. 2020). Different pretreatment techniques employed for bioconversion of LCB components into various biomass-derived compounds have been summarized in Table 2.

4.1 Cellulose

Cellulose (linear homopolysaccharide) is a major structural component of LCB, consisting of monomeric sugar units of D-glucose linked by β-1-4-glycosidic bonds (Rudnik 2012). Sorbitol, lactic acid, succinic acid, levulinic acid, and acetic acid are some of the platform chemicals obtained from cellulose (Devi et al. 2022). LCB is pretreated using different methods for extraction of glucose from cellulose. The pretreatment is generally a two-step sequential treatment, the first step involving structural breakdown to make hemicellulose and cellulose accessible to enzymes, followed by saccharification or hydrolysis to simple sugars (Kucharska et al. 2018). These simple sugars are metabolized by microorganisms for the production of different chemicals.

4.1.1 Lactic Acid

Enantiomers, D or L form of lactic acid, can be synthesized through either fermentation or chemical route. It can be synthesized via chemical synthesis leading to the formation of a racemic mixture, while the fermentative approach using a suitable microorganism produces a pure form of optically active enantiomer (Abdel-Rahman et al. 2013; Daful et al. 2023). Different agro-industrial wastes act as precursors of glucose, sucrose, and lactose that could be used as substrate (Ghaffar et al. 2014). The potential lactic acid-producing microorganisms are *Enterococcus faecalis*, *Lactobacillus rhamnosus*, lactic acid bacteria (LAB), and *Rhizopus* sp. In a study, 36.75 g/L lactic acid was produced using *Lactobacillus plantarum* SKL-22 from imidazolium (ionic liquid) pretreated and enzymatically saccharified rice straw (Yadav et al. 2021). Chatgasem et al. (2023) reported 2.06 g/L of lactic acid produced using thermocellulolytic *Enterococcus* sp. SLC45-3, obtained from rice straw pretreated with 1.5% H_3PO_4 at 121 °C for 15 min.

4.1.2 Succinic Acid (SA)

Succinic acid (C4-dicarboxylic acid) is obtained as a part of tricarboxylic acid cycle-based metabolism. Biomass pretreatment, enzymatic hydrolysis for extraction of valuable sugars, followed by the process of fermentation by a potential succinic acid-producing microorganism are the steps involved in SA production (Ojo 2023; Zhou et al. 2023). SA finds applications in industries involved in chemical production, food, pharmaceuticals, and agriculture. The microbes used for SA

Table 2 Various conversion techniques for the production of different value-added products along with biofuel from LCB components

Lignocellulosic biomass	Pretreatment	Biomass component	Conversion process	Value-added product	Product yield (g/g)	References
Rice straw	Acidic (H_2SO_4)	Hemicellulose	Fermentation	Xylitol	0.61	Kaur et al. (2023)
Mustard biomass	Alkali pretreatment and saccharification (enzyme cocktail)	Cellulose and hemicellulose	Two-step fermentation	Bioethanol and xylitol	Bioethanol: 0.45 Xylitol: 0.62	Pant et al. (2022)
Wheat straw	Ionic liquid (triethylamine hydrogen sulfate)	Cellulose	Fermentation	Ethanol	0.43	Ziaei-Rad et al. (2021)
Rice straw	Acidic (H_2SO_4), saccharification (cellulase)	Cellulose	Fermentation	Bioethanol	0.49	Pooja et al. (2021)
Arundo donax (giant reed)	Dual pretreatment with acid (H_2SO_4) and alkali (NaOH), saccharification (cellulase)	Cellulose	Fermentation	Bioethanol	0.48	Sidana et al. (2022)
Sugarcane bagasse	Alkali (NaOH)	Cellulose and hemicellulose	Co-fermentation	Succinic acid and ethanol	–	Xu et al. (2021)
Hemp hurd	Ionic liquid (1-ethyl-3-methylimidazolium chloride) and acidic (HCl) pretreatment	Cellulose	Glucose dehydration and HMF rehydration	Levulinic acid	0.59	Tulaphol et al. (2019)
Beech wood	Organosolv	Cellulose	Fermentation	Lactic acid	0.69	Karnaouri et al. (2020)
Raw sugarcane bagasse	Ultrasound-ionic liquid pretreated sugarcane bagasse in a microwave-solid acid-ionic liquid system	Cellulose	Catalytic conversion (D001-cc ion-exchange resin)	5-HMF	0.65	Li et al. (2020)
Oil palm fronds	Acidic (HNO_3)	Hemicellulose	Fermentation	Xylitol	0.35	Manaf et al. (2018)
Sugarcane bagasse	Ultrasound-assisted metal salt (titanium oxide) and hydrogen peroxide, dilute acid hydrolysis	Holocellulose	Fermentation	Bioethanol	0.468	Ramadoss and Muthukumar (2016)
Corn cob	Porous polytriphenylamine-SO_3H (SPTPA) solid acid catalyst in lactone solvent	Cellulose and hemicellulose	Catalytic conversion	5-HMF and furfural	5-HMF: 0.32 Furfural: 0.739	Zhang et al. (2016)
Miscanthus giganteus	Dilute oxalic acid	Cellulose	Fermentation	Bioethanol	–	Scordia et al. (2013)
Kenaf stalk	Microwave-assisted catalytic conversion mixed metal oxide citrate complex $CeZrO_2$–CA	Cellulose	Oxidative depolymerization	Vanillin	0.143	Rouf et al. (2023)

production include *Saccharomyces cerevisiae*, *Anaerobiospirillum succiniciproducens*, *Basfia succiniciproducens*, and *Actinobacillus succinogenes* (Pais et al. 2016). In a study, researchers conducted ultrasound-assisted sequential pretreatment using peracetic acid and alkaline peroxide on rice straw and sugarcane bagasse. Subsequently, SA was produced via the step involving simultaneous saccharification along with fermentation resulting in observed yields of 3.64 g/L from rice straw and 5.1 g/L from sugarcane bagasse (Putri et al. 2023).

4.1.3 Acetic Acid

Acetic acid (AA) holds significance as a crucial carboxylic acid with diverse applications in both industrial and household settings. AA acts as precursor for the synthesis of polyvinyl and cellulose acetate, which are used in the production of fibers and fabrics (Ehsanipour et al. 2016). *Gluconacetobacter* sp. and *Acetobacter* sp. are involved in the production of AA (Kim et al. 2019). A study conducted by Ehsanipour et al. (2016) reported the production of acetic acid involved in the fermentation of steam-pretreated LCB, including wheat straw, sugarcane bagasse, forest residue, and switchgrass. The study demonstrated efficient fermentation of both xylose and glucose to acetic acid by *Moorella thermoacetica* ATCC 39073.

4.1.4 Levulinic Acid

Levulinic acid (LA) also known as 4-oxopentanoic acid is prepared from the cellulosic fraction of LCB, which is extracted via a pretreatment technique (Mthembu et al. 2021). Currently, LA is commercially produced using sulfuric acid, which offers certain disadvantages; hence, research is now focused on environment-friendly alternatives such as ionic liquids (Liu et al. 2018). Ionic liquids, often referred to as designer solvents, are highly sought after, due to their modifiable characteristics obtained via characteristic alterations of anionic or cationic groups. Incorporation of sulfonic acid and carboxylic acid groups has been reported to enhance acidity and water solubility, contributing to the creation of environmentally friendly acidic catalysts (Shen et al. 2015; Cole et al. 2002). Other approaches used to synthesize LA include utilization of metal chloride and organic acid catalysts (Zhao et al. 2019; Park et al. 2018).

4.1.5 Sorbitol

Sorbitol, derived from biomass sugars through fermentation by specific microorganisms, stands as a valuable product. As a natural sugar alcohol found in fruits, it holds diverse applications in food, pharmaceutical, cosmetic, and textile industries (Ahmad and Munaim 2018). Chamnipa et al. (2022) reported that producing sorbitol involves utilizing a blend of sugarcane bagasse and cassava pulp hydrolysates with the assistance of thermotolerant *Zymomonas mobilis* TISTR548.

4.1.6 5-Hydroxymethylfurfural

5-Hydroxymethylfurfural (HMF), a promising compound for bio-based plastics, poses challenges in biomass synthesis due to its tendency to undergo degradation into by-products. High-yield synthesis of HMF is predominantly achieved from fructose, and various catalysts are employed for its production from mono and polysaccharides of pretreated biomass (Rout et al. 2016). HMF synthesis is much easier when fructose is used as a substrate of choice in comparison to glucose because of simpler reaction steps with high product yield and purity (Deng et al. 2023). The AVA Biochem company, based in Switzerland, stands as the world's pioneering industrial plant for producing HMF from a fructose substrate. The company uses a hydrothermal pretreatment technique for the conversion of C6 sugars to 5-HMF (Kläusli 2014; avabiochem.com).

4.2 *Hemicellulose*

Hemicellulose, a heteropolymer found in plant cell wall, is composed of various C5 sugars such as xylose, arabinose, and C6 sugars including mannose, galactose, and rhamnose. It comes next to cellulose in terms of abundance (Huang et al. 2021). Numerous products, such as furfural, xylitol, and xylo-oligosaccharides, can be obtained from hemicellulosic fraction of LCB (Kumar and Dixit 2021).

4.2.1 Furfural

Furfural is considered a vital constituent for biorefineries commonly found in LCB such as oat hulls, corn cobs, sugarcane bagasse, etc. (Mika et al. 2018; Cai et al. 2014). It serves as an important precursor for a wide array of versatile bulk chemicals, polymers, hydrocarbon fuels, and its intermediates (Luo et al. 2019; Mika et al. 2018; Mariscal et al. 2016). Both HMF and furfural are the products of acid-catalyzed dehydration (Perez et al. 2019). The production of furfural involved the use of H_2SO_4 (acid catalyst) and steam (stripping agent) with quite low industrial yields, approximately <50% of theoretical yield. However, several other methods like the use of heterogenous catalysts have been proposed in literature to improve furfural yield (Machado et al. 2016; Verma et al. 2017). Bhaumik and Dhepe (2017) reported furfural yield varying from 72% to 87% by transforming various crop wastes (bagasse, wheat straw, and rice husk) utilizing WO_3/SiO_2 in a water/toluene mixture.

4.2.2 Xylitol

In addition to HMF and furfural, xylitol, a favored low-calorie sweetener, can be manufactured from LCB. Xylitol finds extensive applications in food and pharmaceutical industries, owing to its notable properties, including high chemical stability, gastrointestinal tolerance, and insulin-independent metabolism (Kaur et al. 2022). Bacteria and yeasts are examples of microorganisms that ferment xylose. *Candida* sp., in particular, stands out as the most widely utilized strain for xylitol production due to its ability to thrive in toxic conditions (Guleria et al. 2022).

4.3 Lignin

Lignin, a complex non-carbohydrate polymer consisting of aromatic and aliphatic compounds, enhances the hydrophobic nature and rigidity of the plant cell wall, making it a valuable raw material for producing biofuels, platform chemicals, and biomaterials (Yoo et al. 2020; Watkins et al. 2015). Platform chemicals can be produced from biomass lignin through photocatalytic, enzymatic, or chemical pathways (Reddy et al. 2021). Lignin-derived value-added products are of great interest to biorefineries. A US-based company, i.e., Lignol Innovations Inc., produces high purity lignin from different lignocellulosic biomass feedstocks. Lignin possesses a great potential to substitute the petroleum-based products that are used in the manufacturing of industrial coatings, emulsifiers, and gels (De Corato et al. 2018).

4.3.1 Vanillin

Vanillin is generated from lignin and is an essential phenolic compound synthesized on pilot scale from the LCB. It finds huge applications in food and pharmaceutical industries, owing to its flavoring, antioxidant, antimicrobial, anticarcinogenic, and antislicking nature (Devi et al. 2022). The depolymerization of lignin is complex and can be achieved using different technologies, each having certain advantages and disadvantages. Thermal, chemical, biological, lignin-first biorefining, and microwave-assisted depolymerization are some examples of lignin depolymerization techniques. In a study, *Bambusa tulda* was biologically pretreated with a bacterial consortium to degrade lignin, thereby producing vanillin along with other minor compounds (Harshvardhan et al. 2017).

4.3.2 Benzoquinone and Carboxylic Acid

Oxidation of lignin produces benzoquinone and carboxylic acid. Benzoquinone is a fermentative inhibitor synthesized during pretreatment to LCB (Yan et al. 2019). It is a pale yellow-colored crystalline solid widely used in manufacturing of dyes.

Carboxylic acid acts as the precursor for the production of biopolymers, drugs, coatings, adhesives, etc. (Badea and Radu 2018).

5 Techno-Economic and Environmental Assessments of LCB-Based Biorefinery

The incorporation of a lignocellulosic biorefinery into the circular economy concept centers on maximizing the value of LCB, generating a variety of bio-based products of commercial importance. Although biorefineries are progressing toward commercial scale, its implementation still faces many hurdles, such as technological advancements, economic viability, and environmental implications. Understanding the interplay between economic and environmental performance is essential for developing and expanding sustainable biorefinery processes. Comprehensive techno-economic and environmental assessments serve as important frameworks that merge technical, economic, and environmental factors, which explain the broad ramifications of biorefineries (Almada et al. 2023). Conducting techno-economic analysis is crucial for evaluating biorefinery systems, offering vital decision-making information for the investment process. This analysis spans various phases, including assessing technical viability, capital and operational costs, return on investment (RoI), and payback period (Mussatto and Bikaki 2016). Estimation of capital cost comprises expenses related to construction, equipment, installation, and engineering, whereas operational costs include fixed costs, direct production costs, plant general expenses, and overhead expenses (Gezae Daful and Gorgens 2017). Return on investment (RoI) serves as a straightforward metric for evaluating the profitability of biorefinery plants, determined by the annual interest rate generated from the initial investment. Alternatively, the payback period, a subset of the profitability index, offers a fundamental insight into a biorefinery plant. These indicators, along with risk considerations, establish distinct criteria for assessment (Okolie et al. 2021).

Life cycle assessment scrutinizes the environmental impacts and repercussions of the production system from cradle-to-gate (cultivation to usage) or cradle-to-grave (cultivation to industrial operation, usage, and waste disposal). It is commonly employed to assess how entire value chains involved in the manufacture of valuable products influence the environment. The biorefinery, guided by life cycle assessment, strives to scrutinize the ecological impacts of diverse processes (designs and technologies) or ascertain the optimal utilization of biomass feedstock. For conducting life cycle assessment studies, several platforms like open LCA, GaBi, SimaPro, and Ecoinvent databases aid in compiling the inventory and generating the impact assessment results (Almada et al. 2023). Integrating both techno-economics and life cycle assessment offers a potent tool that can assist a variety of stakeholders in establishing economical and sustainable biorefinery.

6 Recent Advances and Future Perspectives

The commercial manufacturing of biofuels and value-added chemicals from lignocellulose-based biorefinery is still lagging in the developing nations potentially due to several difficulties. These difficulties include the high pretreatment costs, production of inhibitors during pretreatment process, expensive enzymes, and challenges incurred during the production of enzyme cocktails to effectively hydrolyze biomass components (Ojo 2023; Singh et al. 2022; Abdel-Rahman et al. 2013). Moreover, the established procedures in use might not result in satisfactory returns despite significant investments. The lignocellulose biorefinery can be operated efficiently if these challenges are tackled appropriately and all the three components of LCB are completely utilized (Sidana and Yadav 2022; Zabaniotou and Kamaterou 2019). The pretreatment techniques play a vital part in utilization of LCB as a feedstock for their conversion to value-added products. Various pretreatment techniques have been developed and used for depolymerization of LCB in the last three decades (Cheah et al. 2020). Optimization of different pretreatment strategies through response surface methodologies is an essential way of improving the process efficiency (Basak et al. 2015; Behera et al. 2015). Research is currently being conducted to improve pretreatment process, and the use of nonconventional pretreatment techniques in combination with the conventional ones has been found promising at lab scale. These methods present numerous benefits, including more environmentally friendly and cost-effective biofuel and value-added chemical production. The integration of nanotechnology may emerge as an optimal pretreatment technique for delignification of LCB and addressing the challenges associated with by-product inhibition (Ingle et al. 2019; Srivastava et al. 2023). Further, the genetic engineering techniques to improve fermentation efficiency and for higher product recovery has also gained impressive attention (Pooja and Yadav 2022). Future aspects of utilization of LCB are focused mostly on improving product yield, selectivity, and economic bioconversion pathways and achieving widespread commercial implementation (Blasi et al. 2023). Continuous progress has also been made to overcome the challenges. Rapid advancement in thermochemical and biochemical methods is currently undertaken for efficient scaleup of valorization of different lignocellulosic wastes. Further directional research on real-time scenarios is essential to make the lignocellulose-based biorefinery an enviro-economically sustainable option in the near future.

7 Conclusion

LCB-based biorefinery has received substantial interest in meeting the increasing energy demands of the expanding population. These biorefineries use either thermochemical or biochemical processes to convert LCB into second-generation biofuels and other bio-based compounds. Thermochemical approaches include torrefaction, pyrolysis, liquefaction, and gasification, whereas biochemical approaches include anaerobic digestion and fermentation. Selection of suitable conversion pathway acts as the governing factor in determining the end products. Extensive studies explore bioethanol, biodiesel, jet fuel, and biogas as alternatives to fossil-based energy for promoting a circular economy. Additionally, various value-added chemicals such as furans, xylitol, succinic acid, sorbitol, lactic acid, and levulinic acid can be derived from LCB. In addition to advantages, these biorefineries often face challenges from technical, economic, and environmental feasibility. Understanding the techno-economic and environmental assessment aids in optimizing the biorefineries to accomplish sustainable development goals. Integrating several biorefineries for complete valorization of LCB into multifaceted bio-based products emerges as an efficient strategy for economic viability and sustainability.

Acknowledgment The authors acknowledge the Department of Biotechnology, Thapar Institute of Engineering and Technology, Patiala, and the Department of Biotechnology, B. R. Ambedkar National Institute of Technology Jalandhar, for providing the requisite infrastructural facilities to conduct the work.

References

Abdel-Rahman MA, Tashiro Y, Sonomoto K (2013) Recent advances in lactic acid production by microbial fermentation processes. Biotechnol Adv 31(6):877–902

Agrawal K, Ruhil T, Verma P (2023) Microalgae: a way toward sustainable development of a society. In: Srivastava N, Mishra P (eds) Basic research advancement for algal biofuels production. Clean energy production technologies. Springer, Singapore

Ahmad ZS, Munaim MSA (2018) Effect of time, moisture content, and substrate amount on sorbitol production using entrapment of *Lactobacillus plantarum* (BAA-793) in sodium alginate beads. Food Biosci 21:27–33

Akbarian A, Andooz A, Kowsari E, Ramakrishna S, Asgari S, Cheshmeh ZA (2022) Challenges and opportunities of lignocellulosic biomass gasification in the path of circular bioeconomy. Bioresour Technol 362:127774

Akhtar N, Gupta K, Goyal D, Goyal A (2016) Recent advances in pretreatment technologies for efficient hydrolysis of lignocellulosic biomass. Environ Prog Sustain Energy 35(2):489–511

Akubo K, Nahil MA, Williams PT (2019) Pyrolysis-catalytic steam reforming of agricultural biomass wastes and biomass components for production of hydrogen/syngas. J Energy Inst 92(6):1987–1996

Almada DP, Galán-Martín Á, Gámez MDMC, Galiano EC (2023) Integrated techno-economic and environmental assessment of biorefineries: review and future research directions. Sustain Energy Fuels 7:4031–4050

Amer M, Nour M, Ahmed M, Ookawara S, Nada S, Elwardany A (2019) The effect of microwave drying pretreatment on dry torrefaction of agricultural biomasses. Bioresour Technol 286:121400

Ashokkumar V, Venkatkarthick R, Jayashree S, Chuetor S, Dharmaraj S, Kumar G, Ngamcharussrivichai C (2022) Recent advances in lignocellulosic biomass for biofuels and value-added bioproducts-a critical review. Bioresour Technol 344:126195

Badea GI, Radu GL (2018) Carboxylic acid: key role in life sciences. In: BoD–books on demand. IntechOpen, London

Banu JR, Sugitha S, Kavitha S, Kannah RY, Merrylin J, Kumar G (2021) Lignocellulosic biomass pretreatment for enhanced bioenergy recovery: effect of lignocelluloses recalcitrance and enhancement strategies. Front Energy Res 9:646057

Basak AK, Behera B, Mukherjee A, Aikat K, Ganguly A, Chatterjee PK (2015) Comparative study of using sodium chlorite and tartaric acid pretreatment on waste rice straw for enhanced production of bioethanol. IOSR J Pharm Biol Sci 10:37–45

Bazaluk O, Havrysh V, Nitsenko V, Balezentis T, Streimikiene D, Tarkhanova EA (2020) Assessment of green methanol production potential and related economic and environmental benefits: the case of China. Energies 13(12):3113

Behera B, Mukherjee A, Basak AK, Aikat K, Ganguly A, Chatterjee PK (2015) Utilisation of the spent biomass from the old thatched roofs for production of bioethanol. Eur J Biotechnol Biosci 3(4):49–55

Bhaumik P, Dhepe PL (2017) From Lignocellulosic biomass to furfural: insight into the active species of a silica-supported tungsten oxide catalyst. Chem Cat Chem 9(14):2709–2716

Bianchini IDA, Sene L, da Cunha MAA, Felipe MDGDA (2022) Short-term adaptation strategy improved xylitol production by *Candida guilliermondii* on sugarcane bagasse hemicellulosic hydrolysate. Bioenergy Res 15(2):1182–1194

Blasi A, Verardi A, Lopresto CG, Siciliano S, Sangiorgio P (2023) Lignocellulosic agricultural waste valorization to obtain valuable products: an overview. Recycling 8(4):61

Cai CM, Zhang T, Kumar R, Wyman CE (2014) Integrated furfural production as a renewable fuel and chemical platform from lignocellulosic biomass. J Chem Technol Biotechnol 89(1):2–10

Chamnipa N, Klanrit P, Thanonkeo S, Thanonkeo P (2022) Sorbitol production from a mixture of sugarcane bagasse and cassava pulp hydrolysates using thermotolerant *Zymomonas mobilis* TISTR548. Ind Crop Prod 188:115741

Chatgasem C, Suwan W, Attapong M, Siripornadulsil W, Siripornadulsil S (2023) Single-step conversion of rice straw to lactic acid by thermotolerant cellulolytic lactic acid bacteria. Biocatal Agric Biotechnol 47:102546

Cheah WY, Sankaran R, Show PL, Ibrahim T, Baizura TN, Chew KW, Culaba A, Chang JS (2020) Pretreatment methods for lignocellulosic biofuels production: current advances, challenges and future prospects. Biofuel Res J 7(1):1115–1127

Chen Z, Chen L, Khoo KS, Gupta VK, Sharma M, Show PL, Yap PS (2023) Exploitation of lignocellulosic-based biomass biorefinery: a critical review of renewable bioresource, sustainability, and economic views. Biotechnol Adv 69:108265

Cheng F, Brewer CE (2017) Producing jet fuel from biomass lignin: potential pathways to alkylbenzenes and cycloalkanes. Renew Sust Energ Rev 72:673–722

Cole AC, Jensen JL, Ntai I, Tran KLT, Weaver KJ, Forbes DC, Davis JH (2002) Novel brønsted acidic ionic liquids and their use as dual solvent-catalysts. J Am Chem Soc 124(21):5962–5963

Daful AG, Loridon M, Chandraratne MR (2023) Lactic acid production from lignocellulosic biomass. IntechOpen, London, p 112739

Dahmen N, Lewandowski I, Zibek S, Weidtmann A (2019) Integrated lignocellulosic value chains in a growing bioeconomy: status quo and perspectives. Glob Change Biol Bioenergy 11(1):107–117

De Corato U, De Bari I, Viola E, Pugliese M (2018) Assessing the main opportunities of integrated biorefining from agro-bioenergy co/by-products and agroindustrial residues into high-value added products associated to some emerging markets: a review. Renew Sust Energ Rev 88:326–346

Deng W, Feng Y, Fu J, Guo H, Guo Y, Han B, Jiang Z, Kong L, Li C, Liu H, Nguyen PT (2023) Catalytic conversion of lignocellulosic biomass into chemicals and fuels. Green Energy Environ 8(1):10–114

Devaraja UMA, Dissanayake CLW, Gunarathne DS, Chen WH (2022) Oxidative torrefaction and torrefaction-based biorefining of biomass: a critical review. Biofuel Res J 9:1672–1696

Devi A, Bajar S, Kour H, Kothari R, Pant D, Singh A (2022) Lignocellulosic biomass valorization for bioethanol production: a circular bioeconomy approach. Bioenergy Res 15(4):1820–1841

Ehsanipour M, Suko AV, Bura R (2016) Fermentation of lignocellulosic sugars to acetic acid by *Moorella thermoacetica*. J Ind Microbiol Biotechnol 43(6):807–816

Elgarahy AM, Eloffy MG, Alengebawy A, El-Sherif DM, Gaballah MS, Elwakeel KZ, El-Qelish M (2023) Sustainable management of food waste; pre-treatment strategies, techno-economic assessment, bibliometric analysis, and potential utilizations: a systematic review. Environ Res 225:115558

Eswaran S, Subramaniam S, Geleynse S, Brandt K, Wolcott M, Zhang X (2021) Techno-economic analysis of catalytic hydrothermolysis pathway for jet fuel production. Renew Sust Energ Rev 151:111516

Garlapati VK, Chandel AK, Kumar SJ, Sharma S, Sevda S, Ingle AP, Pant D (2020) Circular economy aspects of lignin: towards a lignocellulose biorefinery. Renew Sust Energ Rev 130:10997

Gezae Daful A, Gorgens JF (2017) Techno-economic analysis and environmental impact assessment of lignocellulosic lactic acid production. Chem Eng Sci 162:53–65

Ghaffar T, Irshad M, Anwar Z, Aqil T, Zulifqar Z, Tariq A, Kamran M, Ehsan N, Mehmood S (2014) Recent trends in lactic acid biotechnology: a brief review on production to purification. J Radiat Res Appl Sci 7(2):222–229

Goswami RK, Agrawal K, Upadhyaya HM, Gupta VK, Verma P (2022) Microalgae conversion to alternative energy, operating environment and economic footprint: an influential approach towards energy conversion, and management. Energy Convers Manag 269:1–27

Guleria P, Kaur S, Sidana A, Yadav SK (2022) Xylitol production from rice straw hemicellulosic hydrolysate by *Candida tropicalis* GS18 immobilized on bacterial cellulose-sodium alginate matrix. Biomass Convers Biorefinery 14:11

Hai IU, Sher F, Yaqoob A, Liu H (2019) Assessment of biomass energy potential for SRC willow woodchips in a pilot scale bubbling fluidized bed gasifier. Fuel 258:116143

Harshvardhan K, Suri M, Goswami A, Goswami T (2017) Biological approach for the production of vanillin from lignocellulosic biomass (*Bambusa tulda*). J Clean Prod 149:485–490

Hazrat MA, Rasul MG, Khan MMK, Ashwath N, Rufford TE (2019) Emission characteristics of waste tallow and waste cooking oil-based ternary biodiesel fuels. Energy Procedia 160:842–847

Huang LZ, Ma MG, Ji XX, Choi SE, Si C (2021) Recent developments and applications of hemicellulose from wheat straw: a review. Front Bioeng Biotechnol 9:690773

Ingle AP, Chandel AK, Antunes FA, Rai M, da Silva SS (2019) New trends in application of nanotechnology for the pretreatment of lignocellulosic biomass. Biofuels Bioprod Biorefin 13(3):776–788

Isikgor FH, Becer CR (2015) Lignocellulosic biomass: a sustainable platform for the production of bio-based chemicals and polymers. Polym Chem 6(25):4497–4559

Jönsson LJ, Martín C (2016) Pretreatment of lignocellulose: formation of inhibitory by-products and strategies for minimizing their effects. Bioresour Technol 199:103–112

Karnaouri A, Asimakopoulou G, Kalogiannis KG, Lappas A, Topakas E (2020) Efficient d-lactic acid production by *Lactobacillus delbrueckii* subsp. *bulgaricus* through conversion of organosolv pretreated lignocellulosic biomass. Biomass Bioenergy 140:105672

Kaur S, Guleria P, Sidana A, Yadav SK (2022) Efficient process for xylitol production from nitric acid pretreated rice straw derived pentosans by *Candida tropicalis* GS18. Biomass Bioenergy 166:106618

Kaur S, Guleria P, Yadav SK (2023) Evaluation of fermentative xylitol production potential of adapted strains of *Meyerozyma caribbica* and *Candida tropicalis* from rice straw hemicellulosic hydrolysate. Fermentation 9(2):181

Kim HM, Choi IS, Lee S, Yang JE, Jeong SG, Park JH, Ko SH, Hwang IM, Chun HH, Wi SG, Kim JC (2019) Biorefining process of carbohydrate feedstock (agricultural onion waste) to acetic acid. ACS Omega 4(27):22438–22444

Kläusli T (2014) AVA Biochem: commercialising renewable platform chemical 5-HMF. Green Process Synth 3(3):235–236

Kucharska K, Rybarczyk P, Hołowacz I, Łukajtis R, Glinka M, Kamiński M (2018) Pretreatment of lignocellulosic materials as substrates for fermentation processes. Molecules 23(11):2937

Kumar N, Dixit A (2021) Chapter 4: management of biomass. In: Micro and nano technologies, nanotechnology for rural development. Elsevier, Amsterdam, pp 97–140
Kumar P, Barrett DM, Delwiche MJ, Stroeve P (2009) Methods for pretreatment of lignocellulosic biomass for efficient hydrolysis and biofuel production. Ind Eng Chem Res 48(8):3713–3729
Kumar B, Bhardwaj N, Agrawal K, Chaturvedi V, Verma P (2020) Current perspective on pretreatment technologies using lignocellulosic biomass: an emerging biorefinery concept. Fuel Process Technol 199:106244
Li M, Jiang H, Zhang L, Yu X, Liu H, Yagoub AEA, Zhou C (2020) Synthesis of 5-HMF from an ultrasound-ionic liquid pretreated sugarcane bagasse by using a microwave-solid acid/ionic liquid system. Ind Crop Prod 149:112361
Limayem A, Ricke SC (2012) Lignocellulosic biomass for bioethanol production: current perspectives, potential issues and future prospects. Prog Energy Combust Sci 38(4):449–467
Liu L, Li Z, Hou W, Shen H (2018) Direct conversion of lignocellulose to levulinic acid catalysed by ionic liquid. Carbohydr Polym 181:778–784
Luo Y, Li Z, Li X, Liu X, Fan J, Clark JH, Hu C (2019) The production of furfural directly from hemicellulose in lignocellulosic biomass: a review. Catal Today 319:14–24
Machado G, Leon S, Santos F, Lourega R, Dullius J, Mollmann ME, Eichler P (2016) Literature review on furfural production from lignocellulosic biomass. Nat Res Forum 7(3):115–129
Manaf SFA, Jahim JM, Harun S, Luthfi AAI (2018) Fractionation of oil palm fronds (OPF) hemicellulose using dilute nitric acid for fermentative production of xylitol. Ind Crop Prod 115:6–15
Manikandan S, Vickram S, Sirohi R, Subbaiya R, Krishnan RY, Karmegam N, Awasthi MK (2023) Critical review of biochemical pathways to transformation of waste and biomass into bioenergy. Bioresour Technol 372:128679
Mariscal R, Maireles-Torres P, Ojeda M, Sádaba I, Granados ML (2016) Furfural: a renewable and versatile platform molecule for the synthesis of chemicals and fuels. Energy Environ Sci 9(4):1144–1189
Mathanker A, Pudasainee D, Kumar A, Gupta R (2020) Hydrothermal liquefaction of lignocellulosic biomass feedstock to produce biofuels: parametric study and products characterization. Fuel 271:117534
Messineo A, Maniscalco MP, Volpe R (2020) Biomethane recovery from olive mill residues through anaerobic digestion: a review of the state-of-the-art technology. Sci Total Environ 703:135508
Mika LT, Cséfalvay E, Németh Á (2018) Catalytic conversion of carbohydrates to initial platform chemicals: chemistry and sustainability. Chem Rev 118(2):505–613
Mirmohamadsadeghi S, Karimi K, Azarbaijani R, Yeganeh LP, Angelidaki I, Nizami AS, Bhat R, Dashora K, Vijay VK, Aghbashlo M, Gupta VK (2021) Pretreatment of lignocelluloses for enhanced biogas production: a review on influencing mechanisms and the importance of microbial diversity. Renew Sust Energ Rev 135:110173
Moradian JM, Fang Z, Yong YC (2021) Recent advances on biomass-fueled microbial fuel cell. Bioresour Bioprocess 8:1–13
Mosier N, Wyman C, Dale B, Elander R, Lee YY, Holtzapple M, Ladisch M (2005) Features of promising technologies for pretreatment of lignocellulosic biomass. Bioresour Technol 96(6):673–686
Mthembu LD, Gupta R, Deenadayalu N (2021) Conversion of cellulose into value-added products. In: Cellulose science and derivatives. IntechOpen, London, p 125
Mussatto SI, Bikaki N (2016) Chapter 25: technoeconomic considerations for biomass fractionation in a biorefinery context. In: Biomass fractionation technologies for a lignocellulosic feedstock based biorefinery. Elsevier, Amsterdam, pp 587–610
Ojo AO (2023) An overview of lignocellulose and its biotechnological importance in high-value product production. Fermentation 9(11):990
Okolie JA, Nanda S, Dalai AK, Kozinski JA (2021) Techno-economic evaluation and sensitivity analysis of a conceptual design for supercritical water gasification of soybean straw to produce hydrogen. Bioresour Technol 331:125005

Pais C, Franco-Duarte R, Sampaio P, Wildner J, Carolas A, Figueira D, Ferreira BS (2016) Production of dicarboxylic acid platform chemicals using yeasts: focus on succinic acid. In: Biotransformation of agricultural waste and by-products. Elsevier, Amsterdam, pp 237–269
Pant S, Prakash A, Kuila A (2022) Integrated production of ethanol and xylitol from *Brassica juncea* using *Candida sojae* JCM 1644. Bioresour Technol 351:126903
Park MR, Kim SK, Jeong GT (2018) Optimisation of the levulinic acid production from the red macroalga, *Gracilaria verrucosa* using methanesulfonic acid. Algal Res 31:116–121
Perez GP, Mukherjee A, Dumont MJ (2019) Insights into HMF catalysis. J Ind Eng Chem 70:1–34
Pooja, Yadav SK (2022) Diversity and use of genetically modified microorganisms for second-generation ethanol production. In: Liquid biofuels: bioethanol. Springer, Cham, pp 187–207
Pooja, Purohit A, Kaur S, Yadav SK (2021) Identification of a yeast Meyerozyma caribbica M72 from mahua flower for efficient transformation of rice straw into ethanol. Biomass Convers Biorefinery 13:5
Prakesh A, Dave V, Sur S, Sharma P (2021) Vivid techniques of pretreatment showing promising results in biofuel production and food processing. J Food Process Eng 44(2):13580
Pratto B, Chandgude V, de Sousa Junior R, Cruz AJG, Bankar S (2020) Biobutanol production from sugarcane straw: defining optimal biomass loading for improved ABE fermentation. Ind Crop Prod 148:112265
Putri DN, Pratiwi SF, Perdani MS, Rosarina D, Utami TS, Sahlan M, Hermansyah H (2023) Utilizing rice straw and sugarcane bagasse as low-cost feedstocks towards sustainable production of succinic acid. Sci Total Environ 862:160719
Ramadoss G, Muthukumar K (2016) Mechanistic study on ultrasound assisted pretreatment of sugarcane bagasse using metal salt with hydrogen peroxide for bioethanol production. Ultrason Sonochem 28:207–217
Reddy YN, Gogde K, Paul S, Bhaumik J (2021) Lignin to platform chemicals and biomaterials: chemical and biological perspectives. In: Biomass for bioenergy and biomaterials. CRC Press, Boca Raton, pp 31–64
Rouf H, Ramli A, Anuar NASIK, Yunus NM (2023) Ce–Zr-based mixed oxide catalyst for oxidative depolymerization of kenaf stalk (biomass) into vanillin. Bioresour Bioprocess 10(1):76
Rout PK, Nannaware AD, Prakash O, Kalra A, Rajasekharan R (2016) Synthesis of hydroxymethylfurfural from cellulose using green processes: a promising biochemical and biofuel feedstock. Chem Eng Sci 142:318–346
Roy S, Dikshit PK, Sherpa KC, Singh A, Jacob S, Chandra Rajak R (2021) Recent nanobiotechnological advancements in lignocellulosic biomass valorization: a review. J Environ Manag 297:113422
Rudnik E (2012) Compostable polymer materials: definitions, structures, and methods of preparation. In: Handbook of biopolymers and biodegradable plastics, vol 10. Elsevier, Amsterdam, pp 189–211
Sahoo SS, Vijay VK, Chandra R, Kumar H (2021) Production and characterization of biochar produced from slow pyrolysis of pigeon pea stalk and bamboo. Clean Eng Technol 3:100101
Scordia D, Cosentino SL, Jeffries TW (2013) Effectiveness of dilute oxalic acid pretreatment of *Miscanthus giganteus* biomass for ethanol production. Biomass Bioenergy 59:540–548
Shafizadeh A, Rastegari H, Shahbeik H, Mobli H, Pan J, Peng W, Aghbashlo MA (2023) Critical review of the use of nanomaterials in the biomass pyrolysis process. J Clean Prod 400, 136705
Shahbeik H, Peng W, Panahi HKS, Dehhaghi M, Guillemin GJ, Fallahi A, Aghbashlo M (2022) Synthesis of liquid biofuels from biomass by hydrothermal gasification: a critical review. Renew Sust Energ Rev 167:112833
Sharma S, Tsai ML, Sharma V, Sun PP, Nargotra P, Bajaj BK, Chen CW, Dong CD (2023) Environment friendly pretreatment approaches for the bioconversion of lignocellulosic biomass into biofuels and value-added products. Enviro 10(1):6
Shen Y, Sun JK, Yi YX, Wang B, Xu F, Sun RC (2015) One-pot synthesis of levulinic acid from cellulose in ionic liquids. Bioresour Technol 192:812–816
Sidana A, Yadav SK (2022) Recent developments in lignocellulosic biomass pretreatment with a focus on eco-friendly, non-conventional methods. J Clean Prod 335:130286

Sidana A, Kaur S, Yadav SK (2022) Assessment of the ability of *Meyerozyma guilliermondii* P14 to produce second-generation bioethanol from giant reed (*Arundo donax*) biomass. Biomass Convers Biorefinery 13(7):1–13

Sidana A, Guleria P, Kaur S, Yadav SK (2023) Exploration of hemicellulosic hydrolysates derived from *Arundo donax* and *Cynodon dactylon* as potential substrates for microbial protein production by *Candida tropicalis*. Bioenergy Res 17:369–382

Silitonga AS, Masjuki HH, Mahlia TMI, Ong HC, Chong WT (2013) Experimental study on performance and exhaust emissions of a diesel engine fuelled with Ceiba pentandra biodiesel blends. Energy Convers Manag 76:828–836

Singh A, Srivastava S, Rathore D, Pant D (2021) Environmental microbiology and biotechnology: volume 2: bioenergy and environmental health. Springer, Cham

Singh B, Korstad J, Guldhe A, Kothari R (2022) Emerging feedstocks and clean technologies for lignocellulosic biofuel. Front Energy Res 10:1–2

Singh A, Prajapati P, Vyas S, Gaur VK, Sindhu R, Binod P, Varjani S (2023) A comprehensive review of feedstocks as sustainable substrates for next-generation biofuels. Bioenergy Res 16(1):105–122

Srivastava N, Singh R, Srivastava M, Mohammad A, Harakeh S, Singh RP, Pal DB, Haque S, Tayeb HH, Moulay M, Gupta VK (2023) Impact of nanomaterials on sustainable pretreatment of lignocellulosic biomass for biofuels production: an advanced approach. Bioresour Technol 369:128471

Surendra KC, Ogoshi R, Zaleski HM, Hashimoto AG, Khanal SK (2018) High yielding tropical energy crops for bioenergy production: effects of plant components, harvest years and locations on biomass composition. Bioresour Technol 251:218–229

Thilo Z (2018) New oil plants and their potential as feedstock for biokerosene production. In: Kaltschmitt M, Neuling U (eds) Biokerosene: status and prospects. Status and prospects. Springer, Berlin

Tulaphol S, Hossain MA, Rahaman MS, Liu LY, Phung TK, Renneckar S, Grisdanurak N, Sathitsuksanoh N (2019) Direct production of levulinic acid in one pot from hemp hurd by dilute acid in ionic liquids. Energy Fuel 34(2):1764–1772

Usmani Z, Sharma M, Awasthi AK, Lukk T, Tuohy MG, Gong L, Nguyen-Tri P, Goddard AD, Bill RM, Nayak SC, Gupta VK (2021) Lignocellulosic biorefineries: the current state of challenges and strategies for efficient commercialization. Renew Sust Energ Rev 148:111258

Verma S, Baig RN, Nadagouda MN, Len C, Varma RS (2017) Sustainable pathway to furanics from biomass via heterogeneous organo-catalysis. Green Chem 19(1):164–168

Veza I, Zainuddin Z, Tamaldin N, Idris M, Irianto I, Fattah IMR (2022) Effect of palm oil biodiesel blends (B10 and B20) on physical and mechanical properties of nitrile rubber elastomer. Results Eng 16:100787

Wang WC, Lee AC (2019) The study of producing "drop-in" fuels from agricultural waste through fast pyrolysis and catalytic hydro-processing. Renew Energy 133:1–10

Watkins D, Nuruddin M, Hosur M, Tcherbi-Narteh A, Jeelani S (2015) Extraction and characterization of lignin from different biomass resources. J Mater Res Technol 4(1):26–32

Werpy T, Petersen G (2004) Top value added chemicals from biomass: volume I-results of screening for potential candidates from sugars and synthesis gas (No. DOE/GO-102004-1992). National Renewable Energy Lab, Golden

Xu N, Liu S, Xin F, Zhou J, Jia H, Xu J, Jiang M, Dong W (2019) Biomethane production from lignocellulose: biomass recalcitrance and its impacts on anaerobic digestion. Front Bioeng Biotechnol 7:191

Xu C, Alam MA, Wang Z, Peng Y, Xie C, Gong W, Yang Q, Huang S, Zhuang W, Xu J (2021) Co-fermentation of succinic acid and ethanol from sugarcane bagasse based on full hexose and pentose utilization and carbon dioxide reduction. Bioresour Technol 339:125578

Yadav N, Nain L, Khare SK (2021) One-pot production of lactic acid from rice straw pretreated with ionic liquid. Bioresour Technol 323:124563

Yan Z, Gao X, Gao Q, Bao J (2019) Mechanism of tolerance to the lignin-derived inhibitor p-benzoquinone and metabolic modification of biorefinery fermentation strains. Appl Environ Microbiol 85(22):01443
Yoo CG, Meng X, Pu Y, Ragauskas AJ (2020) The critical role of lignin in lignocellulosic biomass conversion and recent pretreatment strategies: a comprehensive review. Bioresour Technol 301:122784
Zabaniotou A, Kamaterou P (2019) Food waste valorization advocating Circular Bioeconomy-A critical review of potentialities and perspectives of spent coffee grounds biorefinery. J Clean Prod 211:1553–1566
Zaimes GG, Soratana K, Harden CL, Landis AE, Khanna V (2015) Biofuels via fast pyrolysis of perennial grasses: a life cycle evaluation of energy consumption and greenhouse gas emissions. Environ Sci Technol 49(16):10007–10018
Zhang L, Xi G, Zhang J, Yu H, Wang X (2016) Efficient catalytic system for the direct transformation of lignocellulosic biomass to furfural and 5-hydroxymethylfurfural. Bioresour Technol 224:656–661
Zhao P, Zhou C, Li J, Xu S, Hu C (2019) Synergistic effect of different species in stannic chloride solution on the production of levulinic acid from biomass. ACS Sustain Chem Eng 7:5176–5183
Zhou S, Zhang M, Zhu L, Zhao X, Chen J, Chen W, Chang C (2023) Hydrolysis of lignocellulose to succinic acid: a review of treatment methods and succinic acid applications. Biotechnol Biofuels Bioprod 16(1):1–17
Ziaei-Rad Z, Fooladi J, Pazouki M, Gummadi SN (2021) Lignocellulosic biomass pre-treatment using low-cost ionic liquid for bioethanol production: an economically viable method for wheat straw fractionation. Biomass Bioenergy 151:106140
Zoghlami A, Paës G (2019) Lignocellulosic biomass: understanding recalcitrance and predicting hydrolysis. Front Chem 7:874

Valorization of the Lignocellulosic Biomass via Bacterial Enzymatic Machinery for Sustainable Biorefinery

Nisha Gupta, Esmil Beliya, Jai Shankar Paul, and S. K. Jadhav

1 Introduction

In the perspective of economy, the valorization of lignocellulosic biomass (LB) holds promising opportunities due to its heterogeneous structure, widespread availability, cost-effectiveness, and nontoxic nature. By modifying all pertinent systems design and research-based business ideas, the circular bioeconomy seeks to replace the conventional approaches through different resource retrieval processes, green energy use, exclusion of toxic chemicals, and negligible discharge idea (Barbosa et al. 2020; Sharma et al. 2022a). In a circular economy, it is emphasized that changes should be made in existing business models and research procedures in such a way that there is no harm to the natural biodiversity present in the environment (Manisha and Yadav 2017; de Almeida et al. 2022). The main objective of circular economy depends majorly on waste generation cutting, reuse and reprocessing of resources, as well as creation of a closed-loop system to get the most value-added commodities out of waste materials (Devi et al. 2022). The International Energy Agency (IEA) states that the goal of biorefineries is "the sustainable transformation of different raw materials or biomasses available in the nature into biological energy and varied bio-products (feed, food, and chemicals), and biological power (heat, power, and biological fuels)" (Saha et al. 2005; Wicker et al. 2021).

N. Gupta · J. S. Paul (✉) · S. K. Jadhav
School of Studies in Biotechnology, Pt. Ravishankar Shukla University, Raipur, Chhattisgarh, India

E. Beliya
School of Studies in Biotechnology, Pt. Ravishankar Shukla University, Raipur, Chhattisgarh, India

Department of Botany, Government College, Bichhua, Chhindwara, Madhya Pradesh, India

K. Agrawal, P. Verma (eds.), *Biotechnological Advances in Biorefinery*, Interdisciplinary Biotechnological Advances,
https://doi.org/10.1007/978-981-97-5544-8_2

Nature's gift of LB is an endless or inexhaustible organic carbon source for the environmentally friendly synthesis of various bio-based products including bioenergy and various chemicals with extremely low or no carbon and sulfur emissions. In biorefinery, attention is paid to the use of cellulose, hemicellulose, and lignin, the three main components of LB in every possible way (Manisha and Yadav 2017). Due to the presence of valuable components in LB, it represents the most intriguing alternative to the various inorganic substrates for generating value-added products (Fig. 1) including biofuels, biomolecules, organic acids, pigments, enzymes, and other industrially important chemicals (Nargotra et al. 2022). The most commonly used LB includes field or crop waste, forest wastes, agro-industrial waste, and weed biomass (Saha et al. 2005; Nargotra et al. 2022). Almost all the biorefineries of today's decade are based on cheap and potential alternatives like LB. However, a bottleneck for biorefineries is the use of recalcitrant LB to satiate the burgeoning demand of the ever-increasing population. However, microorganisms have a great potential for producing extracellular lignocellulolytic enzymes (hydrolases) through their enzymatic machinery and are therefore essential tools for the utilization of recalcitrant feedstocks in biorefineries.

2 Structural Features of Lignocellulosic Biomass

LB constitutes about 32–54% cellulose, 11–37% hemicellulose, and 17–32% lignin (non-carbohydrate polymer) and some amount of fatty acids, amino acids, gums, diterpenes, terpenoids, chlorophyll, phenolic compounds, mineral-enriched ash, resins, and protein, which differ from one variety to another (Sharma et al. 2022b; de Almeida et al. 2022; Wicker et al. 2021; Nargotra et al. 2022). The various components of LB are aligned in such a way as to make a completely rigid and complex structure. Cellulose makes the structure of cell walls, while hemicellulose aids in covalently connecting non-cellulosic and cellulosic polymers to generate

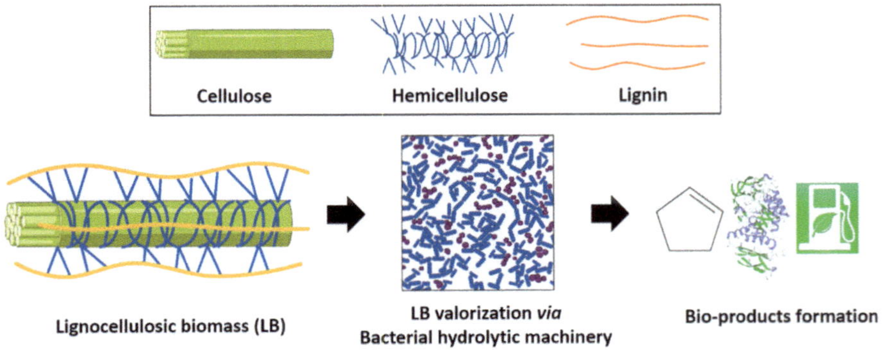

Fig. 1 Valorization of lignocellulosic biomass into various bio-products via bacterial hydrolytic machinery

cross-linked polymers (Fig. 1). Contrarily, lignin, which resembles phenol formaldehyde resin, encloses these polysaccharide polymers (cellulose and hemicellulose) and therefore serves as a "glue," which grip the structure of lignocellulose together, giving the cell walls more mechanical strength to withstand damp circumstances and to fend off insects (Manisha and Yadav 2017; Sankaran et al. 2021; Sharma et al. 2022a).

2.1 *Cellulose*

Being a linear polysaccharide, cellulose molecules are generated by glucose units, which are connected together through four glycosidic linkages (Manisha and Yadav 2017; Sankaran et al. 2021; Wicker et al. 2021). Disaccharide cellobiose, which is composed of two glucose combined molecules, is the repeating unit of cellulose. The main building block of cellulose is a six-carbon (C6) sugar known as glucopyranosyl monomeric units (β-1,4-linked D-glucopyranose residues), which are joined by 1-4-β-glycosidic linkages (Manisha and Yadav 2017; Barbosa et al. 2020; Vélez-Mercado et al. 2021). Cellulose has a flat sheet structure comprising units of glucose monomer, which are connected strongly to numerous intra and intermolecular hydrogen bonds (de Almeida et al. 2022; Devi et al. 2022; Sharma et al. 2022b). Furthermore, van der Waals forces are the primary interfacial forces between cellulose (Barbosa et al. 2020; Devi et al. 2022).

2.2 *Hemicellulose*

On the contrary, hemicellulose is a heteropolysaccharide that constitutes many distinct monosaccharides like galactomannan, xylan, xyloglucan, and arabinoxylan (Saha et al. 2005; Barbosa et al. 2020; Wicker et al. 2021), and due to the presence of diverse functional groups including acetyl, hemicellulose, methyl, galacturonic acid, and glucuronic acid seems to have a branching structure (Vélez-Mercado et al. 2021; Wicker et al. 2021; Devi et al. 2022). Hemicellulose is easier to depolymerize than cellulose due to the lesser degree of polymerization (Rabemanolontsoa and Saka 2016). Hemicellulose acts as an essential source of pentose sugars. These pentose sugars are the building blocks of various bio-products like sugar alcohols, prebiotic oligomers, biofuels, and organic acids (Manisha and Yadav 2017; de Almeida et al. 2022).

2.3 Lignin

A plant's cell wall stiffness and defense are provided by lignin, a complex aromatic polymer with substantial heterogeneity that is connected to carbohydrates via an ester linkage (Rabemanolontsoa and Saka 2016; Lee et al. 2019). Lignin is a three-dimensional, phenolic non-carbohydrate polymer containing phenylpropanoid units. It functions as a cellular adhesive and is located in the secondary cell wall. The three primary categories of phenolic compounds found in lignin are guaiacyl (G), syringyl (S), and 4-hydroxyphenyl (H) (Ralph et al. 2004; Saha et al. 2005; Devi et al. 2022). These compounds are polymerized through cross-linking mediated by an enzymatic radical reaction, which mostly results in β-O-4 bonding and additional C-C and C-O bonding (Ralph et al. 2004). The lignin structure also includes other subunits in addition to these three commonly occurring ones. These are acylated monolignols containing acetate, p-coumarate, sinapaldehyde, ferulic acid, ferulates, coniferaldehyde, and 5-hydroxyconiferyl alcohol (Ralph et al. 2004; Zhang et al. 2020). The β–β, β–5, and β–O–4 bonds in the lignin structure have frequently occurring linkages (Zhang et al. 2020). The major obstacle in LB breakdown is due to the presence of lignin, which is the second most abundantly available biomacromolecule in LB (Saha et al. 2005; Barbosa et al. 2020; Vélez-Mercado et al. 2021). Lignin is a highly branched, aromatic polymer, which makes it resistant to degradation.

3 Hydrolysis of Lignocellulosic Biomass

The structural complexity and heterogeneity of LB, the coverage of cellulose by hemicellulose and lignin, the high lignin content, the crystallinity of cellulose, and the propensity of lignin to repolymerize are some of the challenges that the widespread use of LB still faces (Rabemanolontsoa and Saka 2016; Sethupathy et al. 2021). The highly complex linkage of cellulose, lignin, and hemicellulose gives plant cell walls their hard, recalcitrant, and robust characteristics, making them difficult to use as raw materials for biorefinery by the researchers (Chukwuma et al. 2021; Wicker et al. 2021). The extraction of a specific biopolymer, through its breakdown and eventual conversion into the wanted product, requires a prior processing step known as the pretreatment steps, in which the success rate of LB valorization depends in biorefineries. The pretreatment reduces the recalcitrance and complexity of the LB and makes it accessible for the desired bio-product formation (Barbosa et al. 2020; Vélez-Mercado et al. 2021). Different pretreatment techniques have been developed over the past few years to deal with technological obstacles for valorizing LB such as biomass resistance, high energy consumption, and high operating costs (Baptista and Domingues 2022; Sharma et al. 2022a).

Numerous techniques have been used to break down LB and lessen its naturally tough nature, including physical, chemical, biological, and other integrative techniques (Rabemanolontsoa and Saka 2016; Sethupathy et al. 2021). Out of all the strategies available, biological pretreatment via microbial hydrolases is one of the benign techniques used these days due to its high efficiency, environmental friendliness, and exceptionally high success rate (Rabemanolontsoa and Saka 2016; Baptista and Domingues 2022). Numerous microorganisms existing in nature can contribute to the breakdown of LB, with numerous bacterial and fungal genera producing cellulolytic and hemicellulolytic enzymes (Barbosa et al. 2020; Vélez-Mercado et al. 2021).

4 Biological Pretreatment Methods for Lignocellulosic Biomass Hydrolysis

Biological methods are preferred over other methods because the physical, chemical, and physiological pretreatment strategies generate various types of inhibitory compounds like furans (5-hydroxymethylfurfural (HMF) and furfural), weak acids (acetic acid), and different compounds of phenol like 4-hydroxybenzaldehyde, catechol, and vanillin that are toxic to fermentative organisms (Sethupathy et al. 2021; Baptista and Domingues 2022). In recent eras, the research has turned toward the exploitation of bacterial communities for LB valorization into various value-added products on a sustainable scale (Barbosa et al. 2020; Chukwuma et al. 2021; Wicker et al. 2021).

The specialty of bacteria and fungi is that they act as good decomposers, and therefore, this characteristic is used to hydrolyze and break the complex linkages of LB. The LB is converted by bacteria and fungi into fermentable sugars and aliphatic/aromatic monomers that can be easily utilized by the fermentative organism (Wicker et al. 2021; Baptista and Domingues 2022). Fungi have been used naturally for the degradation of biomass, but in recent research, bacterial communities are being cultivated due to their many features such as the facility of heterologous expression of their genome or the entire genome can be edited and reconstructed as needed through genetic engineering and diverse occurrence, capable of withstanding exceptional environment and shorter incubation time (Sethupathy et al. 2021; Vélez-Mercado et al. 2021). All these features motivate the scientific community to explore the promising property of biomass degradation in bacterial communities as well, which can further ease the way for biorefineries. Bacteria possess a wide variety of hydrolases for the degradation or cleavage of the internal linkages between the monomers of lignin, cellulose, and hemicellulose in the LB (Barbosa et al. 2020; Chukwuma et al. 2021; Sethupathy et al. 2021).

Because lignin serves as a barrier to preserve cellulose during the transformation of biomass, its removal is both crucial and challenging. Extreme conditions of reaction (high pressure, temperature, and acidic/alkaline pH) are needed for the energy-intensive chemical breakdown of lignin (Berg and Laskowski 2005; Lee et al. 2019; Chukwuma et al. 2021). Microbes are thought to be the greatest option for lignin degradation because they can break down lignin both aerobically and anaerobically. For bacterial delignification, *Streptomyces* from the order *Actinomycetales* are frequently chosen species. Apart from *Streptomyces*, *Pseudomonas*, *Sphingomonas*, *Nocardia*, and *Rhodococcus* have also been reported for lignin degradation (Berg and Laskowski 2005; Xie et al. 2016). The sequential step of bacterial delignification involves tunnelling, erosion, and cavitation (Xie et al. 2016; Barbosa et al. 2020; Vélez-Mercado et al. 2021).

5 Bacterial Hydrolytic Machinery for LB Hydrolysis

The recalcitrant nature of LB is a chief obstacle for its use on an industrial level. However, the microbial diversity available in nature has also provided a solution to this problem for researchers. The most primitive and simple life originated on earth is bacteria, which are capable of maintaining the nutrient cycle and are ubiquitous (Manisha and Yadav 2017; Vélez-Mercado et al. 2021). There is a huge metabolic diversity of bacteria involved in breaking the recalcitrance of complex LB via hydrolyzing hemicellulose and cellulose into their simple sugar counterparts. The bacterial genera usually used to hydrolyze the cellulosic content of LB includes *Actinomycetales* (aerobic) from the phylum *Actinobacteria* and *Clostridiales* (anaerobic) from the phylum *Firmicutes* (Manisha and Yadav 2017; Chukwuma et al. 2021). *Clostridium* sp. is rich in a cocktail of enzymes popularly known as cellulosome, which consists of cellulases, hemicellulases, glucanases, chitinases, laccases, peroxidases, pectate lyases, and xylanases (Barbosa et al. 2020). Other reported bacterial genera for LB hydrolysis are *Bacillus*, *Microbispora*, *Acetivibrio*, *Thermomonospora*, *Cellulomonas*, *Bacteroides*, *Erwinia*, *Streptomyces*, and *Ruminococcus* (Xie et al. 2016; Barbosa et al. 2020; Chukwuma et al. 2021). Biomass-degrading enzymes are potent biotechnology tools in the present notion of biorefining (Rabemanolontsoa and Saka 2016; Chukwuma et al. 2021; Vélez-Mercado et al. 2021).

The microorganisms including the potent fungi and bacteria produce lignocellulolytic enzymes, which can be isolated from the site where LB degradation is going on. To effectively and specifically degrade the lignocellulose raw material source to produce diverse value-added products, the expression of these enzymes can be optimized along with their detailed characterization, structure, and reaction mechanism analysis (Lombard et al. 2014; Vélez-Mercado et al. 2021). The carbohydrate-active enzyme database classifies the different enzymes needed for LB to break into simple forms as CAZymes, or carbohydrate-active catalytic enzymes (Vélez-Mercado et al. 2021; Gupta et al. 2022). The family categorization of CAZymes is accessible online through the CAZy database (CAZy, www.cazy.org/) and is continuously updated as per the latest discoveries (Manisha and Yadav 2017; Nargotra et al. 2022; Gupta et al. 2022). Cellulolytic and hemicellulolytic enzymes are glycoside hydrolases (GH), which are classed in CAZy, and are hydrolytic enzymes that can act on cellulose and hemicellulose, respectively (Lombard et al. 2014). Using omics technologies like metagenomics, proteomics, and transcriptomics, which do not require isolating and cultivating microbes to produce enzymes, it is possible to find new CAZymes for hydrolyzing the LB (Manisha and Yadav 2017; Vélez-Mercado et al. 2021).

Apart from the use of various hydrolytic enzymes, lytic polysaccharide monooxygenases (LPMOs) are also employed for LB hydrolysis (Manisha and Yadav 2017; Houfani et al. 2020). The discovery of LPMO had broken the paradigm that the LB is hydrolyzed only through the hydrolytic enzymes. LPMOs have the potential to improve fermentation efficiency and save expenses. LPMO is an oxidative way of breaking the recalcitrance of LB, which is mediated via copper ion (Manisha and Yadav 2017; Valenzuela et al. 2017; Houfani et al. 2020). In the existence of an electron donor like cellobiose dehydrogenase (CDH) and ascorbic acid, copper ions reduce oxygen. Reduced oxygen then consumes the ions of hydrogen from substrate, which causes the β-1,4-glycosidic link to break (Valenzuela et al. 2017). Cellulases can act through LPMOs by oxidizing the glycosidic bond (Barbosa et al. 2020). Histidine and copper are combined in the N- and C-termini of LPMOs' active site (Valenzuela et al. 2017).

The LPMO is capable of cleaving the glycosidic bond of LB, which fails to be hydrolyzed by the application of various hydrolases (Manisha and Yadav 2017). Under the CAZy database, all LPMOs described till now are categorized as auxiliary activities (AA), under families AA9, AA10, AA11, and AA13 with low sequence identity among them (Forsberg et al. 2014; Lombard et al. 2014; Manisha and Yadav 2017). However, they share a common structural architecture (Valenzuela et al. 2017; Barbosa et al. 2020). The only AA family that has bacterial LPMOs that breaks down substrates like cellulose and chitin is family AA10, which is formerly CBM33 (Forsberg et al. 2014). The best-known AA10 enzyme, CBP21 from *Serratia marcescens*, works on chitin (Lombard et al. 2014; Valenzuela et al. 2017).

There are several ways of LB hydrolysis; however, the free enzyme system is the frequently adopted measure by aerobic bacteria. On the contrary, more complex proteinaceous structures in the form of cellulosomes and xylanosomes are the frequently used methods for LB hydrolysis (Chukwuma et al. 2021; Vélez-Mercado et al. 2021). The application of various hydrolytic enzymes for the breakdown of LB is an efficient green approach as well as a cost-effective strategy.

5.1 *Cellulose Hydrolysis*

The key player for the cellulose hydrolysis via enzymatic saccharification involves the application of cellulases, which predominantly contains endo-1,4-β-glucanases (EC 3.2.1.4), exo-1,4-β-glucanases (EC 3.2.1.91), and β-glucosidase (EC 3.2.1.21) (Table 1), which is commonly known as cellobiase (Rabemanolontsoa and Saka 2016; Manisha and Yadav 2017; Wicker et al. 2021; Nargotra et al. 2022). β-1,4 glycosidic linkages in cellulose are sliced by a set of hydrolytic different enzymes known as cellulases, which yield glucose, cellobiose, and cellooligosaccharides as

Table 1 Bacterial hydrolases (lignocellulolytic enzyme) for the hydrolysis of lignocellulosic biomass

Bacteria	Enzyme	Raw material	Result	References
Bacillus halodurans FNP 135	Xylanase, laccase	Wheat bran	Enzyme activity: xylanase, 1685 IU/g, laccase, 2270 nkat/g	Gupta et al. (2015)
Ruminiclostridium thermocellum M3	β-glucoside	Poplar sawdust, corn cobs, rice straw, cornstalk	Oligosaccharide yield of 52.7 ± 2.77, 77.8 ± 5.9, 89.9 ± 9.3 and 107.8 ± 5.88 mg/g respectively	Sheng et al. (2016)
Bacillus aerius and *Bacillus anthracis* (gram-positive bacteria)	Cellulase	Oil palm + rice residue	Degradation of raw material up to 21.35%	Oke et al. (2017)
Cellvibrio japonicus, Klebsiella pneumoniae (gram-negative bacteria)			Degradation of raw material up to 52%	
Micromonospora sp. G7	Xylanase, CMCase, LiP, MnP, laccase	Rice straw, corn straw, wheat straw, soyabean	Degradation ratio increase: cellulose, 46.2%, hemicellulose, 22.5%, lignin, 28.9%	Wei et al. (2019)
Streptomyces sp. H1				
Saccharomonospora sp.				
Mycobacterium sp.				
Bacillus ligniniphilus L1	Laccase	Rice straw	Lignin content decrease, 8.93%; phenolic content decrease, 44.8%	Nazar et al. (2022)
Acinetobacter sp. B213	MnP, LiP	Corn straw	98.51% cellulose retention rate and 12.02% lignin degradation rate	Li et al. (2022)

LiP lignin peroxidase, *MnP* manganese peroxidase, *CMCase* carboxymethyl cellulase

their main by-products after hydrolysis (Manisha and Yadav 2017; Wicker et al. 2021; Tiwari et al. 2022). These can attach to carbohydrates and modify, degrade, or produce glycosidic linkages, which classifies them as CAZymes. Cellulases, such as cellobiohydrolases (CBHs), exoglucanases, endoglucanases, and β-glucosidases, are categorized based on the mechanism of catalysis that they use (Houfani et al. 2020; Vélez-Mercado et al. 2021).

The catalysis of β-1,4-glycosidic bonds of cellulose is catalyzed by endoglucanases in a random fashion. Few of the functional endoglucanases contain a catalytic domain from the glycoside hydrolase family 5 (GH5) family, whereas most of them are in family GH9. Though their enzymatic pathways are still not fully understood, endoglucanases of GH5 appear to have a more complicated process of action for glycosidic bond cleavage than GH9 endoglucanases (Rabemanolontsoa and Saka 2016; Wicker et al. 2021; Vélez-Mercado et al. 2021). In addition to the GH9 catalytic domain, GH9 processive endoglucanases also contain an auxiliary carbohydrate-binding module (CBM), which has a major impression on the activity of the enzyme. The glycosidic bond of nonreducing point is cleaved by the exoglucanases, whereas β-glucanases cleave cellobiose and therefore release two glucose units (Saha et al. 2005; Barbosa et al. 2020; Devi et al. 2022). For the cellulose degradation, a symbiosis is observed between two types of microorganisms. The one is the cellulolytic (primary microorganisms), which are capable of breaking off cellulose to release monomeric sugar. On the contrary, the secondary microorganisms utilize these hydrolytic products as they are unable to utilize complex sugar. For efficient cellulosic breakdown, both of these microorganisms show symbiosis (Rabemanolontsoa and Saka 2016; Chukwuma et al. 2021; Vélez-Mercado et al. 2021).

5.2 Hemicellulose Hydrolysis

Apart from the GH, hemicellulases also comprise carbohydrate esterases (CEs). The hemicellulose structure is mostly made up of xylan residue, and therefore xylanases (EC 3.2.1.8) are vital enzymes in the process of depolymerizing hemicellulose by cleaving the β-1,4 linkages in xylan backbone, forming xylooligomers such as xylose and xylobiose (Vélez-Mercado et al. 2021).

For hemicellulose decomposition, xylanases, glucomannanase, β-xylosidase, galactomannanase, acetylesterase, and glucuronidase are chief enzymes used (Table 1) (Rabemanolontsoa and Saka 2016; Wicker et al. 2021; Vélez-Mercado et al. 2021). The xylan backbone is first depolymerized into xylooligosaccharides by endo-β-1,4-xylanase, which are then further degraded by β-xylosidase (Vélez-Mercado et al. 2021). Meanwhile, side groups are hydrolyzed by various glycosidases, including α-arabinofuranosidase, β-glucuronidases, ferulic acid esterase, and acetyl xylan esterase (Saha et al. 2005; Rabemanolontsoa and Saka 2016). Numerous aerobic and anaerobic mesophilic, thermophilic, and extremophilic microbes produce hemicellulases (Houfani et al. 2020; Vélez-Mercado et al. 2021).

5.3 Lignin Hydrolysis

For the degradation of lignin of LB, white-rot fungi are prominently used. However, several bacterial genera like *Streptomyces*, *Pseudomonas,* and *Actinomycetes* are capable of delignification by generating several enzymes including laccases (E.C 1.10.3.2) and peroxidase like phenol oxidase, catalase, demethylase, lignin peroxidase (LiP; E.C 1.11.1.7), and manganese peroxidase (MnP; E.C 1.11.1.13) (Hatakka 2005; Houfani et al. 2020; Wicker et al. 2021). Aryl-alcohol oxidase, aryl alcohol dehydrogenase, glyoxal oxidase, glucose oxidase, and cellobiose dehydrogenase are a few more enzymes that are responsible for the breakdown of lignin (Saha et al. 2005; Zhang et al. 2020; Vélez-Mercado et al. 2021). These enzymes catalyze a number of chemical changes in lignin, including aliphatic side-chain and aromatic ring breakage, side-chain oxidation, decarboxylation, demethoxylation, and single-bond chain breakage, which leads to the monomerization of the lignin macromolecule (Bugg et al. 2011; Houfani et al. 2020; Zhang et al. 2020). For the delignification,

Microbacterium phyllosphaerae, *Actinobacteria*, *Rhodococcus jostii* RHA1, *Bacillus pumilus*, *Paenibacillus* sp., *Pseudomonas putida* KT2440, and *Ochrobactrum* sp. are reported by several researchers (Manisha and Yadav 2017; Nargotra et al. 2022).

According to the CAZy database, a variety of enzymes, which are redox-active categorized under families of auxiliary activity (AA), are involved in the bacterial delignification. For instance, AA1 and AA2 families, containing laccases and lignin-active class-II peroxidases respectively, are engaged in modification and degradation of lignin and are of the two main subfamilies of these auxiliary activities (Zhang et al. 2020; Nargotra et al. 2022). Together, these enzymes and glycoside-hydrolases (GHs) breakdown lignocellulose. Multi-copper oxidases known as laccases are active against monophenols and substances with p-diphenol structures (Bugg et al. 2011). A heme structure with a Mn^{2+} in the center is present in manganese peroxidases (MnP) (Manisha and Yadav 2017; Barbosa et al. 2020). These enzymes catalyze the oxidation of lignin by utilizing the electron of H_2O_2 to reduce substrates (Bugg et al. 2011; Houfani et al. 2020). Other families, like the AA4 containing vanillyl-alcohol oxidases, the AA5 of glyoxal oxidases/alcohol oxidases, and the family AA6 consisting 1,4-benzoquinone reductases, are also involved in the reduction or oxidation of lignin-derived components (Saha et al. 2005; Zhang et al. 2020). Furthermore, members of the families AA3 and AA7 exhibit possible involvement in lignin modification and degradation via Fenton reactions.

6 Conclusion

The recalcitrance of LB becomes a very challenging matter when we try to use it on a large scale in biorefineries. Different pretreatment strategies are used in the laboratory to break down/hydrolyze the complex linkages between lignin, hemicellulose, and cellulose. The major issues on physical and chemical pretreatment are on the requirements of sophisticated instrumentation as well as the formation of inhibitors during the treatment process that hinder the formation of the desired product and their expensiveness. Keeping all these shortcomings in mind, biological treatment, which involves breaking down their robustness and recalcitrance using lignocellulolytic enzymes made by microorganisms such as bacteria and fungi, is an exceptional alternative. Due to their widespread use in modern industrial processes, hydrolytic enzymes are essential tools in the manufacturing of numerous foods, medicines, and consumables.

The transformation of lignocellulosic biomass into value-based products is one of the main applications of hydrolytic enzymes in this field. Cellulases, hemicellulases, and ligninolytic enzymes are the primary lignocellulose-degrading enzymes. In today's decade, different microorganisms and their hydrolases are being used for hydrolysis of lignocellulose biomass. In current biorefineries, various types of LB are used to produce various value-added products. The current chapter offers insights into a viable alternative for using bacterial hydrolytic machinery to generate value-added goods like organic acids, biofuels, antibiotics, proteins, enzymes, and pigments through the hydrolysis of complex lignocellulosic biomass, which will expand the potential of biorefineries in terms of their ability to be economically and environmentally sustainable.

References

Baptista M, Domingues L (2022) *Kluyveromyces marxianus* as a microbial cell factory for lignocellulosic biomass valorisation. Biotechnol Adv 60:108027. https://doi.org/10.1016/j.biotechadv.2022.108027

Barbosa FC, Silvello MA, Goldbeck R (2020) Cellulase and oxidative enzymes: new approaches, challenges and perspectives on cellulose degradation for bioethanol production. Biotechnol Lett 42:875–884. https://doi.org/10.1007/s10529-020-02875-4

Berg B, Laskowski R (2005) Decomposers: soil microorganisms and animals. Adv Ecol Res 38:73–100. https://doi.org/10.1016/S0065-2504(05)38003-2

Bugg TD, Ahmad M, Hardiman EM, Rahmanpour R (2011) Pathways for degradation of lignin in bacteria and fungi. Nat Prod Rep 28(12):1883–1896. https://doi.org/10.1039/C1NP00042J

Chukwuma OB, Rafatullah M, Tajarudin HA, Ismail N (2021) A review on bacterial contribution to lignocellulose breakdown into useful bio-products. Int J Environ Res Public Health 18(11):6001. https://doi.org/10.3390/ijerph18116001

de Almeida AP, Vargas IP, Marciano CL, Zanoelo FF, Giannesi GC, de Moraes MDLT, Masui DC (2022) Investigation of biochemical and biotechnological potential of a thermo-halo-alkali-tolerant endo-xylanase (GH11) from *Humicola brevis* var. *thermoidea* for lignocellulosic valorization of sugarcane biomass. Biocatal Agric Biotechnol 44:102424. https://doi.org/10.1016/j.bcab.2022.102424

Devi A, Bajar S, Kour H, Kothari R, Pant D, Singh A (2022) Lignocellulosic biomass valorization for bioethanol production: a circular bioeconomy approach. Bioenergy Res 15(4):1820–1841. https://doi.org/10.1007/s12155-022-10401-9

Forsberg Z, Røhr ÅK, Mekasha S, Andersson KK, Eijsink VG, Vaaje-Kolstad G, Sørlie M (2014) Comparative study of two chitin-active and two cellulose-active AA10-type lytic polysaccharide monooxygenases. Biochemist 53(10):1647–1656. https://doi.org/10.1021/bi5000433

Gupta V, Garg S, Capalash N, Gupta N, Sharma P (2015) Production of thermo-alkali-stable laccase and xylanase by co-culturing of *Bacillus* sp. and *B. halodurans* for biobleaching of kraft pulp and deinking of waste paper. Bioprocess Biosyst Eng 38:947–956. https://doi.org/10.1007/s00449-014-1340-0

Gupta N, Beliya E, Paul JS, Jadhav SK (2022) Nanoarmoured α-amylase: a route leading to exceptional stability, catalysis and reusability for industrial applications. Coord Chem Rev 464:214557. https://doi.org/10.1016/j.ccr.2022.214557

Hatakka A (2005) Biodegradation of lignin. In: Biopolymers. Wiley, Hoboken. https://doi.org/10.1002/3527600035.bpol1005

Houfani AA, Anders N, Spiess AC, Baldrian P, Benallaoua S (2020) Insights from enzymatic degradation of cellulose and hemicellulose to fermentable sugars–a review. Biomass Bioenergy 134:105481. https://doi.org/10.1016/j.biombioe.2020.105481

Lee S, Kang M, Bae JH, Sohn JH, Sung BH (2019) Bacterial valorization of lignin: strains, enzymes, conversion pathways, biosensors, and perspectives. Front Bioeng Biotechnol 7:209. https://doi.org/10.3389/fbioe.2019.00209

Li H, Zhang X, Liu J, Chen B, Ran Q, Wang R, Yu C, Ma X, Song H, Jiang Z (2022) A newly isolated lignin degradation bacterium *Acinetobacter* sp. B213 pretreatment effectively improved hydrolysis of corn straw by enhance cellulose retention and reduce lignin degradation inhibitor. https://doi.org/10.21203/rs.3.rs-2060011/v1

Lombard V, Golaconda Ramulu H, Drula E, Coutinho PM, Henrissat B (2014) The carbohydrate-active enzymes database (CAZy) in 2013. Nucleic Acids Res 42(D1):D490–D495. https://doi.org/10.1093/nar/gkt1178
Manisha, Yadav SK (2017) Technological advances and applications of hydrolytic enzymes for valorization of lignocellulosic biomass. Bioresour Technol 245:1727–1739. https://doi.org/10.1016/j.biortech.2017.05.066
Nargotra P, Sharma V, Lee YC, Tsai YH, Liu YC, Shieh CJ, Kuo CH (2022) Microbial Lignocellulolytic enzymes for the effective valorization of lignocellulosic biomass: a review. Catalysts 13(1):83. https://doi.org/10.3390/catal13010083
Nazar M, Lingxia X, Ullah MW, Moradian JM, Wang Y, Sethupathy S, Iqbal B, Nawaz MZ, Zhu D (2022) Biological delignification of rice straw using laccase from *Bacillus ligniniphilus* L1 for bioethanol production: a clean approach for agro-biomass utilization. J Clean Prod 360:132171. https://doi.org/10.1016/j.jclepro.2022.132171
Oke MA, Annuar MSM, Simarani K (2017) Mixed lignocellulosic biomass degradation and utilization for bacterial cellulase production. Waste Biomass Valoriz 8:893–903. https://doi.org/10.1007/s12649-016-9595-0
Rabemanolontsoa H, Saka S (2016) Various pretreatments of lignocellulosic. Bioresour Technol 199:83–91. https://doi.org/10.1016/j.biortech.2015.08.029
Ralph J, Lundquist K, Brunow G, Lu F, Kim H, Schatz PF, Boerjan W (2004) Lignins: natural polymers from oxidative coupling of 4-hydroxyphenyl-propanoids. Phytochem Rev 3:29–60. https://doi.org/10.1023/B:PHYT.0000047809.65444.a4
Saha BC, Iten LB, Cotta MA, Wu YV (2005) Dilute acid pretreatment, enzymatic saccharification and fermentation of wheat straw to ethanol. Process Biochem 40(12):3693–3700. https://doi.org/10.1016/j.procbio.2005.04.006
Sankaran R, Markandan K, Khoo KS, Cheng CK, Ashokkumar V, Deepanraj B, Show PL (2021) The expansion of lignocellulose biomass conversion into bioenergy via nanobiotechnology. Front Nanotechnol 3:96. https://doi.org/10.3389/fnano.2021.793528
Sethupathy S, Morales GM, Li Y, Wang Y, Jiang J, Sun J, Zhu D (2021) Harnessing microbial wealth for lignocellulose biomass valorization through secretomics: a review. Biotechnol Biofuels 14:1–31. https://doi.org/10.1186/s13068-021-02006-9
Sharma V, Tsai ML, Chen CW, Sun PP, Patel AK, Singhania RR, Dong CD (2022a) Deep eutectic solvents as promising pretreatment agents for sustainable lignocellulosic biorefineries: a review. Bioresour Technol 360:127631. https://doi.org/10.1016/j.biortech.2022.127631
Sharma S, Tsai ML, Sharma V, Sun PP, Nargotra P, Bajaj BK, Dong CD (2022b) Environment friendly pretreatment approaches for the bioconversion of lignocellulosic biomass into biofuels and value-added products. Environment 10(1):6. https://doi.org/10.3390/environments10010006
Sheng T, Zhao L, Gao L-F, Liu W-Z, Cui M-H, Guo Z-C, Ma X-D, Ho S-H, Wang A-J (2016) Lignocellulosic saccharification by a newly isolated bacterium, *Ruminiclostridium thermocellum* M3 and cellular cellulase activities for high ratio of glucose to cellobiose. Biotechnol Biofuels 9:1–11. https://doi.org/10.1186/s13068-016-0585-z
Tiwari S, Beliya E, Vaswani M, Khawase K, Verma D, Gupta N, Jadhav SK (2022) Rice husk: a potent lignocellulosic biomass for second generation bioethanol production from *Klebsiella oxytoca* ATCC 13182. Waste Biomass Valoriz 13:2749–2767. https://doi.org/10.1007/s12649-022-01681-5
Valenzuela SV, Ferreres G, Margalef G, Pastor FJ (2017) Fast purification method of functional LPMOs from *Streptomyces ambofaciens* by affinity adsorption. Carbohydr Res 448:205–211. https://doi.org/10.1016/j.carres.2017.02.004
Vélez-Mercado MI, Talavera-Caro AG, Escobedo-Uribe KM, Sánchez-Muñoz S, Luévanos-Escareño MP, Hernández-Terán F, Balagurusamy N (2021) Bioconversion of lignocellulosic biomass into value added products under anaerobic conditions: insight into proteomic studies. Int J Mol Sci 22(22):12249. https://doi.org/10.3390/ijms222212249
Wei Y, Di W, Wei D, Zhao Y, Junqiu W, Xie X, Zhang R, Wei Z (2019) Improved lignocellulose-degrading performance during straw composting from diverse sources with actinomycetes

inoculation by regulating the key enzyme activities. Bioresour Technol 271:66–74. https://doi.org/10.1016/j.biortech.2018.09.081

Wicker RJ, Kumar G, Khan E, Bhatnagar A (2021) Emergent green technologies for cost-effective valorization of microalgal biomass to renewable fuel products under a biorefinery scheme. Chem Eng J 415:128932. https://doi.org/10.1016/j.cej.2021.128932

Xie S, Ragauskas AJ, Yuan JS (2016) Lignin conversion: opportunities and challenges for the integrated biorefinery. Ind Biotechnol 12(3):161–167. https://doi.org/10.1089/ind.2016.0007

Zhang S, Xiao J, Wang G, Chen G (2020) Enzymatic hydrolysis of lignin by ligninolytic enzymes and analysis of the hydrolyzed lignin products. Bioresour Technol 304:122975. https://doi.org/10.1016/j.biortech.2020.122975

The Contribution of Fungal and Its Enzymatic System Toward Multiproduct Biorefinery and Global Biofuel Research

Amisha Patel and Ravi Patel

1 Introduction

The unpredictable and inconsistent availability of petroleum resources, as well as the increasing cost of energy, has necessitated a global shift toward renewable energy sources to produce sustainable and alternative energy that can also fulfill the world's energy crisis (Patel et al. 2019a, b). Biofuel can help the economy and the environment by helping to reduce climate change because biofuels are a C-neutral alternative (Mohanram et al. 2013). In 2018, India's National Biofuels Policy announced that it would step up its efforts to promote biofuels, with the goal of getting 20% ethanol blended into petrol by 2030. Nowadays, oil manufacturing industries are adding 10% ethanol in petrol to meet a target, in accordance with the BIS criteria. To satisfy this demand, biofuels must therefore be produced in considerably greater quantities (Patel and Shah 2021a, b).

Nowadays, the generation of waste has become a major global concern as it has been increased with both population and economic growth (Awasthi et al. 2022). The two sectors that produce the most waste are agriculture and agri-food processing, with lignocellulosic waste being the predominant type (Gil et al. 2015). The lignocellulosic biomass (LCB) found in agricultural waste is mostly composed of cellulose, hemicellulose, and lignin, all of which can be utilized to generate biofuel and value-added products (VAP) (Awasthi et al. 2020; Wainaina et al. 2019). Around

A. Patel (✉)
P. G. Department of Biosciences, Sardar Patel University, Satellite Campus, Anand, Gujarat, India

R. Patel
Department of Botany, Faculty of Science, The Maharaja Sayajirao University of Baroda, Vadodara, Gujarat, India

K. Agrawal, P. Verma (eds.), *Biotechnological Advances in Biorefinery*, Interdisciplinary Biotechnological Advances,
https://doi.org/10.1007/978-981-97-5544-8_3

70% of all renewable energy generated worldwide is produced from LCB (Raud et al. 2019). The biorefinery offers a sustainable and environmentally friendly process for utilizing this waste and producing a large amount of marketable fuel and products. Furthermore, the production of biofuels and bioproducts holds significant potential to fulfill the growing demand for petroleum products (Awasthi et al. 2022; Vlachokostas et al. 2021).

The recalcitrant nature, abrasive structure, and diverse biochemical makeup of LCB have become a mounting task to surmount for bioconversion of LCB into biofuels, notwithstanding the massive global efforts to make biorefinery process more competitive in the energy sector (Onwosi et al. 2024). In nature, lignocellulose-degrading fungi and bacteria use a wide range of holocellulolytic and lignin-derived enzymes to depolymerize this biomass. The utilization of these enzymes to convert LCB into biofuel and bioproducts in a biorefinery process is highly appealing as it enhanced profitability (Østby and Várnai 2023). Much attention is being paid on research for biofuel production using lignocellulolytic fungi as it tackles up the energy demand and eliminates dangerous environmental pollutants, thereby assisting in global warming issues (Saini and Sharma 2021).

In light of these, this chapter aims to review the benefits and scope of employing fungi and their enzyme repertoire for biofuel and VAP generation in a biorefinery system.

2 Lignocellulose: An Entangled Composite Network

LCB is being considered as the most abundant source of carbon-based biomass on Earth and is comprised of three primary components, i.e., the structural polysaccharides portion, which is made up of cellulose and hemicellulose (i.e. holocellulose), lignin, and pectin (Østby and Várnai 2023; Sperandio and Ferreira Filho 2019).

Cellulose is a homopolymer composed of several glucose monosaccharides linked via β-1,4 configuration. Single chains create a long chain polymer by van der Waals force and hydrogen bonding (Sunardi 2020). It exists in two distinct forms, namely, amorphous and crystalline, and is the more prevalent biopolymer (Sperandio and Ferreira Filho 2019). Higher levels of crystallinity make the substrate more recalcitrant, which impedes the process of enzymatic hydrolysis (Bragatto et al. 2012; Howell et al. 2011). The content of cellulose typically accounts for 35–50% of LCB. There are three types of cellulose, i.e., α, β, and γ, in accordance with the degree of polymerization (DP) and ability to dissolve in NaOH (17.5%). The α-cellulose is a polymer with longer chain length and DP of 600–1500 that is unable to dissolve in NaOH, while the β-cellulose is a polymer with shorter chain length and DP of 15–90 that dissolves in NaOH. The γ-cellulose is reminiscent of β-cellulose, but it has DP <15 with hemicellulose as the main content (Suryadi et al. 2022).

Hemicellulose, the second polysaccharide component of LCB, accounts for 15–30% of the dry weight of biomass (Suryadi et al. 2022). Hemicellulose connects directly with the cellulose fibers through an intricate and multifaceted web of non-cellulosic polymers. Considering all of its constituents, hemicellulose ranks as the second most prevalent polymer on Earth (Saha 2003). Hemicellulose and pectin combine to form a structure resembling a gel, which envelops cellulose in an array. Furthermore, the hemicelluloses are the polymers with sugar units like pentoses (arabinose, xylose) and hexoses (glucose, galactose, mannose), as well as substituted side chains of a low molecular weight branched or linear or branched structures (Sunardi 2020). Because of this additional complexity, a variety of hydrolytic enzymes are required to break down the hemicellulosic component into its individual monomers (Sperandio and Ferreira Filho 2019). Moreover, lignin-carbohydrate complexes are formed by interactions between lignin and hemicelluloses in covalent bonds with ester or ether linkages, and it has been proposed that breaking these covalent bonds in lignin-carbohydrate complexes is critical for lignin degradation (Sunardi 2020).

Lignin, the main structural non-polysaccharides in LCB, is an aromatic biopolymer that is created by oxidatively coupling non-regularly arranged monolignols. Sinapyl alcohol, *p*-coumaryl alcohol, and coniferyl alcohol are the three foremost and ubiquitous monolignols. These monolignols become Syringyl (S), *p*-hydroxyphenyl (H), and guaiacyl (G) residues as these are incorporated into the developing lignin polymer (Østby and Várnai 2023). The amount of the third most prevalent component lignin varies among the different LCB, and the lignin content is >30% in hardwood as compared with herbaceous plants or softwood (Ishak et al. 2019).

Pectin is an extremely versatile polysaccharide, consisting of a galacturonic acid-rich backbone bound together via α-(1,4) bonds. There are five pectic domains: homogalacturonan, apiogalacturonan, rhamnogalacturonan I, rhamnogalacturonan II, and xylogalacturonan (Sperandio and Ferreira Filho 2019). The percentage of three main constituents in some LCB of biorefineries interest has been shown in Table 1.

Despite cellulose, the longest-utilized fraction of LCB, being historically used in pulp and paper industries, the hemicellulosic portion is still comparatively underutilized in contemporary biorefineries due to its complex chemical structure (Østby and Várnai 2023). The biomass is more resilient to enzymatic hydrolysis in the

Table 1 Composition of important LCB

Lignocellulosic biomass	Cellulose (%)	Hemicellulose (%)	Lignin (%)	References
Sugarcane bagasse	42–48	19–25	20–42	Kim and Day (2011)
Switchgrass	17–36	20–28	18–26	Yadav et al. (2023)
Corn cobs	42–45	35–39	14–15	Patel and Shah (2021a, b)
Corn stover	38–40	24–26	7–19	Rahikainen et al. (2013)
Rice straw	28–36	23–28	12–14	Rahikainen et al. (2013)
Wheat straw	33–38	26–32	17–19	Rahikainen et al. (2013)
Cotton	85–95	5–15	0	Patel and Shah (2021a, b)

presence of lignin because it inhibits the enzymatic access to carbohydrates and encourages the nonproductive bonds with cellulases, which lowers the hydrolytic efficiency (Rahikainen et al. 2013). Therefore, lignin is considered to be a barrier to biorefining procedures and is handled like a waste product that can be ignited to produce heat or energy (Ten and Vermerris 2015). In this regard, a number of pretreatment processes were intended to delignify the LCB or alter the lignin framework to some degree prior to the utilization of biomass for enzymatic hydrolysis (Chen and Dixon 2007; Ko et al. 2015; Yang et al. 2016). Lately, new methods have changed this "waste" viewpoint toward lignin, and this is currently being evaluated for use in commercial sectors like the creation of bioplastics and wood adhesives (Ten and Vermerris 2015; Saito et al. 2012; Khan et al. 2004).

To utilize the cellulose fraction for biofuel biorefinery, pretreatment is being considered as an indispensable step to disrupt the complex and recalcitrant lignocellulosic matrix (Patel and Shah 2021b). Pretreatment aims to achieve four main goals: (1) alter the crystallinity and degree of polymerization of cellulose, (2) break down the lignin-carbohydrate linkage, (3) remove lignin and/or hemicellulose, and (4) improve porosity (Moreno et al. 2019; Dharmaraja et al. 2023). These aims can be achieved by various physical, chemical, physicochemical, and biological methods (Baruah et al. 2018; Prasad et al. 2023). Despite being environmentally benign and not producing any hazardous compounds, physical treatment is not commonly utilized because it necessitates additional pretreatment (Zabed et al. 2019). The remarkable efficacy and recovery of spent chemicals make the chemical treatment beneficial. However, it has drawbacks because of its costly and toxic chemicals as well as environmental pollution (Baruah et al. 2018). The most widely used method for delignifying LCB is physicochemical treatment due to its advantages of time efficiency and high lignin removal rate. Nonetheless, cellulose degradation, high energy consumption, and generation of inhibitory compounds detrimental to subsequent process of enzymatic hydrolysis may restrict the application of this pretreatment method (Rajendran et al. 2018). In contrary, biological treatment using fungi, bacteria, or the enzyme secreted by lignin- decomposing microorganisms can diminish the negative aspects of chemical and physicochemical approaches (Tsegaye et al. 2019; Nadir et al. 2019; Kumar and Chandra 2020). Biological pretreatment encompasses the primary benefits of being safe and sustainable because it is free from toxic or reactive chemicals, thus diminishing the production of waste streams and the operational expenses. Biological pretreatment also has the advantages of minimal generation of inhibitors and by-products that do not inhibit or impede the process of hydrolysis and allow microbial enzymes to work with high specificity (Rezania et al. 2020; Ponnusamy et al. 2019). Notwithstanding the numerous advantages that the biological pretreatment method offers, it still has a number of shortcomings, including a high spatial requirement, lengthy processing times, and continuous microbial growth monitoring (Suryadi et al. 2022; Nadir et al. 2019).

3 Fungal Cell Factories: A Comprehensive Framework for Lignocellulolytic Enzyme

The main objective of the bioeconomy is the development of bioprocesses that are both environmentally friendly and economically feasible. The application of effective, versatile, and resilient cell factories for formation of VAP is a prerequisite for a productive biorefinery (Chatterjee and Mohan 2022). Fungi are deemed to be the most attractive source for lignocellulosic biorefinery among all microorganisms, owing to their robustness, higher growth rate, and ease of separating biomass and products, all of which lower the production costs (Althuri and Mohan 2019).

The biorefinery combines various strategies to deconstruct the complex and recalcitrant LCB into its principal constituents. One of the attractive ways is to use an array of LCB degrading enzymes derived from fungi (Gupta et al. 2016). However, one organism cannot possibly have all of the enzymes required for complete degradation of LCB. For instance, strains of the *Trichoderma* and *Aspergillus* genera meager in β-glucosidase and endoglucanase activities, respectively, while ascomycetes are unable to metabolize and break down the lignin, which is more preferably done by wood-rot basidiomycetes (Patel et al. 2019a, b; Maciel and Ribeiro 2010). Therefore, selecting a suitable type of fungi to use on a LCB depends on the makeup of the plant cell walls (Cragg et al. 2015). It is often not appropriate to view the breakdown of LCB by wood-rot fungi as a strictly harmful process. It also positively impacts the natural environment's nutrient cycling activities. Wood-rot fungi have evolved a vast array of oxidative and hydrolytic enzymes to attack the lignin and holocellulose components of LCB. These fungi are categorized as soft-rot (SRF), brown-rot (BRF), and white-rot (WRF) according to the type of degradation they have carried out. While SRF belong to ascomycetes, WRF and BRF are basidiomycetes (Sunardi 2020; Goodell 2020).

3.1 White-Rot Fungi: A Biodegradative Source

WRF have received great interest as they efficiently degrade the lignin component, which causes the wood that remains after decay to become more vividly colored (white) (Sunardi 2020). The whitish (white) appearance on wood is linked with the term "white-rot." This kind of fungi grows on dead wood from temperate to tropical regions, or on live trees under growing stage. They secrete numerous extracellular ligninolytic enzymes like manganese peroxidase, lignin peroxidase, and laccase (Suryadi et al. 2022). In accordance with the species of fungi and the amount of carbohydrates in the LCB, WRF can delignify to varying degrees. The utmost examined species of WRF are classified to sub families: *Polyporaceae* (*Pycnoporus sanguineus* and *Trametes versicolor*), *Pleurotaceae* (*Pleurotus pulmonarius* and *Pleurotus ostreatus*), *Phanerochaetaceae* (*Phanerochaete chrysosporium*), *Ganodermataceae* (*Ganoderma applanatum* and *Ganoderma lucidum*),

Hymenochaetaceae (*Phellinus igniarius* and *Inonotus hispidus*), *Meruliaceae* (*Phlebia radiata*, *Irpex lacteus*, and *Bjerkandera adusta*), and *Marasmiaceae* (*Lentinula edodes*) (Peralta et al. 2017). It was observed that certain WRF exhibit greater selectivity for lignin degradation, whereas others are nonselective (Rodríguez-Couto 2017). WRF such as *Phlebia radiata*, *Phanerochaete chrysosporium*, *Dichomitus squalens*, and *Ceriporiopsis subvermispora* have the capacity to degrade both hemicellulose and lignin more than cellulose, which belongs to a selective type of degradation mechanism. Conversely, *Fomes fomentarius* and *Trametes versicolor* are the examples of WRF belonging to the nonselective type of degradation as capable of degrading all components of LCB equally. These fungi are the only known organisms capable of fully breaking down lignin into CO_2 and water-soluble compounds (Suryadi et al. 2022). In recent studies, the WRF *Ganoderma lucidum* MDU-7 was utilized for fungal integrated biorefinery process. The study revealed maximum lignin degradation of 49.16% (Shankar et al. 2022). Conventional delignifications that rely solely on single fungal cultures are not economically feasible due to their extended incubation periods, and therefore, mixed fungal co-culture or consortia have been utilized to enhance the delignification and reduce the processing time. Furthermore, gene editing and multi-omics technologies are being explored as potential alternatives to improve the suitability of fungal cell factories for lignocellulosic biorefinery (Saini and Sharma 2021).

3.2 Brown-Rot Fungi: Deterministic But Enigmatic Source

Fungi that cause brown-rot break down cellulose and hemicelluloses while changing lignin to some extent. Consequently, lignin alteration, such as demethylation and oxidation, causes the rotting wood to shrink and turn brown (Sunardi 2020). The brown color of decayed wood following fungal deterioration has progressed to advanced stages ion, which the fungi termed BRF (Goodell 2020). Though they make up approximately 6% of all identified species of *Basidiomycota*, BRF in the northern hemisphere and in the rest of the world degrade about 80% of wood by mass (Eastwood 2014). Following the oxidation of LCB by hydroxyl radicals by means of Fenton reaction, cellulose and hemicellulose are broken down enzymatically as a consequence of BRF-induced wood decay (Patel et al. 2021b). Numerous studies are available on the production of biofuel and VAP from LCB using BRF like *Fomitopsis pinicola*, *Daedalea quercina*, *Serpula lacrymans*, *Laetiporus sulphureus*, *Postia placenta*, *Oligoporus placenta*, and *Fomitopsis meliae* (Patel et al. 2021b; Vaidya and Singh 2012; Schilling et al. 2009; Nurika et al. 2020; Monrroy et al. 2011; Lu et al. 2017; Giles et al. 2015). The collaborative action of *Fomitopsis pinicola* (BRF) and *Trametes orientalis* (WRF) has demonstrated a considerable shortening of pretreatment time and a rise of glucose yield up to 80% using corn cob (Wang et al. 2017). Holocellulolytic from *Fomitopsis meliae* has been observed to hydrolyze pretreated rice and wheat straws with 38.41% and 24.02% saccharification yield, respectively (Patel et al. 2021b). Unfortunately, researchers have never paid much attention to the BRF's degrading mechanism, which limits the large-scale applications of these fungi in multiproduct biorefinery. The potential of contemporary structural methodologies and multi-omics approaches to yield profound insights into the BRF degradation mechanism could lead to the commercial viability of these cultures (Saini and Sharma 2021).

3.3 *Soft-Rot Fungi: Economic Viable Source*

The cellulose-degrading enzymes secreted by SRF break down cellulose causing minute clefts and holes in the wood, as well as sporadic browning and cracking that resemble those caused by BRF (Sunardi 2020). *Lecythophora*, *Chaetomium*, and *Humicola* are examples of SRF (Goodell 2020). Numerous cellulase, laccase, hemicellulose, and pectinases enzymes have been found to be produced by 27 freshwater soft-rot ascomycetes using agar tests (Simonis et al. 2008). Despite being efficient, SRF are strong producers of extracellular peroxidases and oxidases, but they cannot oxidize the guaiacyl lignin. This is in contrast to their capacity to readily oxidize the syringyl units in lignin (Nargotra et al. 2022). SRF, unlike BRF and WRF, have not been directly exploited to produce biofuels; nonetheless, their potent extracellular holocellulolytic enzymes are widely used to produce important chemicals and biofuels on a commercial scale. The potential of SRF to withstand a broad range of temperatures and moisture levels could lead to their potential use in fungal-integrated multiproduct biorefineries in forthcoming years (Saini and Sharma 2021).

4 Lignocellulolytic Enzymes: A Predominant Path for Integrated Biorefinery

Lignocellulolytic enzymes are biological catalysts that break down or fractionate holocellulose and lignin fractions of LCB into their key constituents for subsequent hydrolysis (Chukwuma et al. 2020). Lignocellulolytic enzymes can be further characterized as ligninolytic enzymes and hydrolytic enzymes (Mtui 2012). The synergistic production of lignocellulolytic enzymes has been investigated using several fungi and bacteria. For effective breakdown of LCB, fungal enzyme systems have garnered greater interest than bacterial enzyme systems due to their proficiency and resilience (Saini and Sharma 2021). These enzymes serve as prerequisites for efficient enzymatic hydrolysis and pretreatment steps in lignocellulosic biorefinery to access their prospects for generation of biofuel and VAP using LCB (Koupaie et al. 2019). However, the impediments such as cost linked to the production and purification of these enzymes and finding robust biocatalysts are being considered as major prerequisites for their utilization in a lignocellulosic biorefinery. These can be circumvented by finding a hyper lignocellulolytic enzyme producing fungal strains and developing robust enzyme cocktails with greater efficiencies, enhanced stability at elevated temperatures, and increased tolerance capacity toward the inhibition by end products and process inhibitors.

4.1 Ligninolytic Enzymes: A Pretreatment Novice

Ligninolytic enzymes are the enzymes that enable the delignification in a variety of organisms, including fungus, bacteria, and insects. However, fungi, especially WRF, are the most efficient decomposers of lignin. The extracellular, nonspecific ligninolytic enzymes that have been identified thus far degrade the aromatic framework of lignin and cleave the connection across the lignocellulosic matrix through oxidative processes (Sunardi 2020). Ligninolytic enzymes are classified into manganese peroxidase (MnP), laccase, lignin peroxidase (LiP), dye-decolorizing peroxidase (DyP), and versatile peroxidase (VP) (Dashora et al. 2023).

Laccases (EC 1.10.3.2), the multicopper oxidases, oxidize the substrate by transferring the electron from a mononuclear copper core to a trinuclear copper center (Weng et al. 2021). The glycoprotein dimers or tetramers that make up the laccase holoenzyme have four copper atoms per monomer linked to redox sites at types 1, 2, and 3 (Binod et al. 2019). They oxidize aromatic compounds by utilizing the redox properties of copper (Nargotra et al. 2022). Laccase oxidizes phenolic substrates to form the unsteady intermediate phenoxyl- free radical that then promotes Cα oxidation and polymer cleavage, including Cα-Cβ and alkyl-aryl cleavage (Weng et al. 2021). Numerous substrates, including aromatic and aliphatic amines, polyphenols, and diphenols, can be hydrolyzed by laccases (Binod et al. 2019). In addition to phenolic substrates, laccases are also able to oxidize non-phenolic substrates by mediator and can assist in Cα-Cβ cleavage, aromatic ring cleavage, ether cleavage, and Cα oxidation (Nargotra et al. 2022). In fungi, laccase is secreted by Basidiomycetes (*Thelephora terrestris*, *Lenzites true*, *Phanerochaete chrysosporium*), WRF (*Pleurotus ostreatus*, *Trametes versicolor*, *Phlebia radiata*), and some *Trichoderma* species (*Trichoderma harzianum*, *Trichoderma longibrachiatum*, *Trichoderma atroviride*) (Suryadi et al. 2022). Fungi-derived laccase has the ability of oxidizing substrates having high E (E > 400 mV) (Agrawal et al. 2018).

Lignin peroxidase (LiP, EC.1.11.1.14), a glycosylated enzyme, has an iron protoporphyrin prosthetic group and a heme cofactor in its active core (Suryadi et al. 2022; Nargotra et al. 2022). LiP was the first discovered ligninolytic enzyme from the WRF, *Phanerochaete chrysosporium* (Dashora et al. 2023). By virtue of their substantial redox potential and ability to produce cation radicals, LiPs may function on both phenolic and non-phenolic molecules without requiring a mediator (Binod et al. 2019). LiPs rely on H_2O_2 and veratryl alcohol for initiation of catalytic cycle, where veratryl alcohol acts as a co-factor and electron donor (Chio et al. 2019). Some examples of fungi that secrete LiPs include *Phlebia tremellosa*, *Phanerochaete sordida*, *Trichoderma versicolor*, *Pleurotus ostreatus*, and *Phlebia radiata* (Suryadi et al. 2022).

Manganese peroxidases (MnPs, EC 1.11.1.13), also called glycosylated heme proteins, oxidize Mn^{2+} as a reducing substrate to produce Mn^{3+} ions, which then form linkages with alpha hydroxy acids and dicarboxylic acids (Saldarriaga-Hernández et al. 2020). The enzyme MnP helps H_2O_2 in binding to the native ferric enzyme for the initiation of catalytic cycle. After that, Mn^{2+} is oxidized to Mn^{3+} ions in the presence of chelators. Being a strong oxidant, Mn^{3+} disintegrates from the enzyme and oxidizes the phenolic segment in lignin to produce phenoxy radicals that depolymerize the lignin. The chelators and organic acid, such as malonate and oxalate, help to stabilize Mn^{3+} and increase the enzyme activity (Nargotra et al. 2022; Weng et al. 2021). Some examples of potent producers of MnPs include *Phanerochaete chrysosporium*, *Phanerochaete sordida*, *Physisporinus rivulosus*, *Ceriporiopsis subvermispora*, *Dichomeris squalens*, and *Phlebia radiata* (Dashora et al. 2023).

Versatile peroxidase (VP, EC 1.11.1.16), a distinct form of MnP and LiP, produced by a small number of WRF, combines the molecular structures of MnP and LiP and displays traits common to both such as the oxidation of phenolic and non-phenolic lignin structures. Like MnPs, VPs can oxidize Mn^{2+} to Mn^{3+}, and similar to LiPs, VPs can oxidize non-phenolic lignin substrate veratryl alcohol without adding Mn (Singhania et al. 2022). Due to their special mechanisms, VPs are effective industrial biocatalysts for the detoxification of biomass and the degradation of lignin (Tsegaye et al. 2019).

Dye-decolorizing peroxidase (DyP, EC 1.11. 1.19), a heme-based peroxidase, is a recently discovered peroxidase and can break down lignin through an oxidation process driven by radicals. Despite structural and sequence differences, these DyPs have identical H_2O_2- and mediator-dependent catalytic mechanism for lignin breakdown as other class II heme peroxidases (Nargotra et al. 2022). Unlike other peroxidases found in fungi, DyPs are widely present in a range of bacteria. DyPs can decarboxylate non-phenolic lignin and yield *p*-anisaldehyde as their primary product.

4.2 *Holocellulolytic Enzymes: The Complete Fate for Hydrolysis*

The complete breakdown of structured carbohydrates or holocellulose into their basic monomeric elements is necessary for generation of biofuel and VAP. A synergistic application of holocellulolytic enzymes (cocktails of cellulases and hemicellulases) is being considered as the most effective method to depolymerize the key carbohydrate polymers into simpler sugars such as glucose, mannose, xylose, galactose, arabinose, and others (Chiranjeevi et al. 2012). The core group of enzymes for the hydrolysis of cellulose is composed of three cellulases: endoglucanases (EG), exoglucanases (cellobiohydrolase, CBH), and β-glucosidases. Endoglucanase (EC 3.2.1.4) attacks the internal glycosidic linkages to form oligosaccharides with different chain lengths and creates new chain ends. Exoglucanases are further classified into two groups: glucan glucanohydrolases (EC 3.2.1.74) and glucan cellobiohydrolases (EC 3.2.1.91). These groups function progressively on the reducing or nonreducing ends of cellulosic polymer chains, secreting cellobiose (cellobiohydrolase) or glucose (glucanohydrolases). β-Glucosidases (EC 3.2.1.21) secrete glucose by acting on cellobiose and cellodextrins (Patel and Shah 2021b).

Hemicellulases, a broad class of hydrolytic and nonhydrolytic enzymes with distinct specificities that work jointly, break down the heterogeneous hemicellulose network (Champreda et al. 2019). They encompass main-chain-degrading enzymes, which are endo- or exo-acting (e.g., xylanase, mannanase, *β*-xylosidase, *β*-mannosidase), and side-chain-acting enzymes (e.g., *α*-L-arabinofuranosidase, *α*-glucuronidase, esterases), which cleave various adorned groups attached to the main chain (Patel and Shah 2021b). Endoxylanase (EC 3.2.1.8) attacks on β-1,4 glycosidic bonds in xylan chain of hemicellulose and generates xylooligosaccharides (Saldarriaga-Hernández et al. 2020). β- Xylosidase (EC 3.2.1.37), which acts after the subsequent action of endoxylanases, serves as a key for catalysis of xylan by acting on the nonreducing end of xylooligosaccharides and converting them into monomers (Binod et al. 2019; Patel et al. 2018a). α-L-Arabinofuranosidases (E.C. 3.2.1.55) cleave α-1,5-, 1,3-, and 1,2-arabinofuranosyl residues of arabinose containing polymers and secrete arabino-oligomers and arabinose (Patel et al. 2015). Esterases are classified into three classes: acetyl xylan esterase (E.C. 3.1.1.72), ferulic acid esterase (E.C.3.1.1.73), and *p*-coumaric acid esterase. Acetyl xylan esterase acts on acetyl-ester bond between xylose units of xylan and acetic acid and releases o-acetyl groups. Ferulic acid esterase can remove ferulic acid from O-5 of arabinose of glucuronoarabinoxylan by disrupting the feruloyl ester bonds with xylan. *p*-Coumaric acid esterase catalyzes the hydrolysis of *p*-coumaroyl ester bonds in xylan backbone (Nargotra et al. 2022; Shah et al. 2017). α-Glucuronidases (E.C.3.2.1.131) act on α-1,2-glycosidic bond between the nonreducing end xylopyranosyl unit of xylooligosaccharides and D-glucuronic acid or its 4-O-methyl ether (Malgas et al. 2019). β-Mannanase (E.C. 3.2.1.78) catalyze the hydrolysis of 1,4-β-mannosidic linkage randomly in the backbone of mannan and heteromannan (Bangoria et al. 2021). α-Galactosidase (E.C. 3.2.1.22), an exo-acting enzyme, hydrolyzes galactose containing oligomers (raffinose, melibiose, verbascose, stachyose) or polymers (galactomannan, galactoglucomannan) by cleaving terminal α-1,6-linked galactose moiety (Bangoria et al. 2023a). β-Mannosidase (EC 3.2.1.25), an enzyme that works after subsequent action of β-mannanase, acts on the nonreducing end of manno-oligosaccharides and release the mannose units (Shah et al. 2017).

In addition to holocellulolytic enzymes, auxiliary enzymes such as cellobiose dehydrogenases (CDH) and lytic polysaccharide monooxygenases (LPMO) work collaboratively with cellulolytic enzymes for efficient cellulose hydrolysis (Barbosa et al. 2020). LPMO, a unique copper-dependent enzyme for Fenton- like chemistry, oxidizes C-H bonds at the C1 and/or C4 carbon of glycosidic bond observed in recalcitrant and crystalline polysaccharides (Bissaro and Eijsink 2023). The glycosylated extracellular proteins called CDH are known to lower Fe^{3+} levels, generate

H_2O_2, and provide electrons to the LPMO. This is an add-on mechanism necessary for LPMO to initiate the redox processes for lignocellulosic hydrolysis (Patel and Shah 2021b). The important characteristics of lignocellulolytic enzymes produced using different fungi have been exemplified in Table 2.

5 Tailor-Made Fungal Enzymes: A Progress in Multiproduct Biorefinery

The use of lignocellulolytic enzymes in multiproduct biorefinery is thought to be more advantageous than using fungi directly because it could address issues with large-scale growth requirements, mycelia penetration into the substrate, extended incubation times, fungal overconsumption of polysaccharides that leads to excess destruction of carbohydrates, and contamination issues (Suryadi et al. 2022). Enzyme-based biorefinery shows promising approach owing to their advantages over conventional methods, including improved catalytic efficiency, reduced waste generation, milder conditions for processing conditions, increased reaction specificity, and energy economy (Okeke et al. 2022). Enzyme activity on LCB is dependent on both the kind of enzyme and the chemistry of biomass (Suryadi et al. 2022). Furthermore, because of compositional diversity, a single enzyme blend may fail to function efficiently for every type of biomass (Patel and Shah 2021b). Thus, commercially marketed lignocellulolytic enzymes are the enzymes, cocktails, or blends obtained using two or more fungal strains (Saini and Sharma 2021). Recently, Novozymes (Denmark), Bio-Cat (India), Genencor (USA), Infinita Biotech Pvt. Ltd. (India), Creative Enzymes (USA), BASF (Japan and Germany), Megazyme (Ireland), Advanced enzymes (India), and Sunson Enzymes (China) are some of the prominent players notorious for lignocellulolytic enzyme manufacturing used in biomass conversion across the globe (Patel and Shah 2021b; Saini and Sharma 2021).

Genencor has marketed a series of Accellerase enzymes such as Accellerase®1000, Accellerase®1500, Accellerase®XY, and Accellerase®BG comprising of single or blends of holocellulolytic enzymes necessary for efficient enzymatic saccharification of LCB for biofuel and VAP (Patel and Shah 2021b). A widely accepted enzyme cocktail Cellic CTec3, developed by Novozymes, has a mixture of cellulases and hemicellulase. All these enzymes have been produced using filamentous fungi like *Trichoderma* and *Aspergillus* species. In India, the lignocellulolytic enzyme preparations widely marketed are BioConvert-ACEL (Noor Enzymes Pvt. Ltd.), Neutrapol Max P (Aumgene Biosciences), EBZYME–AC (Enzyme Bioscience Pvt. Ltd.), SacchariSEB C6L (Advanced Enzymes), and Bio Multizyme (Aum Enzymes), comprising either single or enzyme blend of cellulolytic or hemicellulolytic enzymes mostly produced using *Aspergillus* and *Trichoderma* (Saini and Sharma 2021). Novozymes and Genencor have accomplished a 30-fold reduction in cost with enhanced enzyme activity (Saini et al. 2015). The primary variables influencing the price of enzymes are the source, degree of purity, and packaging of the

Table 2 Characteristics of important lignocellulolytic enzymes produced using different fungi

Lignocellulolytic enzyme	Fungal source	Optimum temperature (°C)	Optimum pH	Molecular weight (kDa)	Km (μM)	Vmax (μM/min)	References
Laccase	*Agrocybe pediades*	45	5.0	55–60	100	285	González-González et al. (2023)
	Trichoderma harzianum	50	3.0	48, 63	100, 65	603, 182	Elsayed et al. (2023)
LiP	*Pleurotus pulmonarius*	30	3.0	40	25	–	Giap et al. (2022)
	Lentinus squarrosulus	35	5.0	55	2.45	285	Dinh Giap et al. (2022)
MnP	*Trametes pubescens*	40	5.0	55	4.61	12.99	Rekik et al. (2019)
	Cerrena unicolor	60	4.5	45	46.7	–	Zhang et al. (2020)
VP	*Phlebia radiata*	80	3.0	40	–	–	Bouacem et al. (2022)
	Lentinus squarrosulus	30	5.0	49	6.84	–	Ravichandran et al. (2019)
DyP	*Xylaria grammica*	–	3.5	49	41, 12	–	Kimani et al. (2021)
	Irpex lacteus	30	4.0	74	3.0	–	Qin et al. (2018)
Endoglucanase	*Fomitopsis meliae*	70	4.8	37.87	12.0	556.58	Patel and Shah (2021a)
	Pycnoporus sanguineus	50	5.0	–	3.18	–	dos Santos et al. (2023)
Exoglucanase	*Thielavia terrestris*	60	5.5	71	3.12	50	López-López et al. (2023)
	Trichoderma virens	60	4.0	54	1.88	41.67	Ab Wahab et al. (2019)
β-Glucosidase	*Neofusicoccum parvum*	65	4.0	65	1.18	28.08	Singh et al. (2023)
	Talaromyces amestolkiae	60	4.0	68.05	3.36	–	Méndez-Líter et al. (2020)
Endoxylanase	*Aspergillus tubingensis*	50	5.0	40–45	32.4	2602.8	Intasit et al. (2022)
	Trichoderma harzianum	65	6.0	72	5.56	1052.63	Dhaver et al. (2022)

β-Xylosidases	*Aspergillus niger*	65	4.0	120.48	1.17	24.39	Patel et al. (2018b)
	Aspergillus fumigatus	75	4.5	74.5	0.25	150.25	Jin et al. (2020)
α-L-Arabinofuranosidase	*Trichoderma harzianum*	55	6.5	53.44	3.30	15.47	Motta et al. (2021)
	Coriolopsis byrsina	50	3.5–4.5	55	3.45	198.2	de Oliveira Nascimento et al. (2023)
Acetyl xylan esterase	*Aspergillus oryzae*	50	7.0	23	2.3	–	Kato et al. (2021)
	Irpex lacteus	30	6.0	60	0.4	0.057	Koh et al. (2019)
Ferulic acid esterase	*Penicillium chrysogenum*	50	6.0–7.0	75	–	–	Phuengmaung et al. (2019)
	Aspergillus aculeatus	50, 60	7.0	55, 57	79.85, 105.98	6.19, 6.53	Zhang et al. (2021)
α-Glucuronidase	*Aspergillus niger*	40	3.5	58	4.4	50	Kiryu et al. (2005)
	Phlebia radiata	60	3.8	110	0.18	0.13	Mierzwa et al. (2005)
β-Mannanase	*Penicillium aculeatum*	70	5.5	40.6	0.90	483	Bangoria et al. (2023b)
	Lentinula edodes	60	6.5	58	–	–	Alaubidi and Abed (2023)
α-Galactosidase	*Penicillium aculeatum*	55	4.5	65.4	2.5	507.8	Bangoria et al. (2021)
	Aspergillus niger	60	3.5	64	0.401	14.65	Othman et al. (2023)
β-Mannosidase	*Aspergillus niger*	65	4.5–5.5	130	0.28	–	Fliedrová et al. (2012)
	Lichtheimia ramosa	65	6.0	51	1.377	–	Xie et al. (2019)

finished product–factors that frequently differ between industries. The mechanism governing the creation and assessment of commercial lignocellulolytic concoctions is still unclear. Thus, the creation of a profitable and cost-competitive enzyme cocktail is important to efficiently address the issue of energy scarcity (Saini and Sharma 2021). Furthermore, most manufacturers exclusively concentrate on the use of fungi, *Aspergillus*, and *Trichoderma* in creating the tailor-made fungal holocellulolytic enzymes individually or in cocktail. Therefore, there is a need to emphasize on other beneficial fungal sources such as BRF and WRF, which are known to secrete a substantial quantity of extracellular lignocellulolytic enzymes that are incredibly resilient and highly thermostable.

6 Fungal Biorefinery: An Innovative Perspective

The substantial expenses associated with the production of biofuel from alternative and renewable resources have obliged to apply innovative and revolutionary methods such as lignocellulosic biorefinery (Maleki et al. 2020). The term "biorefinery" signifies the sustainable processing of native LCB for integrated production of chemicals, energy, and biomaterials using microorganisms and their enzymes (Patel and Shah 2021b). The ideal biocatalysts for biorefinery technologies are nevertheless lignocellulolytic fungi because they spontaneously release a range of cell wall-degrading enzymes that may fractionate lignocellulosic building blocks into a number of industrial products (Troiano et al. 2020). In essence, WRF pretreat the LCB by secreting a wide variety of extracellular ligninolytic enzymes in large quantities as well as noteworthy hydrolytic combinations that break down the polymeric biomass into its structural components (Vasco-Correa et al. 2019). The pretreatment of LCB with the crude enzyme from WRF may generate the carbohydrates that are free from lignin (Kumar and Chandra 2020). Consequently, the liberated carbohydrates are further broken down using the crude enzyme of the same other SRF or BRF to produce the biofuels (Bioethanol, biobutanol) and VAP (xylitol, xylooligosaccharides, furfural, organic acids) (Kumar and Chandra 2020; Champreda et al. 2019).

Researchers are proactively fractionating the holocellulosic portion of LCB using various physicochemical methods and subsequently utilizing these fractionates in the production of biofuel and VAP through the action of hydrolytic enzymes from fungi. For instance, the integrated production of high titer ethanol and VAP like xylooligosaccharides and xylitol has been reported from alkali- treated wheat straw and sugarcane bagasse (Patel et al. 2021a; 2023). The lignin fraction, on the other hand, can be valorized to generate a variety of industrial products, including adhesive, aromatics, vanillin, resins, polyurethane, biopesticides, natural binders, and polyhydroxyalkanoates (PHAs) (Cao et al. 2019; Mandal et al. 2023). With advances in technology, WRF can now be utilized to delignify LCB, and the processed residual biomass can be used to generate biofuel utilizing hydrolytic enzymes derived from other fungi in a biorefinery concept. In a recent study, the WRF,

Ganoderma lucidum, was cultivated on LCB for the production of laccase. The leftover biomass was subsequently saccharified using tailored cellulase consortium of *Trichoderma citrinoviride* and *Aspergillus flavus*. The hydrolyste thus obtained was further fermented using *Saccharomyces cerevisiae* for ethanol production (Shankar et al. 2022). Similarly, it has been discovered that the active consortium of—*Trametes versicolor*, *Pleurotus eryngii*, and *Pleurotus ostreatus*—can produce biogas from wheat straw via simultaneous pretreatment and enzyme extraction (Albornoz et al. 2018). *Auricularia auricula-judae* was utilized for making food, fermentable sugar, bioethanol, and solid fuels using LCB by Chen et al. (2021). The study reported 200 g mushroom/kg of raw substrate, 46% glucan digestibility, 37 ml ethanol/kg of raw substrate, and 280–320 gm DM combustible fuels of 4639–5295 kJ. While much research is being done in the biorefinery concept globally, full-fledged commercialization of biofuels and VAP is often labor intensive, and fungal integrated biorefinery processes are still in their early stages in the current scenario. The product has a very high capital and investment cost due to numerous technological and operational drawbacks, which has reduced its competitiveness in the market. Therefore, to make these processes economically and industrially viable, forthcoming studies must concentrate on technological and operational aspects of biorefineries (Saini and Sharma 2021).

7 Present Difficulties and Future Opportunities

Lignocellulolytic fungi have attracted a lot of attention in the quest to overcome the recalcitrance of LCB due to their robust extracellular lignocellulolytic enzymes and sophisticated metabolic pathways. The heterogenic nature of lignin is the main obstacle being considered for the utilization of LCB in a multiproduct biorefinery. A useful strategy might be the fungal bio-pretreatment with WRF, which has the ability to secrete enzymes that break down lignin, to increase the variety of enzyme activities in the co-culture and aid in the delignification of the substrate. Because lignin-degrading enzymes remove a component of the lignin that causes lignocellulosic recalcitrance, they may have a synergistic impact by allowing other enzymes to access the portion of the carbohydrates. However, fungal pretreatment necessitates a lengthy incubation period and progresses more slowly for efficient delignification. Contemporary gene engineering tools like zinc-finger nucleases, CRISPR-Cas9 and TALENs, and RNA interference (RNAi) can be utilized to overcome these obstacles by creating lignin-reduced wood, in which ligninolytic fungi can quickly and readily break down. Bio-pretreatment leveraging enzymes extracted from ligninolytic fungi are currently receiving more attention from researchers and may be a topic for further investigation. This will also make it necessary to concentrate on a few factors, such as the cost of producing and purifying enzymes and the intricacy of degradable pathways that may be endorsed. Consequently, there is a need to do additional research to understand the ligninolytic enzyme's degradation paths and the proper interactions between various enzymes to develop a predominant lignin degradation system that will allow biomass to be fractionated into several VAP.

Enzymatic saccharification of holocellulosic portion of pretreated biomass is being considered as a vital step for the profitability of multiproduct biorefinery. Despite the fact that successive pretreatment advances have led to a substantial decrease in enzyme loading, the effective enzymatic saccharification of holocellulose remains a key bottleneck to the use of LCB for generation of biofuel and platform chemicals. As previously discussed, the enzyme inadequacies associated with commonly utilized fungal strains (*Trichoderma* and *Aspergillus*) necessitate the acquisition of two or more different complimentary cocktails for a successful biorefinery process. Looking toward the enormous diversity, fungi have the potential to be a valuable asset for their utilization during enzymatic hydrolysis. Unpretentiously, a substantial proportion of fungi, which are still unidentified or uncultivable, may have novel catalyst or metabolic pathways with potential commercial use. It is imperative to explore potential new strains and/or species of fungi to potentially find new lignocellulolytic enzymes. For instances, novel basidiomycetes fungi have been reported by Patel and coworkers, and there are possibilities of hyper active lignocellulolytic strains if explored further (Patel and Rajput 2023; Patel et al. 2018c, d, e; 2019b; 2020). Furthermore, knowledge of the natural relationships among fungi may offer insightful ideas regarding the ways in which these relationships will affect the expression, secretion, and activity of lignocellulolytic enzymes. These could offer the potential to boost the lignocellulolytic enzyme manufacturing and acquire more diverse and robust enzyme blends in a single process. Because this is an expensive approach, producing enzyme mixtures on site ought to be offered meticulous thought to lower the associated costs. An additional consideration that must be implemented when formulating an enzyme blend to prevail over the significant and technical bottleneck is the fact that different types of biomass will not respond optimally to the same enzyme cocktail due to compositional diversity. Advances in the formulation of enzymes have resulted from an improved understanding of the synergy of lignocellulolytic enzymes, which has been acquired through omics studies of specific microorganisms and meta-omics studies of microbiomes. The discovery of new lignocellulolytic enzymes and novel auxiliary elements from fungal sources employing more sophisticated high-throughput genomic and pertinent multi-omics strategies from untapped sources, such as thermophilic fungi, will drive the development of more efficient enzyme cocktail for the biorefinery process. The pursuit of efficient and robust enzymes will be substantially enhanced by developments in enzyme engineering and the use of heterologous hosts for large-scale manufacturing. The effectiveness of recycling enzymes in terms of both quantity and activity may be enhanced by developments in enzyme immobilization technology, perhaps surpassing the amount of fungal biomass.

8 Conclusion

The potential of fungi and their enzyme system have been emphasized in this chapter, particularly in the context of multiproduct lignocellulosic biorefineries. The development of a bio-economy based on biorefineries will undoubtedly face many obstacles, but greater research into the science of fungi is necessary because it may be able to resolve or at least diminish some of the current problems with enzyme costs. There are still a lot of unsolved concerns in this area, which will present novel and fascinating possibilities for additional research and method advancements.

References

Ab Wahab AFF, Karim NAA, Ling JG, Hasan NS, Yong HY, Bharudin I et al (2019) Functional characterisation of cellobiohydrolase I (Cbh1) from *Trichoderma virens* UKM1 expressed in *Aspergillus niger*. Protein Expr Purif 154:52–61

Agrawal K, Chaturvedi V, Verma P (2018) Fungal laccase discovered but yet undiscovered. Bioresour Bioprocess 5(1):1–12

Alaubidi N, Abed FNAM (2023) Purification and characterization of mannanase enzyme from *Lentinula edodes* isolate and evaluation of its activity in removing blood stains from cotton fabrics. Bionatura 8:1–12

Albornoz S, Wyman V, Palma C, Carvajal A (2018) Understanding of the contribution of the fungal treatment conditions in a wheat straw biorefinery that produces enzymes and biogas. Biochem Eng J 140:140–147

Althuri A, Mohan SV (2019) Single pot bioprocessing for ethanol production from biogenic municipal solid waste. Bioresour Technol 283:159–167

Awasthi MK, Sarsaiya S, Patel A, Juneja A, Singh RP, Yan B et al (2020) Refining biomass residues for sustainable energy and bio-products: an assessment of technology, its importance, and strategic applications in circular bio-economy. Renew Sust Energ Rev 127:109876

Awasthi MK, Sindhu R, Sirohi R, Kumar V, Ahluwalia V, Binod P et al (2022) Agricultural waste biorefinery development towards circular bioeconomy. Renew Sust Energ Rev 158:112122

Bangoria P, Divecha J, Shah AR (2021) Production of mannooligosaccharides producing β-Mannanase by newly isolated *Penicillium aculeatum* APS1 using oil seed residues under solid state fermentation. Biocatal Agric Biotechnol 34:102023

Bangoria P, Patel A, Shah AR (2023a) Characterization of a fungal α-galactosidase and its synergistic effect with β-mannanase for hydrolysis of galactomannan. Carbohydr Res 531:108893

Bangoria P, Patel A, Shah AR (2023b) Thermotolerant and protease-resistant GH5 family β-mannanase with CBM1 from *Penicillium aculeatum* APS1: purification and characterization. 3 Biotech 13:107

Barbosa FC, Silvello MA, Goldbeck R (2020) Cellulase and oxidative enzymes: new approaches, challenges and perspectives on cellulose degradation for bioethanol production. Biotechnol Lett 42:875–884

Baruah J, Nath BK, Sharma R, Kumar S, Deka RC, Baruah DC et al (2018) Recent trends in the pretreatment of lignocellulosic biomass for value-added products. Front Energy Res 6:141

Binod P, Gnansounou E, Sindhu R, Pandey A (2019) Enzymes for second generation biofuels: recent developments and future perspectives. Bioresour Technol Rep 5:317–325

Bissaro B, Eijsink VGH (2023) Lytic polysaccharide monooxygenases: enzymes for controlled and site-specific Fenton-like chemistry. Essays Biochem 67(3):575–584

Bouacem K, Allala F, Zaraî Jaouadi N, Hamdi S, Mechri S, Ighilahriz K et al (2022) A novel peroxidase from white-rot Agaricomycetes fungus *Phlebia radiata* strain KB-DZ15: Its purification, characterisation, and potential application for dye-decolorisation and lignin-biodegradation. Biocatal Biotransformation 40(5):365–377

Bragatto J, Segato F, Cota J, Mello DB, Oliveira MM, Buckeridge MS et al (2012) Insights on how the activity of an endoglucanase is affected by physical properties of insoluble celluloses. J Phys Chem B 116(21):6128–6136

Cao Y, Chen SS, Zhang S, Ok YS, Matsagar BM, Wu KC-W et al (2019) Advances in lignin valorization towards bio-based chemicals and fuels: Lignin biorefinery. Bioresour Technol 291:121878

Champreda V, Mhuantong W, Lekakarn H, Bunterngsook B, Kanokratana P, Zhao X-Q et al (2019) Designing cellulolytic enzyme systems for biorefinery: from nature to application. J Biosci Bioeng 128(6):637–654

Chatterjee S, Mohan SV (2022) Fungal biorefinery for sustainable resource recovery from waste. Bioresour Technol 345:126443

Chen F, Dixon RA (2007) Lignin modification improves fermentable sugar yields for biofuel production. Nat Biotechnol 25(7):759–761

Chen F, Grimm A, Eilertsen L, Martín C, Arshadi M, Xiong S (2021) Integrated production of edible mushroom (*Auricularia auricular-judae*), fermentable sugar and solid biofuel. Renew Energy 170:172–180

Chio C, Sain M, Qin W (2019) Lignin utilization: a review of lignin depolymerization from various aspects. Renew Sust Energ Rev 107:232–249

Chiranjeevi T, Rani G, Chandel AK, Sekhar PV, Prakasham RS, Addepally U (2012) Optimization of holocellulolytic enzymes production by *Cladosporium Cladosporioides* using taguchi-L'16 orthogonal array. J Biobaased Mater Bioenergy 6(2):148–157

Chukwuma OB, Rafatullah M, Tajarudin HA, Ismail N (2020) Lignocellulolytic enzymes in biotechnological and industrial processes: a review. Sustainability 12(18):7282

Cragg SM, Beckham GT, Bruce NC, Bugg TDH, Distel DL, Dupree P et al (2015) Lignocellulose degradation mechanisms across the tree of life. Curr Opin Chem Biol 29:108–119

Dashora K, Gattupalli M, Tripathi GD, Javed Z, Singh S, Tuohy M et al (2023) Fungal assisted valorisation of polymeric lignin: mechanism, enzymes and perspectives. Catalysts 13(1):149

de Oliveira Nascimento CE, de Menezes CLA, Boscolo M, da Silva R, Gomes E, da Silva RR (2023) Functional characteristics of an α-l-arabinofuranosidase secreted by the basidiomycete *Coriolopsis byrsina*. Biologia 78:1919–1928

Dharmaraja J, Shobana S, Arvindnarayan S, Francis RR, Jeyakumar RB, Saratale RG et al (2023) Lignocellulosic biomass conversion via greener pretreatment methods towards biorefinery applications. Bioresour Technol 369:128328

Dhaver P, Pletschke B, Sithole B, Govinden R (2022) Optimization, purification, and characterization of xylanase production by a newly isolated *Trichoderma harzianum* strain by a two-step statistical experimental design strategy. Sci Rep 12(1):17791

Dinh Giap V, Nghi DH, Cuong LH, Quynh DT (2022) Lignin peroxidase from the white-rot fungus *Lentinus squarrosulus* MPN12 and its application in the biodegradation of synthetic dyes and lignin. Bioresources 17(3):4480

dos Santos DMRC, Albuquerque F, Silva TP, Ferreira AN, Machado SS, da Luze JMR et al (2023) Production, purification, characterization, and application of halotolerant and thermostable endoglucanase isolated from *Pycnoporus sanguineus*. Waste Biomass Valoriz 14(10):1–12

Eastwood DC (2014) Evolution of fungal wood decay. In: Deterioration and protection of sustainable biomaterials. ACS Publications, Washington, pp 93–112

Elsayed AM, Mahmoud M, Abdel Karim GSA, Abdelraof M, Othman AM (2023) Purification and biochemical characterization of two laccase isoenzymes isolated from *Trichoderma harzianum* S7113 and its application for bisphenol A degradation. Microb Cell Factories 22(1):1–12

Fliedrová B, Gerstorferová D, Křen V, Weignerová L (2012) Production of *Aspergillus niger* β-mannosidase in Pichia pastoris. Protein Expr Purif 85(2):159–164

Giap VD, Duc HT, Huong PTM, Do TH, Do HN, Duy VD et al (2022) Purification and characterization of lignin peroxidase from white-rot fungi *Pleurotus pulmonarius* CPG6 and its application in decolorization of synthetic textile dyes. J Gen Appl Microbiol 68(6):262–269

Gil A, Siles JÁ, Serrano A, Martín MÁ (2015) Mixture optimization of anaerobic co-digestion of tomato and cucumber waste. Environ Technol 36(20):2628–2636

Giles RL, Zackeru JC, Galloway ER, Elliott GD, Parrow MW (2015) Single versus simultaneous species treatment of wood with *Ceriporiopsis subvermispora and Postia placenta* for ethanol applications, with observations on interspecific growth inhibition. Int Biodeterior Biodegradation 99:66–72

González-González P, Gómez-Manzo S, Tomasini A, Martínez Pérez JL, García Nieto E, Anaya-Hernández A et al (2023) Laccase production from *Agrocybe pediades*: Purification and functional characterization of a consistent laccase isoenzyme in liquid culture. Microorganisms 11(3):568

Goodell B (2020) Fungi involved in the biodeterioration and bioconversion of lignocellulose substrates. In: Genetics and biotechnology. Springer, Cham, pp 369–397

Gupta VK, Kubicek CP, Berrin J-G, Wilson DW, Couturier M, Berlin A et al (2016) Fungal enzymes for bio-products from sustainable and waste biomass. Trends Biochem Sci 41(7):633–645

Howell C, Hastrup ACS, Jara R, Larsen FH, Goodell B, Jellison J (2011) Effects of hot water extraction and fungal decay on wood crystalline cellulose structure. Cellulose 18:1179–1190

Intasit R, Cheirsilp B, Suyotha W, Boonsawang P (2022) Purification and characterization of a highly-stable fungal xylanase from *Aspergillus tubingensis* cultivated on palm wastes through combined solid-state and submerged fermentation. Prep Biochem Biotechnol 52(3):311–317

Ishak N, Kassim ASM, Aripin AM, Sharifah SM, Oluwatosin AF (2019) A review on lignin and biodelignification. J Eng Heal Sci 3(1):41–72

Jin X, Song J, Ma J, Liu G-Q (2020) Thermostable β-xylosidase from *Aspergillus fumigatus*: purification, characterization and potential application in lignocellulose bioethanol production. Renew Energy 155:1425–1431

Kato T, Shiono Y, Koseki T (2021) Identification and characterization of an acetyl xylan esterase from *Aspergillus oryzae*. J Biosci Bioeng 132(4):337–342

Khan MA, Ashraf SM, Malhotra VP (2004) Development and characterization of a wood adhesive using bagasse lignin. Int J Adhes Adhes 24(6):485–493

Kim M, Day DF (2011) Composition of sugar cane, energy cane, and sweet sorghum suitable for ethanol production at Louisiana sugar mills. J Ind Microbiol Biotechnol 38(7):803–807

Kimani V, Ullrich R, Büttner E, Herzog R, Kellner H, Jehmlich N et al (2021) First dye-decolorizing peroxidase from an ascomycetous fungus secreted by *Xylaria grammica*. Biomolecules 11(9):1391

Kiryu T, Nakano H, Kiso T, Murakami H (2005) Purification and characterization of a novel α-glucuronidase from *Aspergillus niger* specific for O-α-D-glucosyluronic acid α-D-glucosiduronic acid. Biosci Biotechnol Biochem 69(3):522–529

Ko JK, Kim Y, Ximenes E, Ladisch MR (2015) Effect of liquid hot water pretreatment severity on properties of hardwood lignin and enzymatic hydrolysis of cellulose. Biotechnol Bioeng 112(2):252–262

Koh S, Imamura S, Fujino N, Mizuno M, Sato N, Makishima S et al (2019) Characterization of acetylxylan esterase from white-rot fungus *Irpex lacteus*. J Appl Glycosci 66(4):131–137

Koupaie EH, Dahadha S, Lakeh AAB, Azizi A, Elbeshbishy E (2019) Enzymatic pretreatment of lignocellulosic biomass for enhanced biomethane production-A review. J Environ Manag 233:774–784

Kumar A, Chandra R (2020) Ligninolytic enzymes and its mechanisms for degradation of lignocellulosic waste in environment. Heliyon 6(2):e03170

López-López A, Santiago-Hernández A, Cayetano-Cruz M, García-Huante Y, Campos JE, Bustos-Jaimes I et al (2023) TtCel7A: a native thermophilic bifunctional cellulose/xylanase exoglucanase from the thermophilic biomass-degrading fungus *Thielavia terrestris* Co3Bag1, and its application in enzymatic hydrolysis of agroindustrial derivatives. J Fungi 9(2):152

Lu H, Zhang X, Wu A, Deng X, Ren J, Kong F et al (2017) Comparison of dilute acid, alkali, and biological pretreatments for reducing sugar production from eucalyptus. Bioresources 12(3):6353–6365

Maciel MJM, Ribeiro HCT (2010) Industrial and biotechnological applications of ligninolytic enzymes of the basidiomycota: a review. Electron J Biotechnol 13(6):14–15

Maleki M, Shahraki MF, Kavousi K, Ariaeenejad S, Salekdeh GH (2020) A novel thermostable cellulase cocktail enhances lignocellulosic bioconversion and biorefining in a broad range of pH. Int J Biol Macromol 154:349–360

Malgas S, Mafa MS, Mkabayi L, Pletschke BI (2019) A mini review of xylanolytic enzymes with regards to their synergistic interactions during hetero-xylan degradation. World J Microbiol Biotechnol 35(12):187

Mandal DD, Singh G, Majumdar S, Chanda P (2023) Challenges in developing strategies for the valorization of lignin—a major pollutant of the paper mill industry. Environ Sci Pollut Res 30(5):11119–11140

Méndez-Líter JA, Nieto-Domínguez M, Fernández de Toro B, González Santana A, Prieto A, Asensio JL et al (2020) A glucotolerant β-glucosidase from the fungus *Talaromyces amestolkiae* and its conversion into a glycosynthase for glycosylation of phenolic compounds. Microb Cell Factories 19:1–13

Mierzwa M, Tokarzewska-Zadora J, Deptuła T, Rogalski J, Szczodrak J (2005) Purification and characterization of an extracellular α-D-glucuronidase from *Phlebia radiata*. Prep Biochem Biotechnol 35(3):243–256

Mohanram S, Amat D, Choudhary J, Arora A, Nain L (2013) Novel perspectives for evolving enzyme cocktails for lignocellulose hydrolysis in biorefineries. Sustain Chem Process 1(1):1–12

Monrroy M, Ortega I, Ramírez M, Baeza J, Freer J (2011) Structural change in wood by brown rot fungi and effect on enzymatic hydrolysis. Enzym Microb Technol 49(5):472–477

Moreno AD, Tomás-Pejó E, Ballesteros M, Negro MJ (2019) Pretreatment technologies for lignocellulosic biomass deconstruction within a biorefinery perspective. In: Biofuels: alternative feedstocks and conversion processes for the production of liquid and gaseous biofuels. Elsevier, Amsterdam, pp 379–399

Motta MLL, Filho JAF, de Melo RR, Zanphorlin LM, Dos Santos CA, de Souza AP (2021) A novel fungal metal-dependent α-l-arabinofuranosidase of family 54 glycoside hydrolase shows expanded substrate specificity. Sci Rep 11(1):10961

Mtui GYS (2012) Lignocellulolytic enzymes from tropical fungi: types, substrates and applications. Sci Res Essays 7(15):1544–1555

Nadir N, Ismail NL, Hussain AS (2019) Fungal pretreatment of lignocellulosic materials. In: Biomass for bioenergy-recent trends and future challenges. IntechOpen, London

Nargotra P, Sharma V, Lee Y-C, Tsai Y-H, Liu Y-C, Shieh C-J et al (2022) Microbial lignocellulolytic enzymes for the effective valorization of lignocellulosic biomass: a review. Catalysts 13(1):83

Nurika I, Suhartini S, Barker GC (2020) Biotransformation of tropical lignocellulosic feedstock using the brown rot fungus *Serpula lacrymans*. Waste Biomass Valoriz 11:2689–2700

Okeke ES, Ezugwu AL, Anaduaka EG, Mayel MH, Ezike TC, Ossai EC (2022) Ligninolytic and cellulolytic enzymes—biocatalysts for green agenda. Biomass Convers Biorefinery 14:3031–3055

Onwosi CO, Ezugworie FN, Onyishi CL, Igbokwe VC (2024) Lignocellulose biomass pretreatment for efficient hydrolysis and biofuel production. In: Advances in biofuels production, optimization and applications. Elsevier, Amsterdam, pp 1–19

Østby H, Várnai A (2023) Hemicellulolytic enzymes in lignocellulose processing. Essays Biochem 67(3):533–550

Othman AM, Elshafei AM, Elsayed MA, Ibrahim GE, Hassan MM, Mehanna NS (2023) Biochemical characterization and insights into the potency of the acidic *Aspergillus niger*

NRC114 purified α-galactosidase in removing raffinose family oligosaccharides from soymilk yogurt. BMC Biotechnol 23(1):1–14

Patel RS, Rajput KS (2023) Molecular and morphological studies on some gasteroid (Basidiomycota) fungi from Western India. Asian J Mycol 6(2):72–88

Patel A, Shah A (2021a) Purification and characterization of novel, thermostable and non-processive GH5 family endoglucanase from *Fomitopsis meliae* CFA 2. Int J Biol Macromol 182:1161–1169

Patel A, Shah AR (2021b) Integrated lignocellulosic biorefinery: gateway for production of second generation ethanol and value added products. J Bioresour Bioprod 6(2):108–128

Patel H, Chapla D, Divecha J, Shah A (2015) Improved yield of α-L-arabinofuranosidase by newly isolated *Aspergillus niger* ADH-11 and synergistic effect of crude enzyme on saccharification of maize stover. Bioresour Bioprocess 2(1):1–14

Patel H, Divecha J, Shah A (2018a) Enhanced production of β-xylosidase from *Aspergillus niger* ADH-11 and development of synergistic enzyme cocktail for saccharification of sugarcane bagasse. Curr Biotechnol 7(4):273–287

Patel H, Kumar AK, Shah A (2018b) Purification and characterization of novel bi-functional GH3 family β-xylosidase/β-glucosidase from *Aspergillus niger* ADH-11. Int J Biol Macromol 109:1260–1269

Patel RS, Vasava AM, Rajput KS (2018c) Morphological and molecular evidence for the occurrence of *Itajahya galericulata* (Basidiomycota, Phallales) in India. Plant Fungal Syst 63(2):39–44

Patel RS, Vasava AM, Rajput KS (2018d) New distribution record of *Clathrus delicatus* Berk. & Broome (Phallaceae) from Gujarat. J Indian Bot Soc 97(1):54–57

Patel RS, Vasava AM, Rajput KS (2018e) New distribution record of *Cyathus stercoreus* (Schwein.) De Toni (Nidulariaceae) for India from Gujarat state. Stud Fungi 3(1):227–233

Patel A, Patel H, Divecha J, Shah AR (2019a) Enhanced production of ethanol from enzymatic hydrolysate of microwave-treated wheat straw by statistical optimization and mass balance analysis of bioconversion process. Biofuels 12:1251–1258

Patel RS, Vasava AM, Rajput KS (2019b) Distribution of *Trichaleurina javanica* (Rehm) M. Carbone, Agnello & P. Alvarado (Chorioactidaceae) in India. Plant Biosyst Int J 153(2):231–234

Patel RS, Vasava AM, Rajput KS (2020) Studies on morphological and molecular identification of *Geastrum* from Gujarat, India. Kavaka 54:50–54

Patel A, Patel H, Shah A (2021a) Production of cellulosic ethanol from alkali-treated wheat straw using P-SSF process and bioconversion of hemicellulosic fraction into high-value products. Curr Biotechnol 10(2):122–132

Patel A, Divecha J, Shah A (2021b) *Fomitopsis meliae* CFA 2, a novel brown rot for endoglucanase: emphasis towards enhanced endoglucanase production by statistical approach. Mycology 12:325–340

Patel A, Divecha J, Shah A (2023) A sustainable process for co-production of xylooligosaccharides and ethanol from alkali treated sugarcane bagasse: a strategy towards waste management. Prep Biochem Biotechnol 53(6):599–609

Peralta RM, da Silva BP, Côrrea RCG, Kato CG, Seixas FAV, Bracht A (2017) Enzymes from basidiomycetes—peculiar and efficient tools for biotechnology. In: Biotechnology of microbial enzymes. Elsevier, Amsterdam, pp 119–149

Phuengmaung P, Sunagawa Y, Makino Y, Kusumoto T, Handa S, Sukhumsirichart W et al (2019) Identification and characterization of ferulic acid esterase from *Penicillium chrysogenum* 31B: de-esterification of ferulic acid decorated with l-arabinofuranoses and d-galactopyranoses in sugar beet pectin. Enzym Microb Technol 131:109380

Ponnusamy VK, Nguyen DD, Dharmaraja J, Shobana S, Banu JR, Saratale RG et al (2019) A review on lignin structure, pretreatments, fermentation reactions and biorefinery potential. Bioresour Technol 271:462–472

Prasad BR, Padhi RK, Ghosh G (2023) A review on key pretreatment approaches for lignocellulosic biomass to produce biofuel and value-added products. Int J Environ Sci Technol 20(6):6929–6944
Qin X, Luo H, Zhang X, Yao B, Ma F, Su X (2018) Dye-decolorizing peroxidases in *Irpex lacteus* combining the catalytic properties of heme peroxidases and laccase play important roles in ligninolytic system. Biotechnol Biofuels 11:1–11
Rahikainen JL, Evans JD, Mikander S, Kalliola A, Puranen T, Tamminen T et al (2013) Cellulase–lignin interactions—the role of carbohydrate-binding module and pH in non-productive binding. Enzym Microb Technol 53(5):315–321
Rajendran K, Drielak E, Sudarshan Varma V, Muthusamy S, Kumar G (2018) Updates on the pretreatment of lignocellulosic feedstocks for bioenergy production–a review. Biomass Convers Biorefinery 8:471–483
Raud M, Kikas T, Sippula O, Shurpali NJ (2019) Potentials and challenges in lignocellulosic biofuel production technology. Renew Sust Energ Rev 111:44–56
Ravichandran A, Rao RG, Gopinath MS, Sridhar M (2019) Purification and characterization of versatile peroxidase from *Lentinus squarrosulus* and its application in biodegradation of lignocellulosics. J Appl Biotechnol Bioeng 6(6):280–286
Rekik H, Jaouadi NZ, Bouacem K, Zenati B, Kourdali S, Badis A et al (2019) Physical and enzymatic properties of a new manganese peroxidase from the white-rot fungus *Trametes pubescens* strain i8 for lignin biodegradation and textile-dyes biodecolorization. Int J Biol Macromol 125:514–525
Rezania S, Oryani B, Cho J, Talaiekhozani A, Sabbagh F, Hashemi B et al (2020) Different pretreatment technologies of lignocellulosic biomass for bioethanol production: an overview. Energy 199:117457
Rodríguez-Couto S (2017) Industrial and environmental applications of white-rot fungi. Mycosphere 8(3):456–466
Saha BC (2003) Hemicellulose bioconversion. J Ind Microbiol Biotechnol 30(5):279–291
Saini S, Sharma KK (2021) Fungal lignocellulolytic enzymes and lignocellulose: a critical review on their contribution to multiproduct biorefinery and global biofuel research. Int J Biol Macromol 193:2304–2319
Saini JK, Saini R, Tewari L (2015) Lignocellulosic agriculture wastes as biomass feedstocks for second-generation bioethanol production: concepts and recent developments. 3 Biotech 5(4):337–353
Saito T, Brown RH, Hunt MA, Pickel DL, Pickel JM, Messman JM et al (2012) Turning renewable resources into value-added polymer: development of lignin-based thermoplastic. Green Chem 14(12):3295–3303
Saldarriaga-Hernández S, Velasco-Ayala C, Flores PL-I, de Jesús Rostro-Alanis M, Parra-Saldivar R, Iqbal HMN et al (2020) Biotransformation of lignocellulosic biomass into industrially relevant products with the aid of fungi-derived lignocellulolytic enzymes. Int J Biol Macromol 161:1099–1116
Schilling JS, Tewalt JP, Duncan SM (2009) Synergy between pretreatment lignocellulose modifications and saccharification efficiency in two brown rot fungal systems. Appl Microbiol Biotechnol 84:465–475
Shah A, Patel H, Narra M (2017) Bioproduction of fungal cellulases and hemicellulases through solid state fermentation. In: Fungal metabolites. Springer, Cham, pp 349–393
Shankar A, Saini S, Sharma KK (2022) Fungal-integrated second-generation lignocellulosic biorefinery: utilization of agricultural biomass for co-production of lignocellulolytic enzymes, mushroom, fungal polysaccharides, and bioethanol. Biomass Convers Biorefinery 14:1117–1131
Simonis JL, Raja HA, Shearer CA (2008) Extracellular enzymes and soft rot decay: are ascomycetes important degraders in fresh water. Fungal Divers 31(1):135–146
Singh N, Sithole B, Kumar A, Govinden R (2023) A glucose tolerant β-glucosidase from a newly isolated *Neofusicoccum parvum* strain F7: production, purification, and characterization. Sci Rep 13(1):5134

Singhania RR, Patel AK, Raj T, Chen C-W, Ponnusamy VK, Tahir N et al (2022) Lignin valorisation via enzymes: a sustainable approach. Fuel 311:122608

Sperandio GB, Ferreira Filho EX (2019) Fungal co-cultures in the lignocellulosic biorefinery context: a review. Int Biodeterior Biodegrad 142:109–123

Sunardi S (2020) Fungal glycoside hydrolase of white-rot fungi for cellulosic biofuels production: a review. Asian J Chem 32(8):1815–1823

Suryadi H, Judono JJ, Putri MR, Eclessia AD, Ulhaq JM, Agustina DN et al (2022) Biodelignification of lignocellulose using ligninolytic enzymes from white-rot fungi. Heliyon 8(2):e08865

Ten E, Vermerris W (2015) Recent developments in polymers derived from industrial lignin. J Appl Polym Sci 132(24):42069

Troiano D, Orsat V, Dumont MJ (2020) Status of filamentous fungi in integrated biorefineries. Renew Sust Energ Rev 117:109472

Tsegaye B, Balomajumder C, Roy P (2019) Microbial delignification and hydrolysis of lignocellulosic biomass to enhance biofuel production: an overview and future prospect. Bull Natl Res Cent 43(1):1–16

Vaidya A, Singh T (2012) Pre-treatment of *Pinus radiata* substrates by basidiomycetes fungi to enhance enzymatic hydrolysis. Biotechnol Lett 34:1263–1267

Vasco-Correa J, Shah A (2019) Techno-economic bottlenecks of the fungal pretreatment of lignocellulosic biomass. Fermentation 5(2):30

Vlachokostas C, Achillas C, Diamantis V, Michailidou AV, Baginetas K, Aidonis D (2021) Supporting decision making to achieve circularity via a biodegradable waste-to-bioenergy and compost facility. J Environ Manag 285:112215

Wainaina S, Lukitawesa, Kumar Awasthi M, Taherzadeh MJ (2019) Bioengineering of anaerobic digestion for volatile fatty acids, hydrogen or methane production: a critical review. Bioengineered 10(1):437–458

Wang R, You T, Yang G, Xu F (2017) Efficient short time white rot–brown rot fungal pretreatments for the enhancement of enzymatic saccharification of corn cobs. ACS Sustain Chem Eng 5(11):10849–10857

Weng C, Peng X, Han Y (2021) Depolymerization and conversion of lignin to value-added bioproducts by microbial and enzymatic catalysis. Biotechnol Biofuels 14(1):1–22

Xie J, He Z, Wang Z, Wang B, Pan L (2019) Efficient expression of a novel thermophilic fungal β-mannosidase from *Lichtheimia ramosa* with broad-range pH stability and its synergistic hydrolysis of locust bean gum. J Biosci Bioeng 128(4):416–423

Yadav M, Balan V, Varjani S, Tyagi VK, Chaudhary G, Pareek N et al (2023) Multidisciplinary pretreatment approaches to improve the bio-methane production from lignocellulosic biomass. Bioenergy Res 16(1):228–247

Yang H, Xie Y, Zheng X, Pu Y, Huang F, Meng X et al (2016) Comparative study of lignin characteristics from wheat straw obtained by soda-AQ and kraft pretreatment and effect on the following enzymatic hydrolysis process. Bioresour Technol 207:361–369

Zabed HM, Akter S, Yun J, Zhang G, Awad FN, Qi X et al (2019) Recent advances in biological pretreatment of microalgae and lignocellulosic biomass for biofuel production. Renew Sust Energ Rev 105:105–128

Zhang H, Zhang X, Geng A (2020) Expression of a novel manganese peroxidase from Cerrena unicolor BBP6 in *Pichia pastoris* and its application in dye decolorization and PAH degradation. Biochem Eng J 153:107402

Zhang W, Wang J, Liu Y, Leng J (2021) Heterologous expression and characterization of a ferulic acid esterase from *Aspergillus aculeatus* with potential use in sunflower seed processing. Appl Sci 11(10):4453

An Overview of Fungal Enzymes in the Generation of Value-Added Product of Industrial Significance

Deepshikha, Pradeep Verma, and Komal Agrawal

1 Introduction

In response to the expanding consumption for energy and chemicals and the pressing need to address challenges related to gradual diminishing of fossil fuels and their environmental impact, there is a critical requirement to explore alternative sources for generating energy and production of chemical. One of the foremost challenges confronting humanity involves striving to establish a more sustainable society for succeeding generations (Celiktas and Alptekin 2017). The reduction in primary resources and escalating volume of waste are two critical concerns that provide a significant impediment to achieving the core aim of sustainability. In this context, biomass has been identified as a sustainable source of energy and chemicals while also addressing waste production difficulties. It may be generated from a range of waste sources, including agricultural waste and municipal solid waste, in addition to natural resources (Pandey et al. 2015).

Bioenergy and biofuels are produced from byproducts of other manufacturing processes. This technique improves resource efficiency while also increasing the value of used biomass and harmonizing with the ideals of a circular economy (Balina et al. 2017). Increased resource efficiency not only conserves raw material supplies but also allows for the reuse of biomass. Although transferring residual

The original version of the chapter has been revised. A correction to this chapter can be found at https://doi.org/10.1007/978-981-97-5544-8_18

Deepshikha · K. Agrawal (✉)
Department of Microbiology, School of Bio Engineering and Biosciences, Lovely Professional University, Phagwara, Punjab, India

P. Verma
Bioprocess and Bioenergy Laboratory (BPEL), Department of Microbiology, Central University of Rajasthan, Kishangarh, Ajmer, Rajasthan, India
e-mail: pradeepverma@curaj.ac.in

K. Agrawal, P. Verma (eds.), *Biotechnological Advances in Biorefinery*, Interdisciplinary Biotechnological Advances, https://doi.org/10.1007/978-981-97-5544-8_4

biomass between sites for subsequent production stages is a possible barrier, the biorefinery design mitigates this issue by condensing the whole process inside a single location. A cascade strategy that prioritizes the production of the higher-value product first and lower-value goods later in turn has been developed to ensure the sustainable exploration of biomass (Kamm et al. 2016). The notion of biorefinery first evolved in the late 1990s, with an emphasis on energy production and the manufacturing of biofuels. Biorefineries are defined (as per US Department of Energy) as complete facilities that convert and/or extract biomass feedstocks to generate a variety of valued products. These make it easier to convert diverse biomass processes and technologies into fuels, electricity, and chemicals. The use of combined heat, power, and biofuels in combination with chemical manufacturing is an interesting option for energy generation that has the potential to minimize environmental impact (Celiktas and Alptekin 2017).

The product made should have substantial production volume and the feasibility of biological synthesis from common sugars derived from diverse forms of biomass. These could be achieved through the exploration of different feedstocks combined with advanced biological techniques. To determine an appropriate product portfolio, specific criteria have been established. These criteria include high theoretical yields of the product from the substrate and market demand for the product either as a final product or as a crucial industrial intermediate (Liguori and Faraco 2016).

In this current chapter, the various generations of biorefineries will be briefly discussed. The significance of LCB as the main feedstock for second-generation biorefineries will then be thoroughly examined. Following that, the discussions will expand into the complexity of harnessing the potential of LCB in the context of sustainable and advanced biorefinery processes. Given its heterogeneous cellulose, hemicellulose, and lignin composition, the study will necessitate into consideration the special qualities and difficulties related to LCB. This chapter intends to present an understanding of the significance of LCB as a crucial resource in improving the ideas and practices of second-generation biorefinery technology.

2 Phases in Evolution of Biorefineries

Biofuel gains prominence as a promising sustainable energy option due to its origin from replenishable resources or biological sources, such as agricultural residues, municipal wastes, and other bio-based materials (Ramos et al. 2017). BE, biodiesel (BD), and biohydrogen (BH) exhibit distinctive and intriguing properties, finding widespread utilization across various applications (Papong et al. 2017). Based on the industrial processes and raw materials used in their development, biofuels are separated into four generations (Papong et al. 2017).

The first-generation biorefineries employ sugarcane, maize, barley, wheat, and other crops to generate BE, BD, and biogas from starch (Azad et al. 2015). First-generation biorefineries are primarily advantageous due to their high sugar content feedstocks, which facilitates the process of turning raw materials into biofuels

(Gasol et al. 2007). A life cycle analysis focused on first-generation biofuels demonstrated a net reduction in GHG emissions and a decrease in fossil fuel consumption. This is attributed to the effective substitution of conventional diesel and gasoline by biofuels in transportation, particularly in the case of BE and BD (Cherubini 2010). Within the spectrum of biofuels, the production of BE entails the utilization of starch-based substrates such as barley, potatoes, maize, wheat, and cassava, along with sucrose-rich botanical sources like sweet sorghum, sugarcane, and sugar beet. Among cereal grains, barley and wheat have been especially renowned. Barley, distinguished for its versatility as a winter crop, emerges as the foremost choice due to its pronounced resistance to drought (Arun and Dalai 2020). The process of BE production comprises of five primary stages, which includes (1) purification through water-washing; (2) perforation, defibration, crushing, and soaking; (3) extraction and processing of raw broth; (4) fermentation of processed solution; and (5) distillation of the BE (de Almeida and Colombo 2023). However, these biorefineries face significant challenges due to competition for biomass resources and agricultural land from the food and animal feed industries. The utilization of food resources and the exploration of agricultural land give rise to ethical concerns such as food versus fuel (Dutta et al. 2014).

Second-generation biofuels, as opposed to first-generation biofuels derived from cereals, are made from nonfood crops and primarily utilize feedstock residues. These feedstocks are referred to as LCB. Straws, forage, bagasse, stover, and municipal solid waste are all possible sources of LCB suitable for conversion into fuels or value-added goods (Geddes et al. 2011). Second-generation liquid biofuels such as ethanol, butanol, and other gaseous biofuels such as methane, hydrogen, and syngas may be produced from the abovementioned feedstock utilizing various methods, e.g., fungal enzymes (Singh et al. 2023). The production of BE from agricultural residues, for instance, rape straw and rice hulls, has witnessed a significant rise in recent years. *Wickerhamomyces anomalus* employs potato peel waste as a primary ingredient for BD production, utilizing saccharification and fermentation through EH (Wu et al. 2021). *Trichoderma citrinoviride* C1 is involved in the production of BE from sweet sorghum waste (Ruan et al. 2019). Additionally, the efficiency of solid-state fermentation (SSF) using cellulase and *S. cerevisiae* has been carried out for the conversion of pomegranate peel into BE. A thermochemical process has been applied to convert coffee residue waste into bioenergy. Citrus peelings using a solvent-free microwave extraction also turned out to be an excellent approach for producing biofuel. To manufacture cellulosic methanol from oil palm trunk sap, *S. cerevisiae* uses the mill effluent from palm oil mills as fertilizer (Vaish et al. 2019).

While the simultaneous presence of first-generation and second-generation biorefineries is apparent, the widespread adoption of second-generation biorefineries faces challenges linked to the necessary advancements in processing technology and infrastructure designed for these particular feedstocks (Dutta et al. 2014). Nevertheless, second-generation biorefineries are commonly perceived as being more ecologically sustainable, economically efficient, and socially acceptable. This is especially notable in situations where competition for food supplies and fertile land presents a significant challenge (Azad et al. 2015).

Later, the third-generation biorefineries evolved focusing on using aquatic biomass, notably algae (Agrawal et al., 2020), as their principal feedstock. This algal biomass is strong in oils, proteins, and carbohydrates, making it ideal for biorefinery operations (Agrawal et al. 2022, 2023a, b; Goswami et al. 2021, 2022a, b; Mehariya et al. 2021). Algal biorefineries hold considerable potential in the future for optimizing biomass utilization and generating a diverse range of developed products. The National Renewable Energy Laboratory's (NREL) combined algal processing (CAP) is one example of a fuel-based biorefinery (di Visconte et al. 2019). These biorefineries are not limited by seasonal growth, resulting in larger oil production than conventional oilseed crops. In addition, these biorefineries have great carbon dioxide tolerance and quick growth rates when compared with other biomass sources (Ho et al. 2011). The key obstacle to microalgae advancement is obtaining a suitable quantity of sunlight while regulating the pH levels of inorganic salts. On the contrary, the prior can be improved using cost-effective acrylic materials in system manufacturing (Anto et al. 2020). By supplying artificial lighting, it is possible to compensate for the absence of sunlight throughout the night, which will eventually increase the cultivation system's total production. Gaseous CO_2 sparging prevents algal cells from absorbing HCO^{3-} ions, ensuring a stable pH in the system. Achieving a profitable and environmentally friendly third-generation biorefinery necessitates the synergistic integration of various technologies encompassing material sciences, biological processes, and chemical, which has recently been achieved with advances in technology (Bhatia et al. 2022).

Additionally, fourth-generation biofuels include the study of the production of biofuel through genetic engineering of microorganisms. The development of complex genome editing tools, such as MAGE/eMAGE (Das et al. 2020) and CRISPR/Cas9 (Lakhawat et al. 2022), has the potential to increase the rate of genetic engineering comprising both traditional and unconventional hosts. The goal of this method is to create extremely efficient organisms for biofuel generation (Aamer Mehmood et al. 2021). The engineering of *S. cerevisiae* to generate bioactive substances can be facilitated by various factors such as the following: (1) due to the presence of genetic tools for manipulation, its genome is easily accessible in comparison to other oleaginous yeast strains; (2) isolation, production, and analysis of mutant strains are uncomplicated processes; (3) readily grows in specified medium; (4) rapid rate of growth; and (5) integrated expanding techniques ensuring efficient production on a larger scale (Malik et al. 2021). Among yeasts, *Y. lipolytica* is also considered a potentially suitable model species for producing long-chain hydrocarbons. Beyond yeasts, various bacterial strains, including *Z. mobilis*, *C. glutamicum*, *B. subtilis*, and *L. lactis*, have also been investigated for potential applications in biofuel generation (Aamer Mehmood et al. 2021).

3 Renewable Energy Resource: Lignocellulosic Biomass

Within the area of renewable resources, LCB derived from plants is the most easily accessible raw material for supporting the expansion of a sustainable energy while simultaneously reducing reliance on fossil fuels. In terms of dry matter composition, a typical LCB consists of 15–25% lignin, 20–40% cellulose, and 20–40% hemicellulose, which can vary according to the biomass under consideration (Østby et al. 2020). Moreover, covalent interactions, intermolecular connections, and van der Waals forces sustain the multilayer structure of lignocellulose, making it insoluble in water and incredibly resistant to enzyme breakdown (Nanda et al. 2014).

Lignin, a complex constituent present in plant cell walls, holds significance among the components of LCB due to its abundant aromatic group composition (Baeyens et al. 2015). It is characterized as a phenyl propane polymer connected through ester bonds and serves as a binding agent, establishing a robust connection with cellulose and hemicellulose. The key structural components of lignin include coniferyl alcohol, sinapyl alcohol, and p-coumaryl alcohol. It comprises macromolecules with highly branching phenolic compounds. Several phenyl propane subunits are connected via C-O-C and C-C connections (Nanda et al. 2014). Its side chain contains methoxyl, hydroxyl, phenolic, and terminal aldehyde groups, and it is insoluble in the majority of solvents. Notably, the polydispersity index of lignin is estimated to be more than that of cellulose (Demirbas 2006).

Another complex in LCB is cellulose, a linear polymer that results from the removal of water from glucose molecules, and its fundamental repeating unit is cellobiose. This polymer is made up of glucose subunits connected by β-(1→4) linkages (Bajpai 2016). Cellulose exhibits a structural pattern characterized by alternating amorphous and crystalline regions, forming microfibrils. Numerous cross-connections, particularly interactions with van der Waals and hydrogen bonds, increase cellulose chain cohesion, resulting in bundle formation. These bundles, in turn, aggregate to create microfibrils with diameters ranging from 5 to 15 nm (Sajith et al. 2016). Each cellulose chain plays an essential role in composing the microfibril, analogous to the contribution of individual threads to a rope. Typically, cellulose chains align parallelly within microfibrils, forming a highly ordered crystalline structure. In contrast, the less organized regions contribute to the amorphous aspects of the microfibrils (Sajith et al. 2016).

Hemicellulose, the second most abundant component within LCB, is characterized by its diverse composition of sugar acids, including pentoses and hexoses. These polymers exhibit a high degree of branching and lack a crystalline structure. A homopolymer or heteropolymer provides the hemicellulose's backbone, and along with β-(1,3)-glycosidic connections, its short branches are also connected by β-(1,4)-glycosidic linkages. Furthermore, hemicelluloses may exhibit acetylation to some extent, as seen in heteroxylan. In contrast to cellulose, hemicelluloses possess a lower molecular weight and shorter lateral chains, making them more readily susceptible to hydrolysis (Sajith et al. 2016).

In its native state, renewable plant biomass exhibits limited enzymatic decomposability due to its low reactivity (Bomble et al. 2017). The primary protective agent against enzymatic degradation of plant polysaccharides is lignin, which functions as both a physical barrier and an enzyme adsorbent, thereby restricting enzymatic action (Zeng et al. 2017). Cellulose displays resistance to EH owing to its insolubility and a stable structure characterized by a high degree of polymerization (DP) and crystallinity index (CI). Furthermore, hemicelluloses have a role in the decreased reactivity of renewable plant biomass through their formation of lignin-carbohydrate complexes (LCCs). Generally, the factors contributing to the resistance of renewable plant biomass against enzymes encompass the high DP and CI of cellulose, the limited surface area accessible for enzyme attack, cellulose protection facilitated by lignin, and the shielding effect of hemicelluloses surrounding cellulose microfibrils (Tarasov et al. 2018). To augment the enzymatic conversion potential of renewable plant biomass, pretreatment becomes imperative to eliminate lignin and disrupt the structure of cellulose fibers. This pretreatment results in a reduction in the CI and DP of cellulose, a reduction in particle size of the raw material, and an increase in cellulose porosity. Consequently, this process expands the surface area available for enzyme action (Haldar and Purkait 2020).

4 Fungal Enzyme in Biorefinery

Enzymes play a vital role in catalyzing various biological reactions essential for sustaining life by reducing the activation energy without causing any permanent alterations. They are ubiquitously found in all living organisms, encompassing animals, plants, and microorganisms (Guerrand 2018). Microbial enzyme production offers several advantages, including acceleration and the potential for genetic modifications. Among microbial sources, fungi are particularly noteworthy for their industrial enzyme production, constituting a significant portion of the enzyme market (Kango et al. 2019). Fungal enzymes are known for their high efficiency in production, relatively uncomplicated purification, and separation processes, especially in the case of filamentous fungi. Moreover, they demonstrate effective catalytic activity and robust stability even in challenging environmental conditions (Singh and Mahapatra 2019).

Biological breakdown of raw materials containing lignocellulose relies on fungi, predominantly falling into the categories of white-rot (*Phanerochaete chrysosporium*) (Wan and Li 2012), *Pleurotus ostreatus* (Suryadi et al. 2022), and brown-rot basidiomycetes (*Gloeophyllum trabeum*) (Umezawa et al. 2020). Fungi exhibit both intracellular and extracellular enzymatic activities. Due to their natural role as decomposers, fungi can deliver a wide array of extracellular enzymes crucial for converting diverse substrates and complexes (Kumla et al. 2020). However, this process demands prolonged implementation periods, and the rate of fungal degradation is insufficient for industrial applications, leading to the utilization of only a fraction of plant polysaccharides (Couturier and Berrin 2013). In contrast, enzymatic degradation offers a more convenient alternative to fungal degradation due to its high selectivity and swiftness, although it comes with a higher cost at an industrial scale. A variety of carbohydrate-active enzymes are needed for the enzymatic breakdown of important materials including lignocellulose; these enzymes typically work in tandem with one another, exhibiting complementary and synergistic properties and modes of action (Payne et al. 2015). Commercially significant fungal enzymes primarily fall into the categories of hydrolases and oxidoreductases (Berbee et al. 2017).

4.1 Enzymatic Degradation of Lignin

Enzymes play vital roles in the breakdown and consumption of lignin, facilitating its conversion into simpler, more accessible compounds. Lignin degradation is an oxidative process predominantly driven by basidiomycetes, with white-rot fungi being particularly proficient in this field (Singhania et al. 2022). The effective utilization of lignin necessitates its conversion into monomeric or oligo-aromatic compounds, with the aim of achieving a consistent and desirable pool of aromatic products (Ren et al. 2019). The enzymatic machinery responsible for lignin degradation includes laccases (Agrawal et al. 2019, 2022; Agrawal and Verma 2020), versatile peroxidase (VP), manganese-dependent peroxidase (MnP), and lignin peroxidase (LiP), all working together to dismantle lignin into more accessible compounds (Ren et al. 2019).

There are two primary avenues for lignin fractionation: thermochemical methods and biochemical methods. Within the thermochemical category, various techniques are employed, including pyrolysis, hydrogenolysis, gasification, hydrolysis, and chemical oxidation (Marinović et al. 2018). It is worth noting that chemical procedures typically involve the utilization of harsh chemicals and necessitate stringent reaction conditions. These conditions result into nonselective degradation of lignin and often lead to high operational costs. Biological agents, for instance, enzymes and microorganisms carrying these enzymes, display a notable capacity to efficiently deconstruct lignin under moderate reaction conditions. These exhibit a high level of specificity in targeting structural and stereochemical aspects of lignin (Cragg et al. 2015).

The enzymes that break down lignin are classified into two types: (1) lignin-modifying enzymes (LMEs), which serve as the fundamental components in the lignin degradation process, and (2) lignin-degrading auxiliary (LDA) enzymes, which, while unable to independently break down lignin, are essential for the successful completion of lignin degradation (Levasseur et al. 2008). Auxiliary enzymes perform a vital role in facilitating the depolymerization of lignin through the sequential action of various proteins, including the generation of H_2O_2 oxidative species. This group contains a variety of enzymes, such as pyranose 2-oxidase, glucose oxidase, aryl alcohol oxidases (AAO), glyoxal oxidase, and cellobiose dehydrogenase. The secretomes of white-rot fungus and basidiomycetes are commonly found to include these enzymes (Levasseur et al. 2008). Aerobic white-rot fungi have the extraordinary capacity to completely break down lignin in their native habitat. Anaerobic fungi, on the contrary, lack the essential enzymatic machinery for this activity because the enzymatic processes rely heavily on the presence of oxygen for separating the aromatic ring. Consequently, the complete degradation of lignin does not take place in an anaerobic environment (Berg and McClaugherty 2008). LiPs, which are oxidoreductases, are specially present to deal with the lignin biopolymer's intrinsic recalcitrance and heterogeneity. This biopolymer is composed of diverse phenylpropanoid units, forming a complex 3D network with various substructures (Ruiz-Dueñas and Martínez 2009). A recently published study presented a unique technique that combines the use of biocompatible ionic liquids to facilitate enzymatic lignin depolymerization with aqueous lignin extraction. This depolymerization process utilized a bi-enzyme system consisting of AAO and LiP. LiP and AAO's pH and temperature characteristics were determined as part of the study, and these enzymes' biocompatibilities in different solvents were also investigated. Interestingly, it was found that aqueous cholinium glycinate was an appropriate and efficient solvent for facilitating enzyme activity and solubilizing lignin. During the enzymatic lignin depolymerization process, the simultaenous action of AAO and LiP resulted in a significant decrease in lignin molecular weight in liquid and solid phases (Liu et al. 2021). In another investigation, lignin-degrading fungal species and mild chemical treatment were used in combination to function as agents for the effective synthesis of LMEs, which are intended to depolymerize lignin inside biomasses. *Phanerochaete chrysosporium* emerged as the most effective lignin degrader, achieving a substantial lignin degradation rate of 47.11% in paddy straw (PS) + pine needle (PN) over a span of 28 days, with the peak production of LMEs observed between the 10th and 17th days. The effective lignin breakdown in PS and PN biomass holds promise for its prospective application in pulp production, thereby contributing to the realization of a circular economy in an environmentally sustainable manner (Gupta et al. 2023).

Heme-containing peroxidases, including multifunctional VP, LiP, and MnP, and laccase are classified as LMEs. LiP and VP may efficiently oxidize benzene rings of lignin due to their significant redox potential, which is caused by their architectural features, including a heme pocket. This capability remains unaffected by the degree of methoxylation and interunit linkage within the lignin structure. Additionally, a recent addition to this enzyme repertoire is the heme-thiolate haloperoxidase group, which has been proposed to contribute to the process of lignin degradation (Fernández-Fueyo et al. 2012). The absence of experimental evidence supporting the autonomous lignin-degrading capabilities of these enzymes categorizes them as LDA enzymes by definition. Moreover, numerous studies have showcased various other lignolytic enzymes, including feruloyl esterase (Liu et al. 2023), lipases (Šibalić et al. 2023), catechol 2,3-dioxygenase (Zhang et al. 2021), xylanase (Bhardwaj et al. 2019, 2021a, b), and quinone reductases (Majeke et al. 2021), in the natural degradation of lignin. The hydroxycinnamic acid produced during the reduction process serves as a mediator for the laccase (Janusz et al. 2017).

LMEs integral to the peroxidase-catalase superfamily, which encompasses LiP, MnP, and VP, are predominantly present in white-rot fungi. These enzymes perform a vital function in lignin degradation, utilizing H_2O_2 as a cosubstrate. All enzymes classified within this superfamily share to porphyrin IX as a cofactor (Zámocký et al. 2015). These three peroxidases require highly conserved catalytic residues for oxidative degradation. In these peroxidases, oxidation begins with H_2O_2 deprotonation assisted by a distal base, resulting in the production of a short-lived [Fe^{4+}-O-O-H] intermediate. Subsequently, H_2O_2 undergoes heterolytic cleavage of the dioxygen bond, resulting in the generation of a high-valent species [Fe^{4+}-O]$^+$, known as compound I. After attaching to the substrate, compound I oxidizes with the assistance of a porphyrin radical or specific amino acid radical a, donating two electrons to form a stable Fe^{3+} state (Singhania et al. 2022).

4.2 *Enzymatic Degradation of Cellulose*

Cellulase constitutes a sophisticated multi-enzyme network consisting of various hydrolytic enzymes that display significant synergy. The efficient degradation of cellulose proves challenging for any single cellulase (Agrawal et al. 2023a, b; Bhardwaj et al. 2021a, b; Kumar et al. 2018). It may be categorized into three types based on the unique role of each enzyme. Exo β-1,4-glucanase (EC 3.2.1.91), commonly known as cellobiohydrolase, belongs to the first group. Cellobiose is produced when this enzyme catalyst, also known as β-1,4-glucan fiber two carbohydrase, promotes the breakdown of β-1,4-glucoside bonds located on the nonreducing sugar end of cellulose (Jayasekara and Ratnayake 2019). When exonucleases act solely on natural cellulose, the generation of reducing sugars is challenging to detect. However, the synergistic collaboration with endoglucanases proves effective in the efficient decomposition of natural cellulose. Endo-β-1,4-glucanase (EC 3.2.1.4) hydrolyzes cellulose's glycosidic linkages to produce glucose, cellobiose, and cello-oligosaccharides (Šuchová et al. 2022). Cello-oligosaccharides, β-glucosides, and cellobiose are broken down into glucose by the enzyme β-1,4-glucosidase (EC 3.2.1.21). Also referred to as cellobiase, this enzyme, present in enzyme preparations, demonstrates a minimal protein content, comprising approximately 1% of the total composition. β-glucosidase breaks down cello-oligosaccharides and cellobiose to generate glucose during the saccharification of cellulose (Giwa et al. 2023).

The glycosidic bond cleavage by cellulases is achieved through acid-based catalysis and involves two different catalytic mechanisms: inverting mechanisms and retaining mechanisms. The two catalytic residues of the enzyme, a nucleophile/base and a generic acid (proton donor), assist in the hydrolysis process. The particular catalytic mechanism that is utilized is contingent upon the spatial orientation of individual residues (Saini and Sharma 2021). Cellulases conserve the original configuration of the anomeric carbon carrying the unique glycosidic bond in the "retaining" mechanism, even after a double-displacement hydrolysis involving two critical deglycosylation or glycosylation stages. In contrast, in the "inverting" process, cellulases change the conformation of the anomeric carbon following a single nucleophilic displacement hydrolysis (Giwa et al. 2023). During the screening phase of a study, ten fungal isolates were identified using carboxymethyl cellulose-Congo Red agar plates with substantial endoglucanase activity. Six (6) of the selected isolates were tested based on their higher total cellulase activity and were identified to be *Aspergillus* and *Trichoderma* members. In submerged fermentation (SmF) using synthetic media with a pH of 6.5 at 30 °C, *T. longibrachiatum* LMLSAUL 14-1 demonstrated a noteworthy total cellulase activity of 8 FPU/mL and an endoglucanase activity of 23 U/mL. Similarly, *T. harzianum* LMLBP07 13-5 exhibited a total cellulase activity of 6 FPU/mL and an endoglucanase activity of 16 U/mL. Notably, the levels of cellulases and endoglucanase produced by *Trichoderma* species surpassed those observed in strains of *A. fumigatus*. The β-glucosidase activity of *A. fumigatus* LMLPS 13-4 was higher at 38 U/mL in comparison to *Trichoderma* species (Legodi et al. 2019). In a separate investigation, 37 species had been isolated and evaluated for their capability of degrading rice straw cellulose. The isolates *Rhizopus oryzae* (AUMC 14361), *Fusarium verticillioides* (AUMC 14360), and *Neurospora intermedia* (AUMC 14359) showed the greatest cellulase activity. Optimal conditions for maximal activity were determined for each strain, which varied as well. *N. intermedia* exhibited maximum activity at 2% carboxymethylcellulose (CMC) concentration, beef extract and peptone primary nitrogen sources, pH 7, 30 °C, and an inoculum size of 2.5(v/v). *R. oryzae* exhibited maximal activity at 2% CMC concentration, beef extract providing a nitrogen source, pH 7, 30 °C, and an inoculum size of 2.5(v/v). *F. verticillioides* thrived under circumstances with 2.5% CMC, using yeast extract as a source of nitrogen, pH 6, 30 °C, and an inoculum size of 2(v/v). The strains were considered unique starting strains because of their great effectiveness under ideal circumstances (Helal et al. 2021).

4.3 Enzymatic Degradation of Hemicellulose

Debranching enzymes, responsible for removing diverse side chains from hemicellulose polysaccharides and contributing to their chemical heterogeneity, are classified into two groups based on their target bonds: glycosidic bond-acting enzymes (GHs) and ester bond-acting enzymes (carbohydrate esterases or CEs) (Terrett et al. 2019). GHs encompass xylosidases, arabinosidases, glucuronidases, and galactosidases, while CEs include feruloyl, glucuronoyl, acetyl, and p-coumaroyl esterases. The elimination of substituting groups, such as acetyl, (methyl) glucuronoyl, and galactosyl, is crucial for xylanases and mannanases (Busse-Wicher et al. 2014). To fully convert biomass or generate cellulose fibers, this method makes it easier to penetrate the underlying cellulosic fibers. Specialized CEs like feruloyl, p-coumaroyl esterases (Monrad et al. 2018), and glucuronoyl esterases (Ernst et al. 2020) play a pivotal role in separating hemicellulose fragments from residual lignin through ester bond cleavage in LCCs (Várnai et al. 2014). The dual benefit of CEs in lignocellulose analysis lies in their ability to assist hemicellulose-lignin separation and concurrently promote cellulose exposure by facilitating the depolymerization of hemicellulose fractions. Organisms that break down lignocellulose also employ oxidative enzymes like lytic polysaccharide monooxygenases (LPMOs) for the depolymerization of hemicellulosic polysaccharides. The significance of these enzymes in lignocellulose degradation lies in their ability, akin to endo-acting glycoside hydrolases, to generate new chain ends that serve as substrates for exo-acting enzymes (Østby and Várnai 2023).

Enzymes with hemicellulose activity are used in lignocellulose processing for two reasons: (1) to purify cellulose fibers with the aim of augmenting the availability of cellulose microfibrils, facilitating comprehensive saccharification, or enabling fiber engineering within lignocellulose biorefineries, and (2) to transform separated hemicellulose fractions into separate streams of polymeric, oligomeric, or monomeric products (Østby and Várnai 2023). Among the enzymes classified under auxiliary activities (AA), some enzymes catalyze the oxidation of poly-, oligo-, and monosaccharides derived from hemicellulose. Specifically, galactose oxidases such as AA5_2 have the capability to oxidize the C6-hydroxyl group of terminals (non-reducing) galactosyl substitutions in galactoglucomannan and xyloglucan (Parikka et al. 2010). Certain oligosaccharides can have their terminal nonsubstituted reducing-end sugar oxidized by hemicellulose-specific oxidases such as xylooligosaccharide oxidase (AA7) (Vuong et al. 2013). By giving hemicellulose-derived biomaterials a functional transformation, these enzymes enhance the value generation of biomass even if they do not directly depolymerize hemicelluloses. These eliminate hemicellulose to purify cellulose fibers, improving the availability of cellulose microfibrils for thorough saccharification or fiber engineering in LCB biorefineries. Furthermore, the separated hemicellulose fractions are converted into discrete streams of poly-, oligo-, or monomeric products. In an investigation, pineapple peel waste, containing 31.8 ± 1.9% hemicellulose, was explored for value addition. Using an alkali-based method, optimal conditions for hemicellulose recovery (95.9 ± 2.0%) were observed at 15% (w/v) alkali concentration, 16 h, and 45 °C. Hydrothermal-assisted alkali extraction achieved a maximum relative recovery of approximately 87.6% with 10% (w/v) alkali after 1.5 h at 121 °C and 15 psi pressure. Xylo-oligosaccharides (XOS) were generated by the EH of extracted hemicellulose, whereas a xylose-rich liquid (91% yield) with the potential to be converted to compounds such as xylitol was obtained by direct hydrolysis with diluted nitric acid (Banerjee et al. 2019).

5 Application

A platform chemical is one that may function as a substrate to provide a range of higher value-added goods. The Department of Energy (DOE) recognized 12 chemical compounds that are produced from biomass as possible platform chemicals in

2004 (Takkellapati et al. 2018). The DOE upgraded the Platform Chemical List in 2010, in which they included furfural, ethanol, 3-hydroxypropionic acid/aldehyde, hydroxymethylfurfural, isoprene, glycerol, succinic acid, lactic acid, 2,5-furandicarboxylic acid, levulinic acid, xylitol, and sorbitol. Except for isoprene and glycerol, all of the listed platform chemicals may be produced from carbohydrate sources generated from biomass (Singh et al. 2015). Alcohols of low molecular weight, for instance, methanol, ethanol, propanol, and butanol, have significant potential as alternatives to fossil fuels in the transportation sector (Demain 2009). Alcohol fuel derived from biomass has the ability to lower GHG emissions, decrease fuel costs, improve energy efficiency overall, and provide employment in the agriculture sector. Recently, there has been a heightened focus on the production pathways of biomass-derived ethanol, generating substantial interest as a biofuel (Nanda et al. 2014) (Fig. 1) (Table 1).

5.1 *Bioethanol*

BE possesses the capacity to serve as a long-term source of butadiene, propylene, and ethylene—all essential for the production of polymers. Numerous types of microorganisms are employed in the fermentation process of LCB to produce BE. A combination of critical features is required for the selection of an appropriate microbe for effective ethanol production: (1) an ethanol yield ≤90% of theoretical estimation; (2) an ethanol productivity greater than 1 g/L/h; (3) an ethanol tolerance greater than 40 g/L; (4) strong metabolic properties and minimal growth needs, preferably achieved through cost-effective media formulation; (5) growth potential in undiluted hydrolyzates; (6) growth conditions that inhibit the growth of

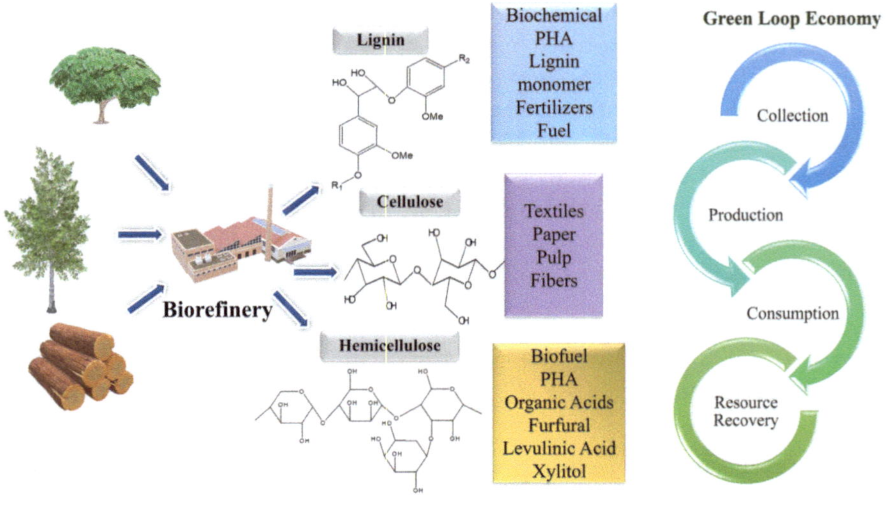

Fig. 1 Diagrammatic illustration of various value-added product forms of LCB

Table 1 Compilation of biorefineries employing LCB

Biorefinery route	Residue	Pretreatment	Process	Products	Yield	References
Lignin	Organosolv lignin	Mechanochemical	Oxidation	Phenolic compounds	95%	Dabral et al. (2018)
	LCB	Hydrogenation	Extraction by biomass derived solvent, γ-valerolactone	Phenolic compounds	76.41%	Ji et al. (2018)
	Alkali lignin	Fast pyrolysis	Fixed bed reactor (Ni formate catalyst)	Liquid yield alkyl phenols	30.20 wt%	Geng et al. (2017)
	Kraft lignin	Catalytic hydrotreatment	Depolymerization	Monomer yield alkyl phenolics and guaiacolics	28 wt% 76%	Narani et al. (2015)
	Lignin	Depolymerization	ES	Monomeric phenols	47%	Xiao et al. (2017)
Cellulose	Corn stover	Diluted pretreatment	ES	BD and BE	54.47%	Su et al. (2020)
	Sugarcane bagasse	One-pot hydrothermal precipitation sulfonation	H	Fermentable sugars	74.5%	Alavijeh et al. (2020)
	Sugarcane bagasse	Hydrodynamic cavitation-assisted alkaline H_2O_2	H	Ethanol	84.31%	Terán Hilares et al. (2020)
	Rice hull	Extrusion and ultrasound	EH	Fermentable sugars	381.59 mg/g	Zang et al. (2020)
	Rice straw	Steam explosion	EH	Fermentable sugars	81.1%	Amoah et al. (2019)
Hemicellulose	Sugarcane bagasse	Microwave hydrothermal liquefaction	H	Food additives	8.1 wt%	Shao et al. (2020)
	LCB	Ultrasound-assisted acid		Furfural	3.64%	Bizzi et al. (2019)
					4.99%	
	Wheat straw	Microwave-assisted acid pretreatment		Furfural and HMF	84.01% 76.42%	Prasad et al. (2018)
	Aspen-wood sawdust,	Microwave-assisted pressurized hot water system		Gels, films, adhesives, coating	66%	Mihiretu et al. (2017)
	Sugarcane trash				50%	
	Rice straw	Steam explosion		Furfural	375 g/kg	Chen et al. (2015)

H hydrolysis, *EH* enzymatic hydrolysis, *ES* enzyme saccharification contaminants; and (7) resistance to inhibitors, acidic pH, and higher temperature (Dien et al. 2003). There are two distinct approaches available for ethanol production from LCB. The initial method employs a thermal gasification process to produce syngas, subsequently transformed into ethanol through either a chemical catalytic process or a biochemical fermentative technique (Fang et al. 2008). The alternative method relies on biochemical fermentation, following the saccharification process, in which complex carbohydrates are transformed to simple monomers through hydrolysis of cellulose, hemicellulose forms fungal enzyme, cellulose produces glucose, while hemicellulose produces different pentose and hexose isomers (Schwab et al. 2016). The monomers, notably pentose and hexose sugars, are subsequently fermented by microorganisms to make ethanol. Since normally existing microorganisms cannot successfully digest both pentose and hexose sugars, genetically engineered microbes are required for complete fermentation, resulting in increased ethanol yields (Fig. 2). There are two basic fermentation techniques that are routinely used: separate hydrolysis and fermentation (SHF) and simultaneous saccharification and fermentation (SSF) (Takkellapati et al. 2018).

5.2 *Biodiesel*

BD represents a crucial environmentally friendly fuel sourced from diverse biomass types. It can be used either independently or blended with petro-diesel, leading to a reduction in gas emissions and combustion-related impurities. Key physical attributes including cetane number, flash point, specific gravity, and sulfur content are therefore improved by this method (Ahorsu et al. 2018). When edible, nonedible, and waste oils are esterified or trans-esterified with alcohol using a certain

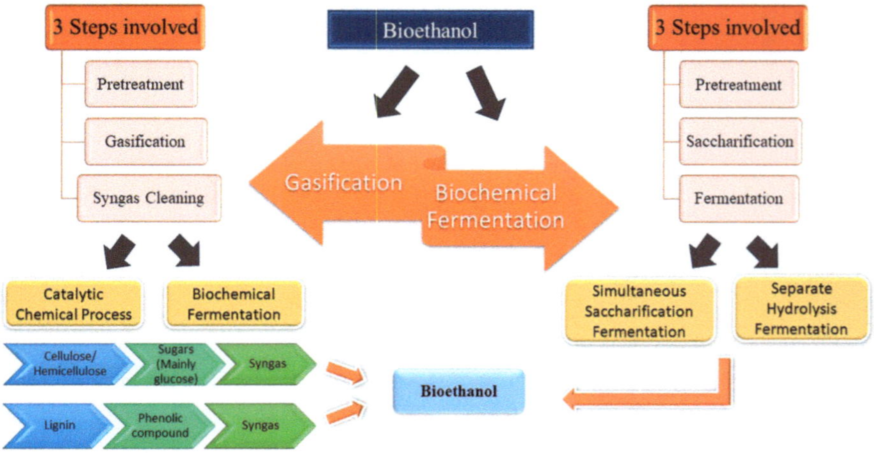

Fig. 2 Diagrammatic representation of two methods for BE production form of LCB

catalyst—which may be basic or acidic in nature—BD, a mono-alkyl ester, is produced (Singh et al. 2021). The BD production process from biowaste consists of two main stages. The initial stage involves the extraction of lipids, a process that is conducted using mechanical, chemical, or enzymatic methods. Subsequently, the extracted lipids undergo esterification with alcohols (Olkiewicz et al. 2016). In a specific investigation, lipids were extracted from dewatered sludge (with a total suspended solids content of 15 wt.%) in the initial step using hexane. This was accompanied by the second step, which involved esterification via methanol in presence of an acid catalyst (H_2SO_4) (Pastore et al. 2013). In another study, a lipid content of approximately 37.99% was obtained from treated LCB (sugarcane bagasse) through steam explosion at 121 °C for 2 h. The lipid content is acknowledged as a critical feedstock in BD production (Ananthi et al. 2019).

5.3 *Furfural*

Furfural is derived from diverse renewable agricultural sources in a LCB biorefinery. Acid-catalyzed xylose dehydration is the process used in the commercial manufacturing of bio-based furfural (Dutta et al. 2012). Furfural manufacturing involves two unique processes: one-step and two-step. The one-step process uses acids to assist the combined dehydration to furfural and depolymerization of pentosans in xylose. In comparison, the second approach is a two-step process that begins with the dissolving and depolymerization of pentosans under moderate circumstances, followed by the dehydration of xylose to yield furfural (Ribeiro et al. 2016). Approximately 70% of the manufactured furfural serves as a chemical feedstock for furfuryl alcohol production. Furthermore, furfural can be converted into 5-membered oxygen heterocycles, levulinic acid, and succinic acid. Furfural finds extensive applications in industries such as plastics, pharmaceuticals, agrochemicals, adhesives, and flavor enhancers (Liu et al. 2021). Furfuryl acid-derived resins have remarkable thermal, chemical, and physical characteristics, including resistance to solvents and caustic substances. When producing fiber-reinforced polymers, furfuryl acid is used; high-performance procedures utilizing oxygenated organic solvents and chlorinated aromatics are particularly well-suited for its application (Yan et al. 2014).

5.4 *Biohydrogen*

BH is one possible renewable energy source that may be developed effectively with the use of renewable energy resources. It shows significant potential as an alternative energy opportunity as a clean fuel, addressing the negative environmental consequences connected with the use of fossil fuels (Khan et al. 2018). A vast array of limitless biological sources may be employed to generate BH, such as organic wastes, biomass, and wastewater from industries that produce sugar, beverages, and chemicals, in addition to palm oil effluent and distillery sector effluent (Preethi et al. 2019). In addition to these sources, LCB stands out as especially appropriate for BH generation owing to its high cellulose concentration, sustainability, and availability of access (Wang et al. 2017). Furthermore, it functions as a carbon-neutral energy carrier, offering an energy yield 2.75 times greater than that of fossil fuels (Singh and Mahapatra 2019). Its broad range of applications encompasses the production of fertilizers (such as ammonia), impurity removal in methanol synthesis, oil refineries, pharmaceutical industries, cryogenics, rotor cooling, petroleum recovery, utilization as a rocket engine fuel, and its role as a hydrogenating and reducing agent. The biochemical production of hydrogen is demonstrated by a variety of species, including bacteria, green algae, blue-green algae, and higher plants. This makes using biological methods for hydrogen generation an ecologically beneficial method. Microorganisms accelerate the photoproduction of oxygen and hydrogen at ambient pressures and temperatures (Khetkorn et al. 2017).

5.5 *Biogas*

Biogas is a gas mixture predominantly composed of carbon dioxide (CO_2) and methane, along with trace amounts of water vapor, hydrogen peroxide (H_2O_2), and hydrogen sulfide (Mirmohamadsadeghi et al. 2019). It is produced by the breakdown of organic materials anaerobically, which involves several steps of substrate-bacteria interaction. The four steps of this biological process are acetogenesis, methanogenesis, hydrolysis, and acidogenesis (Silva dos Santos et al. 2018). Biogas is considered an economically viable and environmentally friendly renewable energy source because its carbon content is sourced from organic matter, making it carbon neutral as it captures carbon from atmospheric CO_2 (Awe et al. 2017).

In addition to its use in electricity generation, biogas serves various purposes such as automotive fuel, residential heating, combined heat and power generation, and applications in the chemical industry (Zhu et al. 2019). Notably, this process contributes to the elimination of GHG emissions and maintains a nonpolluting profile. Oxidative treatment with H_2O_2 is considered as an additional pretreatment method for breaking complex framework of biomass. This method holds the capacity to enhance the cellulose proportion through the delignification of LCB (Kapoor et al. 2020).

6 Challenges

The current state of biorefineries encounters significant challenges in gaining acceptance within the established market dominated by the fossil-based economy. These significant challenges encompass concerns pertaining to the accessibility and composition of feedstock, the ability to fulfill market demands, the efficacy of resource recovery, the techno-economic viability of the endeavor, and adherence to environmental sustainability standards (Mohan et al. 2019). Optimizing processes, developing techno-economic models, encouraging technical advances, running simulations with cost and risk analysis, and putting life cycle evaluations into practice are all necessary for successfully transferring laboratory-scale technologies into commercial operations (Katakojwala and Mohan 2021). Essential considerations for scaling up biorefinery operations include emphasizing product diversification, minimizing waste discharge, optimizing water utilization, deploying cost-effective equipment and labor, and selecting the appropriate combination of pretreatment methods and biomass types (Sharma et al. 2020). Efficient commercialization is dependent upon the automation of critical biorefinery procedures, which includes the transportation, sorting, and management of biomass prior to its arrival at processing facilities. Furthermore, by automating the commencement of enzyme hydrolysis and subsequent phases of product separation and packaging, substantial operational efficiencies can be achieved, energy consumption can be reduced, and the requirement for manual interventions can be eliminated (Ubando et al. 2020).

7 Conclusion

Enzymes have a central and indispensable place in the biomass-based biorefinery paradigm, coordinating a series of essential processes ranging from the hydrolysis of polysaccharides to the intricate transformation of biomass into an array of valuable products. These include platform chemicals, BE, intermediate chemical compounds, and a variety of additional value-added goods. However, the multidimensional nature of these conversions often results in a difficult and multistep procedure that raises manufacturing costs. The complex transformation of these items involves a multistep procedure, which automatically leads to increased prices. Consequently, there is a strong need to reduce the number of processing stages using novel methodologies. This requirement emphasizes the need of having an improved comprehension of the biochemical and kinetic features of diverse hydrolytic enzymes. When combined with sophisticated technology, this enhanced understanding has the potential to improve the generation of microorganisms suited for integrated biorefinery systems.

References

Aamer Mehmood M, Shahid A, Malik S, Wang N, Rizwan Javed M, Nabeel Haider M, Verma P, Umer Farooq Ashraf M, Habib N, Syafiuddin A, Boopathy R (2021) Advances in developing metabolically engineered microbial platforms to produce fourth-generation biofuels and high-value biochemicals. Bioresour Technol 337:125510. https://doi.org/10.1016/j.biortech.2021.125510

Agrawal K, Verma P (2020) Production optimization of yellow laccase from *Stropharia* sp. ITCC 8422 and enzyme-mediated depolymerization and hydrolysis of lignocellulosic biomass for biorefinery application. Biomass Convers Biorefinery 12:4939–4958. https://doi.org/10.1007/s13399-020-00869-w

Agrawal K, Bhardwaj N, Kumar B, Chaturvedi V, Verma P (2019) Process optimization, purification and characterization of alkaline stable white laccase from Myrothecium verrucaria ITCC-8447 and its application in delignification of agroresidues. Int J Biol Macromol 125:1042–1055. https://doi.org/10.1016/j.ijbiomac.2018.12.108

Agrawal K, Bhatt A, Bhardwaj N, Kumar B, Verma P (2020) Algal biomass: potential renewable feedstock for biofuels production–part I. In: Biofuel production technologies: critical analysis for sustainability. Springer, Cham, pp 203–237. https://doi.org/10.1007/978-981-13-8637-4_8

Agrawal K, Vaishnavi S, Verma P (2022) Laccase mediated green composite synthesis: a name synonymous with each other. In: Deshmukh SK, Deshpande MV, Sridhar KR (eds) Fungal biopolymers and biocomposites: prospects and avenues. Springer, Singapore, pp 295–312. https://doi.org/10.1007/978-981-19-1000-5_14

Agrawal K, Nair LG, Chaturvedi V, Verma P (2023a) Designing microbial cellulases using genetic engineering approach: a promising strategy towards zero-waste cellulosic biorefinery. Biocatal Agric Biotechnol 52:102830. https://doi.org/10.1016/j.bcab.2023.102830

Agrawal K, Ruhil T, Verma P (2023b) Microalgae: a way toward sustainable development of a society. In: Basic research advancement for algal biofuels production. Springer, Singapore, pp 259–277. https://doi.org/10.1007/978-981-19-6810-5_10

Ahorsu R, Medina F, Constantí M (2018) Significance and challenges of biomass as a suitable feedstock for bioenergy and biochemical production: a review. Energies 11(12):3366

Alavijeh RS, Karimi K, van den Berg C (2020) An integrated and optimized process for cleaner production of ethanol and biodiesel from corn stover by Mucor indicus. J Clean Prod 249:119321

Amoah J, Ogura K, Schmetz Q, Kondo A, Ogino C (2019) Co-fermentation of xylose and glucose from ionic liquid pretreated sugar cane bagasse for bioethanol production using engineered xylose assimilating yeast. Biomass Bioenergy 128:105283. https://doi.org/10.1016/j.biombioe.2019.105283

Ananthi V, Prakash GS, Chang SW, Ravindran B, Nguyen DD, Vo D-VN, La DD, Bach Q-V, Wong JWC, Gupta SK (2019) Enhanced microbial biodiesel production from lignocellulosic hydrolysates using yeast isolates. Fuel 256:115932

Anto S, Mukherjee SS, Muthappa R, Mathimani T, Deviram G, Kumar SS, Verma TN, Pugazhendhi A (2020) Algae as green energy reserve: technological outlook on biofuel production. Chemosphere 242:125079. https://doi.org/10.1016/j.chemosphere.2019.125079

Arun N, Dalai AK (2020) Chapter 9: Environmental and socioeconomic impact assessment of biofuels from lignocellulosic biomass. In: Yousuf A, Pirozzi D, Sannino F (eds) Lignocellulosic biomass to liquid biofuels. Academic, New York, pp 283–299. https://doi.org/10.1016/B978-0-12-815936-1.00009-5

Awe OW, Zhao Y, Nzihou A, Minh DP, Lyczko N (2017) A review of biogas utilisation, purification and upgrading technologies. Waste Biomass Valoriz 8(2):267–283. https://doi.org/10.1007/s12649-016-9826-4

Azad AK, Rasul MG, Khan MMK, Sharma SC, Hazrat MA (2015) Prospect of biofuels as an alternative transport fuel in Australia. Renew Sust Energ Rev 43:331–351. https://doi.org/10.1016/j.rser.2014.11.047

Baeyens J, Kang Q, Appels L, Dewil R, Lv Y, Tan T (2015) Challenges and opportunities in improving the production of bio-ethanol. Prog Energy Combust Sci 47:60–88. https://doi.org/10.1016/j.pecs.2014.10.003

Bajpai P (2016) Structure of lignocellulosic biomass. In: Pretreatment of lignocellulosic biomass for biofuel production. Springer, Singapore, pp 7–12. https://doi.org/10.1007/978-981-10-0687-6_2

Balina K, Romagnoli F, Blumberga D (2017) Seaweed biorefinery concept for sustainable use of marine resources. Energy Procedia 128:504–511

Banerjee S, Patti AF, Ranganathan V, Arora A (2019) Hemicellulose based biorefinery from pineapple peel waste: xylan extraction and its conversion into xylooligosaccharides. Food Bioprod Process 117:38–50. https://doi.org/10.1016/j.fbp.2019.06.012

Berbee ML, James TY, Strullu-Derrien C (2017) Early diverging fungi: diversity and impact at the dawn of terrestrial life. Ann Rev Microbiol 71:41–60. https://doi.org/10.1146/annurev-micro-030117

Berg B, McClaugherty C (2008) Plant litter: decomposition, humus formation, carbon sequestration. Springer, Cham

Bhardwaj N, Kumar B, Agarwal K, Chaturvedi V, Verma P (2019) Purification and characterization of a thermo-acid/alkali stable xylanases from Aspergillus oryzae LC1 and its application in Xylo-oligosaccharides production from lignocellulosic agricultural wastes. Int J Biol Macromol 122:1191–1202. https://doi.org/10.1016/j.ijbiomac.2018.09.070

Bhardwaj N, Agrawal K, Verma P (2021a) Xylanases: an overview of its diverse function in the field of biorefinery. In: Bioenergy research: commercial opportunities & challenges. Springer, Cham, pp 295–317

Bhardwaj N, Kumar B, Agrawal K, Verma P (2021b) Current perspective on production and applications of microbial cellulases: a review. Bioresour Bioprocess 8:1–34. https://doi.org/10.1186/s40643-021-00447-6

Bhatia L, Bachheti RK, Garlapati VK, Chandel AK (2022) Third-generation biorefineries: a sustainable platform for food, clean energy, and nutraceuticals production. Biomass Convers Biorefinery 12(9):4215–4230. https://doi.org/10.1007/s13399-020-00843-6

Bizzi CA, Santos D, Sieben TC, Motta GV, Mello PA, Flores EMM (2019) Furfural production from lignocellulosic biomass by ultrasound-assisted acid hydrolysis. Ultrason Sonochem 51:332–339. https://doi.org/10.1016/j.ultsonch.2018.09.011

Bomble YJ, Lin C-Y, Amore A, Wei H, Holwerda EK, Ciesielski PN, Donohoe BS, Decker SR, Lynd LR, Himmel ME (2017) Lignocellulose deconstruction in the biosphere. Curr Opin Chem Biol 41:61–70

Busse-Wicher M, Gomes TCF, Tryfona T, Nikolovski N, Stott K, Grantham NJ, Bolam DN, Skaf MS, Dupree P (2014) The pattern of xylan acetylation suggests xylan may interact with cellulose microfibrils as a twofold helical screw in the secondary plant cell wall of Arabidopsis thaliana. Plant J 79(3):492–506

Celiktas MS, Alptekin FM (2017) Biorefinery concept: current status and future prospects. Int Conf Eng Technol 7:9

Chen H, Qin L, Yu B (2015) Furfural production from steam explosion liquor of rice straw by solid acid catalysts (HZSM-5). Biomass Bioenergy 73:77–83. https://doi.org/10.1016/j.biombioe.2014.12.013

Cherubini F (2010) The biorefinery concept: using biomass instead of oil for producing energy and chemicals. Energy Convers Manag 51(7):1412–1421. https://doi.org/10.1016/j.enconman.2010.01.015

Couturier M, Berrin J-G (2013) The saccharification step: the main enzymatic components. In: Faraco V (ed) Lignocellulose conversion: enzymatic and microbial tools for bioethanol production. Springer, Berlin, pp 93–110. https://doi.org/10.1007/978-3-642-37861-4_5

Cragg SM, Beckham GT, Bruce NC, Bugg TDH, Distel DL, Dupree P, Etxabe AG, Goodell BS, Jellison J, McGeehan JE, McQueen-Mason SJ, Schnorr K, Walton PH, Watts JEM, Zimmer M (2015) Lignocellulose degradation mechanisms across the tree of life. Curr Opin Chem Biol 29:108–119. https://doi.org/10.1016/j.cbpa.2015.10.018

Dabral S, Wotruba H, Hernández JG, Bolm C (2018) Mechanochemical oxidation and cleavage of lignin β-O-4 model compounds and lignin. ACS Sustain Chem Eng 6(3):3242–3254. https://doi.org/10.1021/acssuschemeng.7b03418

Das M, Patra P, Ghosh A (2020) Metabolic engineering for enhancing microbial biosynthesis of advanced biofuels. Renew Sust Energ Rev 119:109562

de Almeida MA, Colombo R (2023) Production chain of first-generation sugarcane bioethanol: characterization and value-added application of wastes. Bioenergy Res 16(2):924–939

Demain AL (2009) Biosolutions to the energy problem. J Ind Microbiol Biotechnol 36(3):319–332. https://doi.org/10.1007/s10295-008-0521-8

Demirbas MF (2006) Current technologies for biomass conversion into chemicals and fuels. Energy Sour Part A Recov Utiliz Environ Effects 28(13):1181–1188

di Visconte GS, Spicer A, Chuck CJ, Allen MJ (2019) The microalgae biorefinery: a perspective on the current status and future opportunities using genetic modification. Appl Sci 9(22):4793. https://doi.org/10.3390/app9224793

Dien BS, Cotta MA, Jeffries TW (2003) Bacteria engineered for fuel ethanol production: current status. Appl Microbiol Biotechnol 63(3):258–266. https://doi.org/10.1007/s00253-003-1444-y

Dutta S, De S, Saha B (2012) A brief summary of the synthesis of polyester building-block chemicals and biofuels from 5-hydroxymethylfurfural. ChemPlusChem 77(4):259–272

Dutta K, Daverey A, Lin J-G (2014) Evolution retrospective for alternative fuels: first to fourth generation. Renew Energy 69:114–122. https://doi.org/10.1016/j.renene.2014.02.044

Ernst HA, Mosbech C, Langkilde AE, Westh P, Meyer AS, Agger JW, Larsen S (2020) The structural basis of fungal glucuronoyl esterase activity on natural substrates. Nat Commun 11(1):1026

Fang Z, Sato T, Smith RL Jr, Inomata H, Arai K, Kozinski JA (2008) Reaction chemistry and phase behavior of lignin in high-temperature and supercritical water. Bioresour Technol 99(9):3424–3430

Fernández-Fueyo E, Ruiz-Dueñas FJ, Miki Y, Martínez MJ, Hammel KE, Martínez AT (2012) Lignin-degrading peroxidases from genome of selective ligninolytic fungus Ceriporiopsis subvermispora. J Biol Chem 287(20):16903–16916

Gasol CM, Gabarrell X, Anton A, Rigola M, Carrasco J, Ciria P, Solano ML, Rieradevall J (2007) Life cycle assessment of a Brassica carinata bioenergy cropping system in southern Europe. Biomass Bioenergy 31(8):543–555. https://doi.org/10.1016/j.biombioe.2007.01.026

Geddes CC, Nieves IU, Ingram LO (2011) Advances in ethanol production. Curr Opin Biotechnol 22(3):312–319. https://doi.org/10.1016/j.copbio.2011.04.012

Geng J, Wang W-L, Yu Y-X, Chang J-M, Cai L, Shi SQ (2017) Adding nickel formate in alkali lignin to increase contents of alkylphenols and aromatics during fast pyrolysis. Bioresour Technol 227:1–6. https://doi.org/10.1016/j.biortech.2016.11.036

Giwa AS, Ali N, Akhter MS (2023) Cellulose degradation enzymes in filamentous fungi, a bioprocessing approach towards biorefinery. Mol Biotechnol. https://doi.org/10.1007/s12033-023-00900-1

Goswami RK, Agrawal K, Verma P (2021) Microalgae-based biofuel-integrated biorefinery approach as sustainable feedstock for resolving energy crisis. In: Bioenergy research: commercial opportunities & challenges. Springer, Cham, pp 267–293. https://doi.org/10.1007/978-981-16-1190-2_9

Goswami RK, Agrawal K, Verma P (2022a) Microalgae dunaliella as biofuel feedstock and β-carotene production: an influential step towards environmental sustainability. Energy Convers Manag 13:100154. https://doi.org/10.1016/j.ecmx.2021.100154

Goswami RK, Agrawal K, Upadhyaya HM, Gupta VK, Verma P (2022b) Microalgae conversion to alternative energy, operating environment and economic footprint: an influential approach towards energy conversion, and management. Energy Convers Manag 269:116118. https://doi.org/10.1016/j.enconman.2022.116118

Guerrand D (2018) Chapter 26: Economics of food and feed enzymes: status and prospectives. In: Nunes CS, Kumar V (eds) Enzymes in human and animal nutrition. Academic, New York, pp 487–514. https://doi.org/10.1016/B978-0-12-805419-2.00026-5

Gupta A, Tiwari A, Ghosh P, Arora K, Sharma S (2023) Enhanced lignin degradation of paddy straw and pine needle biomass by combinatorial approach of chemical treatment and fungal enzymes for pulp making. Bioresour Technol 368:128314. https://doi.org/10.1016/j.biortech.2022.128314

Haldar D, Purkait MK (2020) Lignocellulosic conversion into value-added products: a review. Process Biochem 89:110–133

Helal GA, Khalil RR, Galal YG, Soliman SM, Abd Elkader RS (2021) Studies on cellulases of some cellulose-degrading soil fungi. Arch Microbiol 204(1):65. https://doi.org/10.1007/s00203-021-02705-9

Ho S-H, Chen C-Y, Lee D-J, Chang J-S (2011) Perspectives on microalgal CO_2-emission mitigation systems – a review. Biotechnol Adv 29(2):189–198. https://doi.org/10.1016/j.biotechadv.2010.11.001

Janusz G, Pawlik A, Sulej J, Świderska-Burek U, Jarosz-Wilkołazka A, Paszczyński A (2017) Lignin degradation: microorganisms, enzymes involved, genomes analysis and evolution. FEMS Microbiol Rev 41(6):941–962. https://doi.org/10.1093/femsre/fux049

Jayasekara S, Ratnayake R (2019) Microbial cellulases: an overview and applications. In: Cellulose. IntechOpen, London. https://doi.org/10.5772/intechopen.84531

Ji H, Dong C, Yang G, Pang Z (2018) Valorization of lignocellulosic biomass toward multipurpose fractionation: furfural, phenolic compounds, and ethanol. ACS Sustain Chem Eng 6(11):15306–15315. https://doi.org/10.1021/acssuschemeng.8b03766

Kamm B, Schönicke P, Hille C (2016) Green biorefinery - industrial implementation. Food Chem 197:1341–1345. https://doi.org/10.1016/j.foodchem.2015.11.088

Kango N, Jana UK, Choukade R (2019) Fungal enzymes: sources and biotechnological applications. In: Satyanarayana T, Deshmukh SK, Deshpande MV (eds) Advancing frontiers in mycology & mycotechnology: basic and applied aspects of fungi. Springer, Singapore, pp 515–538. https://doi.org/10.1007/978-981-13-9349-5_21

Kapoor R, Ghosh P, Kumar M, Sengupta S, Gupta A, Kumar SS, Vijay V, Kumar V, Kumar Vijay V, Pant D (2020) Valorization of agricultural waste for biogas based circular economy in India: a research outlook. Bioresour Technol 304:123036. https://doi.org/10.1016/j.biortech.2020.123036

Katakojwala R, Mohan SV (2021) A critical view on the environmental sustainability of biorefinery systems. Curr Opin Green Sustainable Chem 27:100392

Khan MA, Ngo HH, Guo W, Liu Y, Zhang X, Guo J, Chang SW, Nguyen DD, Wang J (2018) Biohydrogen production from anaerobic digestion and its potential as renewable energy. Renew Energy 129:754–768. https://doi.org/10.1016/j.renene.2017.04.029

Khetkorn W, Rastogi RP, Incharoensakdi A, Lindblad P, Madamwar D, Pandey A, Larroche C (2017) Microalgal hydrogen production – a review. Bioresour Technol 243:1194–1206. https://doi.org/10.1016/j.biortech.2017.07.085

Kumar B, Bhardwaj N, Alam A, Agrawal K, Prasad H, Verma P (2018) Production, purification and characterization of an acid/alkali and thermo tolerant cellulase from Schizophyllum commune NAIMCC-F-03379 and its application in hydrolysis of lignocellulosic wastes. AMB Express 8:1–16. https://doi.org/10.1186/s13568-018-0696-y

Kumla J, Suwannarach N, Sujarit K, Penkhrue W, Kakumyan P, Jatuwong K, Vadthanarat S, Lumyong S (2020) Cultivation of mushrooms and their lignocellulolytic enzyme production through the utilization of agro-industrial waste. Molecules 25:12. https://doi.org/10.3390/molecules25122811

Lakhawat SS, Malik N, Kumar V, Kumar S, Sharma PK (2022) Implications of CRISPR-Cas9 in developing next generation biofuel: a mini-review. Curr Protein Pept Sci 23(9):574–584

Legodi LM, La Grange D, Van Rensburg EL, Ncube I (2019) Isolation of cellulose degrading fungi from decaying banana pseudostem and Strelitzia alba. Enzyme Res 2019:1390890

Levasseur A, Piumi F, Coutinho PM, Rancurel C, Asther M, Delattre M, Henrissat B, Pontarotti P, Asther M, Record E (2008) FOLy: an integrated database for the classification and functional annotation of fungal oxidoreductases potentially involved in the degradation of lignin and related aromatic compounds. Fungal Genet Biol 45(5):638–645

Liguori R, Faraco V (2016) Biological processes for advancing lignocellulosic waste biorefinery by advocating circular economy. Bioresour Technol 215:13–20. https://doi.org/10.1016/j.biortech.2016.04.054

Liu E, Segato F, Prade RA, Wilkins MR (2021) Exploring lignin depolymerization by a bi-enzyme system containing aryl alcohol oxidase and lignin peroxidase in aqueous biocompatible ionic liquids. Bioresour Technol 338:125564. https://doi.org/10.1016/j.biortech.2021.125564

Liu X, Jiang Y, Liu H, Yuan H, Huang D, Wang T (2023) Research progress and biotechnological applications of feruloyl esterases. Biocatal Biotransformation 41(6):421–439
Majeke BM, Collard F-X, Tyhoda L, Görgens JF (2021) The synergistic application of quinone reductase and lignin peroxidase for the deconstruction of industrial (technical) lignins and analysis of the degraded lignin products. Bioresour Technol 319:124152
Malik S, Shahid A, Liu C-G, Zafar Khan A, Nawaz MZ, Zhu H, Mehmood MA (2021) Developing fourth-generation biofuels secreting microbial cell factories for enhanced productivity and efficient product recovery; a review. Fuel 298:120858. https://doi.org/10.1016/j.fuel.2021.120858
Marinović M, Nousiainen P, Dilokpimol A, Kontro J, Moore R, Sipilä J, de Vries RP, Mäkelä MR, Hildén K (2018) Selective cleavage of lignin β-O-4 aryl ether bond by β-etherase of the white-rot fungus Dichomitus squalens. ACS Sustain Chem Eng 6(3):2878–2882. https://doi.org/10.1021/acssuschemeng.7b03619
Mehariya S, Goswami RK, Verma P, Lavecchia R, Zuorro A (2021) Integrated approach for wastewater treatment and biofuel production in microalgae biorefineries. Energies 14(8):2282. https://doi.org/10.3390/en14082282
Mihiretu GT, Brodin M, Chimphango AF, Øyaas K, Hoff BH, Görgens JF (2017) Single-step microwave-assisted hot water extraction of hemicelluloses from selected lignocellulosic materials – a biorefinery approach. Bioresour Technol 241:669–680. https://doi.org/10.1016/j.biortech.2017.05.159
Mirmohamadsadeghi S, Karimi K, Tabatabaei M, Aghbashlo M (2019) Biogas production from food wastes: a review on recent developments and future perspectives. Bioresour Technol Rep 7:100202. https://doi.org/10.1016/j.biteb.2019.100202
Mohan SV, Dahiya S, Amulya K, Katakojwala R, Vanitha TK (2019) Can circular bioeconomy be fueled by waste biorefineries—a closer look. Bioresour Technol Rep 7:100277
Monrad RN, Eklöf J, Krogh KBRM, Biely P (2018) Glucuronoyl esterases: diversity, properties and biotechnological potential. A review. Crit Rev Biotechnol 38(7):1121–1136
Nanda S, Mohammad J, Reddy SN, Kozinski JA, Dalai AK (2014) Pathways of lignocellulosic biomass conversion to renewable fuels. Biomass Convers Biorefinery 4:157–191
Narani A, Chowdari RK, Cannilla C, Bonura G, Frusteri F, Heeres HJ, Barta K (2015) Efficient catalytic hydrotreatment of Kraft lignin to alkylphenolics using supported NiW and NiMo catalysts in supercritical methanol. Green Chem 17(11):5046–5057. https://doi.org/10.1039/C5GC01643F
Olkiewicz M, Torres CM, Jiménez L, Font J, Bengoa C (2016) Scale-up and economic analysis of biodiesel production from municipal primary sewage sludge. Bioresour Technol 214:122–131
Østby H, Várnai A (2023) Hemicellulolytic enzymes in lignocellulose processing. Essays Biochem 67(3):533–550. https://doi.org/10.1042/EBC20220154
Østby H, Hansen LD, Horn SJ, Eijsink VGH, Várnai A (2020) Enzymatic processing of lignocellulosic biomass: principles, recent advances and perspectives. J Ind Microbiol Biotechnol 47(9-10):623–657. https://doi.org/10.1007/s10295-020-02301-8
Pandey A, Höfer R, Taherzadeh M, Nampoothiri M, Larroche C (2015) Industrial biorefineries and white biotechnology. Elsevier, Amsterdam
Papong S, Rewlay-ngoen C, Itsubo N, Malakul P (2017) Environmental life cycle assessment and social impacts of bioethanol production in Thailand. J Clean Prod 157:254–266
Parikka K, Leppanen A-S, Pitkanen L, Reunanen M, Willfor S, Tenkanen M (2010) Oxidation of polysaccharides by galactose oxidase. J Agric Food Chem 58(1):262–271
Pastore C, Lopez A, Lotito V, Mascolo G (2013) Biodiesel from dewatered wastewater sludge: a two-step process for a more advantageous production. Chemosphere 92(6):667–673
Payne CM, Knott BC, Mayes HB, Hansson H, Himmel ME, Sandgren M, Ståhlberg J, Beckham GT (2015) Fungal cellulases. Chem Rev 115(3):1308–1448. https://doi.org/10.1021/cr500351c
Prasad S, Malav MK, Kumar S, Singh A, Pant D, Radhakrishnan S (2018) Enhancement of bioethanol production potential of wheat straw by reducing furfural and 5-hydroxymethylfurfural (HMF). Bioresour Technol Rep 4:50–56. https://doi.org/10.1016/j.biteb.2018.09.007
Preethi, Usman TMM, Rajesh Banu J, Gunasekaran M, Kumar G (2019) Biohydrogen production from industrial wastewater: an overview. Bioresour Technol Rep 7:100287. https://doi.org/10.1016/j.biteb.2019.100287

Ramos JL, García-Lorente F, Valdivia M, Duque E (2017) Green biofuels and bioproducts: bases for sustainability analysis. Microb Biotechnol 10(5):1111–1113

Ren T, Qi W, Su R, He Z (2019) Front cover: promising techniques for depolymerization of lignin into value-added chemicals. ChemCatChem 11(2):636. https://doi.org/10.1002/cctc.201900018

Ribeiro LS, Delgado JJ, de Melo Órfão JJ, Pereira MFR (2016) A one-pot method for the enhanced production of xylitol directly from hemicellulose (corncob xylan). RSC Adv 6(97):95320–95327

Ruan D, Zhou Z, Pang H, Yao J, Chen G, Qiu Z (2019) Enhancing methane production of anaerobic sludge digestion by microaeration: enzyme activity stimulation, semi-continuous reactor validation and microbial community analysis. Bioresour Technol 289:121643. https://doi.org/10.1016/j.biortech.2019.121643

Ruiz-Dueñas FJ, Martínez ÁT (2009) Microbial degradation of lignin: how a bulky recalcitrant polymer is efficiently recycled in nature and how we can take advantage of this. Microb Biotechnol 2(2):164–177

Saini S, Sharma KK (2021) Fungal lignocellulolytic enzymes and lignocellulose: a critical review on their contribution to multiproduct biorefinery and global biofuel research. Int J Biol Macromol 193:2304–2319. https://doi.org/10.1016/j.ijbiomac.2021.11.063

Sajith S, Priji P, Sreedevi S, Benjamin S (2016) An overview on fungal cellulases with an industrial perspective. J Nutr Food Sci 6(1):461

Schwab A, Warner E, Lewis J (2016) 2015 survey of non-starch ethanol and renewable hydrocarbon biofuels producers. National Renewable Energy Lab, Golden

Shao Y, Tsang DCW, Shen D, Zhou Y, Jin Z, Zhou D, Lu W, Long Y (2020) Acidic seawater improved 5-hydroxymethylfurfural yield from sugarcane bagasse under microwave hydrothermal liquefaction. Environ Res 184:109340. https://doi.org/10.1016/j.envres.2020.109340

Sharma B, Larroche C, Dussap C-G (2020) Comprehensive assessment of 2G bioethanol production. Bioresour Technol 313:123630. https://doi.org/10.1016/j.biortech.2020.123630

Šibalić D, Šalić A, Zelić B, Tran NN, Hessel V, Nigam KDP, Tišma M (2023) Synergism of ionic liquids and lipases for lignocellulosic biomass valorization. Chem Eng J 461:142011

Silva dos Santos IF, Braz Vieira ND, de Nóbrega LGB, Barros RM, Tiago Filho GL (2018) Assessment of potential biogas production from multiple organic wastes in Brazil: Impact on energy generation, use, and emissions abatement. Resour Conserv Recycl 131:54–63. https://doi.org/10.1016/j.resconrec.2017.12.012

Singh L, Mahapatra DM (2019) Waste to sustainable energy: MFCs–prospects through prognosis. CRC Press, Boca Raton

Singh J, Suhag M, Dhaka A (2015) Augmented digestion of lignocellulose by steam explosion, acid and alkaline pretreatment methods: a review. Carbohydr Polym 117:624–631. https://doi.org/10.1016/j.carbpol.2014.10.012

Singh D, Sharma D, Soni SL, Inda CS, Sharma S, Sharma PK, Jhalani A (2021) A comprehensive review on 1st-generation biodiesel feedstock palm oil: production, engine performance, and exhaust emissions. Bioenergy Res 14(1):1–22. https://doi.org/10.1007/s12155-020-10171-2

Singh A, Prajapati P, Vyas S, Gaur VK, Sindhu R, Binod P, Kumar V, Singhania RR, Awasthi MK, Zhang Z, Varjani S (2023) A comprehensive review of feedstocks as sustainable substrates for next-generation biofuels. Bioenergy Res 16(1):105–122. https://doi.org/10.1007/s12155-022-10440-2

Singhania RR, Patel AK, Raj T, Chen C-W, Ponnusamy VK, Tahir N, Kim S-H, Dong C-D (2022) Lignin valorisation via enzymes: a sustainable approach. Fuel 311:122608. https://doi.org/10.1016/j.fuel.2021.122608

Su T, Zeng J, Gao H, Jiang L, Bai X, Zhou H, Xu F (2020) One-pot synthesis of a chemically functional magnetic carbonaceous acid catalyst for fermentable sugars production from sugarcane bagasse. Fuel 262:116512

Šuchová K, Fehér C, Ravn JL, Bedő S, Biely P, Geijer C (2022) Cellulose- and xylan-degrading yeasts: enzymes, applications and biotechnological potential. Biotechnol Adv 59:107981. https://doi.org/10.1016/j.biotechadv.2022.107981

Suryadi H, Judono JJ, Putri MR, Eclessia AD, Ulhaq JM, Agustina DN, Sumiati T (2022) Biodelignification of lignocellulose using ligninolytic enzymes from white-rot fungi. Heliyon 8:e08865

Takkellapati S, Li T, Gonzalez MA (2018) An overview of biorefinery-derived platform chemicals from a cellulose and hemicellulose biorefinery. Clean Techn Environ Policy 20(7):1615–1630. https://doi.org/10.1007/s10098-018-1568-5

Tarasov D, Leitch M, Fatehi P (2018) Lignin–carbohydrate complexes: properties, applications, analyses, and methods of extraction: a review. Biotechnol Biofuels 11(1):1–28

Terán Hilares R, Dionízio RM, Sánchez Muñoz S, Prado CA, de Sousa Júnior R, da Silva SS, Santos JC (2020) Hydrodynamic cavitation-assisted continuous pre-treatment of sugarcane bagasse for ethanol production: effects of geometric parameters of the cavitation device. Ultrason Sonochem 63:104931. https://doi.org/10.1016/j.ultsonch.2019.104931

Terrett OM, Lyczakowski JJ, Yu L, Iuga D, Franks WT, Brown SP, Dupree R, Dupree P (2019) Molecular architecture of softwood revealed by solid-state NMR. Nat Commun 10(1):4978

Ubando AT, Felix CB, Chen W-H (2020) Biorefineries in circular bioeconomy: a comprehensive review. Bioresour Technol 299:122585. https://doi.org/10.1016/j.biortech.2019.122585

Umezawa K, Niikura M, Kojima Y, Goodell B, Yoshida M (2020) Transcriptome analysis of the brown rot fungus Gloeophyllum trabeum during lignocellulose degradation. PLoS One 15(12):e0243984

Vaish B, Sharma B, Srivastava V, Singh P, Ibrahim MH, Singh RP (2019) Energy recovery potential and environmental impact of gasification for municipal solid waste. Biofuels 10(1):87–100

Várnai A, Costa THF, Faulds CB, Milagres AMF, Siika-Aho M, Ferraz A (2014) Effects of enzymatic removal of plant cell wall acylation (acetylation, p-coumaroylation, and feruloylation) on accessibility of cellulose and xylan in natural (non-pretreated) sugar cane fractions. Biotechnol Biofuels 7(1):1–11

Vuong TV, Vesterinen A-H, Foumani M, Juvonen M, Seppälä J, Tenkanen M, Master ER (2013) Xylo-and cello-oligosaccharide oxidation by gluco-oligosaccharide oxidase from Sarocladium strictum and variants with reduced substrate inhibition. Biotechnol Biofuels 6:1–14

Wan C, Li Y (2012) Fungal pretreatment of lignocellulosic biomass. Biotechnol Adv 30(6):1447–1457. https://doi.org/10.1016/j.biotechadv.2012.03.003

Wang S, Dai G, Yang H, Luo Z (2017) Lignocellulosic biomass pyrolysis mechanism: a state-of-the-art review. Prog Energy Combust Sci 62:33–86. https://doi.org/10.1016/j.pecs.2017.05.004

Wu Y, Ge S, Xia C, Mei C, Kim K-H, Cai L, Smith LM, Lee J, Shi SQ (2021) Application of intermittent ball milling to enzymatic hydrolysis for efficient conversion of lignocellulosic biomass into glucose. Renew Sust Energ Rev 136:110442. https://doi.org/10.1016/j.rser.2020.110442

Xiao L-P, Wang S, Li H, Li Z, Shi Z-J, Xiao L, Sun R-C, Fang Y, Song G (2017) Catalytic hydrogenolysis of lignins into phenolic compounds over carbon nanotube supported molybdenum oxide. ACS Catal 7(11):7535–7542. https://doi.org/10.1021/acscatal.7b02563

Yan K, Wu G, Lafleur T, Jarvis C (2014) Production, properties and catalytic hydrogenation of furfural to fuel additives and value-added chemicals. Renew Sust Energ Rev 38:663–676

Zámocký M, Hofbauer S, Schaffner I, Gasselhuber B, Nicolussi A, Soudi M, Pirker KF, Furtmüller PG, Obinger C (2015) Independent evolution of four heme peroxidase superfamilies. Arch Biochem Biophys 574:108–119

Zang G, Shah A, Wan C (2020) Techno-economic analysis of an integrated biorefinery strategy based on one-pot biomass fractionation and furfural production. J Clean Prod 260:120837. https://doi.org/10.1016/j.jclepro.2020.120837

Zeng Y, Himmel ME, Ding S-Y (2017) Visualizing chemical functionality in plant cell walls. Biotechnol Biofuels 10(1):1–16

Zhang R, Wang J, Milligan S, Yan Y (2021) Microbial utilization of lignin-derived aromatics via a synthetic catechol meta-cleavage pathway. Green Chem 23(20):8238–8250

Zhu T, Curtis J, Clancy M (2019) Promoting agricultural biogas and biomethane production: lessons from cross-country studies. Renew Sust Energ Rev 114:109332. https://doi.org/10.1016/j.rser.2019.109332

Bioprocessing of Biomass to Value-Added Products Using Enzymes

Swagat Kumar Das and Ashwini Kumar Dash

1 Introduction

As per the European Renewable Energy Directive, biomass can be defined as "the biodegradable fraction of products, waste, and residues of biological origin from agricultural, including plant and animal substances, from forestry and related industries, including fisheries and aquaculture, as well as the biodegradable fraction of waste, including industrial and municipal waste of biological origin." The biomass can be yoked with the production of several energy-related products like biodiesel, bio jet fuel, biogas, etc. The production of energy from biomass offers several advantages like sustainability, cost-effectiveness, etc. The lignocellulosic biomass that consists of lignin, cellulose, and hemicellulose needs to be degraded to separate cellulose-hemicellulose components from lignin. The cellulose-hemicellulose and lignin components are further utilized to produce different products. Therefore, it is of utmost importance to remove, degrade, and isolate the different components of lignocellulosic biomass because the proper utilization of biomass can save waste management cost significantly. The biomass can be utilized for the production of several products such as biofuel (bioethanol, 1-butanol, bio-methane, bio-hydrogen), industrially important chemicals (citric acid, fumaric acid, lactic acid, succinic acid, schizophyllan, pullulan, xylitol), single-cell protein, curdlan, dextran, etc. (Kumar et al. 2019). One of the major steps in the conversion of biomass conversion involves

S. K. Das
Department of Biotechnology, Odisha University of Technology and Research, Bhubaneswar, Odisha, India

A. K. Dash (✉)
Department of Textile Engineering, Odisha University of Technology and Research, Bhubaneswar, Odisha, India
e-mail: akdash@outr.ac.in

K. Agrawal, P. Verma (eds.), *Biotechnological Advances in Biorefinery*, Interdisciplinary Biotechnological Advances,
https://doi.org/10.1007/978-981-97-5544-8_5

hydrolysis of cellulose, but the recalcitrant nature of lignocellulosic biomass makes this process difficult. Hence, pretreatment method is conducted to remove lignin and hemicellulose before hydrolysis using different methods that include physical, chemical, or biological (Singh et al. 2015). Enzymes play a key role in the conversion of lignocellulosic biomass impacting the overall time and cost of the biomass conversion. The present study summarizes the role of enzymes used for the conversion of lignocellulosic biomass to value-added products.

2 Techniques of Biomass Conversion

The plant biomass usually consists of polymeric substances like cellulose, hemicellulose, and lignin. The biomass in its original form is difficult to process due to several factors that affect the process such as high humidity, low density, and difficulties in milling. The biomass conversion irrespective of their sources can be conducted in two ways, namely, thermochemical conversion and biochemical conversion. The thermochemical conversion technique involves several steps namely, combustion, pyrolysis, gasification, liquefaction, etc. On the contrary, biochemical conversion employes different enzymes to break down the biomass. The strategies of biochemical conversion technique include anaerobic digestion and fermentation (McKendry 2002). An alternative to the above two techniques is hybrid processing, which is a combination of biological, thermal, and/or catalytic processes. However, the hybrid conversion technique of biomass conversion showed more advantages over the above two traditional methods in terms of robustness, selectivity, fast pyrolysis, reduction in recalcitrant biomass, etc. (Shen et al. 2015). The conversion of lignocellulosic biomass comprises three steps: pretreatment, hydrolysis of the pretreated biomass, and fermentation of reducing sugars into fuels and biochemicals (Fig. 1).

Pretreatment is a very important step in lignocellulosic biomass conversion in which the outermost seal of lignin is broken, leading to the subsequent exposure of cellulose and hemicellulose for hydrolysis (Pandiyan et al. 2019). Therefore,

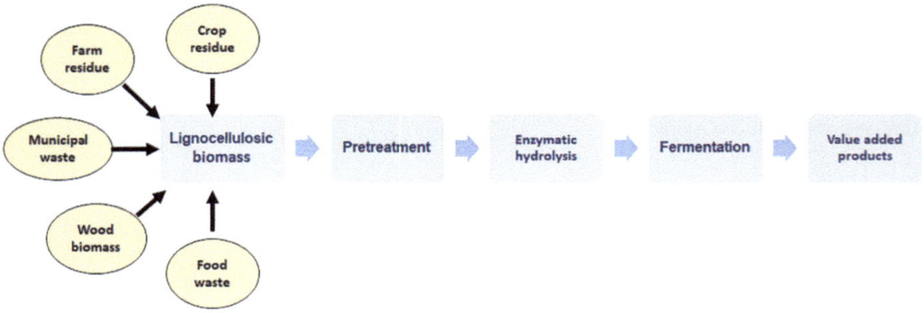

Fig. 1 Major steps involved in lignocellulosic biomass conversion to value-added products

different pretreatment techniques have been studied to optimize the condition for individual feedstock materials. The pretreatment procedure can be categorized into physical, chemical, physicochemical, and biological processes along with a combination of the above methods (Nitsos et al. 2017). The physical pretreatment leads to size reduction of biomass along with an increase in effective surface area. The different physical methods involve milling, freezing, ultrasonication, etc. (Fig. 2). The chemical pretreatment involves treatment with different chemicals like acid, alkali, organic solvents, liquid mixture of ions, etc. The physicochemical pretreatment involves a combination of physical processes such as steam explosion, liquid hot water, or wet oxidation along with chemical treatment. On the contrary, the biological pretreatment encompasses the use of enzymes for deconstruction of the rigid structure of biomass.

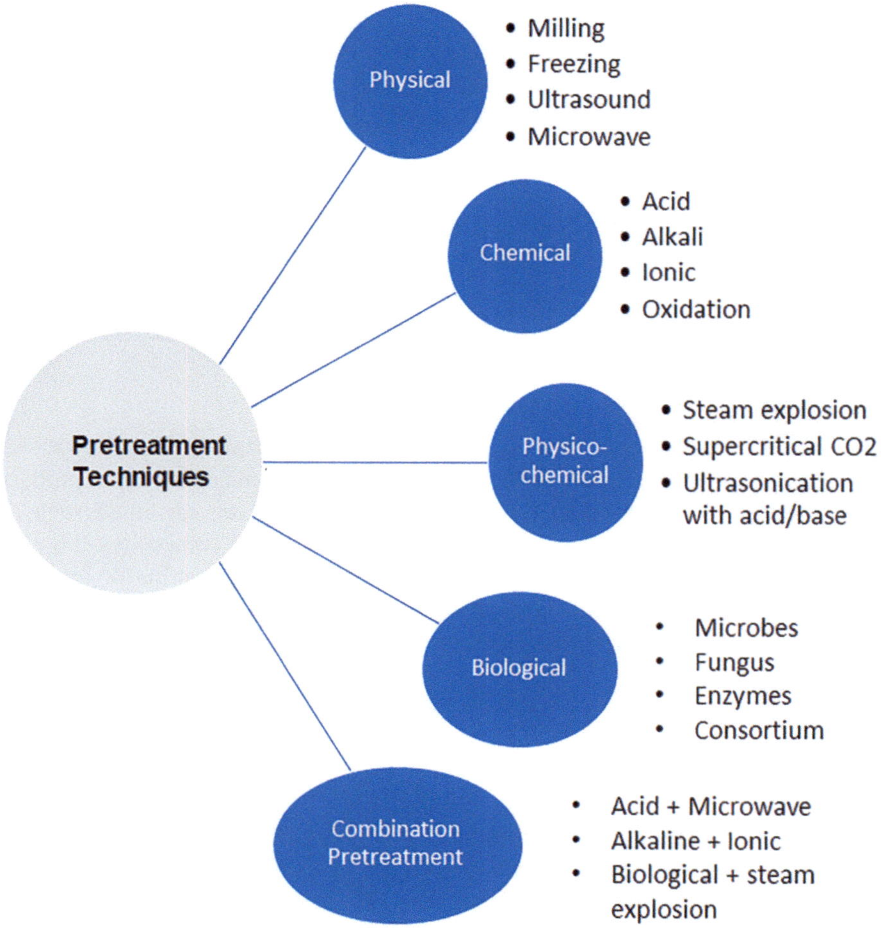

Fig. 2 Different types of pretreatment methods employed in lignocellulosic biomass treatment

The hydrolysis is a limiting step in the conversion of lignocellulosic biomass to biofuel, where a number of hydrolytic enzymes played a substantial role. The lignocellulosic biomass of a plant usually consists of 40–60% cellulose, 20–40% hemicellulose, and 10–24% lignin; however, the composition may vary depending on the nature of different plant species (Sharma et al. 2019). Besides, pectin forms the base matrix in the middle plate and cell walls together with hemicelluloses and proteins. The pectin mostly consists of alpha-D-galacturonic acid (Stoffels et al. 2020). The enzymatic hydrolysis of cellulosic and hemicellulosic biomass to fermentable sugars involves a complex cocktail of cellulolytic and hemicellulolytic enzymes. The hydrolysis of cellulosic biomass involves enzymes like exo-1,4-β-glucanase, endo-1,4-β-glucanase, cellobiohydrolase, β-glucosidase, and copper-dependent lytic polysaccharide monooxygenases that catalyze the oxidative cleavage of polysaccharides (Johansen 2016). However, hemicellulose needs a more complex set of enzymes due to its structural complexity as compared with cellulose. The hemicellulolytic enzymes can be broadly categorized into core enzymes and ancillary enzymes. The core enzymes that catalyze the cleavage of the polysaccharide backbone involve endo-β-1,4-xylanase (EC 3.2.1.8), xylan-1,4-β-xylosidase (EC 3.2.1.37), endo-1,4-β-mannanase (EC 3.2.1.78), and β-1,4-mannosidase (EC 3.2.1.25). On the contrary, ancillary enzymes are involved in the removal of functional groups, and they include enzymes like β-glucuronidase (EC 3.2.1.139), acetylxylan esterase (EC 3.2.1.55), ferulic acid esterase (EC 3.1.1.73), and *p*-coumaric acid esterase (EC 3.1.1).

3 Enzymatic Intervention in the Production of Value-Added Products from Lignocellulosic Biomass

The last few decades have witnessed extensive improvements on lignocellulose biomass conversion processes. In addition to the traditional applications in the production of biofuels, new markets for lignocellulosic have been identified such as enzymes, chemicals, pulp and paper, animal feedstuff, and composites (Fig. 3). Potential applications of lignocellulosic materials in the production of biofuels, enzymes, chemicals, the pulp and paper, animal feedstuff, and composites have been summarized in the following section.

3.1 Biofuel Production

Biomass-based biorefineries are considered sustainable alternatives to the fossil fuel and environmental protection. The plant-derived lignocellulosic biomass can have varying contents of cellulose, hemicellulose, and lignin, which can be utilized to produce biofuels like bioethanol, biodiesel, biobutanol, etc. However, it is necessary

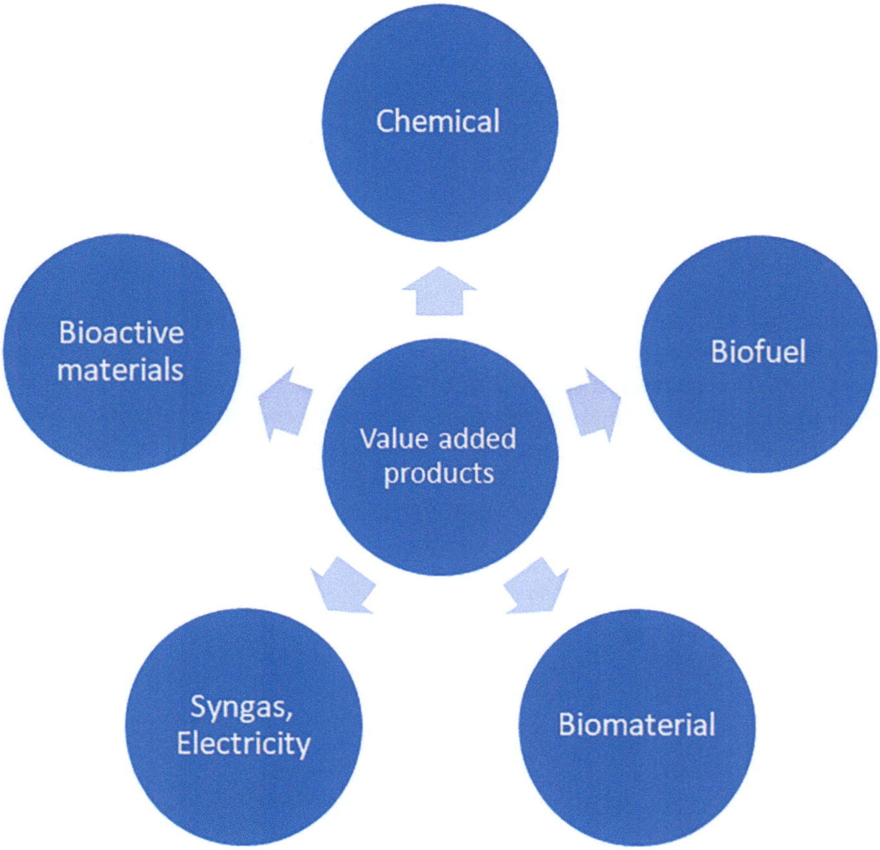

Fig. 3 Different value-added products from lignocellulosic biomass

to pretreat the biomass components for being recalcitrant. Pretreatment is the process of converting a lignocellulosic biomass into a more accessible form for effective enzymatic hydrolysis. The pretreatment of biomass may involve physical, chemical, physicochemical, and biological methods. The biological pretreatment method may use fungi and bacteria or their enzymes. The enzymatic hydrolysis of biomass is one of the most preferable methods; however, it requires a pretreatment step that may involve various steps like steam explosion, ammonia fiber expansion, lime treatment, etc. These treatments help in opening up the recalcitrant lignocellulose structure and facilitate the action of enzyme cocktails (Rabemanolontsoa and Saka 2016; Chen et al. 2017). Different degradative enzymes are used to break down the lignocellulosic structure of biomass to facilitate hydrolysis or oxidation of the polymers present in them (Pellegrini et al. 2020).

3.2 Production of Value-Added Chemicals

The hydrolysis of lignocellulosic biomass can be used to produce other value-added products besides ethanol. Some of the important products produced are discussed below (Table 1).

3.2.1 Xylitol

It is considered as an alternative to natural sweetener and found wide applications in food and pharmaceutical industries and is also recommended for sugar management in diabetic patients. It is produced by hydrotreating xylose, which is obtained from lignocellulosic biomass having rich fraction of hemicellulose such as wood, agricultural wastes, aquatic weeds, etc. (Espinoza-Acosta 2019). The xylose is converted into xylitol using enzymes like the xylose reductase (XR) enzyme (Lugani et al. 2021). The enzymes other than XR include NADPH recycling systems

Table 1 Production of value-added products from biomass using enzymes

Value-added product	Source of biomass	Feed stock	Enzyme(s) used	References
Xylitol	Wood, agricultural wastes, aquatic weeds	Xylan	Xylose reductase, formate dehydrogenase, glucose dehydrogenase, methanogen oxidoreductase	Espinoza-Acosta (2019)
2,3-Butanediol	Sugarcane molasses, sugarcane bagasse, corn stover, Jerusalem artichoke tuber	Lignocellulosic	Cellulases, xylanases, hemicellulases, ligninase	Guo et al. (2022)
Furfural	Bagasse, corncob, and oat husks	Xylose	Cellobiohydrolases, endo-1,4-β-D-glucanases and β-glucosidases	Eriksson et al. (2002)
Vanillin	Wood	Lignin	Laccase, versatile peroxidase	Saikia et al. (2020)
Itaconic acid	Eucalyptus wood	Cellulosic pulp	Cellulase	Kerssemakers et al. (2020)
Adipic acid	Sugarcane bagasse	Hemicellulose	Cellulase	Choe et al. (2021)
Lactic acid	Agricultural straw, distiller's grains, and garden waste	Lignocellulosic biomass	Amylase, cellulases, ligninases, xylanases	Yankov (2022)
Succinic acid	Agricultural waste like corn cob	Lignocellulosic biomass	Cellulase, xylanase, exoglucanase	Khullar et al. (2013)

including formate dehydrogenase and glucose dehydrogenase, methanogen oxidoreductase, etc. (Sharma et al. 2019). The xylitol also serves as a precursor for the production of various other compounds like xylaric acid, ethylene glycol, and propylene glycol.

3.2.2 Furfural

Furfural or 2-furaldehyde is an environmentally friendly alternative to petroleum-derived products and is derived from the hemicellulose fraction in lignocellulosic biomass. The furfural can be produced from biomass using different strategies like acid hydrolysis, solvent-thermal conversion, aqueous acid catalyst, catalytic hydrogenation, catalytic conversion, autocatalytic system, biorefinery process, etc. (Sardar et al. 2023). The process begins with the acid treatment of the hemicellulose biomass to convert it to xylose. The xylose is then reduced further to produce furfural. Common raw materials used for lignocellulosic biomass are obtained from bagasse, corncob, and oat husks. The production of furfural requires enzymatic hydrolysis of cellulose and hemicellulose that involves enzymes like cellobiohydrolases, endo-1,4-β-D-glucanases, and β-glucosidases (Eriksson et al. 2002).

3.2.3 2,3-Butanediol

2,3-Butanediol or 2,3-BD is an important chemical with diverse industrial applications like manufacturing of different solvents, anti-freeze compound, octane booster, etc. The high industrial demand has resulted in the production of 2,3-BD for lignocellulosic biomass and common agro-wastes that include corn stover, wheat straw (WS), rye and rice straw, etc. (Bai et al. 2023). The production of 2,3-BD from lignocellulosic biomass involves enzymatic hydrolysis or saccharification, hydrolyzation, and fermentation. The above process requires the involvement of different enzymes, namely, hydrolases like cellulases, xylanases, and other carbohydrases. Cellulases, hemicellulases, ligninases, and auxiliary enzymes can be applied to the hydrolysis of lignocellulose (Guo et al. 2022). Several studies reported that sugarcane molasses, sugarcane bagasse, corn stover, and stalk of the Jerusalem were used for the production of 2,3-BD using several enzymes like cellulase, hemicellulose, ligninase, inulinase, etc. (Sun et al. 2009; Olowa and Nuñeza 2013; Aghaei et al. 2022).

3.2.4 Lactic Acid

Lactic acid is one of the most popular bio-based chemicals having important applications in various industries including chemical, pharmaceutical, food, cosmetics, etc. Therefore, the steady demand for lactic acid production has resulted in the production of lactic acid from lignocellulosic biomass. The bioconversion of lignocellulosic biomass into lactic acid commenced in four major stages. Pretreatment of lignocellulosic biomass is followed by hydrolysis and fermentation. The hydrolyzed products are then fermented to produce either L- or D-lactic acid (Gezae Daful et al. 2023). The enzymatic hydrolysis involves depolymerizing lignocellulose to fermentable sugars, such as glucose and xylose, by means of hydrolytic enzymes like cellulases, ligninases, xylanases, etc. (Yankov 2022).

3.2.5 Vanillin

Vanillin is a key intermediate for the synthesis of bio-based polymers produced in industrial scale from lignin biomass. It is used as flavoring and fragrance agents in food and cosmetic industries, respectively. The increased demand for vanillin has compelled researchers to produce it from synthetic sources using lignin biomass. One of the important methods for the production of vanillin includes processing of lignocellulose biomass by Kraft process. This process helps in the concentration of lignin-rich biomass, which can be used for vanillin production (Zirbes et al. 2023). The ligninolytic enzyme, e.g., laccase and versatile peroxidase-assisted depolymerization of lignin to vanillin, is one of the most commercially attractive and promising approaches in vanillin production (Saikia et al. 2020).

3.2.6 Itaconic Acid

Itaconic acid is one of the important chemicals that can be produced industrially using cellulose pulp as a feedstock. Itaconic acid has several industrial uses in synthesis of a wide range of polymers through cross-linking and hydrogels in water decontamination, drug delivery, food applications, coatings, elastomers, etc. (Teleky and Vodnar 2019). The bleached cellulose pulp from *Eucalyptus* wood was used for the production of itaconic acid using enzyme concentrate Cellic® that contains cellulase (Kerssemakers et al. 2020).

3.2.7 Adipic Acid

Adipic acid is a straight-chain dicarboxylic acid with a wide utilization in chemical, pharmaceutical, and lubricant manufacturing industries. It can be produced via chemical, biological, or chemical-biological methods. The biological method involves enzymatic saccharification of sugarcane bagasse (SCB) followed by enzymatic hydrolysis of cellulose to ethanol and catalytic conversion of hemicellulose to adipic acid (Choe et al. 2021). The biomass can be pretreated suitably to disrupt crystalline structure for enhancement of enzymatic saccharification. Alkaline pretreatment is reported to be an effective strategy to improve the enzymatic hydrolysis of lignocellulosic biomass (Wu et al. 2023).

3.2.8 Succinic Acid

The microbes can produce succinic acid during conversion of phosphoenolpyruvate (PEP) to oxaloacetate (OAA) with the help of PEP carboxykinase enzyme. The bacteria *Anaerobiospirillum succiniciproducens* can produce succinic acid from different carbohydrate sources like many carbohydrates such as fructose, glucose, glycerol, lactose, and starch (Ahn et al. 2017). Succinic acid can be produced industrially from lignocellulosic biomass like corncob using microbes like fungi, yeast, and gram-positive bacteria, in which pretreatment of lignocellulosic biomass is essential. The pretreatment of lignocellulosic biomass involves enzymatic hydrolysis, in which the complex lignocellulosic structure is destroyed by enzymes, namely, cellulase, xylanase, and exoglucanase, which convert the cellulose in biomass and hemicellulose to fermentable monosaccharides (Khullar et al. 2013).

3.3 Pulp and Paper Industry

The pulp and paper industry is a well-recognized industry that plays an important role in the economic development of any country. The lignocellulosic biomass from agricultural and weed plant residues can cater the growing demand for paper and pulp industry. The pulp and paper industry involves various steps like pulping, bleaching, etc. Pulping involves breaking of bonds between fibers, for which enzymatic treatments can be used. In biological retreatment, microorganisms and their enzymatic complex are employed to break down lignin and modify lignocellulosic structures. Hemicellulolytic enzymes, such as xylanase, are used frequently for pretreatment of pulp bleaching (Bajpai 2004). Studies have shown the use of laccase and lignin peroxidase enzymes produced by white-rot fungi for paper and pulp production (Chandel et al. 2015).

3.4 Production of Animal Feeds

Animal feeds play a significant role in the food industry where the conversion of lignocellulosic biomass through livestock becomes inefficient indicating the need for a new approach. The animal feeds that mostly contain lignocellulosic possess low protein, high fiber, and antinutritional factors. They have low digestibility coefficients leading to reduced meat and/or milk production. The digestibility can be increased by breaking the linkage between cellulose, hemicellulose, and lignin with the help of enzymes. Studies have shown that enzymes like cellulase and fibrinolytic enzymes improve the ruminal enzymatic activity in cows, augmentation in milk production, etc. (Stella et al. 2007; Dean et al. 2013).

3.5 Production of Composites

Several studies have been made toward using lignocellulosic materials in the preparation ofvarious composites with different functionalities of interest, which may be useful from both ecological and environmental perspectives (Bajpai et al. 2013). The development of composites like reinforcing fillers for thermo-plastic polymers from lignocellulosic biomass has received encouraging attention in recent years. Lignocellulosic fibers are flexible and will not damage during processing, which results in high volume fraction filling during processing and high mechanical properties. Polybutylene succinate (PBS) is a commercially available aliphatic polyester having higher flexibility, ease of processability, chemical resistance and thermal stability, better thermoplasticity, superior biodegradability, etc. The PBS can be used for different purposes like food bags, bottles, tableware, agricultural films, fertilizer slow-release materials, scaffolds, etc. Polyhydroxyalkanoates or PHAs are a group of bio-based polyesters derived from lignocellulose biomass, which are biodegradable, renewable, and environment friendly.

4 Future Prospectives and Conclusion

Lignocellulosic materials offer significant opportunities to produce numerous value-added products. The bioconversion of lignocellulosic biomass requires several pretreatment approaches including chemical, mechanical, and enzymatic treatments or a combination of the above methods. Though enzymatic hydrolysis is costly, the application of integrated enzymes can increase the production rate. Developing enzyme with combined activity of cellulase, hemicellulose, and ligninase and integrating the process of pretreatment and hydrolysis into a single step would drastically decrease the production cost and make the process more practical and efficient. Therefore, identifying enzymes used in the breakdown of lignocellulosic feedstocks could be used as targets to increase the production of value-added products.

References

Aghaei S et al (2022) A comprehensive review on bioethanol production from corn stover: worldwide potential, environmental importance, and perspectives. Biomass Bioenergy 161:106447. https://doi.org/10.1016/j.biombioe.2022.106447

Ahn JH, Jang Y, Yup Lee S (2017) Succinic acid. In: Industrial biotechnology. Wiley, New York, pp 505–544. https://doi.org/10.1002/9783527807833.ch17

Bai Y et al (2023) Biomass-derived 2,3-butanediol and its application in biofuels production. Energies 16:5802. https://doi.org/10.3390/en16155802

Bajpai P (2004) Biological bleaching of chemical pulps. Crit Rev Biotechnol 24(1):1–58. https://doi.org/10.1080/07388550490465817

Bajpai PK, Singh I, Madaan J (2013) Tribological behavior of natural fiber reinforced PLA composites. Wear 297(1):829–840. https://doi.org/10.1016/j.wear.2012.10.019

Chandel AK et al (2015) Biodelignification of lignocellulose substrates: an intrinsic and sustainable pretreatment strategy for clean energy production. Crit Rev Biotechnol 35(3):281–293. https://doi.org/10.3109/07388551.2013.841638

Chen H et al (2017) A review on the pretreatment of lignocellulose for high-value chemicals. Fuel Process Technol 160:196–206. https://doi.org/10.1016/j.fuproc.2016.12.007

Choe B et al (2021) Integrated strategy for coproducing bioethanol and adipic acid from lignocellulosic biomass. J Clean Prod 311:127849. https://doi.org/10.1016/j.jclepro.2021.127849

Dean D et al (2013) Effect of method of adding a fibrolytic enzyme to dairy cow diets on feed intake digestibility, milk production, ruminal fermentation, and blood metabolites. Anim Nutr Feed Technol 13:337–353

Eriksson T, Karlsson J, Tjerneld F (2002) A model explaining declining rate in hydrolysis of lignocellulose substrates with cellobiohydrolase I (Cel7A) and endoglucanase I (Cel7B) of Trichoderma reesei. Appl Biochem Biotechnol 101(1):41–60. https://doi.org/10.1385/ABAB:101:1:41

Espinoza-Acosta JL (2019) Biotechnological production of xylitol from agricultural waste// Producción biotecnológica de xilitol a partir de residuos agrícolas. Biotecnia 22(1):126–134. https://doi.org/10.18633/biotecnia.v22i1.1160

Gezae Daful A, Loridon M, Chandraratne MR (2023) Lactic acid production from lignocellulosic biomass. In: From biomass to biobased products [working title]. IntechOpen, London. https://doi.org/10.5772/intechopen.112739

Guo H, He T, Lee D-J (2022) Contemporary proteomic research on lignocellulosic enzymes and enzymolysis: a review. Bioresour Technol 344:126263. https://doi.org/10.1016/j.biortech.2021.126263

Johansen KS (2016) Discovery and industrial applications of lytic polysaccharide monooxygenases. Biochem Soc Trans 44(1):143–149. https://doi.org/10.1042/BST20150204

Kerssemakers AAJ et al (2020) Production of Itaconic acid from cellulose pulp: feedstock feasibility and process strategies for an efficient microbial performance. Energies 13(7):1654. https://doi.org/10.3390/en13071654

Khullar E et al (2013) Effect of particle size on enzymatic hydrolysis of pretreated Miscanthus. Ind Crop Prod 44:11–17. https://doi.org/10.1016/j.indcrop.2012.10.015

Kumar A et al (2019) 3. Biotechnological conversion of plant biomass into value-added products. In: Industrial biotechnology. De Gruyter, Berlin, pp 51–72. https://doi.org/10.1515/9783110563337-003

Lugani Y, Puri M, Sooch BS (2021) Recent insights, applications and prospects of xylose reductase: a futuristic enzyme for xylitol production. Eur Food Res Technol 247(4):921–946. https://doi.org/10.1007/s00217-020-03674-x

McKendry P (2002) Energy production from biomass (part 2): conversion technologies. Bioresour Technol 83(1):47–54. https://doi.org/10.1016/S0960-8524(01)00119-5

Nitsos C et al (2017) Investigation of different pretreatment methods of Mediterranean-type ecosystem agricultural residues: characterisation of pretreatment products, high-solids enzymatic hydrolysis and bioethanol production. https://doi.org/10.6084/m9.figshare.5446804.v1

Olowa LF, Nuñeza OM (2013) Brine shrimp lethality assay of the ethanolic extracts of three selected species of medicinal plants from Iligan City, Philippines. Int Res J Biol Sci 2(11):74–77

Pandiyan K et al (2019) Technological interventions for utilization of crop residues and weedy biomass for second generation bio-ethanol production. Renew Energy 132:723–741. https://doi.org/10.1016/j.renene.2018.08.049

Pellegrini VOA, Sepulchro AGV, Polikarpov I (2020) Enzymes for lignocellulosic biomass polysaccharide valorization and production of nanomaterials. Curr Opin Green Sustain Chem 26:100397. https://doi.org/10.1016/j.cogsc.2020.100397

Rabemanolontsoa H, Saka S (2016) Various pretreatments of lignocellulosics. Bioresour Technol 199:83–91. https://doi.org/10.1016/j.biortech.2015.08.029

Saikia K et al (2020) Development of a magnetically separable co-immobilized laccase and versatile peroxidase system for the conversion of lignocellulosic biomass to vanillin. J Air Waste Manage Assoc 70(12):1252–1259. https://doi.org/10.1080/10962247.2020.1760958
Sardar MRI et al (2023) A brief review of recent furfural production from lignocellulosic biomass: state of the art of processes, technologies, and optimization. J Appl Organomet Chem 3(2):108–122. https://doi.org/10.22034/jaoc.2023.394685.1078
Sharma HK, Xu C, Qin W (2019) Biological pretreatment of lignocellulosic biomass for biofuels and bioproducts: an overview. Waste Biomass Valoriz 10(2):235–251. https://doi.org/10.1007/s12649-017-0059-y
Shen Y et al (2015) A thermochemical–biochemical hybrid processing of lignocellulosic biomass for producing fuels and chemicals. Biotechnol Adv 33(8):1799–1813. https://doi.org/10.1016/j.biotechadv.2015.10.006
Singh S et al (2015) Comparison of different biomass pretreatment techniques and their impact on chemistry and structure. Front Energy Res 3:62. https://doi.org/10.3389/fenrg.2014.00062
Stella AV et al (2007) Effect of administration of live Saccharomyces cerevisiae on milk production, milk composition, blood metabolites, and faecal flora in early lactating dairy goats. Small Rumin Res 67(1):7–13. https://doi.org/10.1016/j.smallrumres.2005.08.024
Stoffels P et al (2020) Complementing the intrinsic repertoire of Ustilago maydis for degradation of the pectin backbone polygalacturonic acid. J Biotechnol 307:148–163. https://doi.org/10.1016/j.jbiotec.2019.10.022
Sun L-H et al (2009) Microbial production of 2,3-butanediol from Jerusalem artichoke tubers by Klebsiella pneumoniae. Appl Microbiol Biotechnol 82(5):847–852. https://doi.org/10.1007/s00253-008-1823-5
Teleky B-E, Vodnar D (2019) Biomass-derived production of itaconic acid as a building block in specialty polymers. Polymers 11(6):1035. https://doi.org/10.3390/polym11061035
Wu M et al (2023) Enhanced adipic acid production from sugarcane bagasse by a rapid room temperature pretreatment. Chem Eng J 452:139320. https://doi.org/10.1016/j.cej.2022.139320
Yankov D (2022) Fermentative lactic acid production from lignocellulosic feedstocks: from source to purified product. Front Chem 10:823005. https://doi.org/10.3389/fchem.2022.823005
Zirbes M et al (2023) Peroxodicarbonate as a green oxidizer for the selective degradation of Kraft lignin into vanillin. Angew Chem Int Ed 62(14):e202219217. https://doi.org/10.1002/anie.202219217

Production of Value-Added Products Using Microalgae: A Zero-Waste Biorefinery Approach

Nirmalya Haldar, Kolli Venkata Supraja, Anamika, Mansi Achhoda, Mrigank Mayank, Mansi Sharma, Nandini Thakur, Anee Mohanty, Sumer Singh Meena, Prangya Ranjan Rout, and Bunushree Behera

1 Introduction

Over the years, the exploration of fossil-based resources for fuel as well as other much-needed commodities has led to severe environmental pollution and global warming. Nevertheless, as per the United Nations (UN) report, the population is expected to reach 9.7 million from 7.7 million by the year 2050 (UN 2019). The growing population and the urge to meet their burgeoning demands concomitant with the need to meet the enviroeconomic and sustainability aspects have led toward the search for alternative biomass-based resources (Dessie et al. 2020). In lieu of the abovementioned issues, the use of third-generation algal biomass with the capability to grow in wastewater resources, sequestering carbon dioxide (CO_2) for high

N. Haldar · M. Sharma · N. Thakur · B. Behera (✉)
Bioprocess Laboratory, Department of Biotechnology, Thapar Institute of Engineering and Technology, Patiala, Punjab, India
e-mail: bunushree.behera@thapar.edu

K. V. Supraja
Waste Treatment Laboratory, Department of Biochemical Engineering and Biotechnology, Indian Institute of Technology Delhi, Hauz Khas, New Delhi, India

Anamika · M. Mayank · A. Mohanty · S. S. Meena · P. R. Rout (✉)
Department of Biotechnology, Dr. B. R. Ambedkar National Institute of Technology, Jalandhar, Punjab, India
e-mail: routpr@nitj.ac.in

M. Achhoda
Bioprocess Laboratory, Department of Biotechnology, Thapar Institute of Engineering and Technology, Patiala, Punjab, India

Department of Civil Engineering , University of Manitoba, Winnipeg, Canada

K. Agrawal, P. Verma (eds.), *Biotechnological Advances in Biorefinery*, Interdisciplinary Biotechnological Advances,
https://doi.org/10.1007/978-981-97-5544-8_6

value commodities, apart from its exploration for biofuel has received attention (Behera et al. 2019).

Algal biomass, regarded as photosynthetic bio-factories, are rich in bioactive chemicals like polysaccharides, proteins, lipids, pigments, and antioxidants produced via primary and secondary metabolism that could possibly be transformed into value-added products (VAPs) (Paliwal et al. 2017; Uma et al. 2023). Many influencing factors inherent to specific algal species, such as their cultivation conditions, also influence their potential utilization for high-value commodities (Li et al. 2019; Behera et al. 2021a). *Spirulina* sp., *Chlorella* sp., and *Dunaliella* sp., are commonly supplemented as additives because of their high protein, antioxidants, and pigment content, thereby improvising the overall food quality and providing substantial health benefits (Mobin and Alam 2017). Omega 3 fatty acids like eicosapentaenoic acid (EPA) and docosahexaenoic acid (DHA) supplemented in nutraceuticals, bakery, infant formulas, as well as animal feed are commonly obtained from *Isochrysis* sp., *Chaetoceros* sp., and *Nannochloropsis* sp. (Chew et al. 2017). Extracts from *Polysiphonia* sp., *Spirulina* sp., and *Ulva lactuca* are utilized in cosmetics as anti-aging, moisturizing, and skin softening agents (Martins et al. 2020). Algal biomass-based metabolites find applications as food supplements, nutraceuticals, pharmaceuticals, and cosmetics, with their antimicrobial, anti-inflammatory, and anticancerous properties that have a huge market impact (Costa et al. 2019). Several researchers like Bhattacharya et al. (2017), Costa et al. (2020), and Siddiki et al. (2022) have reviewed the status of VAPs from microalgae. Since algal yield and metabolite content, which governs the amount and quality of VAPs, are dependent on the cultivation mode along with the extraction process, it is essential to understand and comprehend their fundamental effects.

Based on the abovementioned context, the present chapter elucidates the potential of microalgae for green synthesis of VAPs. Different photobioreactors used for algal cultivation and the extraction methodologies are elaborated in detail. Further, the potential applications of algal biomass, along with their environmental impacts and economic status, have been delineated. The chapter could act as a guideline for researchers and stakeholders to scale up algal biorefineries for sustainable fuel and value-added bio-commodities.

2 Algae as Bio-Factories for Value-Added Products

Algae is a viable feedstock for the long-term production of a variety of VAPs as presented in Fig. 1. Thermochemical and biochemical conversion techniques are often employed to produce fuel and other value-added bio-commodities from algal biomass (Dutta et al. 2023). Algal extracts have a potent therapeutic and biostimulatory functions, thus having a massive market demand. Successfully converting algal biomass to VAPs requires optimization of several factors including the choice of production system, nutrient supply, aeration, mixing, harvesting, and processing of biomass.

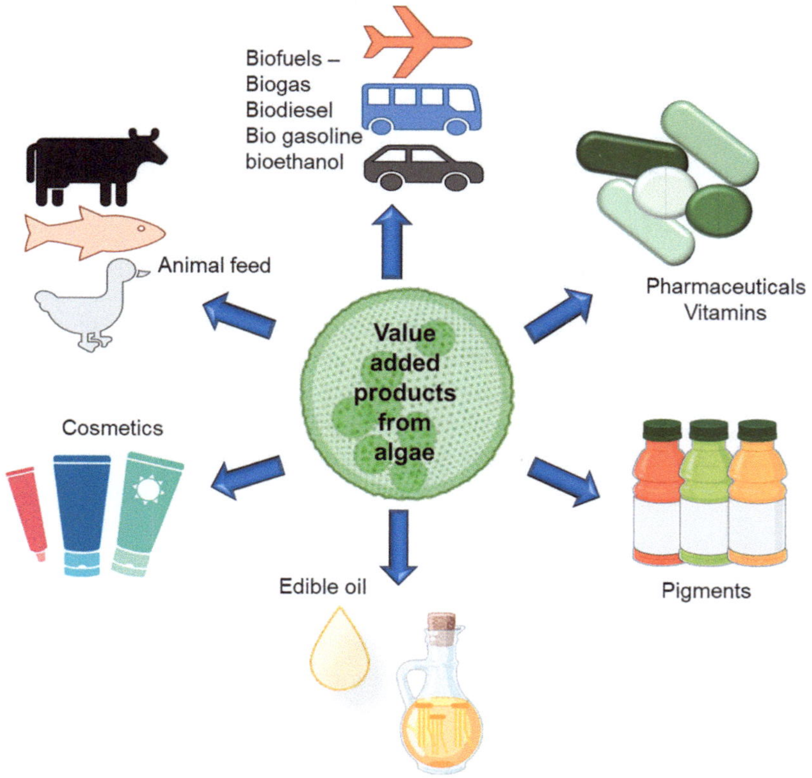

Fig. 1 VAPs from microalgae

Algae-based membrane lipids and monounsaturated fatty acid derivatives have diverse applications as oleochemicals in food, feed, and dietary supplements. The fatty acids containing C16 and C18 carbon chains are commonly used in vegetable oils. Omega 3 and omega 9 fatty acids in algal oil/pigments have antioxidant and medicinal properties, especially for cardiovascular ailments; thus, often finding applications as dietary supplements (Tang et al. 2015). Algae-based fatty alcohol derivatives comprising of sulfates, alkoxylates, and amines are used as versatile surfactants, emulsifier, and thickeners in several industries (Pfleger et al. 2015).

Algal biomass is a rich source of pigments like chlorophylls, phycobiliproteins, and xanthophylls having potential applications in the food industry as natural food colorants. Several cyanobacterial species are wide sources of water-soluble phycobiliproteins (Patel et al. 2022). Xanthophylls are yellow pigments that are found in algae. Carotenoid is a natural pigment that is lipid soluble. Its roles include accessory pigment for light collection during photosynthesis, and it also acts as cell antioxidants and pro-vitamins. Carotenoids produced by algae comprising of β-carotene, lutein, fucoxanthin, zeaxanthin, astaxanthin, echinenone, and violaxanthin are often employed in nutraceuticals due to their health-promoting properties (Generalić Mekinić et al. 2023).

Algae has been recognized as a promising and sustainable source of animal feed due to the presence of protein along with fiber, antioxidants, fatty acids, vitamins (B12, K, D, A, and C), and minerals like phosphorus, zinc, magnesium, and iron. It improves the overall immune system, intestinal health, skin, coat, and oral hygiene (Bratosin et al. 2021).

Bioplastics derived from algae have emerged as a sustainable alternative to standard plastics, with biodegradability and strong mechanical characteristics. With the aim to reduce waste generation and greenhouse gas emissions, research is currently focused on identifying different physiochemical properties of various biopolymers generated from marine and freshwater microalgae for packaging applications (Yap et al. 2023). Undoubtedly, the exploration of algal biomass through different strategies would aid in improvising process economics by reducing waste generation, facilitating the use of algal residues, and generating multiple products of societal importance.

3 Cultivation Techniques for Microalgae-based Bioproducts

The interdependency of various cultivation parameters, like pH, temperature, light intensity, and photo-period during scale-up of the microalgae cultivation process, has made it imperative to implement an intricately designed strategy to account for operation efficiency and product quality (Behera et al. 2021b).

3.1 Open Photobioreactors

Open system photobioreactors (PBRs) are versatile and cost-effective platforms designed for the large-scale cultivation of microalgae. Unlike closed systems, open PBRs are characterized by their direct exposure to natural sunlight and their exchange of carbon dioxide with the surrounding atmosphere (Behera et al. 2019). The shallow design of open PBRs allows sunlight to penetrate the entire culture volume, supporting efficient photosynthesis. Due to their exposure to the external environment, open systems can harbor contaminants including viruses, bacteria, and other microbes (Marsullo et al. 2015). Additionally, evaporation may cause open systems to use more water than closed systems. Water seepage in such systems can be prevented by lining with high-density polyethylene (HDPE) (Scofield et al. 2022). Raceway ponds are distinguished by its oval or round form, and the culture medium is enabled to flow either through an incline or by introducing paddlewheels (Rogers et al. 2014). For optimum yield, the depth of a raceway pond is around 0.15–0.3 m (Sawant et al. 2019). This relatively shallow design allows for good exposure to sunlight but has certain disadvantages pertaining to the hydraulic jump effects of the paddlewheels such as sheer stress and damage to the microalgae cells.

Another form of open PBR are floating bioreactors, which are designed in a way that allows the microalgae to float on the water surface, eliminating the requirement of land resources (Khor et al. 2022). These PBRs enable efficient mixing by virtue of natural turbulence and waves in the water body and makes for a really cost-efficient open system for microalgae cultivation. Photodynamic movement makes for a great deal of performance parameters in the form of wave period, PBR configuration, and wave height. Studies have shown that a square model has better movement than a rectangular model (Zhu et al. 2019). Floating bioreactors can be built on freshwater as well as on saline water bodies depending on algal species and could generate biochemically rich biomass to be utilized for several VAPs (Ma et al. 2016).

3.2 *Closed Photobioreactors*

Closed PBRs, as opposed to open systems, are enclosed structures that offer a well-regulated environment with conditions specifically catered to the needs of the cultured microalgae by precisely manipulating variables including temperature, light intensity, and nutrient concentrations (Razzak et al. 2013). These photobioreactors can have complex designs and usually require a beforehand technical knowledge (Hossain et al. 2018).

Flat-panel PBRs are shaped like a cuboid and consist of flat, transparent, or semi-transparent panels/plates made up of materials like plexiglass or polycarbonate, which allow optimum light penetration (Singh and Sharma 2012). Microalgae are circulated as a thin film over these panels, and the liquid medium is made to pump around these panels (Razzak et al. 2023). Flat-panel PBRs are further employed in various configurations depending on the strain and byproduct. For instance, alveolar panel configuration is used in cases where photoinhibition is the main problem, and microalgae like *Scenedesmus obliquus* greatly benefit from this configuration. On the contrary, U-shaped configuration is helpful in reducing sedimentation and works well for *Chlorella sorokiniana* (Carone et al. 2022; Fuchs et al. 2021).

Closed tubular photobioreactors are cylinder-shaped vessels with transparent walls that allow maximum light exposure, usually composed of glass or acrylic (Chanquia et al. 2022). The tube diameter is preferably kept in range from a few millimeters to centimeters allowing for maximum light penetration and, in certain cases, maintaining high cell concentrations. The tubes' length, which can reach hundreds or even thousands of meters, can be placed in a horizontal and parallel fashion or in a curved manner. The medium is allowed to be circulated within the tube. Certain disadvantages of using these types of systems come in the form of mass transfer problems in relatively larger length tubes. The variations in the amounts of dissolved oxygen and carbon dioxide throughout the length of the tube lead to oxygen toxicity in certain areas, limiting its use during scale-up (Huang et al. 2017). Bubble-column PBRs employ rising gas bubbles in a vertical column to circulate microalgae. The bubbles provide a mixing effect and facilitate nutrient exchange. These PBRs have a relatively simple design, do not require a lot of space for setup, and allow good heat transfer and light exposure (Cui et al. 2020). However, one of the main areas of concern with this type of PBR is the microalgae cell rupture that extensively takes place because of bubble rupture. This rupture has been observed near the sparger, which has led to research in trying to determine the critical gas entrance velocity for mitigating such adverse effects and better performance of bubble-column PBR (López-Rosales et al. 2019).

Annular PBR consists of a ring-shaped enclosed structure that supports the growth of microalgae. There is an artificial light source present at the very center of the ring-shaped system that enables a uniform light exposure (Masojídek and Torzillo 2014). Various variations have been proposed in this type of system depending on the specific requirements of different microalgae. For instance, alternate light and dark regions in the ring have been introduced for the cultivation of microalgae like *Nannochloropsis* sp. (Taisir et al. 2016). Such systems are gaining attention because they are easy to set up without the requirement of much land area.

4 Algal Cell Disruption and Extraction Methods

The robust algal cell wall often hinders the extraction of VAPs from microalgae. Cell disruption involving a series of mechanical and nonmechanical strategies is a crucial step in the extraction process as it allows for the release of intracellular products. Mechanical cell disruption strategies including high-pressure homogenization, bead milling, and ultrasonication often require significant energy input and may lead to the degradation of heat-sensitive compounds. Nonmechanical methods involving the use of chemical solvents or enzymes are generally less energy-intensive but might endure additional costs. Advanced methods like microwave-assisted extraction (MAE) and pulsed-electric field (PEF) are currently explored for enhanced product efficacy and selectivity. The following sections elucidate the

Table 1 Different extraction methods of microalgae for VAPs

VAPs	Microalgae	Extraction technique	References
Biosurfactants	*Scenedesmus acutus*	Acid treatment followed by extraction with hexane	Fioroni et al. (2022)
Carotenoids	*Tetradesmus obliquus*	Supercritical CO_2-based extraction	Chronopoulou et al. (2019)
Vitamins	*Dunaliella, Spirulina platensis,* and *Skeletonema marinoi*	Solvent extraction	Del Mondo et al. (2020)
Edible oil	*Chlorella vulgaris, Scenedesmus obliquus,* and *Nannochloropsis oceanica*	Hexane-based solvent extraction	Huang et al. (2016)

principle as well as its potential application during the extraction of VAPs from microalgae. Few VAPs along with their algal source, underlying extraction principle, are enlisted in Table 1.

4.1 Bead Milling

Bead milling is a mechanical technique involving the use of glass, ceramic, or metal beads agitated within a closed chamber with the material to be processed. Agitation is typically achieved through stirring arms, rotating discs, or high-speed impellers. Agitation causes the beads to collide with each other and the material, facilitating grinding and particle size reduction. It aids in the effective release of intracellular contents due to the impact and shear forces generated. The milling conditions, including agitation speed, time, and bead size, can be adjusted to achieve specific outcomes. Schwenzfeier et al. (2011) demonstrated the bead-milling process with *Tetraselmis* sp. for the extraction of protein. The study reported that with the use of 3.3–4 kW power, ceramic beads of 0.3–0.6 mm in diameter (65% bead filling ratio) along with 12% dry cell weight (DCW), a flow rate of 1.5 L min^{-1}, and a treatment duration of 30 min, 21% protein could be extracted.

4.2 High-Speed Homogenization

High-speed homogenization involves the application of mechanical force to break down cell structures. Operating at elevated speeds, typically ranging from 20,000 to 30,000 revolutions per minute (rpm), the homogenizer utilizes a rapidly rotating blade or rotor-stator assembly. As the cell suspension passes through the narrow gap between the rotor and stator, intense shear forces and turbulence are generated, which effectively rupture cell walls and membranes, leading to the release of intracellular contents into the surrounding medium. Variations in design in terms of the applied voltage and distance between the impeller and stator blades could be optimized to achieve efficient results. Wang and Wang (2012) subjected *Nannochloropsis* sp. (6% DCW) to high-speed homogenization at 10,000 rpm for 1 min, resulting in 75.8–78% of lipid yield, indicating the effectiveness of the employed conditions in extraction.

4.3 High-Pressure Homogenization

Algal suspensions are subjected to intense pressures generated by a homogenizer, typically a piston or diaphragm pump, which forces the cells through a narrow orifice, leading to an increase in kinetic energy and subsequent sudden pressure drop.

Pressure differential, along with shear forces, causes the algal cells to undergo mechanical stress, leading to the disruption of cell walls and membranes, thereby releasing intracellular contents. The intense and rapid changes in pressure and flow conditions contribute to the efficient disruption of algal cells, making high-pressure homogenization a valuable technique for downstream applications in various industries. Yap et al. (2015) reported that subjecting *Nannochloropsis* sp. suspensions at pressures varying from 25 to 150 MPa resulted in effective homogenization. *Nannochloropsis oculata* exposed to high pressure of 2760 bars, with 0.023–0.035% DCW, with 4 passes, resulted in 67% cell disintegration, accompanied by 8.5 times more oil extraction compared with undisrupted algae (Samarasinghe et al. 2012).

4.4 Ultrasonication

Sonication involves the utilization of ultrasonic sound energy exceeding 20 kHz to induce cavitation, where air bubbles in a liquid implode, exerting disruptive force on the cell membrane causing cell lysis. The instrument utilizes a fast back-and-forth movement of a probe, which under high acoustic power, produces multiple microbubbles at nucleation sites in the fluid, growing during the rarefying phase and collapsing during compression. The resulting energy input causes motions within the cells, leading to their disintegration when the kinetic energy surpasses the cell wall strength. Additionally, microstreaming near rapidly vibrating gas bubbles, induced by ultrasound, further contributes to cell breakage through high shear stress (Saranya et al. 2014). *Scenedesmus obliquus* (7–10% DCW) subjected to 200 W power for 30 s in 5 cycles with 10 min breaks between cycles resulted in cell lysis yielding 0.025 g of glucose/g of biomass (Miranda et al. 2012).

4.5 Microwave Treatment

Electromagnetic waves with frequencies 300 MHz to 300 GHz in microwave treatment induce selective heating within biological samples. When exposed to microwave, polar molecules within the cells, such as water and certain cellular constituents, rapidly oscillate due to the changing electric field. This oscillation results in frictional heating and a subsequent temperature increase, leading to the rupture of cell membranes. Additionally, the formation and expansion of vapor bubbles within the cells due to the microwave-induced heating contribute to mechanical stress, further aiding in cell disruption. The advantages of microwave treatment include its rapid and selective heating properties, which can lead to efficient cell lysis without the need for mechanical forces or harsh chemicals. However, careful optimization of parameters such as microwave power and exposure time is crucial to ensure effective cell disruption while minimizing the risk of denaturing heat-sensitive biomolecules. Pasquet et al. (2011) demonstrated that 0.16% DCW of *Dunaliella tertiolecta*, exposed to 50 W power (56 °C), revealed the extraction of approximately 4.5 μg/L Chlorophyll *a*, 1.4 μg/L Chlorophyll *b*, and 1.3 μg/L β-carotene.

4.6 Pulsed-Electric Field (PEF)

The application of short high-frequency electric pulses in the range of kV/cm for micro-/milliseconds to algal cells creates pores via electroporation. These pores destabilize the lipid bilayer of the cell membrane, allowing ions to flow in and out of the cell, disrupting the electrochemical balance, thereby resulting in the release of intracellular components into the surrounding medium. Process efficacy can be improvised by optimizing the electric field strength and the strength and duration of electric pulses. Zbinden et al. (2013) projected the utilization of 5.8 kWh/kg PEF over *Ankistrodesmus falcatus* (0.19% DCW), resulting in a twofold increase in extraction efficiency.

4.7 Enzymatic Treatment

Enzymes, such as proteases, lipases, and other hydrolytic enzymes, target the major macromolecules like proteins, lipids, and carbohydrates, respectively. The enzymatic action cleaves the macromolecules, leading to cell wall/membrane disruption. Enzymatic cell disruption is particularly useful when dealing with sensitive biomolecules, as it allows for selective and controlled breakdown without the harsh conditions associated with some physical methods. The efficiency of enzymatic treatment depends on factors such as enzyme specificity, concentration, and reaction time. The study by Prajapati et al. (2015) reported the effectivity of fungal crude enzymes from *Aspergillus lentulus* during the pretreatment of algal biomass.

4.8 Chemical Surfactant Treatment

Chemical surfactant treatment involves the use of surface-active agents for disrupting the macromolecular linkages in cell membrane, facilitating intracellular components to be released. The hydrophobic portions of the surfactant embed themselves into the lipid membrane, while the hydrophilic portions face the external surrounding aqueous medium. The abovementioned interference disrupts the organization of the lipid bilayer, leading to increased membrane permeability and, ultimately, cell lysis. The choice of surfactant and its concentration is crucial in achieving effective disruption while minimizing undesirable side effects, such as protein denaturation or the formation of undesirable byproducts. Huang and Kim (2013) reported approximately 16% lipids extracted out of *Chlorella vulgaris* using 7.5 mg of cationic surfactants (cetyl trimethyl ammonium bromide, CTAB) applied at a dosage of 50 mL to 1 g/L algal slurry.

5 Microalgae-based Value-Added Products

Microalgae have versatile applications in producing an array of value-added products. Among them, biofuel production is the most prominent, contributing to biodiesel and biogas generation through lipid-rich strains. Additionally, certain microalgae species serve as valuable sources of omega 3 fatty acids, proteins, and amino acids, making them suitable for nutritional supplements and food fortification. Their natural pigments find use as food colorants, while bioactive compounds, including antioxidants like astaxanthin, are incorporated into pharmaceuticals and cosmetics for their health and skincare benefits (Behera et al. 2021a). Furthermore, their ability to sequester carbon dioxide during photosynthesis positions them as a tool for carbon capture, while some microalgae produce biopolymers for use in biodegradable plastics. In the field of agriculture, microalgae possess macro and micronutrients and phytohormones, which could accelerate plant growth (Supraja et al. 2020). The ongoing research and innovation surrounding microalgae underscore their potential in advancing environmentally friendly and economically viable solutions across diverse industries.

5.1 Pigments as Food Additives/Colorants

Microalgae-based pigments have gained attention as natural and sustainable alternatives for food additives and colorants. Certain microalgae species are rich sources of pigments, such as chlorophyll, carotenoids, phycobilins, and astaxanthin, offering a diverse range of colors and health benefits (Patel et al. 2022). Chlorophyll, the green pigment found in microalgae, is utilized for imparting green hues to food and beverages (Sun et al. 2023). Carotenoids, including beta-carotene, lutein, and zeaxanthin, contribute yellow to orange shades and serve as antioxidants with potential health-promoting properties. Phycobilins, present in red and blue-green algae, are employed for their vibrant red and blue hues in various food applications. Astaxanthin, a red carotenoid pigment found in certain microalgae, is known for its powerful antioxidant activity and is used to color various food products. These microalgae-derived pigments offer an eco-friendly alternative to synthetic colorants, addressing consumer preferences for natural and sustainable ingredients (Patel et al. 2022). The use of microalgae-based pigments aligns with the growing demand for clean-label products and contributes to the development of a more environmentally conscious and health-oriented food industry.

5.2 *Polysaccharides*

Microalgae-based polysaccharides have emerged as versatile compounds with multifaceted applications across diverse industries (Table 2). Extracted from the cell wall or extracellular matrices of microalgae, these polysaccharides exhibit distinctive chemical compositions that render them valuable for a spectrum of purposes. In the biomedical domain, certain microalgae polysaccharides, particularly sulfated varieties, showcase promising immunomodulatory effects, antiviral, and antitumor properties, suggesting potential therapeutic applications (Qi and Kim 2017). The different functional potentials of algal polysaccharides are presented in Fig. 2. The food and beverage industry benefits from microalgae polysaccharides as thickeners and stabilizers, contributing texture and stability to various products. In cosmetics, these polysaccharides offer emollient and moisturizing attributes, making them suitable for inclusion in skincare formulations. Beyond consumer goods, microalgae polysaccharides play roles in biotechnology, contributing to biofilm production with applications in wastewater treatment and bioremediation (Miranda et al. 2017). Geremia et al. (2021) discussed bio-flocculation as a cost-effective alternative for cell harvesting, focusing on the adhesive properties of polymeric exopolysaccharides (EPS) derived from microalgae. EPS enables microalgae to form biofilm, a solid substrate application, with advantages of biomass retention, nutrient removal, tolerance to environmental changes, shorter hydraulic retention time, resistance to heavy metals, and a more economical harvesting process compared with suspended systems (Geremia et al. 2021). In agriculture, microalgal polysaccharides serve as bio-stimulants, enhancing plant growth, stress tolerance, and crop yields (Rachidi et al. 2021). The ongoing research in microalgae-based polysaccharides underscores their potential to revolutionize various industries, driven by their natural origin, biodegradability, and diverse array of applications.

Several algal species have been studied for extraction of EPS. Exopolysaccharides from *P. marinum* have been reported to possess therapeutic functions and are increasingly employed in food and pharmaceutical industries (Gargouch et al. 2021). Microalgal EPS can also inhibit retroviral malignant cell transformation and replication by murine leukemia virus (MuLV) and murine sarcoma virus (MuSV-124) (Talyshinsky et al. 2002). Algal EPS have also been shown to reduce infection caused by herpes simplex virus and varicella zoster virus (Huleihel et al. 2001). This is attributed to the adsorption of virus over algal EPS. Consumption of *Porphyridium* sp. also significantly improves total serum cholesterol and triglycerides, improvising the metabolic activity (Dvir et al. 2009). The therapeutic effects of *Porphyridium* sp. have been predicted to be similar or far better to those from plant sources (Nikolova et al. 2019). Gardeva et al. (2009) reported antitumorous activity of sulfated polysaccharide of *P. cruentum*, thereby reducing tumor growth and increasing the survival chances of a patient. EPS from *Porphyridium* sp. also exhibited biolubricant properties with stable viscosity across a range of temperature, pH, and salinity, resulting in high stability and reduced friction (Arad et al. 2006). Chemical modification of EPS with (3-chloro-2-hydroxypropyl)trimethylammonium chloride under alkaline conditions produces a gel-like substance with elastic response,

Table 2 Microalgae polysaccharides and its potential applications

Microalgae	Polysaccharides	Applications	References
Chlorella ellipsoidea	–	Immuno-stimulating applications	Qi and Kim (2017)
Chlorella pyrenoidosa	–	Prebiotic applications	Lv et al. (2022)
Dunaliella salina, Desmodesmus sp.	–	Plant growth stimulator	Rachidi et al. (2021)
Graesiella sp.	[a]HAS-EPS	Antioxidant applications, antiproliferative applications	Trabelsi et al. (2016)
Green algae	[b]CP	Biosorption applications	Ali et al. (2018)
Gyrodinium impudium	[c]SP	Antiviral applications	Kim et al. (2012)
Haematococcus pluvialis	–	Anti-aging abilities	Liu et al. (2018)
Isochrysis galbana	–	Physiological health booster	Nuño et al. (2013)
Nannochloropsis oculata	–	Physiological health booster	Nuño et al. (2013)
Nannochloropsis oculata	[d]WSP	Immunostimulatory applications	Pandeirada et al. (2019)
Phaeodactylum tricornutum	–	Plant growth stimulator	Rachidi et al. (2021)
Pavlova viridis		Antitumor applications	Sun et al. (2016)
Phaeodactylum tricornutum	[c]SP	Anticancer applications	Yang et al. (2019)
Porphyridium cruentum	[e]EPS	Immunomodulatory applications	Sun et al. (2012)
Porphyridium cruentum	[c]SP	Immunomodulatory applications, antioxidant applications	Casas-Arrojo et al. (2021)
Porphyridium cruentum	[e]EPS	Antiviral applications	Parra-Riofrio et al. (2023)
Porphyridium marinum	[e]EPS	Antibacterial, antibiofilm applications, anticancer applications	Gargouch et al. (2021)
Porphyridium purpureum	[e]EPS	Antiviral applications	Radonić et al. (2011)
Porphyridium sordidum	[e]EPS	Antitumor applications	Nikolova et al. (2019)
Porphyridium sp.	–	Plant growth stimulator	Rachidi et al. (2021)
Rhodella reticulata	[e]EPS	Free radical scavenging applications, antioxidant applications	Chen et al. (2010)
Tetraselmis sp.	[d]WSP	Antioxidant, antifungal applications, tyrosinase inhibitory activities	Park et al. (2017)

(continued)

Table 2 (continued)

Microalgae	Polysaccharides	Applications	References
Tetraselmis suecica	[e]EPS	Antiviral applications	Parra-Riofrio et al. (2023)
Tribonema sp.	[c]SP	Immunostimulatory applications, anticancer applications	Chen et al. (2019)

[a]*HAS-EPS* Hetero-sulfated-Anionic Extracellular Polysaccharides, [b]*CP* Cellulosic Polysaccharides, [c]*SP* Sulfated polysaccharide, [d]*WSP* Water-Soluble Polysaccharides, [e]*EPS* Exopolysaccharide

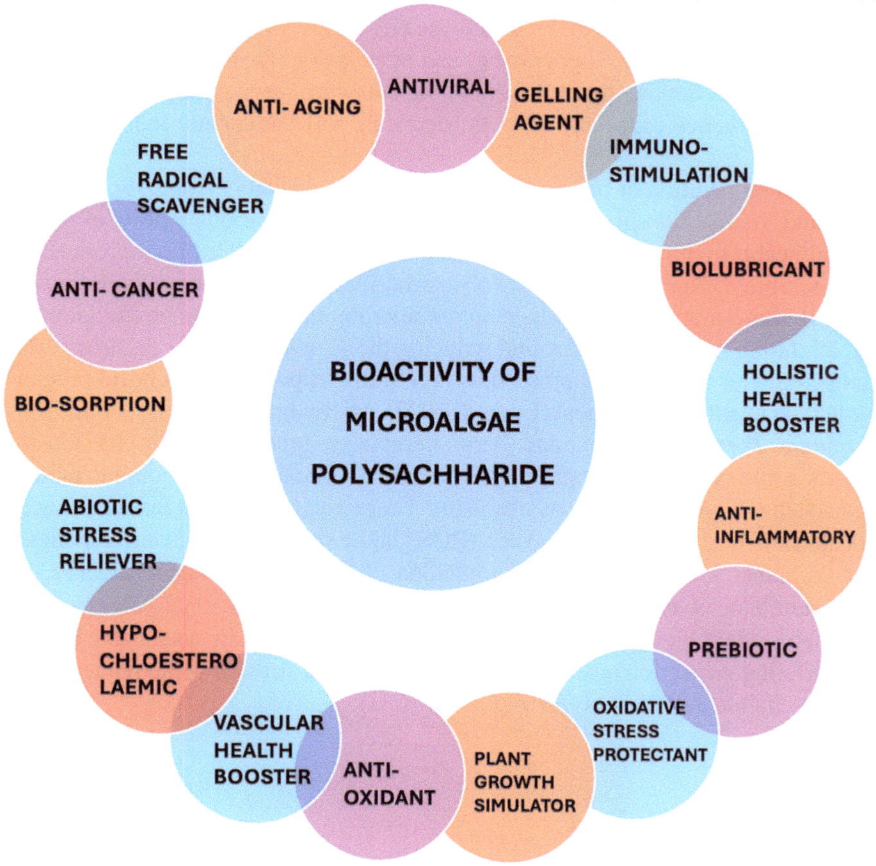

Fig. 2 Bioactivity of microalgal polysaccharides

having potential applications in cosmetic industries, particularly in formulations for hair coloring (Geresh and Dawadi 2000).

Chen et al. (2010) reported that EPS from *Rhodella reticulata* exhibited dose-dependent free radical scavenging and antioxidant activity. Antiviral activity has also been reported via the use of sulfated polysaccharide from *Gyrodinium impudicum* (Yim et al. 2005). Similar to the abovementioned study, sulfated polysaccharide from *Gyrodinium impudicum* have been shown to inhibit influenza A virus (Kim et al. 2012). The use of sulfated polysaccharides is proposed as an alternative therapeutic treatment for marine organisms in culture, addressing the drawbacks of uncontrolled antibiotic use and the associated issues of antibiotic resistance. Microalgal polysaccharides also show anti-inflammatory and vaso-relaxing properties in TNF-α-induced system involving human coronary artery endothelial cells, suggesting potential applications as therapeutic agents or food additives for enhancing vascular health (Levy-Ontman et al. 2017). In another study, polysaccharides from *Chlorella ellipsoidea* were found to act as functional ingredients or immune-stimulating supplements as they can activate NF-κB and MAPK pathways in murine macrophage cells, producing cytokines and nitric oxide (Qi and Kim 2017). Water-soluble polysaccharides derived from *Tetraselmis* sp. serve as a potent source of biological moieties especially in cosmeceuticals and therapeutic industries (Park et al. 2017).

The administration of microalgae polysaccharides, notably from *Isochrysis galbana* and *Nannochloropsis oculata*, appears to influence various aspects of physiological health by improving the body mass, nutrient absorption, lipid profiles, and gut microbiota (Nuño et al. 2013). Polysaccharides produced from *Chlorella pyrenoidosa* showed prebiotic influence on gut microbiota, promoting beneficial bacteria, inhibiting harmful bacteria, and reducing the *Firmicutes* to *Bacteroidetes* ratio compared with conventional prebiotics, suggesting its potential as a novel prebiotic for human health improvement (Lv et al. 2022). Microalgae such as *D. salina*, *P. tricornutum*, *Porphyridium* sp., and *Desmodesmus* sp. produce polysaccharides that can act as bio-elicitors for improving several biochemical and metabolomic markers associated with plant defense mechanisms. These polysaccharides induce the accumulation of reactive oxygen species (ROS) and enhance antioxidant activities in tomato plants while aiding the accumulation of fatty acids, alkanes, alkenes, and phytosterols involved in the wax construction in leaves. Microalgal polysaccharide treatments could be a promising biotechnological approach in sustainable agriculture, offering protection against pathogens and abiotic stresses while reducing reliance on chemical pesticides (Rachidi et al. 2021).

5.3 Protein-Rich Feed

The emergence of microalgae-based protein-rich feed marks a significant stride in the pursuit of sustainable and nutritious alternatives in the food sector. About 30–70% of proteins are present in microalgae, depending on the species and cultivation conditions (Gerde et al. 2013). Notably, these microorganisms offer a complete amino acid profile, including those that animals cannot synthesize independently, thereby enhancing the overall nutritional quality of the feed. Beyond nutritional benefits, animals fed with microalgae-based protein show improved feed conversion ratios and immune responses. While single-cell protein sources like microalgae have been considered suitable for consumption by animals and humans, there is renewed interest in their integration into animal feed (Bratosin et al. 2021). Investigations have revealed that proteins derived from microalgae show promise as alternatives to fishmeal protein or as valuable additives in aquafeeds (Hanel et al. 2007). Notably, the study by Chen et al. (2021) portrayed no adverse effect of algae-based proteinaceous feed on fish. *Chlorella* sp., incorporated into the poultry feed can serve as a partial substitute for soybean owing to its 50% protein content and substantially good amino-acid profile (Lamminen et al. 2019).

Algal residue, left after lipid extraction being rich in protein, could act as a potential alternative to cottonseed meal (Drewery et al. 2014). Ju et al. (2012) investigated the impact of substituting fishmeal protein with 3% of defatted microalgae meal (DMM) derived from *Haematococcus pluvialis* in shrimp diets, which resulted into enhanced shrimp growth and feed utilization efficiency. Even replacing 50% of fishmeal protein with DMM did not adversely affect shrimp growth or its nutritional composition. Additionally, DMM also enriched shrimp quality by increasing astaxanthin levels, which could benefit both animal and human health. Tibaldi et al. (2015) projected that replacing up to 20% crude protein from fish meal co-combined with the freeze-dried biomass of *Isochrysis* sp. (about 36% lipids) did not adversely affect feed intake or growth performance. According to Sarker et al. (2020), substitution of protein-rich defatted biomass of *Nannochloropsis oculata* for fishmeal resulted into significantly better growth metrics, weight gain, specific growth rate, and comparable feed conversion ratio in fishes. Additionally, the fish-free feed resulted in higher fillet lipid, DHA, and protein content with highest degree of in vitro protein hydrolysis and protein digestibility. Utilizing defatted marine microalgae, an underutilized protein-rich biomass, contributes to aquaculture sustainability by reducing dependence on fishmeal from forage fisheries. The approach has been reported to improve tilapia growth and also resulted in an increased DHA levels in the fillet, offering health benefits to consumers (Sarker et al. 2020). *Scenedesmus* sp. and *S. dimorphus* have also been considered as protein-rich supplements in animal feed on a larger scale for fish and shrimp diets (Duong et al. 2015). Microalgae have comparable and permissible levels for protein degree of hydrolysis and predicted protein apparent digestibility coefficients (Tibbetts et al. 2017). These results highlighted the potential application of defatted microalgae as an alternative protein and natural pigment source in aquaculture systems.

5.4 *Biofertilizers*

The rise of microalgae-based biofertilizers signifies a groundbreaking shift toward sustainable and environmentally conscious practices in agriculture (Uysal et al., 2015). Microalgae act as valuable allies in fostering soil fertility and supporting plant growth through their nutrient-rich composition loaded with nitrogen, phosphorus, potassium, and trace elements (Fig. 3). Beyond these basic nutrients, microalgae also constitute bioactive compounds, including growth-promoting substances, vitamins, and hormones, which positively impact plant growth and overall health (Viegas et al. 2021a; Dineshkumar et al. 2020). Their rapid growth and adaptability make them an eco-friendly source of soil enrichment, promoting microbial activity and enhancing soil structure (Marks et al. 2017). Furthermore, microalgae-based biofertilizers have demonstrated their potential to improve nutrient uptake by plants, increase crop yields, and even impart drought tolerance (Rachidi et al. 2021) (Table 3). The environmental sustainability of these biofertilizers is evident in their ability to mitigate nutrient runoff and soil degradation, contributing to more responsible agricultural practices. Notably, certain microalgae species can perform biological nitrogen fixation, reducing the reliance on synthetic nitrogen fertilizers and mitigating associated environmental concerns. The customizable formulations derived from microalgae cultivation offer tailored solutions for diverse crops and soil conditions. As research continues to refine cultivation techniques and explore

Fig. 3 Benefits of microalgae as biofertilizers or bio-stimulants

Table 3 Microalgae as biofertilizer or plant growth bio-stimulant

Microalgae	Nutrient or bio-stimulant source/ role	Plant applied on	Common name	References
Acutodesmus dimorphus	[a]PGP	*Solanum lycopersicum*	Roma tomato	Garcia-Gonzalez and Sommerfeld (2016)
Chlorella vulgaris	[d]N	*Lactuca sativa*	Lettuce	Alvarenga et al. (2023)
Chaetomorpha sp.	[d]N, [f]K, [h]Mg, [e]P	*Hordeum vulgare*	Barley	González et al. (2020)
Chlorella sp.	[d]N, [f]K, [h]Mg, [e]P	*Hordeum vulgare*	Barley	González et al. (2020)
Chlorella fusca	[a]PGP, antifungal agent	*Fragaria* × *ananassa*	Strawberry	Kim et al. (2020)
Chlorella minutissima	[d]N and [e]P	*Zea mays, Spinacia oleracea*	Corn, spinach	Sharma et al. (2021)
Chlorella minutissima	Biofertilizer			Khan et al. (2019)
Chlorella protothecoides	[b]PB, biopesticide			Ferreira et al. (2021)
Chlorella protothecoides	[d]N and [e]P	*Triticum aestivum*	Wheat	Viegas et al. (2021a)
Chlorella sorokiniana	[b]PB	*Triticum aestivum*	Wheat	Kholssi et al. (2019)
Chlorella sp.	Organic carbon and nutrients, growth promoter of rhizosphere microorganism			Marks et al. (2017)
Chlorella sp.	[d]N, [a]PGP	*Triticum aestivum*	Wheat	Mückschel et al. (2023)
Chlorella sp.	[d]N, [e]P, [f]K, [g]Ca, [h]Mg, organic carbon			Slinksienė et al. (2022)
Chlorella variabilis	[a]PGP	*Zea mays*	Corn	Gatamaneni Loganathan et al. (2020)
Chlorella vulgaris	[b]PB	*Triticum aestivum* L. *Nasturtium officinale*	Wheat, watercress	Viegas et al. (2021a)
Chlorella vulgaris	[b]PB, biopesticide			Ferreira et al. (2021)
Chlorella vulgaris	[c]PGR	*Cattleya labiate*	Orchid	Corbellini et al. (2020)
Chlorella vulgaris	[b]PB	Romaine lettuce, leaf lettuce		Kopta et al. (2018)

(continued)

Table 3 (continued)

Microalgae	Nutrient or bio-stimulant source/ role	Plant applied on	Common name	References
Chlorella vulgaris	[d]N, [e]P, Organic carbon, [a]PGP	*Triticum aestivum* L.	Wheat	Schreiber et al. (2018)
Chlorella vulgaris	[d]N and [e]P	*Triticum aestivum* L.	Wheat	Viegas et al. (2021a)
Chlorella vulgaris	[b]PB, antimicrobial activities			Ranglová et al. (2021)
Chlorella vulgaris	[a]PGP	*Brassica oleracea*	Broccoli	Kusvuran (2021)
Chlorella vulgaris	Alternative to chemical fertilizer	*Pennisetum glaucum*	Pearl millet	de Siqueira Castro et al. (2017)
Chlorella vulgaris	[d]N	*Oryza sativa*	Rice	Jochum et al. (2018)
Chlorella vulgaris	[a]PGP	*Vigna mungo*	Black gram	Dineshkumar et al. (2020)
Chlorella vulgaris	[d]N, [e]P, [f]K, [g]Ca, [j]Fe, [i]Zn, [k]Cu, [l]Mn	*Lepidium sativum* L.	Garden cress	Sadak and Şensoy (2022)
Chlorella vulgaris and *Spirulina platensis*	[a]PGP	*Oryza sativa*	Rice	Dineshkumar et al. (2018)
Cystoseira sp.	[d]N, [f]K, [h]Mg, [e]P	*Hordeum vulgare*	Barley	González et al. (2020)
Dunaliella salina and *Desmodesmus* sp.	Elicitor agents against abiotic stress	*Solanum lycopersicum*	Tomato	Rachidi et al. (2021)
Dunaliella salina	[b]PB	*Solanum lycopersicum*	Tomato	El Arroussi et al. (2018)
Messastrum gracile	[c]PGR	*Cattleya labiate*	Orchid	Corbellini et al. (2020)
Mixed algal consortia	[b]PB	*Solanum lycopersicum*	Tomato	Supraja et al. (2020)
Monoraphidium sp.	[d]N and [e]P	*Solanum lycopersicum*	Tomato	Jimenez et al. (2020)
Phaeodactylum tricornutum	Elicitor agents against abiotic stress	*Solanum lycopersicum*	Tomato	Rachidi et al. (2021)
Pediastrum sp.	[d]N, [a]PGP	*Triticum aestivum*	Wheat	Mückschel et al. (2023)
Porphyridium sp.	[b]PB	*Solanum lycopersicum*	Tomato	Rachidi et al. (2020)
Porphyridium sp.	Elicitor agents against abiotic stress	*Solanum lycopersicum*	Tomato	Rachidi et al. (2021)

(continued)

Table 3 (continued)

Microalgae	Nutrient or bio-stimulant source/role	Plant applied on	Common name	References
Scenedesmus obliquus	[d]N	*Lactuca sativa*	Lettuce	Alvarenga et al. (2023)
Spirulina platensis	[a]PGP	*Zea mays* L.	Maize	Dineshkumar et al. (2019)
Chlorella vulgaris	[a]PGP	*Zea mays* L.	Maize	Dineshkumar et al. (2019)
Scenedesmus dimorphus	[d]N	*Oryza sativa*	Rice	Jochum et al. (2018)
Scenedesmus obliquus	[b]PB			Viegas et al. (2021a)
Scenedesmus obliquus	[a]PGP	*Glycine max*	Soybean	Gatamaneni Loganathan et al. (2020)
Scenedesmus quadricauda	[b]PB	*Lactuca sativa*	Lettuce	Puglisi et al. (2020)
Scenedesmus quadricauda	[a]PGP, growth promoter of rhizosphere bacteria and fungi	*Cucumis sativus*	Cucumber	Lv et al. (2020)
Scenedesmus sp.	[d]N and [e]P	*Hordeum vulgare*	Barley	González et al. (2020)
Scenedesmus sp.	Organic fertilizer			Khan et al. (2019)
Scenedesmus sp.	[d]N, [a]PGP	*Triticum aestivum*	Wheat	Mückschel et al. (2023)
Scenedesmus sp.	[a]PGP	*Lolium perenne* *Hordeum vulgare*	Ryegrass, barley	González et al. (2020)
Spirulina sp.	[d]N, [f]K, [h]Mg, [e]P	*Hordeum vulgare*	Barley	González et al. (2020)
Spirulina platensis	[a]PGP	*Allium cepa*	Onion	Dineshkumar et al. (2020)
Tetradesmus obliquus	[b]PB, biopesticide			Ferreira et al. (2021)
Tetradesmus obliquus	[d]N and [e]P	*Triticum aestivum*	Wheat	Viegas et al. (2021a)
Tetraselmis sp.	[a]PGP	*Phoenix dactylifera*	Date palm	Saadaoui et al. (2019)
Unicellular and filamentous microalgae	[d]N, [e]P, and [f]K	*Triticum aestivum*	Wheat	Renuka et al. (2016)
Chlorella sp. and *spirulina* sp.	[a]PGP		Barley	Atzori et al. (2020)

[a] *PGP* Plant growth promoter, [b]*PB* Plant bio-stimulant, [c]*PGR* Plant growth regulators, [d]*N* Nitrogen, [e]*P* Phosphorus, [f]*K* Potassium, [g]*Ca* Calcium, [h]*Mg* Magnesium, [i]*Zn* Zinc, [j]*Fe* Iron, [k]*Cu* Copper, [l]*Mn* Manganese

novel strains, microalgae-based biofertilizers stand as a promising frontier in the pursuit of sustainable and efficient agricultural systems.

González et al. (2020) explored the viability of *Scenedesmus* sp. through twin layer biofilm microalgae cultivation system for removing total soluble and ammoniacal nitrogen (up to 66% and 94%, respectively) in small wastewater treatment plants. The obtained biomass exhibited promising characteristics in formulating solid-state bio-based fertilizers by enhancing the growth and nutrient absorption capabilities of ryegrass and barley crops. The addition of *Chlorella* sp. suspension in the soil is found to increase the activity of heterotrophic microorganisms present in the soil and resulted into a photosynthetic soil biofilm (Marks et al. 2017). In another study by Atzori et al. (2020), *Spirulina* sp.-fertilized barley plants demonstrated high uptake efficiencies of nitrogen (60%) and potassium (41%), whereas *Chlorella* sp.-fertilized plants showed about 30% nutrient uptake efficiency, stating that algal fertilizers present a sustainable alternative to synthetic N-fertilizers. Blending algae with synthetic N fertilizer showed the best results, offering greater synchrony in N release and crop uptake (Rupawalla et al. 2021). Algal fertilizers contributed to lower N loss, reduced CO_2 emissions, and improved crop and soil quality. Moreover, recycled N from wastewater further reduced embodied energy in algal and synthetic fertilizer blends (Rupawalla et al. 2021). According to Viegas et al. (2021b), bio-stimulants produced from *Chlorella vulgaris* and *Scenedesmus obliquus* by remediating raw effluent from brown crab aquaculture showed significant increase in the germination index of wheat and watercress plants.

Other studies include the use of *Tetraselmis* sp. as a biofertilizer for young date palms (*Phoenix dactylifera* L.) (Saadaoui et al. 2019), *Chlorella* sp. as a biofertilizer for wheat growth (Kholssi et al. 2019), *T. obliquus* and *C. vulgaris* as bio-stimulants for *Fusarium oxysporum* (Ferreira et al. 2021), *Scenedesmus quadricauda* extract as bio-stimulant for lettuce seedlings (Puglisi et al. 2020), and *Acutodesmus dimorphus* as seed primer, foliar spray, and biofertilizer for tomato plants (Garcia-Gonzalez and Sommerfeld 2016) and bio-stimulants from mixed microalgal consortia as seed treatment and foliar spray for tomato plants (Supraja et al. 2020), which resulted in significantly higher germination index, plant growth rates, 100% survival rates, improved rooting ability, increased number of leaves, larger stem thickness, longer shoots, higher total chlorophyll content, and plant nutrient utilization efficiency. Hence, microalgal biofertilizers and bio-stimulants pave a way toward circular bio-economy as a cost-effective method in future agronomic applications by improving soil fertility and crop yields while reducing environmental pollution and dependence on chemical fertilizers.

6 Enviro-economic Assessments of Microalgal Biorefinery

The utilization of microalgae in the production of various high-end, value-added products through a zero-waste biorefinery approach has been explored, particularly in the context of the microalgal industry's improvement. Algal-based biorefineries are regarded as a potential solution to enhance the economic viability of microalgal-based biofuels by valorizing multiple co-products (Zhu 2015). Despite the acknowledged potential of microalgae, various economic and environmental challenges hinder its commercialization, including high production costs compared with fossil fuels and substantial water consumption during cultivation (Chisti 2013). To address these challenges, it is essential to understand the environmental and economic impacts of algal biorefineries.

6.1 Environmental Impacts of Microalgae Cultivation

The cultivation of microalgae presents environmental advantages by eliminating the need for agricultural land, thereby reducing strain on land use. Microalgae, with varying protein, carbohydrate, lipid, and nucleic acid contents across species, offer versatility in applications (Breuer et al. 2012). Strategies such as nitrogen depletion, optimized irradiance, temperature, pH variations, and CO_2 enhancement have been explored to enhance microalgal metabolite content. Nitrogen starvation, an efficient approach, leads to the conversion of proteins or peptides to carbohydrates or lipids. The anaerobic digestion of microalgae along with organic wastes produces biogas, primarily methane, with potential applications in energy production (Ward et al. 2014). However, challenges arise from high protein content leading to a low C/N ratio and ammonia production, inhibiting anaerobic performance. Proper management of biochemical conversion residues allows its use as animal feed or biofertilizer contributing to improved animal health and crop fertility (Giwa et al. 2018). Microalgae cultivation necessitates careful consideration of nutrient sources and water quality. Traditional reliance on inorganic fertilizers, with associated environmental concerns and carbon footprint, prompts exploration of wastewater as an alternative nutrient source. Wastewater, rich in phosphate and nitrate, can serve as an economical, environmentally friendly nutrient source, reducing dependence on chemical fertilizers and production costs (Pavithra et al. 2020). Large-scale microalgae cultivation using wastewater holds potential for substantial biomass production, contributing significantly to energy generation However, careful water management is essential to prevent environmental risks associated with nutrient release and water quality degradation, emphasizing the importance of properly constructed ponds with adequate lining and containment measures (Chia et al. 2018).

6.2 Microalgae-Based Commodities: Market Scenarios and Economic Dynamics

In the energy market, the microalgae's role in bio-energy production offers a potential solution to address fluctuating fossil fuel prices. The feed and food market, valued at approximately €113 billion globally, predominantly favors the feed sector due to the growth of aquaculture (Crofcheck and Crocker 2016). The chemicals and materials market, valued at about €13 billion globally, is significantly influenced by biopolymer products in the packaging industry. The pharmaceuticals and personal care market, with a value of €1.2 billion each for beauty-related products and therapeutic compounds, represents a low-volume, high-value sector. Analysis reveals that over 70% of the total energy value is attributed to bioenergy conversion and fertilizer production, playing a crucial role in greenhouse gas emissions (Saral et al. 2022). Cruce and Quinn (2019) highlighted different pathways for biofuel production from microalgae, emphasizing their impact on economic feasibility and public policy. Comparative evaluation of various biorefinery pathways underscores the need for non-fuel co-products or reduced capital costs for economic viability. Various microalgae products, including biomass, primary metabolites, and secondary metabolites, find diverse applications with prices ranging from €500 to €15,000 per ton (Karemore and Sen 2016). Economic opportunities exist, especially in markets with high commercial values, such as chemicals and materials. The exploitation of microalgae biomass presents economic potential, provided that careful attention is given to economic feasibility, public policies, and environmental sustainability (Barsanti and Gualtieri 2018). Market scenarios reveal diverse applications for microalgae-based commodities, offering economic opportunities, particularly in high-value markets. As global efforts intensify toward sustainable alternatives, microalgae emerge as a candidate poised to meet energy and product demands while minimizing environmental impact.

7 Conclusion and Future Perspectives

The algal technologies explored for the production of VAPs undoubtedly provide a safer alternative to currently existing chemical-based commodities. Apart from improvising the overall economics, the algal biorefineries sequester CO_2, remediating pollutants making the environment sustainable to dwell upon. There is a need to optimize strategies for microalgal cultivation mode, targeted for improvising productivity. The algal reactors must be incorporated with easier in-place harvesting and extraction systems facilitating rapid product recovery. The selection of extraction methodology must be done in such a manner that the product viability and metabolic content are being maintained without unwanted contamination. Moreover, optimization of the upstream process of algal cultivation would result in significant improvement in biomass yield and productivity. More studies must be focused on bioprospecting robust strains, optimizing process conditions to maximize algal productivity. Research pursuits in the directions of cost-efficient harvesting and extraction for maximizing concentration and retaining value-added metabolites must be emphasized. Nevertheless, an interdisciplinary collaborative approach including researchers from different professional backgrounds must be encouraged to proficiently work in removing the barriers linked with commercialization of algal products.

Acknowledgments The authors acknowledge the Department of Biotechnology, Thapar Institute of Engineering and Technology, Patiala; the Department of Biochemical Engineering and Biotechnology, Indian Institute of Technology Delhi; and the Department of Biotechnology, B. R. Ambedkar National Institute of Technology Jalandhar for providing the requisite facilities to conduct the study.

References

Ali ME, Abd El-Aty AM, Badawy MI, Ali RK (2018) Removal of pharmaceutical pollutants from synthetic wastewater using chemically modified biomass of green alga *Scenedesmus obliquus*. Ecotoxicol Environ Saf 151:144–152

Alvarenga P, Martins M, Ribeiro H, Mota M, Guerra I, Cardoso H, Silva JL (2023) Evaluation of the fertilizer potential of *Chlorella vulgaris* and *Scenedesmus obliquus* grown in agricultural drainage water from maize fields. Sci Total Environ 861:160670

Arad S, Rapoport L, Moshkovich A, Van Moppes D, Karpasas M, Golan R, Golan Y (2006) Superior biolubricant from a species of red microalga. Langmuir *22*(17):7313–7317

Atzori G, Nissim WG, Rodolfi L, Niccolai A, Biondi N, Mancuso S, Tredici MR (2020) Algae and Bioguano as promising source of organic fertilizers. J Appl Phycol 32:3971–3981

Barsanti L, Gualtieri P (2018) Is exploitation of microalgae economically and energetically sustainable? Algal Res 31:107–115

Behera B, Acharya A, Gargey IA, Aly N, Balasubramanian P (2019) Bioprocess engineering principles of microalgal cultivation for sustainable biofuel production. Bioresour Technol Rep 5:297–316

Behera B, Nayak S, Rangabhashiyam S, Jayabalan R, Balasubramanian P (2021a) Microalgae as a viable bioresource for sustainable biofuel production through biorefinery technologies. In: Bioresource utilization and management: applications in therapeutics, biofuels, agriculture, and environmental science. Routledge, London

Behera B, Unpaprom Y, Ramaraj R, Maniam GP, Govindan N, Paramasivan B (2021b) Integrated biomolecular and bioprocess engineering strategies for enhancing the lipid yield from microalgae. Renew Sust Energ Rev 148:111270

Bhattacharya S, Pramanik SK, Gehlot PS, Patel H, Gajaria T, Mishra S, Kumar A (2017) Process for preparing value-added products from microalgae using textile effluent through a biorefinery approach. ACS Sustain Chem Eng 5(11):10019–10028

Bratosin BC, Darjan S, Vodnar DC (2021) Single cell protein: a potential substitute in human and animal nutrition. Sustainability 13(16):9284

Breuer G, Lamers PP, Martens DE, Draaisma RB, Wijffels RH (2012) The impact of nitrogen starvation on the dynamics of triacylglycerol accumulation in nine microalgae strains. Bioresour Technol 124:217–226

Carone M, Alpe D, Costantino V, Derossi C, Occhipinti A, Zanetti M, Riggio VA (2022) Design and characterization of a new pressurized flat panel photobioreactor for microalgae cultivation and CO2 bio-fixation. Chemosphere 307:135755

Casas-Arrojo V, Decara J, de los Ángeles Arrojo-Agudo M, Pérez-Manríquez C, Abdala-Díaz RT (2021) Immunomodulatory, antioxidant activity and cytotoxic effect of sulfated polysaccharides from *Porphyridium cruentum* (sf Gray) Nägeli. Biomolecules 11(4):488

Chanquia SN, Vernet G, Kara S (2022) Photobioreactors for cultivation and synthesis: specifications, challenges, and perspectives. Eng Life Sci 22(12):712–724

Chen B, You W, Huang J, Yu Y, Chen W (2010) Isolation and antioxidant property of the extracellular polysaccharide from *Rhodella reticulata*. World J Microbiol Biotechnol 26:833–840

Chen X, Song L, Wang H, Liu S, Yu H, Wang X, Li P (2019) Partial characterization, the immune modulation and anticancer activities of sulfated polysaccharides from filamentous microalgae *Tribonema* sp. Molecules 24(2):322
Chen F, Leng Y, Lu Q, Zhou W (2021) The application of microalgae biomass and bio-products as aquafeed for aquaculture. Algal Res 60:102541
Chew KW, Yap JY, Show PL, Suan NH, Juan JC, Ling TC, Chang JS (2017) Microalgae biorefinery: high value products perspectives. Bioresour Technol 229:53–62
Chia SR, Chew KW, Show PL, Yap YJ, Ong HC, Ling TC, Chang JS (2018) Analysis of economic and environmental aspects of microalgae biorefinery for biofuels production: a review. Biotechnol J 13(6):1700618
Chisti Y (2013) Constraints to commercialization of algal fuels. J Biotechnol 167(3):201–214
Chronopoulou L, Dal Bosco C, Di Caprio F, Prosini L, Gentili A, Pagnanelli F, Palocci C (2019) Extraction of carotenoids and fat-soluble vitamins from Tetradesmus obliquus microalgae: an optimized approach by using supercritical CO2. Molecules 24(14):2581
Corbellini JR, Ribas LLF, de Maia FR, Correa DDO, Noseda MD, Suzuki RM, Amano É (2020) Effect of microalgae *Messastrum gracile* and *Chlorella vulgaris* on the in vitro propagation of orchid Cattleya labiata. J Appl Phycol 32:4013–4027
Costa JAV, Freitas BCB, Rosa GM, Moraes L, Morais MG, Mitchell BG (2019) Operational and economic aspects of Spirulina-based biorefinery. Bioresour Technol 292:121946
Costa JAV, Freitas BCB, Moraes L, Zaparoli M, Morais MG (2020) Progress in the physicochemical treatment of microalgae biomass for value-added product recovery. Bioresour Technol 301:122727
Crofcheck C, Crocker M (2016) Application of recycled media and algae-based anaerobic digestate in *Scenedesmus* cultivation. J Renew Sust Energy 8(1):013116
Cruce JR, Quinn JC (2019) Economic viability of multiple algal biorefining pathways and the impact of public policies. Appl Energy 233:735–746
Cui X, Yang J, Feng Y, Zhang W (2020) Simulation of a novel tubular microalgae photobioreactor with aerated tangent inner tubes: improvements in mixing performance and flashing-light effects. Archaea 2020:8815263
de Siqueira Castro J, Calijuri ML, Assemany PP, Cecon PR, de Assis IR, Ribeiro VJ (2017) Microalgae biofilm in soil: greenhouse gas emissions, ammonia volatilization and plant growth. Sci Total Environ 574:1640–1648
Del Mondo A, Smerilli A, Sané E, Sansone C, Brunet C (2020) Challenging microalgal vitamins for human health. Microb Cell Factories 19(1):1–23
Dessie W, Luo X, Wang M, Feng L, Liao Y, Wang Z, Qin Z (2020) Current advances on waste biomass transformation into value-added products. Appl Microbiol Biotechnol 104(11):4757–4770
Dineshkumar R, Kumaravel R, Gopalsamy J, Sikder MNA, Sampathkumar P (2018) Microalgae as bio-fertilizers for rice growth and seed yield productivity. Waste Biomass Valori 9:793–800
Dineshkumar R, Subramanian J, Gopalsamy J, Jayasingam P, Arumugam A, Kannadasan S, Sampathkumar P (2019) The impact of using microalgae as biofertilizer in maize (*Zea mays* L.). Waste Biomass Valori 10:1101–1110
Dineshkumar R, Subramanian J, Sampathkumar P (2020) Prospective of *Chlorella vulgaris* to augment growth and yield parameters along with superior seed qualities in black gram, Vigna mungo (L.). Waste Biomass Valori 11:1279–1287
Drewery ML, Sawyer JE, Pinchak WE, Wickersham TA (2014) Effect of increasing amounts of postextraction algal residue on straw utilization in steers. J Anim Sci 92(10):4642–4649
Duong VT, Ahmed F, Thomas-Hall SR, Quigley S, Nowak E, Schenk PM (2015) High protein- and high lipid-producing microalgae from northern Australia as potential feedstock for animal feed and biodiesel. Front Bioeng Biotechnol 3:53
Dutta N, Kundu P, Lee JTE, Bhattacharya S (2023) Implementation and Optimization of Algal Biomass in Value-Added Products Recovery: A Step towards Algae-Based Green Economy. Hydrobiology 2(2):326–346

Dvir I, Stark AH, Chayoth R, Madar Z, Arad SM (2009) Hypocholesterolemic effects of nutraceuticals produced from the red microalga *Porphyridium* sp in rats. Nutrients 1(2):156–167

El Arroussi H, Benhima R, Elbaouchi A, Sijilmassi B, El Mernissi N, Aafsar A, Smouni A (2018) *Dunaliella salina* exopolysaccharides: a promising biostimulant for salt stress tolerance in tomato (*Solanum lycopersicum*). J Appl Phycol 30:2929–2941

Ferreira A, Melkonyan L, Carapinha S, Ribeiro B, Figueiredo D, Avetisova G, Gouveia L (2021) Biostimulant and biopesticide potential of microalgae growing in piggery wastewater. Environ Adv 4:100062

Fioroni G, Katahira R, Van Wychen S, Rowland SM, Christensen ED, Dong T, Laurens LM (2022) Synthesis of hydrophilic derivative surfactants from algae-derived unsaponifiable lipids. Front Chem Eng 3:768382

Fuchs T, Arnold ND, Garbe D, Deimel S, Lorenzen J, Masri M, Brück TB (2021) A newly designed automatically controlled, sterilizable flat panel photobioreactor for axenic algae culture. Front Bioeng Biotechnol 9:697354

Garcia-Gonzalez J, Sommerfeld M (2016) Biofertilizer and biostimulant properties of the microalga *Acutodesmus dimorphus*. J Appl Phycol 28:1051–1061

Gardeva E, Toshkova R, Minkova K, Gigova L (2009) Cancer protective action of polysaccharide, derived from red microalga *Porphyridium cruentum*—a biological background. Biotechnol Biotechnol Equip 23(Suppl 1):783–787

Gargouch N, Elleuch F, Karkouch I, Tabbene O, Pichon C, Gardarin C, Laroche C (2021) Potential of exopolysaccharide from *Porphyridium marinum* to contend with bacterial proliferation, biofilm formation, and breast cancer. Mar Drugs 19(2):66

Gatamaneni Loganathan B, Orsat V, Lefsrud M (2020) Utilizing the microalgal biomass of *Chlorella variabilis* and *Scenedesmus obliquus* produced from the treatment of synthetic dairy wastewater as a biofertilizer. J Plant Nutr 44(10):1486–1497

Generalić Mekinić I, Šimat V, Rathod NB, Hamed I, Čagalj M (2023) Algal carotenoids: chemistry, sources, and application. Foods 12(14):2768

Gerde JA, Wang T, Yao L, Jung S, Johnson LA, Lamsal B (2013) Optimizing protein isolation from defatted and non-defatted Nannochloropsis microalgae biomass. Algal Res 2(2):145–153

Geremia E, Ripa M, Catone CM, Ulgiati S (2021) A review about microalgae wastewater treatment for bioremediation and biomass production—a new challenge for Europe. Environments 8(12):136

Geresh S, Dawadi RP (2000) Chemical modifications of biopolymers: quaternization of the extracellular polysaccharide of the red microalga *Porphyridium* sp. Carbohydr Polym 43(1):75–80

Giwa A, Adeyemi I, Dindi A, Lopez CGB, Lopresto CG, Curcio S, Chakraborty S (2018) Techno-economic assessment of the sustainability of an integrated biorefinery from microalgae and Jatropha: a review and case study. Renew Sust Energ Rev 88:239–257

González I, Ekelhof A, Herrero N, Siles JÁ, Podola B, Chica AF, Martín AM, Melkonian M, Izquierdo CG, Gómez JM (2020) Wastewater nutrient recovery using twin-layer microalgae technology for biofertilizer production. Water Sci Technol 82(6):1044–1061

Hanel R, Broekman D, De Graaf S, Schnack D (2007) Partial replacement of fishmeal by lyophilized powder of the microalgae Spirulina platensis in Pacific white shrimp diets. Open Mar Biol J 1(1):1–5

Hossain SZ, Alnoaimi A, Razzak SA, Ezuber H, Al-Bastaki N, Safdar M, Hossain MM (2018) Multiobjective optimization of microalgae (Chlorella sp.) growth in a photobioreactor using Box-Behnken design approach. Can J Chem Eng 96(9):1903–1910

Huang WC, Kim JD (2013) Cationic surfactant-based method for simultaneous harvesting and cell disruption of a microalgal biomass. Bioresour Technol 149:579–581

Huang Y, Zhang D, Xue S, Wang M, Cong W (2016) The potential of microalgae lipids for edible oil production. Appl Biochem Biotechnol 180:438–451

Huang Q, Jiang F, Wang L, Yang C (2017) Design of photobioreactors for mass cultivation of photosynthetic organisms. Engineering 3(3):318–329

Huleihel M, Ishanu V, Tal J, Arad S (2001) Antiviral effect of red microalgal polysaccharides on *Herpes simplex* and *Varicella zoster* viruses. J Appl Phycol 13:127–134

Jimenez R, Markou G, Tayibi S, Barakat A, Chapsal C, Monlau F (2020) Production of microalgal slow-release fertilizer by valorizing liquid agricultural digestate: growth experiments with tomatoes. Appl Sci 10(11):3890

Jochum M, Moncayo LP, Jo YK (2018) Microalgal cultivation for biofertilization in rice plants using a vertical semi-closed airlift photobioreactor. PLoS One 13(9):e0203456

Ju ZY, Deng DF, Dominy W (2012) A defatted microalgae (Haematococcus pluvialis) meal as a protein ingredient to partially replace fishmeal in diets of Pacific white shrimp (Litopenaeus vannamei, Boone, 1931). Aquaculture 354:50–55

Karemore A, Sen R (2016) Downstream processing of microalgal feedstock for lipid and carbohydrate in a biorefinery concept: a holistic approach for biofuel applications. RSC Adv 6(35):29486–29496

Khan SA, Sharma GK, Malla FA, Kumar A, Gupta N (2019) Microalgae based biofertilizers: a biorefinery approach to phycoremediate wastewater and harvest biodiesel and manure. J Clean Prod 211:1412–1419

Kholssi R, Marks EA, Miñón J, Montero O, Debdoubi A, Rad C (2019) Biofertilizing effect of *Chlorella sorokiniana* suspensions on wheat growth. J Plant Growth Regul 38:644–649

Khor WH, Kang HS, Lim JW, Iwamoto K, Tang CHH, Goh PS, Lai NYG (2022) Microalgae cultivation in offshore floating photobioreactor: state-of-the-art, opportunities and challenges. Aquac Eng 98:102269

Kim M, Yim JH, Kim SY, Kim HS, Lee WG, Kim SJ, Lee CK (2012) In vitro inhibition of influenza A virus infection by marine microalga-derived sulfated polysaccharide p-KG03. Antivir Res 93(2):253–259

Kim MJ, Shim CK, Ko BG, Kim J (2020) Effect of the microalga *Chlorella fusca* CHK0059 on strawberry PGPR and biological control of *Fusarium* wilt disease in non-pesticide hydroponic strawberry cultivation. J Microbiol Biotechnol 30(5):708

Kopta T, Pavlikova M, Sękara A, Pokluda R, Maršálek B (2018) Effect of bacterial-algal biostimulant on the yield and internal quality of lettuce (*Lactuca sativa* L.) produced for spring and summer crop. Not Bot Horti Agrobot Cluj Napoca 46(2):615–621

Kusvuran S (2021) Microalgae (Chlorella vulgaris Beijerinck) alleviates drought stress of broccoli plants by improving nutrient uptake, secondary metabolites, and antioxidative defense system. Hortic Plant J 7(3):221–231

Lamminen M, Halmemies-Beauchet-Filleau A, Kokkonen T, Jaakkola S, Vanhatalo A (2019) Different microalgae species as a substitutive protein feed for soya bean meal in grass silage based dairy cow diets. Anim Feed Sci Technol 247:112–126

Levy-Ontman O, Huleihel M, Hamias R, Wolak T, Paran E (2017) An anti-inflammatory effect of red microalga polysaccharides in coronary artery endothelial cells. Atherosclerosis 264:11–18

Li X, Liu J, Chen G, Zhang J, Wang C, Liu B (2019) Extraction and purification of eicosapentaenoic acid and docosahexaenoic acid from microalgae: a critical review. Algal Res 43:101619

Liu X, Zhang M, Liu H, Zhou A, Cao Y, Liu X (2018) Preliminary characterization of the structure and immunostimulatory and anti-aging properties of the polysaccharide fraction of *Haematococcus pluvialis*. RSC Adv 8(17):9243–9252

López-Rosales L, Sánchez-Mirón A, Contreras-Gómez A, García-Camacho F, Battaglia F, Zhao L, Molina-Grima E (2019) Characterization of bubble column photobioreactors for shear-sensitive microalgae culture. Bioresour Technol 275:1–9

Lv J, Liu S, Feng J, Liu Q, Guo J, Wang L, Jiao X, Xie S (2020) Effects of microalgal biomass as biofertilizer on the growth of cucumber and microbial communities in the cucumber rhizosphere. Turk J Bot 44(2):167–177

Lv K, Yuan Q, Li H, Li T, Ma H, Gao C, Zhao L (2022) *Chlorella pyrenoidosa* polysaccharides as a prebiotic to modulate gut microbiota: physicochemical properties and fermentation characteristics in vitro. Foods 11(5):725

Ma XN, Chen TP, Yang B, Liu J, Chen F (2016) Lipid production from Nannochloropsis. Mar Drugs 14(4):61

Marks EA, Miñón J, Pascual A, Montero O, Navas LM, Rad C (2017) Application of a microalgal slurry to soil stimulates heterotrophic activity and promotes bacterial growth. Sci Total Environ 605:610–617

Marsullo M, Mian A, Ensinas AV, Manente G, Lazzaretto A, Marechal F (2015) Dynamic modeling of the microalgae cultivation phase for energy production in open raceway ponds and flat panel photobioreactors. Front Energ Res 3:41

Martins AA, Mata TM, de Sá Caetano N, Junior WGM, Gorgich M, Corrêa PS (2020) Microalgae for biotechnological applications: cultivation, harvesting and biomass processing. Aquaculture 528:735562

Masojídek J, Torzillo G (2014) Mass cultivation of freshwater microalgae. Elsevier, Amsterdam

Miranda JR, Passarinho PC, Gouveia L (2012) Pre-treatment optimization of Scenedesmus obliquus microalga for bioethanol production. Bioresource technology 104:342–348

Miranda AF, Ramkumar N, Andriotis C, Höltkemeier T, Yasmin A, Rochfort S, Mouradov A (2017) Applications of microalgal biofilms for wastewater treatment and bioenergy production. Biotechnol Biofuels 10(1):1–23

Mobin S, Alam F (2017) Some promising microalgal species for commercial applications: a review. Energy Procedia 110:510–517

Mückschel F, Ollo E, Glaeser SP, Düring R, Yan F, Velten H, Frei M (2023) Nitrogen use efficiency of microalgae application in wheat compared to mineral fertilizer. J Plant Nutr Soil Sci 186(5):522–531

Nikolova B, Semkova S, Tsoneva I, Antov G, Ivanova J, Vasileva I, Kabaivanova L (2019) Characterization and potential antitumor effect of a heteropolysaccharide produced by the red alga *Porphyridium sordidum*. Eng Life Sci 19(12):978–985

Nuño K, Villarruel-López A, Puebla-Pérez AM, Romero-Velarde E, Puebla-Mora AG, Ascencio F (2013) Effects of the marine microalgae *Isochrysis galbana* and *Nannochloropsis oculata* in diabetic rats. J Funct Foods 5(1):106–115

Paliwal C, Mitra M, Bhayani K, Bharadwaj SV, Ghosh T, Dubey S, Mishra S (2017) Abiotic stresses as tools for metabolites in microalgae. Bioresour Technol 244:1216–1226

Pandeirada CO, Maricato E, Ferreira SS, Correia VG, Pinheiro BA, Evtuguin DV, Palma AS, Correia A, Vilanova M, Coimbra MA, Nunes C (2019) Structural analysis and potential immunostimulatory activity of *Nannochloropsis oculata* polysaccharides. Carbohydr Polym 222:114962

Park GT, Go RE, Lee HM, Lee GA, Kim CW, Seo JW, Hwang KA (2017) Potential anti-proliferative and immunomodulatory effects of marine microalgal exopolysaccharide on various human cancer cells and lymphocytes in vitro. Mar Biotechnol 19:136–146

Parra-Riofrio G, Moreno P, García-Rosado E, Alonso MC, Uribe-Tapia E, Abdala-Diaz RT, Bejar J (2023) *Tetraselmis suecica* and *Porphyridium cruentum* exopolysaccharides show anti-VHSV activity on RTG-2 cells. Aquac Int 31:3145–3157

Pasquet V, Chérouvrier JR, Farhat F, Thiéry V, Piot JM, Bérard JB, Picot L (2011) Study on the microalgal pigments extraction process: performance of microwave assisted extraction. Process Biochem 46(1):59–67

Patel AK, Albarico FPJB, Perumal PK, Vadrale AP, Nian CT, Chau HTB, Singhania RR (2022) Algae as an emerging source of bioactive pigments. Bioresour Technol 351:126910

Pavithra KG, Kumar PS, Jaikumar V, Vardhan KH, SundarRajan P (2020) Microalgae for biofuel production and removal of heavy metals: a review. Environ Chem Lett 18:1905–1923

Pfleger BF, Gossing M, Nielsen J (2015) Metabolic engineering strategies for microbial synthesis of oleochemicals. Metab Eng 29:1–11

Prajapati SK, Bhattacharya A, Malik A, Vijay VK (2015) Pretreatment of algal biomass using fungal crude enzymes. Algal Res 8:8–14

Puglisi I, La Bella E, Rovetto EI, Lo Piero AR, Baglieri A (2020) Biostimulant effect and biochemical response in lettuce seedlings treated with a *Scenedesmus quadricauda* extract. Plants 9(1):123

Qi J, Kim SM (2017) Characterization and immunomodulatory activities of polysaccharides extracted from green alga *Chlorella ellipsoidea*. Int J Biol Macromol 95:106–114

Rachidi F, Benhima R, Sbabou L, El Arroussi H (2020) Microalgae polysaccharides bio-stimulating effect on tomato plants: growth and metabolic distribution. Biotechnol Rep 25:e00426

Rachidi F, Benhima R, Kasmi Y, Sbabou L, Arroussi HE (2021) Evaluation of microalgae polysaccharides as biostimulants of tomato plant defense using metabolomics and biochemical approaches. Sci Rep 11(1):930

Radonić A, Thulke S, Achenbach J, Kurth A, Vreemann A, König T, Nitsche A (2011) Anionic polysaccharides from phototrophic microorganisms exhibit antiviral activities to Vaccinia virus. J Antivir Antiretrovir 2:51–55

Ranglová K, Lakatos GE, Manoel JAC, Grivalský T, Estrella FS, Fernández FGA, Masojídek J (2021) Growth, biostimulant and biopesticide activity of the MACC-1 *Chlorella* strain cultivated outdoors in inorganic medium and wastewater. Algal Res 53:102136

Razzak SA, Hossain MM, Lucky RA, Bassi AS, De Lasa H (2013) Integrated CO2 capture, wastewater treatment and biofuel production by microalgae culturing—a review. Renew Sust Energ Rev 27:622–653

Razzak SA, Bahar K, Islam KO, Haniffa AK, Faruque MO, Hossain SZ, Hossain MM (2023) Microalgae cultivation in photobioreactors: sustainable solutions for a greener future. Green Chem Eng

Renuka N, Prasanna R, Sood A, Ahluwalia AS, Bansal R, Babu S, Nain L (2016) Exploring the efficacy of wastewater-grown microalgal biomass as a biofertilizer for wheat. Environ Sci Pollut Res 23:6608–6620

Rogers JN, Rosenberg JN, Guzman BJ, Oh VH, Mimbela LE, Ghassemi A, Donohue MD (2014) A critical analysis of paddlewheel-driven raceway ponds for algal biofuel production at commercial scales. Algal Res 4:76–88

Rupawalla Z, Robinson N, Schmidt S, Li S, Carruthers S, Buisset E, Wolf J (2021) Algae biofertilisers promote sustainable food production and a circular nutrient economy–an integrated empirical-modelling study. Sci Total Environ 796:148913

Saadaoui I, Sedky R, Rasheed R, Bounnit T, Almahmoud A, Elshekh A, Al Jabri H (2019) Assessment of the algae-based biofertilizer influence on date palm (*Phoenix dactylifera* L.) cultivation. J Appl Phycol 31:457–463

Sadak A, Şensoy S (2022) Utilization of microalgae [*Chlorella vulgaris* (Beijerinck)] on plant growth and nutrient uptake of garden cress (*Lepidium sativum* L.) grown in different fertilizer applications. Int J Agric Environ Food Sci 6(2):240–245

Samarasinghe N, Fernando S, Lacey R, Faulkner WB (2012) Algal cell rupture using high pressure homogenization as a prelude to oil extraction. Renew Energy 48:300–308

Saral JS, Satheesh AR, Ranganathan P (2022) Economic and environmental analysis of algal biorefinery for the production of renewable fuels and co-product. Energ Convers Manage 14:100189

Saranya N, Devi P, Nithiyanantham S, Jeyalaxmi R (2014) Cells disruption by ultrasonication. BioNanoScience 4:335–337

Sarker PK, Kapuscinski AR, McKuin B, Fitzgerald DS, Nash HM, Greenwood C (2020) Microalgae-blend tilapia feed eliminates fishmeal and fish oil, improves growth, and is cost viable. Sci Rep 10(1):19328

Sawant SS, Gosavi SN, Khadamkar HP, Mathpati CS, Pandit R, Lali AM (2019) Energy efficient design of high depth raceway pond using computational fluid dynamics. Renew Energy 133:528–537

Schreiber C, Schiedung H, Harrison L, Briese C, Ackermann B, Kant J, Nedbal L (2018) Evaluating potential of green alga *Chlorella vulgaris* to accumulate phosphorus and to fertilize nutrient-poor soil substrates for crop plants. J Appl Phycol 30:2827–2836

Schwenzfeier A, Wierenga PA, Gruppen H (2011) Isolation and characterization of soluble protein from the green microalgae *Tetraselmis* sp. Bioresour Technol 102(19):9121–9127

Scofield J, Shi S, Prime EL, Kentish SE, Qiao GG, Martin GJ (2022) Evaporation reduction and salinity control in microalgae production ponds using chemical monolayers. Algal Res 66:102783

Sharma GK, Khan SA, Shrivastava M, Bhattacharyya R, Sharma A, Gupta DK, Gupta N (2021) Circular economy fertilization: phycoremediated algal biomass as biofertilizers for sustainable crop production. J Environ Manag 287:112295

Siddiki SYA, Mofijur M, Kumar PS, Ahmed SF, Inayat A, Kusumo F, Mahlia TMI (2022) Microalgae biomass as a sustainable source for biofuel, biochemical and biobased value-added products: an integrated biorefinery concept. Fuel 307:121782

Singh RN, Sharma S (2012) Development of suitable photobioreactor for algae production—a review. Renew Sust Energ Rev 16(4):2347–2353

Slinksienė R, Sendzikiene E, Mikolaitiene A, Makareviciene V, Paleckiene R, Ragauskaite D (2022) Use of microalgae biomass for production of granular nitrogen biofertilizers. Green Chem Lett Rev 15(2):416–426

Sun L, Wang L, Zhou Y (2012) Immunomodulation and antitumor activities of different-molecular-weight polysaccharides from *Porphyridium cruentum*. Carbohydr Polym 87(2):1206–1210

Sun L, Chu J, Sun Z, Chen L (2016) Physicochemical properties, immunomodulation and antitumor activities of polysaccharide from *Pavlova viridis*. Life Sci 144:156–161

Sun H, Wang Y, He Y, Liu B, Mou H, Chen F, Yang S (2023) Microalgae-derived pigments for the food industry. Mar Drugs 21(2):82–109

Supraja KV, Behera B, Balasubramanian P (2020) Efficacy of microalgal extracts as biostimulants through seed treatment and foliar spray for tomato cultivation. Ind Crop Prod 151:112453

Taisir M, Teo CL, Idris A, Yusuf AM (2016) Cultivation of Nannochloropsis sp. using narrow beam angle light emitting diode in an internally illuminated photobioreactor. Bioresour Bioprocess 3:1–8

Talyshinsky MM, Souprun YY, Huleihel MM (2002) Anti-viral activity of red microalgal polysaccharides against retroviruses. Cancer Cell Int 2:1–7

Tang Y, Rosenberg JN, Bohutskyi P, Yu G, Betenbaugh MJ, Wang F (2015) Microalgae as a feedstock for biofuel precursors and value-added products: green fuels and golden opportunities. BioResources 11(1):2850–2885

Tibaldi E, Zittelli GC, Parisi G, Bruno M, Giorgi G, Tulli F, Poli BM (2015) Growth performance and quality traits of European sea bass (D. labrax) fed diets including increasing levels of freeze-dried Isochrysis sp.(T-ISO) biomass as a source of protein and n-3 long chain PUFA in partial substitution of fish derivatives. Aquaculture 440:60–68

Tibbetts SM, Yasumaru F, Lemos D (2017) In vitro prediction of digestible protein content of marine microalgae (Nannochloropsis granulata) meals for Pacific white shrimp (Litopenaeus vannamei) and rainbow trout (Oncorhynchus mykiss). Algal Res 21:76–80

Trabelsi L, Chaieb O, Mnari A, Abid-Essafi S, Aleya L (2016) Partial characterization and antioxidant and antiproliferative activities of the aqueous extracellular polysaccharides from the thermophilic microalgae *Graesiella* sp. BMC Complement Altern Med 16(1):1–10

Uma VS, Usmani Z, Sharma M, Diwan D, Sharma M, Guo M, Gupta VK (2023) Valorisation of algal biomass to value-added metabolites: emerging trends and opportunities. Phytochem Rev 22(4):1015–1040

UN United Nations Department of Economic and Social Affairs, Population Division (2019) World population prospects 2019: volume I: comprehensive tables

Uysal O, Uysal FO, Ekinci K (2015) Evaluation of microalgae as microbial fertilizer. European Journal of Sustainable Development 4(2):77–77

Viegas C, Gouveia L, Gonçalves M (2021a) Evaluation of microalgae as bioremediation agent for poultry effluent and biostimulant for germination. Environ Technol Innov 24:102048

Viegas C, Gouveia L, Goncalves M (2021b) Aquaculture wastewater treatment through microalgal. Biomass potential applications on animal feed, agriculture, and energy. J Environ Manag 286:112187
Wang G, Wang T (2012) Characterization of lipid components in two microalgae for biofuel application. J Am Oil Chem Soc 89(1):135–143
Ward AJ, Lewis DM, Green FB (2014) Anaerobic digestion of algae biomass: a review. Algal Res 5:204–214
Yang S, Wan H, Wang R, Hao D (2019) Sulfated polysaccharides from *Phaeodactylum tricornutum*: isolation, structural characteristics, and inhibiting HepG2 growth activity in vitro. PeerJ 7:e6409
Yap BH, Dumsday GJ, Scales PJ, Martin GJ (2015) Energy evaluation of algal cell disruption by high pressure homogenisation. Bioresour Technol 184:280–285
Yap XY, Gew LT, Khalid M, Yow YY (2023) Algae-based bioplastic for packaging: a decade of development and challenges (2010–2020). J Polym Environ 31(3):833–851
Yim JH, Son E, Pyo S, Lee HK (2005) Novel sulfated polysaccharide derived from red-tide microalga *Gyrodinium impudicum* strain KG03 with immunostimulating activity in vivo. Mar Biotechnol 7:331–338
Zbinden MDA, Sturm BS, Nord RD, Carey WJ, Moore D, Shinogle H, Stagg-Williams SM (2013) Pulsed electric field (PEF) as an intensification pretreatment for greener solvent lipid extraction from microalgae. Biotechnol Bioeng 110(6):1605–1615
Zhu L (2015) Biorefinery as a promising approach to promote microalgae industry: an innovative framework. Renew Sust Energ Rev 41:1376–1384
Zhu C, Chi Z, Bi C, Zhao Y, Cai H (2019) Hydrodynamic performance of floating photobioreactors driven by wave energy. Biotechnol Biofuels 12(1):1–12

Microalgal in the Production of Various High-End Value-Added Products Using Zero-Waste Biorefinery Approach

Jeya. K. R., Sankaranarayanan. A, and Veerapagu. M

1 Introduction

In recent years, the decline of fossil fuels and the environmental consequences of greenhouse gas (GHG) emissions has led to research in carbon-free renewable energy sources at current levels of energy production and consumption. It is estimated that after 2050, conventional oil reserves may be depleted, and even if they are, associated GHG emissions including CO_2 are likely to have a negative impact on the global climate (Morris et al. 2023). Biodiesel from healthy crops is recognized as a potentially sustainable as well as low-carbon substitute to fossil fuel. But oil crops, suitable cooking oil, and animal products are not sufficient to fill the growing gap in transportation fuels (Manaf et al. 2019). Alternative energy sources such as solar, wind, and hydropower have gained increasing interest due to the increasing energy demand and environmental concerns related to fossil fuels (Weliwaththage and Yildirim 2020) to meet the challenges associated with the generation of primary biofuels in transport. Second-generation biofuels derived from lignocellulosic biomass are under investigation (Binod et al. 2019).

Microalgae exhibit the ability to transform nutrients present in wastewater into biomass, potentially reducing manual harvesting costs and aiding in contaminant

J. K. R.
Department of Biotechnology, Government Arts and Science College (Women), Veppur, Perambalur, Tamil Nadu, India

S. A
Department of Life Sciences, Sri Sathya Sai University for Human Excellence, Navanihal, Kalaburagi, Karnataka, India

V. M (✉)
Department of Biotechnology, Microbiology and Bioinformatics, National College (Autonomous), Tiruchirappalli, Tamil Nadu, India

K. Agrawal, P. Verma (eds.), *Biotechnological Advances in Biorefinery*, Interdisciplinary Biotechnological Advances,
https://doi.org/10.1007/978-981-97-5544-8_7

removal. This process aligns with the concept of a biorefinery, where microalgal lipid can be transformed into biodiesel, and solid carbon material derived through pyrolysis contributes to contaminant removal (De Bhowmick et al. 2015). Despite these benefits, microalgae-derived biofuel production faces challenges related to net energy gain, capital investment, and overall cost. Consequently, the algal research community is shifting its focus toward a sustainable processing approach, emphasizing a precise method, which explores equally upstream along with downstream process to optimize efficiency and overcome existing challenges (De Bhowmick et al. 2018).

2 Microalgae and High Value-Added Product

In recent years, there is an increasing global interest in the use of microalgal biomass for micropharmaceuticals and biofuels, aiming to partially replace chemicals from petroleum-based food sources. Microalgae exhibit distinct advantages than other fresh food sources, such as rice straw, plant tubers, and leaves. Notably, microalgae have a flexible structure with low levels of lignin, making them less critical compared with other biomass sources with complex structures such as flexible lignin, cellulose, and hemicellulose (Rodionova et al. 2022).

One of the key advantages of microalgae lies in their microscopic nature and photosynthetic capabilities. They utilize sunlight and carbon dioxide (CO_2) as essential regulators for photosynthesis, presenting an opportunity for CO_2 biosequestration applications. Additionally, microalgae demonstrate efficient nutrient uptake, rapid growth rates, and shorter harvesting times in contrast to plants that are in terrestrial, which often need several months before the harvest of biomass (Dębowski et al. 2020). The cultivated biomass contains important chemical substances, including lipid, carbohydrate, and protein, with diverse types synthesizing distinct metabolic substances on cultivation conditions and population of cell.

Historically, microalgae have been utilized for various purposes, including providing food for native communities over a thousand years ago (Olabi et al. 2023). However, the significant development of microphytic biomass production started in the mid-twentieth century, particularly in Germany, driven by concerns about population expansion and its implications for food resilience.

Microalgae have been more important to the academic and engineering sectors over the previous five decades as ecological sustainability gained greater attention. These algae were living things that take carbon from organic matter and energy from light. They exhibit higher annual photon-biomass fluctuations compared with higher plants, nor do they have any inherent sensitivity to seasonal factors. These efforts, in conjunction with species composition, enable the synthesis of valuable molecules such as proteins, lipids, carbohydrates, and dyes (Perin et al. 2019).

Moreover, microalgae could potentially be cultivated on sewage or land that is not farmland without the need for pesticide, avoiding interference with food or other crop productions. Their ability to recycle atmospheric carbon dioxide further minimizes environmental impacts. Given the current challenges in energy, the environment, and food production, microalgal biotechnology has emerged as a powerful solution (Hamed 2016).

Recognized for their versatile uses, microalgae play a role in the treatment of water (wastewater) and gaseous (smoke gas) emissions, serve as food for human animals, produce biofuels, and serve as sources of bioactive compounds (Chew et al. 2017). The biologically active substances generated by microalgae, including polysaccharides, proteins, polyunsaturated fatty acid, sterols, carotenoid, chlorophyll, and phycobiliprotein, depend on factors such as genus, species, and agricultural conditions (Li et al. 2019a; Paliwal et al. 2017).

2.1 Bioactive Compound Production by Microalgae

Photosynthetic microorganisms, particularly microalgae, are recognized as a potential natural reservoir of natural sources of bioactive substances with extensive applications in various industries, including food, nutraceutical, cosmetic, and pharmaceutical (Gallego et al. 2018, 2019; Ventura et al. 2017). These bioactive compounds exhibit diverse functionalities, serving as antioxidants, antimicrobials, anti-inflammatories, antitumor agents, and anticancer compounds among others (Fu et al. 2017; Montalvo et al. 2019).

Microalgal biomass is inherently rich in various substances such as protein, carbohydrate, lipid, and pigment, making it a valuable resource for the generation of biofuel and valuable bio-derived goods (Costa et al. 2017). Bioactive compounds from microalgae can be obtained either directly from primary metabolism or synthesized through secondary metabolism pathways (Park et al. 2022). The primary bioactive substances produced by microalgae encompass polyunsaturated fatty acid, carotenoid, chlorophyll, phycobiliprotein, polysaccharide, and protein compounds such as peptides and amino acids (Wu et al. 2021). These compounds hold significant potential for various applications, contributing to the development of innovative products in multiple industries.

2.2 Polyunsaturated Fatty Acids (PUFAs)

Fatty acids, integral components of storage lipids and primary metabolites, include a noteworthy subset known as polyunsaturated fatty acids (PUFAs) (Behera et al. 2021). PUFAs are characterized by a long unsaturated hydrocarbon chain with multiple double bonds, making them highly desirable for applications in the food and pharmaceutical industries. From a medical perspective, PUFAs have demonstrated positive effects in reducing the prevalence of long-term conditions and chronic diseases such as diabetes, arthritis, cardiovascular disease, and obesity (Katiyar and Arora 2020). Medical reviews, incorporating in vitro and in vivo studies, have underscored the anti-inflammatory, antiproliferative, and anticachectic effects of ω-3 fatty acid supplementation (Kiran and Venkata Mohan 2021). Microalgae have emerged as promising sources of omega 3 fatty acids, particularly eicosapentaenoic acid (EPA) and docosahexaenoic acid (DHA), and have been extensively studied for PUFA production. Microalgae genera, including *Crypthecodinium*, *Schizochytrium*, and *Ulkenia* from the kingdom Chromista, have demonstrated the capability to accumulate over 40% of total fatty acid content in DHA (Sahin et al. 2018).

Apart from their role as a PUFA source, lipids derived from microalgae are proposed as raw materials for biodiesel production, especially in the context of microalgae being superior oil sources compared with conventional plants. Green microalgae, in particular, offer advantages such as faster growth, adaptability to a broader range of temperature and pH conditions, efficient carbon dioxide utilization, and increased lipid yields (Hussain et al. 2021).

In the context of a microalgal biorefinery, lipids serve a dual purpose by not only as a feedstock for biodiesel but also as sources of highly valuable compounds, including PUFAs (Sivaramakrishnan et al. 2022). Major PUFAs produced by microalgae include eicosapentaenoic acid (EPA), docosahexaenoic acid (DHA), arachidonic acid (ARA), and γ-linolenic acid (GLA). Studies on omega 3 fatty acids, such as EPA and DHA, emphasize their biological activities, contributing to the prevention of cardiovascular disorders, diabetic conditions, obesity, and cancer (Saini and Keum 2018).

2.3 Carbohydrates

Microalgae exhibit the synthesis of general carbohydrates, primarily in the form of starch and cellulose, constituting approximately 20% of dry cell weight. Notably, certain microalgae, such as *Chlorella* sp., produce polysaccharides containing galactose, rhamnose, mannose, arabinose, *N*-acetylglucosamine, *N*-acetylgalactosamine, and β-1,3-glucan (Dourou et al. 2021). These polysaccharides boast beneficial biological activities, including cardiovascular and immune system support, along with antioxidant properties (Saide et al. 2021).

Carbohydrates are produced within the cell and signify a significant portion of substances derived from photosynthesis, constituting up to 50% of the dry weight of several microalga (Yen et al. 2013). Microalgae generally utilized for polysaccharide production are *Tetraselmis* sp., *Isochrysis* sp., *Porphyridium cruentum*, *Porphyridium purpureum*, *Chlorella* sp., and *Rhodella reticulata* (Mourelle et al. 2017). After being produced, carbohydrates play diverse biological roles and are categorized into three classes based on their physiological functions: polysaccharide energy stores, polysaccharide framework contributing to cell wall synthesis, and polysaccharide participating in cellular messaging (Bernaerts et al. 2019).

In a biorefinery system, carbohydrates serve as feedstock for bioethanol production and as suppliers of high-value compound. Polysaccharides from microalgae, extensively studied due to their biologically active nature, find applications in the cosmetic additive, food, and pharmaceutical industries. Sulfated polysaccharides, among the carbohydrates produced by microalgae, stand out for their antioxidant, antitumor, antiviral, antibacterial, anti-inflammatory, and immune system modulating effects (Mohammed et al. 2021). Structural polysaccharides, primarily associated with cell walls, have gained significance not only for their rheological properties but also due to various biological activities. These macromolecules have applications in the food, cosmetics, and pharmaceutical industries. Some polysaccharides are recognized for their utilization of biosurfactant functions in the cleanup of water and soil (Siqueira et al. 2018).

Furthermore, starch and other polysaccharides present in the microalgae may be converted into ethanol through different processes like acid hydrolysis before fermentation. Biochemical conversions, such as anaerobic digestion and photobiological hydrogen production, offer avenues for producing extra biofuels like methane and hydrogen (Deviram et al. 2020).

2.4 Protein

The protein content within microalgae is contingent upon both the taxonomic group and the culturing parameters employed. Cyanobacteria like *Spirulina* can exhibit protein content ranging between 50% and 70% (w/w) (Moraes et al. 2016), whereas *Chlorella* may showcase protein content varying from 30% to 60% (w/w) (Bauer et al. 2017).

Recent studies underscore the potential of microalgal biomass as a source of bioactive peptide. Lisboa et al. (Cristiane et al. 2016) explored the production of bioactive peptides with antioxidant activity through the enzymatic hydrolysis of *Spirulina* sp. LEB 18 biomass. Additionally, Montalvo et al. (2019) discovered that biopeptide fractions from *Arthrospira maxima* OF15 biomass exhibited antimicrobial, anti-inflammatory, antioxidant, chelating, and anti-collagenase activities, holding promise for applications in the pharmaceutical, cosmetic, and food industries.

Microalgal proteins, primarily derived from *Arthrospira*, *Chlorella*, and *Dunaliella salina*, are predominantly used as nutraceuticals or integrated into functional foods (Khanra et al. 2018). The high protein content of microalgae, especially in *Arthrospira, Chlorella*, and *Dunaliella salina*, positions them as an unconventional but valuable protein source for human consumption, especially in developing countries. Microalgal proteins, particularly from the *Arthrospira* and *Chlorella* genera, have made inroads into the cosmetic industry, specifically the skincare market (Apone et al. 2019). Protein extracts were commonly present in cosmetic product formulations due to their compelling biological activities, encompassing collagen stimulation, skin firmness enhancement, reduction of stretch mark morphology, and mitigation of vascular imperfections (Yaakob et al. 2014). *Arthrospira* proteins, recognized for their skin repair properties, are also employed in cosmetics targeting early telltale signs of aging skin. Micosporin-like amino acids derived from microalgae have garnered significant interest as potential cosmetic agents, particularly for their photoprotective and antioxidant actions combating photoinduced skin aging (Berthon et al. 2017).

From a pharmacological perspective, peptides derived from microalgae have gained prominence as alternative bioactive compounds due to their safety profile. These peptides exhibit diverse properties, including anti-inflammatory, antihypertensive, anticancer, antibacterial, and antioxidant activities, making them applicable in promoting human health (Sadeghi et al. 2018).

2.5 Pigments

Within microalgal biomass, the primary pigments encompass chlorophyll, carotenoid, and phycobiliproteins. Carotenoids, exhibiting yellow, orange, or red lipophilic pigments with an aliphatic or alicyclic structure composed of eight isoprenoid units, play a crucial role. A key carotenoid, β-carotene, is predominantly produced by *Dunaliella* and finds application as a food colorant, vitamin A precursor in food and animal feed, additive in cosmetics and multivitamin preparations, and a natural food ingredient with antioxidant activity (Becker 2013). Astaxanthin, another carotenoid with robust antioxidant activity, has diverse applications in food industries, animal feed development, cosmetics, and nutraceuticals (Khoo et al. 2019). Microorganisms like *Chlorella zofingiensis* (Liu et al. 2012), *Haematococcus*, and *Chlorococcum* sp. (Yuan et al. 2011) have been identified as capable of biosynthesizing this natural pigment.

Given the broad spectrum of naturally colored pigment in microalgae, they serve as sustainable sources for food colorants. Microalgal cultivation stands out as an environmentally friendly, safe, and cost-effective alternative to artificial colorants, providing natural food dyes. Carotenoid, chlorophyll, and phycobiliprotein, being natural algal-derived pigments, serve as precursors to vitamins present in food and animal products, as well as coloring agents for biomaterials, cosmetics, and pharmaceuticals (Pagels et al. 2021). Most microalgae species contain approximately 0.5–1% chlorophyll and 0.1–0.2% carotenoid. Chlorophyll, a vital pigment for photosynthesis, is universally present in all microalgae (Nwoba et al. 2020). Pigments, as crucial compounds, find wide usage in the food, cosmetics, and pharmaceutical industries (Sosa-Hernández et al. 2019), holding significant commercial potential.

2.6 Vitamins

Microalgae emerge as valuable sources of various vitamins, including A, B1, B2, B6, B12, C, E, biotin, folic acid, pantothenic acid, and more. *Isochrysis galbana*, for instance, stands out as a significant source of vitamins A and E, along with folic acid, nicotinic acid, pantothenic acid, biotin, thiamine, riboflavin, pyridoxine, cobalamin, chlorophyll (*a* and *c*), fucoxanthin, and diadinoxanthin. Another example is *Euglena gracilis*, an antioxidant-rich microalgae capable of producing vitamins such as β-carotene and vitamins C and E (Mulders et al. 2013). The water-soluble vitamin B12, vital for human health, is typically produced in animal product but notably absent in plant sources. This deficiency poses a common challenge for individuals adhering to a vegan or vegetarian diet. Some microalgae types, like *Chlorella* sp. and *Pleurochrysis carterae*, may contain or synthesize vitamin B12 (Dolganyuk et al. 2020). However, the bioavailability of B12 in these microalgae is not guaranteed, necessitating further studies to identify potential sources of readily available vitamin B12 among microalgae.

3 Microalgae and Value-Added Energy Products

Biofuels, derived from biologically sourced and cultivated materials, offer a sustainable alternative to fossil fuels, which pose environmental and economic challenges. Biomass, categorized into four generations, encompasses sustainably cultivated resources and residues or byproducts. The latter includes materials like wood, dried plants, seeds, and manure, which may or may not be renewable (Mobin and Alam 2018).

3.1 Biodiesel

Fossil fuels, as major contributors to the energy sector, have prompted the search for eco-friendly alternatives due to concerns about climate change, air pollution, and economic implications. Biodiesel, recognized for being nontoxic, biodegradable, carbon-neutral, renewable, and sulfur-free, emerges as a sustainable alternative. Its production has evolved, with changing feedstocks, aiming to reduce greenhouse gas emissions and air pollution. As a biofuel, biodiesel demonstrates a lower environmental impact, contributing to reduced air pollution and mitigating climate change (Shuba and Kifle 2018). The global demand for sustainable and renewable energy solutions has led to the exploration of biomass for power generation. Biomass, considered almost carbon neutral, offers environmental benefits, making it an attractive option for reducing net carbon emissions. The urgency to transition to sustainable and renewable energy sources is driven by economic growth, population expansion, and the demand for oil and gas resources (de Farias Silva and Bertucco 2019).

Over the years, investigators have explored biomass as a primary material for various generations of bio-based energy. Third-generation biofuels, particularly derived from microalgae, are viewed as a promising solution to challenges associated with earlier generations. Microalgae can be employed to produce a range of biofuels, such as biomethane, bioethanol, and biohydrogen, after extraction of lipid from the algae (Hamed et al. 2017; Lam et al. 2018). Biodiesel, a focal point in recent discussions, stands out as a sustainable, biodegradable, and safe energy source with superior properties compared with petro-diesel. It exhibits a superior lubrication, a higher cetane number, decreased sulfur percentage, increased flash point, and better lubricating efficiency. Direct use or blending of biodiesel with regular diesel positively impacts exhaust gases, reducing carbon monoxide emissions and combustion gases (Mostafa and El-Gendy 2017).

3.1.1 Feedstock for Biodiesel Production

In recent decades, the search for suitable feedstocks for biofuel production has led to the identification of four biofuel generations (Enamala et al. 2018). First-generation biofuels, derived from food crops, oil seeds, and animal fats, raised concerns about competition with food and fiber production for land and water resources. Second-generation biofuels, produced from lignocellulosic biomass, faced challenges related to land requirements. To address these issues, third-generation biofuels, using microalgae biomass, emerged as a more efficient and sustainable solution (Voloshin et al. 2016).

Microalgae present several advantages, utilizing atmospheric carbon dioxide for growth, aiding in wastewater bioremediation, accumulating high lipid content, thriving in extreme environmental conditions, requiring less water than terrestrial crops, and not relying on herbicides or pesticides (Mathimani and Pugazhendhi 2019; Suganya et al. 2016). Their versatility extends to different water sources, oxygenic photosynthesis using water, and the production of valuable compounds (Kiran et al. 2014). Microalgae biofuels are very affordable in contrast to other feed materials; microalgae-based biofuels are highly cost-effective, with microalgae exhibiting superior photosynthetic efficiency and growth rates (Sun et al. 2018). Biodiesel production from microalgae has gained attention due to their high oil content (Mani Rathnam and Madras 2019). Microalgal biodiesel is biodegradable, nontoxic, and sulfur-free and has a chemical composition similar to petroleum diesel, contributing to reduced carbon emissions (Alaswad et al. 2015; Scaife et al. 2015).

Microalgae biomass, beyond biodiesel extraction, contains various valuable components, making it a significant feedstock for a biorefinery approach. The biorefinery concept involves the comprehensive utilization of microalgae biomass for various products, potentially reducing the costs associated with biodiesel production (Wicker et al. 2021). The focus is on achieving complete valorization of microalgae biomass to enhance economic.

3.1.2 Microalgae Biomass to Biodiesel

For commercial biofuel and value-added product production using algae as feedstock, a significant requirement is the availability of a large quantity of algal biomass. Selecting a species with high lipid content and efficient biofuel production capabilities is crucial. Cultivation, whether in open ponds or photobioreactors, demands attention to various parameters influencing growth and biofuel quality (Ahmad et al. 2021). Achieving rich algal biomass quickly is essential. Algae's adaptability to altering physical and chemical conditions enhances their ability to produce high-quality biofuels.

Contamination is a critical concern, impacting bioreactor design for economic considerations. Capital costs, associated with land acquisition, bioreactor design, construction, and major equipment, along with maintenance costs, including labor, nutrients, equipment maintenance, and water, contribute to the overall expenses. Costs vary widely between urban and rural areas, including land and labor costs. Developing cost-effective, environmentally friendly, and scalable technologies for large-scale microalgae cultivation is imperative (Udayan et al. 2022). Harvesting efficiency and lipid extraction after harvesting significantly influence biofuel costs.

Biodiesel production involves the direct conversion of solar energy into fuel from raw materials. Microalgae's unique advantages, such as efficient light utilization, rapid growth rates, and the ability to produce energy-rich compounds, contribute to their suitability for biodiesel production. The biodiesel production process includes cultivation, harvesting, oil extraction, and transesterification. Microalgal biodiesel exhibits biodegradability, nontoxicity, sulfur absence, reduced carbon emissions, and a chemical composition resembling petroleum diesel. The diverse components in microalgae biomass make it valuable for the biorefinery approach, supporting the complete valorization of microalgae biomass into various products (Suparmaniam et al. 2019).

3.1.3 Environmental Biorefinery

Microalgae play a crucial role in environmental biorefineries, addressing issues related to CO_2 emissions and wastewater pollution. During their growth and metabolic activities, microalgae require CO_2 and nutrients, which can be sourced from polluted air and wastewater. In this process, microalgae serve a dual purpose by fixing atmospheric CO_2 with an efficiency ranging from 10% to 50%. The presence of a high concentration of CO_2 has been linked to increased lipid content and growth rates. Water pollution, a significant threat, can be mitigated by cultivating microalgae in wastewater. Nutrients present in wastewater, such as nitrogen, phosphorus, and heavy metals, which typically require extensive infrastructures for removal, can be utilized by microalgae for growth (Ahmed et al. 2022). Components like nitrogen, phosphorus, ammonia, and nitrates act as nutrients for microalgae, eventually leading to the production of biodiesel-rich biomass. Microalgae are capable of thriving in wastewater sources with high biological oxygen demand (BOD), chemical oxygen demand (COD), dissolved oxygen (DO), and elevated nitrogen and phosphorus concentrations.

Microalgae can grow in the presence of various emerging pollutants (EPs) found in wastewater, demonstrating potential in wastewater treatment (Rempel et al. 2022). Experiments with different microalgal species, such as *Spirulina platensis*, *Scenedesmus obliquus*, and *Chlorella homosphaera*, show their tolerance to certain concentrations of EPs, making them suitable for wastewater treatment. Microalgae cultivation in various wastewater sources, including sewage, distillery, and dairy wastewater, has been successful in achieving high biomass and lipid yields (Chandra et al. 2021).

The concept of an environmental biorefinery involves the complete valorization of microalgae biomass into various products, including biodiesel. The rich fatty acid content in microalgae, extracted after cultivation in wastewater, makes them ideal raw materials for biodiesel production (Moreno-Garcia et al. 2021). The utilization of wastewater as a growth medium for microalgae enhances the sustainability of the biorefinery approach, utilizing nutrients remaining in the residual streams (Javed et al. 2019). This not only addresses environmental concerns but also contributes to the economic viability of valuable compounds derived from microalgae.

3.2 *Biogas*

Biogas, a promising biofuel, primarily composed of methane (CH_4) and carbon dioxide (CO_2), has the potential to mitigate concerns related to fossil fuels, energy crises, and climate change (Ayala-Parra et al. 2017). Microalgae are considered a viable biomass resource for production of biogas through anaerobic digestion (AD). The mature technology of AD offers advantages such as using wastewater and seawater, nutrient recycling, optimal biomass utilization, minimal sludge production, cost-effective operations, and green biogas generations.

Despite the potential of microalgal biogas production, challenges persist, including the recalcitrance of raw microalgal biomass to biodegradation, low biomass yield, carbon-nitrogen ratio (C/N ratio), and volatile fatty acid (VFA) accumulation in the digester (Bohutskyi et al. 2014). Because different species produce different amounts of biofuel, choosing the right microalgal species for an integrated and sustainable bioprocess is still quite difficult. Operational and environmental factors significantly influence microalgae cultivation and AD kinetics, affecting biomass and biogas yields.

The utilization of wastewater and effluents from AD processes as culture media for microalgae cultivation contributes to the sustainability and cost-effectiveness of biogas production (Rafa et al. 2021). Microalgae's ability to grow in wastewater provides tertiary treatment, taking advantage of remaining carbon and nutrients. This integration of microalgae with AD processes not only enhances overall process efficiency but also makes microalgal AD more sustainable and economically viable (Nayak et al. 2016).

3.2.1 Biorefinery Approach

In the context of biorefineries, the effective utilization of microalgae can be accomplished through coordinated processing, leading to enhanced economic viability and positive energy balances. An example by Mussgnug et al. demonstrated that the biogas yield enhanced by 123% through the anaerobic digestion (AD) of algae biomass following the synthesis of biohydrogen (Mussgnug et al. 2010). In another study, the lipid-extracted biomass of *Scenedesmus* was employed in a two-stage process involving hydrogen and methane production, which was more productive compared with a single-stage methane production process (Yang et al. 2011).

When compared with modern methods, biogas generation in an coordinated process may provide better financial sustainability and an energy surplus (Kumar and Verma 2020). The integration of microalgal biogas production with microalgae-based bioproducts, other biomass-based biorefineries, and wastewater treatment holds promise. In these integration approaches, biogas production is technologically and economically enhanced by eliminating the biomass production step, which constitutes a significant portion of the costs and energy required for microalgal cultivation, harvesting, and dewatering (Cuevas-Castillo et al. 2020). The residues generated after biodiesel and bioethanol production from microalgae can be repurposed for generation of biogas, minimizing the costs and energy needed for standalone biogas production (Salinas-Salazar et al. 2019).

The biorefinery concept, involving the production of high-value bioproducts alongside energy production from biomass, presents an attractive option when dealing with microalgae (Goswami et al. 2022). Co-producing high-value compounds and bioenergy seems to be more economically viable than focusing solely on biofuel generation. For instance, González-González et al. (2018) advocated for integrated biodiesel and biogas production from microalgae. Biorefinery concepts that integrate various valorization pathways, encompassing both bioproducts and bioenergy/biofuels, are of high interest to maximize the economic value of microalgal cultivation systems. The residual biomass of the microalga *Chlorella vulgaris*, after extracting chlorophyll, proteins, or lipid fraction, has been utilized as raw material for biomethane generation (Markou et al. 2022).

3.3 Bioethanol

The ever-increasing demand for energy, coupled with depleting fossil fuel reserves and environmental concerns, has led to a surge in interest in biomass energy, specifically bioethanol (Bibi et al. 2017). Bioethanol production processes rely on energy derived from renewable sources, making it an environmentally beneficial energy source with no net addition of carbon dioxide to the atmosphere (Niphadkar et al. 2018). Microalgae have emerged as a promising reservoir for feedstock in renewable ethanol synthesis, attracting attention from researchers globally (Wusnah et al. 2021).

Microalgae's unique features make them an attractive biomass source for bioethanol. Their cultivation is a crucial step in developing bioethanol from microalgae biomass. Microalgae exhibit superior photosynthetic efficiency, producing around ten times more biomass than traditional plants. Importantly, microalgae lack lignin, a component which may hinder the hydrolytic enzyme treatment, simplifying the bioethanol production process (Mishra et al. 2019). The development of renewable energy sources is imperative for sustainable future energy availability. Bioethanol, as a form of biofuel, has garnered significant attention due to its renewable nature, environmental friendliness, and economic viability (Telussa et al. 2023). It serves as a potential substitute for liquid fuels, addressing the escalating global energy demand and mitigating the environmental impact associated with fossil fuel combustion (Parascanu et al. 2021).

Despite the promise of bioethanol, challenges, particularly high production costs associated with feedstock, have hindered its widespread commercialization (Sundaram et al. 2023). Traditionally, bioethanol is predominantly produced from sugar and starch feedstocks, sourced from fuel crops like maize, corn, and wheat. However, the reliance on food crops for bioethanol production has raised concerns, leading to increased food prices. The need for economically viable bioethanol commercialization has prompted research into alternative, cost-effective raw materials to enhance bioethanol yield (Kossatz et al. 2017; Selvakumar et al. 2019). Implementing a sustainable biorefinery approach, exploring valuable byproducts from microalgae, becomes crucial for addressing these challenges and making bioethanol production economically feasible.

3.3.1 Zero-Waste Biorefinery Approach

In the context of a microalgal bioenergy production system, waste could be extensively classified into liquid effluents and solid residuals (Mishra et al. 2019). Establishing a sustainable process that effectively reduces, recycles, and reuses those waste materials is paramount for achieving a sustainable energy biorefinery. Recent literature has highlighted the potential of a zero-waste microalgal biorefinery approach, emphasizing the production of food, pharmaceutical, and supplementary value-enhanced items from residues of microalgal cultivation (Cheirsilp and Maneechote 2022; Malik et al. 2022; Venkata Mohan et al. 2016). While discussions have primarily focused on these valuable co-products, the recycling and conversion of both solid and liquid residues for biofuel generation are essential components for ensuring the sustainability of commercial biofuel production. Microalgae stand out as an alternative feedstock for biofuel production due to their rapid growth rate, ease of cultivation, and high yield (Carneiro et al. 2017). Notably, microalgae offer the advantage of not requiring soil or freshwater, and some species can thrive in wastewater with a growth rate approximately ten times faster than that of sugar cane (Osundeko et al. 2019). The utilization of wastewater for large-scale microalgae biomass production has become a prominent focus in current research.

The concept of a zero-waste biorefinery aligns with the principles of circular economy and sustainable resource management. It emphasizes minimizing waste, maximizing resource efficiency, and extracting value from all components of the microalgal biomass. Through the integration of various waste streams into bioenergy and value-added product production processes, a zero-waste microalgal biorefinery aims to contribute to both environmental sustainability and economic viability in the bioenergy sector.

3.3.2 Effluent from Microalgal Culture

Microalgae produced in a photobioreactor and open circuit pond have a high medium for cultivation to biomass ratio. Diverse sources of wastewater are used in place of freshwater and commercial media to provide the water and nutrients needed for microalgae cell development. The method of converting waste into biofuels also uses wastewater. The source of the medium affects the collected medium's chemical makeup. Heavy metal concentrations are high in the harvested medium made from wastewater streams (e.g., lead, copper, and zinc). The removal of this collected media may have an adverse effect on the ecosystem. Because there are a lot of leftover nutrients in the harvested media, recycling it is a great way to lessen the impact on the environment as well as the demand for water and nutrients. Using wastewater for microalgae cultivation eliminates competition for freshwater, saves the cost of nutrient supplements, and reduces CO_2 emissions. This is coupled with simultaneous nutrient removal and lipid production during wastewater treatment. Microalgae from wastewater were utilized as a suitable substrate for bioethanol production (Acebu et al. 2022; Jayaseelan et al. 2021). Hydrolyzing with *Aspergillus niger* and optimizing fermentation conditions at pH 6.0 and 40° yielded a higher bioethanol concentration (Kida 2023).

3.4 Bioelectricity

Bioelectricity is essential for the success of various activities dependent on facilities and infrastructures, serving as a vital component for continuity and success. Energy from fossil fuels is now the most common source of energy. Although it is easily accessible, it is not recyclable, takes an extended period to create, and is not good for our surroundings (Rozina et al. 2022). Growing populations and rapid urbanization have raised the demand for energy from fossil fuels, which resulted in higher CO_2 liberation and contributed to ecological upheaval. The academic and professional sectors are becoming more interested in systems that store electrochemical energy as an alternative to the aforementioned difficulties. Notable among these are supercapacitors, micro-supercapacitors (Jiang et al. 2020), two-dimensional materials (Zhang et al. 2018), and bioelectricity devices (Rahimnejad et al. 2020). Bioelectricity devices, particularly those involved in wastewater treatment and CO_2

sequestration (Li et al. 2019b), offer a promising sustainable solution. These devices harness electricity directly from degradable organic matter through various biological processes. Biocathodes, emerging as alternatives to standard chemical or non-biological cathodes, utilize algae as biocatalysts to furnish electron acceptors in the cathode compartment (Enamala et al. 2020). Algal biocathodes present advantages such as oxygen release during photosynthesis, ubiquity in nature, growth in adverse conditions, nutrient and toxic metal removal, and the production of valuable biomass for biofuels (Hou et al. 2016). Currently, microbial fuel cells (MFCs) are grounded in various supporting theories, such as utilizing wastewater, waste from alcoholic beverage manufacture, and domestic wastewater (Sengupta et al. 2020). Reports indicate the potential development of MFC technology using microalgae and wastewater due to the easy availability of these ingredients. Both microalgae and wastewater contain essential nutrients like carbohydrates, glucose, and fats, providing a suitable environment for bacterial growth, while microalgae contribute to oxygen production through photosynthesis (Fallahi et al. 2021). The utilization of algal biomass or living algal cells directly in energy generators is a potential method. Recent advancements in MFCs have broadened the scope for bioelectricity generation through microbial metabolism.

MFCs, relying on biocatalytic reactions performed by electroactive microorganisms like bacteria or yeast, produce clean bioelectricity (Elshobary et al. 2021). One environmentally friendly alternative is the use of microbial fuel cells (MFCs) that utilize organic materials (substrates) as an energy source. *Chlorella pyrenoidosa*, a microalgae species, is introduced into the MFC anode as an electron donor. Microalgae do not need outside substrates to produce energy; they can do so by regulating their oxygen concentration, amount of light, and number of algal cells. Kitchen wastewater treatment has been demonstrated using MFCs consisting of photosynthetic microorganisms like *Synechococcus* sp. and *Chlorococcum* sp. as cathode catalysts, showcasing their potential as biocathodes for enhanced power generation (Naina Mohamed et al. 2020). Mango (*Mangifera indica* L.), a widely consumed tropical fruit, holds significant economic importance globally. Diverse biotechnology options are being put up to repurpose and enhance various mango residuals (Mandha et al. 2021; Manhongo et al. 2021). MFCs can utilize organic substrates, including mango residues, to generate electricity (Rojas-Flores et al. 2020). Mango pulp, rich in reducing sugars, amino acids, aromatic compounds, pectin, vitamins, anthocyanins, and polyphenols, serves as a valuable source oxidized by microorganisms to generate electricity. Studies have demonstrated the production of electrical energy in double-chamber MFCs employing *Chlorella vulgaris* (Huarachi-Olivera et al. 2018). Other researchers have utilized microalgae for both the anodic and cathodic chambers, effectively generating bioelectricity (De La et al. 2022; Nookwam et al. 2022). The green microalgae *Chlamydomonas* sp. TRC-1 has been isolated from the effluent treatment plant of a textile dyeing mill, demonstrating bioelectricity generation from algal biomass after remediation (Behl et al. 2020). A three-chamber microalgal bio-electrochemical system has shown promise for simultaneous cultivation of *C. vulgaris* and treatment of industrial organic wastewater, with increased electric current improving COD and total Kjeldahl nitrogen (TKN) removal (Pan et al. 2021).

4 Microalgae as Food, Pigment, and Biostimulant

4.1 Food

Microalgal farming, encompassing eukaryotes and photosynthetic cyanobacteria, yields potential biomass, and the diverse species allow the production of various products, including pigment, antioxidant, and polyunsaturated fatty acid. The resulting derivatives find applications as food supplement, ingredients, or additives (Uribe-Wandurraga et al. 2021). Developments in the degradation of biomass obtained from microalgae result in intriguing dishes and components that are utilized in bread, dough, and beverage compositions. In particular, microscopic algae are a prospective resource for bioaccessible peptides. Notable dietary sources of omega 3 fatty acids include *Phaeodactylum tricornutum*, *Chlorella*, and *Nannochloropsis oculate.*

Microalgae, with a historical role in human nutrition, have varieties like *Nostoc* and *Spirulina* with records as food sources before 1900 (Borowitzka 2018). Microalgal biomass have been used as foodstuff with health benefits for the past several years, mostly as components, components, and dietary supplementation. For instance, fruit juices, milk, and other liquids that are fortified with docosahexaenoic acid (DHA) are particularly beneficial to babies and kids (Araújo et al. 2021). Around 15,000 novel food items incorporating algae or substances made with algae have been documented globally between 2015 and 2019, while 79% of those items being foods and 21% being drinks (Boukid and Castellari 2021). Spirulina, rich in γ-linolenic acid, is utilized for its nutraceutical properties, including lowering low-density lipoprotein in high-cholesterol individuals and increasing vitamin B1 content (Hayes et al. 2017). The amino acids obtained from microalgae are superior to those from other proteinaceous meals, and microalgae may be manipulated to create different types of protein more effectively than from other protein sources (Kusmayadi et al. 2021).

4.2 Pigments

Recognized for their eco-friendly attributes, microalgal pigments have gained prominence as colorant, nutraceutical, and antioxidant with considerable financial significance. They are extensively applied in diverse sectors such as food, dairy, alternative foods, nutritional supplements, pharmaceutical, cosmetics, aquaculture, textiles, and other industrial sectors (Heydarizadeh et al. 2013). Presently, these pigments have undergone varying degrees of development, with some achieving productive industrial output and substantial financial benefits. The need for colorants in food, whether naturally sourced or artificially produced, is crucial for enhancing the visual appeal of products. Recent trends indicate a growing preference for naturally sourced colorants due to health concerns linked to artificial options, including instances of attention deficit hyperactivity disorder and hyperactivity. Within the production process, food colorants play a pivotal role by masking undesirable attributes. They fall into categories like dye, pigment, or specific substances that give color to foods, medications, cosmetics, or the human body either directly or indirectly. In contrast to plants in the terrestrial, microalgae exhibit higher productivities owing to accelerated growth rate, cost-effectiveness, and sustainable cultivation using seawater or wastewater, particularly given the limited availability of freshwater. The resulting biomass is well-suited for biorefinery applications, concurrently producing several other valuable macromolecules (Nwoba et al. 2019).

Microalgae are distinguished by their unique hue, which is a product of many pigments. These pigments may be categorized into three main groups: fat-soluble carotenoids, water-soluble phycobilin, and fat-soluble chlorophylls. These pigments are crucial to the mechanism of algal photosynthesis. The use of pigments generated from microalgae is growing in popularity as a natural substitute. They are full of health-promoting elements with immune-suppressive, neurologically protective, rejuvenation, and vitamin source properties (Christaki et al. 2015). Due to the health issues associated with artificial pigments, there is an increasing desire for natural colors, and microalgal pigments appear as perfect substitutes (Parmar and Singh 2018). Phycocyanin, a microalgal pigment, is frequently employed as a coloring factor in a variety of culinary products, including dairy products, popsicles, ice creams, chewing gum, and jellies (Begum et al. 2016). Pauline et al. (2006) claimed that the microalgal pigment β-carotene strengthens immune responses and protects against glaucoma and allergy reactions. In addition to being utilized in food, the aforementioned pigments may be found in dietary supplements and nutraceuticals. Particularly chlorophylls are consumed because of their remarkable bioactive properties as dietary supplements and antioxidants (Sun et al. 2023).

4.3 Biostimulants

One key tactic for promoting environmentally friendly agricultural practices is the use of natural plant biostimulants into farming operations. These plant growth enhancers function as signaling molecules and successfully promote the development and growth of plants. Biostimulants, composed of natural elements like polyamines, amino acids, humic compounds, and algal extracts, play a crucial role in enhancing plants' physiological responses, encompassing improved nutrient absorption, stress resilience, yield, and quality, as stated by Kapoore et al. (2021). The rapid growth of the commercial biostimulants market is attributed to the exceptional capabilities of microalgae biostimulants, promoting plant growth and increasing resistance to abiotic stresses, as highlighted by Santini et al. (2021). Microalgae serve as effective biostimulants, generating diverse bioactive compounds, including proteins, vitamins, carbohydrates, amino acids, polysaccharides, and phytohormones, contributing to their positive impact on plant development (Álvarez-González et al. 2023). Recent popularity surrounds natural biostimulants derived from seaweeds, offering growth-promoting components such as vitamins, cytokinins, auxins, amino acids, and carbohydrates. These substances, influencing cellular metabolism, enhance plant resilience to external stresses, resulting in improved nutrient absorption, seed germination, root development, antioxidant properties, high-quality leaves, and resistance to infections (Bello et al. 2021; González-Pérez et al. 2022). While these substances may work synergistically, their precise mechanisms of action are still not well understood (Mazepa et al. 2021). Microalgal concentrates can be utilized as biostimulants to encourage plant development and lessen biotic and abiotic stresses as they are known to produce intrinsic hormones generated from plants and hormone-related components (Lu and Xu 2015). Auxin-like activity and cytokinins have been found in axenic unicellular strains of *Chlorophyta* and *Cyanophyta* (Tan et al. 2021), whereas microalgae strains belonging to the *Protococcus*, *Chlorella*, and *Scenedesmus* genera have been examined for cytokinins. Potential plant biostimulants include chlorophytes and cyanobacteria, which are rich sources of bioactive compounds including proteins, carbohydrates, lipids, amino acids, vitamins, pigments, and phytohormones (Kapoore et al. 2021). A combination of cyanobacteria and chlorophytes, for example, showed notable quantities of protein, fat, and carbohydrates, suggesting their role as early progenitors of biostimulant compounds (Supraja et al. 2020). When algal biomass was added, different effluent conditions revealed different accumulations of proteins (Choudhary et al. 2017). Auxins, cytokinins, abscisic acid, ethylene, and gibberellins are examples of phytohormones that are necessary for low-quantity physiological and biochemical responses in plants (Han et al. 2018). Microalgae contain compounds that promote plant development, such as polyamines, vitamins, cytokinins, betaines, auxins, and amino acids (Munaro et al. 2021; Parmar et al. 2023). Microalgal protein hydrolysates and amino acids positively affect the development and growth of crop plants. Microalgal extracts containing phytohormones offer new prospects for phytohormone applications in crop production, expanding microalgae valorization (Renuka et al. 2018). These small molecules serve as chemical messengers, regulating cellular activities and influencing metabolic processes in crops (Gonçalves 2021). Phytohormone contents in microalgae strains, such as *Scenedesmus* spp. and *Arthrospira* spp., suggest potential applications in increasing plant growth and development (Plaza et al. 2018). Beyond phytohormones, microalgae have the potential for increasing plant production because they contain a variety of compounds that promote plant development, including protein, lipid, polysaccharide, and phytohormones (Suriya et al. 2018). Intracellular extracts from microalgae, obtained through physical, mechanical, and enzymatic methods, can enhance seed pretreatment and foliar spray applications, resulting in improved germination, plant growth, and nutrient content (Chiaiese et al. 2018). Leaf surface application is emerging as a critical substitute for effective nutrient uptake, as it involves the absorption of biostimulants through stomata particularly during daylight hours (Ronga et al. 2019). Noteworthy enhancements in plant growth, shoot and root development, and early flowering have been reported with the foliar application of *Scenedesmus* sp. extracts (Plaza et al. 2018). Investigations are ongoing regarding the diverse substances present in microalgal extracts, encompassing proteins, lipids, polysaccharides, and phytohormones (Corbellini et al. 2020; Garcia-Gonzalez and Sommerfeld 2016). These microalgal extracts serve various purposes, contributing to improved crop growth, yield, and nutritional status. Utilizing these compounds as biostimulants holds the potential to foster efficient and environmentally friendly farming practices.

5 Challenges and Futuristic Perspectives

Microalgal bioenergy production stands out as a clean and sustainable process, yet it grapples with several challenges impeding its commercialization. These challenges encompass strain selection, water requirements, biomass harvesting, feedstock availability during adverse weather conditions, and the choice of a suitable biofuel conversion process. While numerous literatures have explored algal biorefinery approaches addressing integrated algal wastewater treatment, bioenergy production, and waste valorization, most of these approaches remain confined to lab-scale studies, raising uncertainty about their sustainability at a commercial scale (Chandra et al. 2019; De Bhowmick et al. 2019). Beyond feedstock limitations, addressing challenges such as waste recovery and the impact of diverse various harvesting methods on effluent recycling emerges as crucial at a large scale.

In response to these challenges, a futuristic microalgal biorefinery approach is proposed, aiming to overcome energy efficiency and sustainability hurdles. This approach emphasizes the reduction, recycling, and reuse of liquid effluents and solid leftovers within a circular bioeconomy. Research indicate that an advanced biorefinery, combined with integrated wastewater treatment and commercial CO_2 sequestration, might be a reality in the future, holds the potential to facilitate bioremediation, and mitigates greenhouse gas emissions. Utilizing mixed-biomass feedstocks, including lignocellulosic waste, food waste, agricultural waste, and sewage waste, is seen as instrumental for maintaining uninterrupted plant operation (Brilman et al. 2017; Wang et al. 2019). While exploring new potent species may offer promise in terms of growth kinetics and biomass yield, the practicality of these investigations appears less attractive for addressing current needs. The focus should shift toward addressing technological issues with the powerful species now in use, especially considering that only 100 out of approximately 30,000 identified microalgal species are commercially targeted (Su et al. 2017). Optimizing equipment capacity and selecting proper construction materials are crucial to minimizing manufacturing expenses. The impact of water consumption, CO_2, and nutrient sources on profitability underscores the importance of utilizing nutrients and CO_2 from industrial wastes for microalgae biomass production, reducing processing costs and environmental impact (Ganesh Saratale et al. 2022). Additionally, the selection of potential strains with autosedimentation properties is proposed to avoid energy-consuming harvesting methods, enhancing the efficiency of effluent recycling for cultivation and biofuel production.

Improving the performance of algal seeds to increase the accumulation of organic matter for biofuel preparation and developing cost-effective methods for algae collection are deemed essential. The concept of a synthetic "cell factory" comprising photosynthetic microalgae through synthetic biology technologies is considered a pivotal step toward the industrialization of microalgae biofuels. Genetic engineering of microalgae is underway to produce robust strains with enhanced economic viability. Despite advancements in DNA synthesis, tools for genetic manipulation, and functional genomes, the lack of principles for genetic strain design remains a limiting factor in the field's progress. Prospective investigations ought to concentrate on creating strong, purpose-based bioengineered strains with an emphasis on CO_2 fixation, higher biomass earnings, and improved photosynthesis performance. While genetic manipulation holds immense potential, overcoming limitations in genetic knowledge is crucial for advancing this field.

6 Conclusion

In summary, the investigation of microalgae in the manufacture of high-end enhanced value products employing a zero-waste biorefinery technique is a groundbreaking effort toward sustainability. The opening established the tone for the next parts by highlighting the global trend toward eco-friendly methods. The many uses of microalgae were described in detail; they included biofuels, medications, nutraceuticals, and energy sources such as biodiesel, biogas, bioethanol, and bioelectricity. In addition, the organisms demonstrated their adaptability in producing food, pigment, and biostimulants, which supported a circular bioeconomy. Notwithstanding these assurances, difficulties with downstream processing, cost-effectiveness, and scalability were noted. The importance of futuristic viewpoints was underlined in the conclusion, which also highlighted the need for study, technology development, and teamwork to overcome these obstacles. The goal is a sustainable future driven by a zero-waste biorefinery model, in which microalgae are essential to meeting the world's energy and food demands. This expedition represents a meeting point of sustainability and innovation, presenting microalgae as drivers of a robust and ecologically responsible bioeconomy.

References

Acebu PIG, de Luna MDG, Chen CY, Abarca RRM, Chen JH, Chang JS (2022) Bioethanol production from Chlorella vulgaris ESP-31 grown in unsterilized swine wastewater. Bioresour Technol 352:127086. https://doi.org/10.1016/j.biortech.2022.127086

Ahmad I, Abdullah N, Koji I, Yuzir A, Eva Muhammad S (2021) Evolution of photobioreactors: A review based on microalgal perspective. IOP Conf Ser Mater Sci Eng 1142(1):012004. https://doi.org/10.1088/1757-899x/1142/1/012004

Ahmed SF, Mofijur M, Parisa TA, Islam N, Kusumo F, Inayat A, Le VG, Badruddin IA, Khan TMY, Ong HC (2022) Progress and challenges of contaminate removal from wastewater using microalgae biomass. Chemosphere 286:131656. https://doi.org/10.1016/j.chemosphere.2021.131656

Alaswad A, Dassisti M, Prescott T, Olabi AG (2015) Technologies and developments of third generation biofuel production. Renew Sust Energy Rev 51:1446. https://doi.org/10.1016/j.rser.2015.07.058

Álvarez-González A, Greque de Morais E, Planas-Carbonell A, Uggetti E (2023) Enhancing sustainability through microalgae cultivation in urban wastewater for biostimulant production and nutrient recovery. Sci Total Environ 904:166878. https://doi.org/10.1016/j.scitotenv.2023.166878

Apone F, Barbulova A, Colucci MG (2019) Plant and microalgae derived peptides are advantageously employed as bioactive compounds in cosmetics. Front Plant Sci 10:756. https://doi.org/10.3389/fpls.2019.00756

Araújo R, Vázquez Calderón F, Sánchez López J, Azevedo IC, Bruhn A, Fluch S, Garcia Tasende M, Ghaderiardakani F, Ilmjärv T, Laurans M, Mac Monagail M, Mangini S, Peteiro C, Rebours C, Stefansson T, Ullmann J (2021) Current status of the algae production industry in Europe: an emerging sector of the blue bioeconomy. Front Mar Sci 7:626389. https://doi.org/10.3389/fmars.2020.626389

Ayala-Parra P, Liu Y, Field JA, Sierra-Alvarez R (2017) Nutrient recovery and biogas generation from the anaerobic digestion of waste biomass from algal biofuel production. Renew Energy 108:410. https://doi.org/10.1016/j.renene.2017.02.085

Bauer LM, Costa JAV, da Rosa APC, Santos LO (2017) Growth stimulation and synthesis of lipids, pigments and antioxidants with magnetic fields in Chlorella kessleri cultivations. Bioresour Technol 244:1425. https://doi.org/10.1016/j.biortech.2017.06.036

Becker EW (2013) Microalgae for human and animal nutrition. In: Handbook of microalgal culture: applied phycology and biotechnology: second edition. Wiley, New York. https://doi.org/10.1002/9781118567166.ch25

Begum H, Yusoff FMD, Banerjee S, Khatoon H, Shariff M (2016) Availability and utilization of pigments from microalgae. Crit Rev Food Sci Nutr 56(13):2209. https://doi.org/10.1080/10408398.2013.764841

Behera B, Unpaprom Y, Ramaraj R, Maniam GP, Govindan N, Paramasivan B (2021) Integrated biomolecular and bioprocess engineering strategies for enhancing the lipid yield from microalgae. Renew Sust Energ Rev 148:111270. https://doi.org/10.1016/j.rser.2021.111270

Behl K, SeshaCharan P, Joshi M, Sharma M, Mathur A, Kareya MS, Jutur PP, Bhatnagar A, Nigam S (2020) Multifaceted applications of isolated microalgae Chlamydomonas sp. TRC-1 in wastewater remediation, lipid production and bioelectricity generation. Bioresour Technol 304:122993. https://doi.org/10.1016/j.biortech.2020.122993

Bello AS, Saadaoui I, Ben-Hamadou R (2021) "Beyond the source of bioenergy": microalgae in modern agriculture as a biostimulant, biofertilizer, and anti-abiotic stress. Agronomy 11(8):1610. https://doi.org/10.3390/agronomy11081610

Bernaerts TMM, Gheysen L, Foubert I, Hendrickx ME, Van Loey AM (2019) The potential of microalgae and their biopolymers as structuring ingredients in food: a review. Biotechnol Adv 37(8):107419. https://doi.org/10.1016/j.biotechadv.2019.107419

Berthon JY, Nachat-Kappes R, Bey M, Cadoret JP, Renimel I, Filaire E (2017) Marine algae as attractive source to skin care. Free Radic Res 51(6):555–567. https://doi.org/10.1080/10715762.2017.1355550

Bibi R, Ahmad Z, Imran M, Hussain S, Ditta A, Mahmood S, Khalid A (2017) Algal bioethanol production technology: a trend towards sustainable development. Renew Sust Energ Rev 71:976. https://doi.org/10.1016/j.rser.2016.12.126

Binod P, Gnansounou E, Sindhu R, Pandey A (2019) Enzymes for second generation biofuels: recent developments and future perspectives. Bioresour Technol Rep 5:317. https://doi.org/10.1016/j.biteb.2018.06.005

Bohutskyi P, Betenbaugh MJ, Bouwer EJ (2014) The effects of alternative pretreatment strategies on anaerobic digestion and methane production from different algal strains. Bioresour Technol 155:366. https://doi.org/10.1016/j.biortech.2013.12.095

Borowitzka MA (2018) Microalgae in medicine and human health: a historical perspective. In: Microalgae in health and disease prevention. Elsevier, Amsterdam. https://doi.org/10.1016/B978-0-12-811405-6.00009-8

Boukid F, Castellari M (2021) Food and beverages containing algae and derived ingredients launched in the market from 2015 to 2019: a front-of-pack labeling perspective with a special focus on Spain. Food Secur 10(1):173. https://doi.org/10.3390/foods10010173

Brilman DWF, Drabik N, Wądrzyk M (2017) Hydrothermal co-liquefaction of microalgae, wood, and sugar beet pulp. Biomass Convers Biorefinery 7(4):445. https://doi.org/10.1007/s13399-017-0241-2

Carneiro MLNM, Pradelle F, Braga SL, Gomes MSP, Martins ARFA, Turkovics F, Pradelle RNC (2017) Potential of biofuels from algae: comparison with fossil fuels, ethanol and biodiesel in Europe and Brazil through life cycle assessment (LCA). Renew Sust Energ Rev 73:632. https://doi.org/10.1016/j.rser.2017.01.152

Chandra R, Iqbal HMN, Vishal G, Lee HS, Nagra S (2019) Algal biorefinery: a sustainable approach to valorize algal-based biomass towards multiple product recovery. Bioresour Technol 278:346. https://doi.org/10.1016/j.biortech.2019.01.104

Chandra R, Pradhan S, Patel A, Ghosh UK (2021) An approach for dairy wastewater remediation using mixture of microalgae and biodiesel production for sustainable transportation. J Environ Manag 297:113210. https://doi.org/10.1016/j.jenvman.2021.113210

Cheirsilp B, Maneechote W (2022) Insight on zero waste approach for sustainable microalgae biorefinery: sequential fractionation, conversion and applications for high-to-low value-added products. Bioresour Technol Rep 18:101003. https://doi.org/10.1016/j.biteb.2022.101003

Chew KW, Yap JY, Show PL, Suan NH, Juan JC, Ling TC, Lee DJ, Chang JS (2017) Microalgae biorefinery: high value products perspectives. Bioresour Technol 229:53. https://doi.org/10.1016/j.biortech.2017.01.006

Chiaiese P, Corrado G, Colla G, Kyriacou MC, Rouphael Y (2018) Renewable sources of plant biostimulation: microalgae as a sustainable means to improve crop performance. Front Plant Sci 9:1782. https://doi.org/10.3389/fpls.2018.01782

Choudhary P, Prajapati SK, Kumar P, Malik A, Pant KK (2017) Development and performance evaluation of an algal biofilm reactor for treatment of multiple wastewaters and characterization of biomass for diverse applications. Bioresour Technol 224:276. https://doi.org/10.1016/j.biortech.2016.10.078

Christaki E, Bonos E, Florou-Paneri P (2015) Innovative microalgae pigments as functional ingredients in nutrition. In: Handbook of marine microalgae: biotechnology advances. Elsevier, Amsterdam. https://doi.org/10.1016/B978-0-12-800776-1.00014-5

Corbellini JR, Ribas LLF, de Maia FR, Corrêa DO, Noseda MD, Suzuki RM, Amano É (2020) Effect of microalgae Messastrum gracile and Chlorella vulgaris on the in vitro propagation of orchid Cattleya labiata. J Appl Phycol 32(6):4013. https://doi.org/10.1007/s10811-020-02251-9

Costa JAV, Moraes L, Moreira JB, da Rosa GM, Henrard ASA, de Morais MG (2017) Microalgae-based biorefineries as a promising approach to biofuel production. In:

Prospects and challenges in algal biotechnology. Springer Singapore, Singapore. https://doi.org/10.1007/978-981-10-1950-0_4

Cristiane RL, Aline MP, Jorge AVC (2016) Biopeptides with antioxidant activity extracted from the biomass of spirulina sp. LEB 18. Afr J Microbiol Res 10(3):79. https://doi.org/10.5897/ajmr2015.7760

Cuevas-Castillo GA, Navarro-Pineda FS, Baz Rodríguez SA, Sacramento Rivero JC (2020) Advances on the processing of microalgal biomass for energy-driven biorefineries. Renew Sust Energ Rev 125:109606. https://doi.org/10.1016/j.rser.2019.109606

De Bhowmick G, Koduru L, Sen R (2015) Metabolic pathway engineering towards enhancing microalgal lipid biosynthesis for biofuel application—a review. Renew Sust Energ Rev 50:1239. https://doi.org/10.1016/j.rser.2015.04.131

De Bhowmick G, Sarmah AK, Sen R (2018) Lignocellulosic biorefinery as a model for sustainable development of biofuels and value added products. Bioresour Technol 247:1144. https://doi.org/10.1016/j.biortech.2017.09.163

De Bhowmick G, Sarmah AK, Sen R (2019) Zero-waste algal biorefinery for bioenergy and biochar: a green leap towards achieving energy and environmental sustainability. Sci Total Environ 650:2467. https://doi.org/10.1016/j.scitotenv.2018.10.002

de Farias Silva CE, Bertucco A (2019) Bioethanol from microalgal biomass: a promising approach in biorefinery. Braz Arch Biol Technol 62:1–14. https://doi.org/10.1590/1678-4324-2019160816

De La CNM, Nazario-Naveda R, Benites SM, Rojas-Flores S, Delfín-Narciso D, Rojas-Villacorta W, Díaz F (2022) Potential use of mango waste and microalgae spirulina sp. for bioelectricity generation. Environ Res Eng Manag 78(3):129. https://doi.org/10.5755/j01.erem.78.3.31117

Dębowski M, Zieliński M, Kazimierowicz J, Kujawska N, Talbierz S (2020) Microalgae cultivation technologies as an opportunity for bioenergetic system development—advantages and limitations. Sustainability 12(23):9980. https://doi.org/10.3390/su12239980

Deviram G, Mathimani T, Anto S, Ahamed TS, Ananth DA, Pugazhendhi A (2020) Applications of microalgal and cyanobacterial biomass on a way to safe, cleaner and a sustainable environment. J Clean Prod 253:119770. https://doi.org/10.1016/j.jclepro.2019.119770

Dolganyuk V, Belova D, Babich O, Prosekov A, Ivanova S, Katserov D, Patyukov N, Sukhikh S (2020) Microalgae: a promising source of valuable bioproducts. Biomolecules 10(8):1153. https://doi.org/10.3390/biom10081153

Dourou M, Dritsas P, Baeshen MN, Elazzazy A, Al-Farga A, Aggelis G (2021) High-added value products from microalgae and prospects of aquaculture wastewaters as microalgae growth media. FEMS Microbiol Lett 367(12):fnaa081. https://doi.org/10.1093/FEMSLE/FNAA081

Elshobary ME, Zabed HM, Yun J, Zhang G, Qi X (2021) Recent insights into microalgae-assisted microbial fuel cells for generating sustainable bioelectricity. Int J Hydrog Energy 46(4):3135. https://doi.org/10.1016/j.ijhydene.2020.06.251

Enamala MK, Enamala S, Chavali M, Donepudi J, Yadavalli R, Kolapalli B, Aradhyula TV, Velpuri J, Kuppam C (2018) Production of biofuels from microalgae—$ review on cultivation, harvesting, lipid extraction, and numerous applications of microalgae. Renew Sust Energ Rev 94:49. https://doi.org/10.1016/j.rser.2018.05.012

Enamala MK, Dixit R, Tangellapally A, Singh M, Dinakarrao SMP, Chavali M, Pamanji SR, Ashokkumar V, Kadier A, Chandrasekhar K (2020) Photosynthetic microorganisms (algae) mediated bioelectricity generation in microbial fuel cell: concise review. Environ Technol Innov 19:100959. https://doi.org/10.1016/j.eti.2020.100959

Fallahi A, Rezvani F, Asgharnejad H, Khorshidi E, Hajinajaf N, Higgins B (2021) Interactions of microalgae-bacteria consortia for nutrient removal from wastewater: a review. Chemosphere 272:129878. https://doi.org/10.1016/j.chemosphere.2021.129878

Fu W, Nelson DR, Yi Z, Xu M, Khraiwesh B, Jijakli K, Chaiboonchoe A, Alzahmi A, Al-Khairy D, Brynjolfsson S, Salehi-Ashtiani K (2017) Bioactive compounds from microalgae: current development and prospects. Stud Nat Prod Chem 54:199–225. https://doi.org/10.1016/B978-0-444-63929-5.00006-1

Gallego R, Montero L, Cifuentes A, Ibáñez E, Herrero M (2018) Green extraction of bioactive compounds from microalgae. J Anal Testing 2(2):109. https://doi.org/10.1007/s41664-018-0061-9
Gallego R, Martínez M, Cifuentes A, Ibáñez E, Herrero M (2019) Development of a green downstream process for the valorization of Porphyridium cruentum biomass. Molecules 24(8):1564. https://doi.org/10.3390/molecules24081564
Ganesh Saratale R, Ponnusamy VK, Jeyakumar RB, Sirohi R, Piechota G, Shobana S, Dharmaraja J, Lay CH, Dattatraya Saratale G, Seung Shin H, Ashokkumar V (2022) Microalgae cultivation strategies using cost–effective nutrient sources: recent updates and progress towards biofuel production. Bioresour Technol 361:127691. https://doi.org/10.1016/j.biortech.2022.127691
Garcia-Gonzalez J, Sommerfeld M (2016) Biofertilizer and biostimulant properties of the microalga Acutodesmus dimorphus. J Appl Phycol 28(2):1051. https://doi.org/10.1007/s10811-015-0625-2
Gonçalves AL (2021) The use of microalgae and cyanobacteria in the improvement of agricultural practices: a review on their biofertilising, biostimulating and biopesticide roles. Appl Sci 11(2):871. https://doi.org/10.3390/app11020871
González-González LM, Correa DF, Ryan S, Jensen PD, Pratt S, Schenk PM (2018) Integrated biodiesel and biogas production from microalgae: towards a sustainable closed loop through nutrient recycling. Renew Sust Energ Rev 82:1137. https://doi.org/10.1016/j.rser.2017.09.091
González-Pérez BK, Rivas-Castillo AM, Valdez-Calderón A, Gayosso-Morales MA (2022) Microalgae as biostimulants: a new approach in agriculture. World J Microbiol Biotechnol 38(1):4. https://doi.org/10.1007/s11274-021-03192-2
Goswami L, Kayalvizhi R, Dikshit PK, Sherpa KC, Roy S, Kushwaha A, Kim BS, Banerjee R, Jacob S, Rajak RC (2022) A critical review on prospects of bio-refinery products from second and third generation biomasses. Chem Eng J 448:137677. https://doi.org/10.1016/j.cej.2022.137677
Hamed I (2016) The evolution and versatility of microalgal biotechnology: a review. Compr Rev Food Sci Food Saf 15(6):1104. https://doi.org/10.1111/1541-4337.12227
Hamed SM, Raut MP, Jaffé SRP, Wright PC (2017) Evaluation of the effect of aerobic–anaerobic conditions on photohydrogen and chlorophyll a production by environmental Egyptian cyanobacterial and green algal species. Int J Hydrog Energy 42(10):6567. https://doi.org/10.1016/j.ijhydene.2017.01.058
Han X, Zeng H, Bartocci P, Fantozzi F, Yan Y (2018) Phytohormones and effects on growth and metabolites of microalgae: a review. Fermentation 4(2):25. https://doi.org/10.3390/fermentation4020025
Hayes M, Skomedal H, Skjånes K, Mazur-Marzec H, Toruńska-Sitarz A, Catala M, Isleten Hosoglu M, García-Vaquero M (2017) Microalgal proteins for feed, food and health. In: Microalgae-based biofuels and bioproducts: from feedstock cultivation to end-products. Elsevier, Amsterdam. https://doi.org/10.1016/B978-0-08-101023-5.00015-7
Heydarizadeh P, Poirier I, Loizeau D, Ulmann L, Mimouni V, Schoefs B, Bertrand M (2013) Plastids of marine phytoplankton produce bioactive pigments and lipids. Mar Drugs 11(9):3425. https://doi.org/10.3390/md11093425
Hou Q, Nie C, Pei H, Hu W, Jiang L, Yang Z (2016) The effect of algae species on the bioelectricity and biodiesel generation through open-air cathode microbial fuel cell with kitchen waste anaerobically digested effluent as substrate. Bioresour Technol 218:902. https://doi.org/10.1016/j.biortech.2016.07.035
Huarachi-Olivera R, Dueñas-Gonza A, Yapo-Pari U, Vega P, Romero-Ugarte M, Tapia J, Molina L, Lazarte-Rivera A, Pacheco-Salazar DG, Esparza M (2018) Bioelectrogenesis with microbial fuel cells (MFCs) using the microalga Chlorella vulgaris and bacterial communities. Electron J Biotechnol 31:34. https://doi.org/10.1016/j.ejbt.2017.10.013
Hussain F, Shah SZ, Ahmad H, Abubshait SA, Abubshait HA, Laref A, Manikandan A, Kusuma HS, Iqbal M (2021) Microalgae an ecofriendly and sustainable wastewater treatment option: biomass application in biofuel and bio-fertilizer production. A review. Renew Sust Energy Rev 137:110603. https://doi.org/10.1016/j.rser.2020.110603

Javed F, Aslam M, Rashid N, Shamair Z, Khan AL, Yasin M, Fazal T, Hafeez A, Rehman F, Rehman MSU, Khan Z, Iqbal J, Bazmi AA (2019) Microalgae-based biofuels, resource recovery and wastewater treatment: a pathway towards sustainable biorefinery. Fuel 255:115826. https://doi.org/10.1016/j.fuel.2019.115826

Jayaseelan M, Usman M, Somanathan A, Palani S, Muniappan G, Jeyakumar RB (2021) Microalgal production of biofuels integrated with wastewater treatment. Sustainability 13(16):8797. https://doi.org/10.3390/su13168797

Jiang K, Baburin IA, Han P, Yang C, Fu X, Yao Y, Li J, Cánovas E, Seifert G, Chen J, Bonn M, Feng X, Zhuang X (2020) Interfacial approach toward benzene-bridged polypyrrole film-based micro-supercapacitors with ultrahigh volumetric power density. Adv Funct Mater 30(7):1908243. https://doi.org/10.1002/adfm.201908243

Kapoore RV, Wood EE, Llewellyn CA (2021) Algae biostimulants: a critical look at microalgal biostimulants for sustainable agricultural practices. Biotechnol Adv 49:107754. https://doi.org/10.1016/j.biotechadv.2021.107754

Katiyar R, Arora A (2020) Health promoting functional lipids from microalgae pool: a review. Algal Res 46:101800. https://doi.org/10.1016/j.algal.2020.101800

Khanra S, Mondal M, Halder G, Tiwari ON, Gayen K, Bhowmick TK (2018) Downstream processing of microalgae for pigments, protein and carbohydrate in industrial application: a review. Food Bioprod Process 110:60. https://doi.org/10.1016/j.fbp.2018.02.002

Khoo KS, Lee SY, Ooi CW, Fu X, Miao X, Ling TC, Show PL (2019) Recent advances in biorefinery of astaxanthin from Haematococcus pluvialis. Bioresour Technol 288:121606. https://doi.org/10.1016/j.biortech.2019.121606

Kida ZH (2023) Production, characterization and optimization of bioethanol from microalgae obtained from waste water in Maiduguri metropolitan council, Borno state-Nigeria. Arid Zone J Basic Appl Res 2(2):10–22. https://doi.org/10.55639/607.191817

Kiran BR, Venkata Mohan S (2021) Microalgal cell biofactory—therapeutic, nutraceutical and functional food applications. Plants 10(5):836. https://doi.org/10.3390/plants10050836

Kiran B, Kumar R, Deshmukh D (2014) Perspectives of microalgal biofuels as a renewable source of energy. Energy Convers Manag 88:1228. https://doi.org/10.1016/j.enconman.2014.06.022

Kossatz HL, Rose SH, Viljoen-Bloom M, van Zyl WH (2017) Production of ethanol from steam exploded triticale straw in a simultaneous saccharification and fermentation process. Process Biochem 53:10. https://doi.org/10.1016/j.procbio.2016.11.023

Kumar B, Verma P (2020) Techno-economic assessment of biomass-based integrated biorefinery for energy and value-added product. In: Biorefineries: a step towards renewable and clean energy. clean energy production technologies. Springer, Singapore. https://doi.org/10.1007/978-981-15-9593-6_23

Kusmayadi A, Leong YK, Yen H-W, Huang C-Y, Chang J-S (2021) Microalgae as sustainable food and feed sources for animals and humans—biotechnological and environmental aspects. Chemosphere 271:129800. https://doi.org/10.1016/j.chemosphere.2021.129800

Lam MK, Khoo CG, Lee KT (2018) Scale-up and commercialization of algal cultivation and biofuels production. In: Biomass, biofuels, biochemicals: biofuels from algae, second edition. Elsevier, Amsterdam. https://doi.org/10.1016/B978-0-444-64192-2.00019-6

Li S, Ji L, Shi Q, Wu H, Fan J (2019a) Advances in the production of bioactive substances from marine unicellular microalgae Porphyridium spp. Bioresour Technol 292:122048. https://doi.org/10.1016/j.biortech.2019.122048

Li M, Zhou M, Tan C, Tian X (2019b) Enhancement of CO2 biofixation and bioenergy generation using a novel airlift type photosynthetic microbial fuel cell. Bioresour Technol 272:501. https://doi.org/10.1016/j.biortech.2018.10.078

Liu J, Huang J, Jiang Y, Chen F (2012) Molasses-based growth and production of oil and astaxanthin by Chlorella zofingiensis. Bioresour Technol 107:393. https://doi.org/10.1016/j.biortech.2011.12.047

Lu Y, Xu J (2015) Phytohormones in microalgae: a new opportunity for microalgal biotechnology? Trends Plant Sci 20(5):273–282. https://doi.org/10.1016/j.tplants.2015.01.006

Malik S, Shahid A, Haider MN, Amin M, Betenbaugh MJ, Mehmood MA, Haq MAU, Syafiuddin A, Boopathy R (2022) Prospects of multiproduct algal biorefineries involving cascading processing of the biomass employing a zero-waste approach. Curr Pollut Rep 8(2):147. https://doi.org/10.1007/s40726-022-00213-y

Manaf ISA, Embong NH, Khazaai SNM, Rahim MHA, Yusoff MM, Lee KT, Maniam GP (2019) A review for key challenges of the development of biodiesel industry. Energy Convers Manag 185:508. https://doi.org/10.1016/j.enconman.2019.02.019

Mandha J, Shumoy H, Matemu AO, Raes K (2021) Valorization of mango by-products to enhance the nutritional content of maize complementary porridges. Food Secur 10(7):1635. https://doi.org/10.3390/foods10071635

Manhongo TT, Chimphango A, Thornley P, Röder M (2021) Techno-economic and environmental evaluation of integrated mango waste biorefineries. J Clean Prod 325:129335. https://doi.org/10.1016/j.jclepro.2021.129335

Mani Rathnam V, Madras G (2019) Conversion of Shizochitrium limacinum microalgae to biodiesel by non-catalytic transesterification using various supercritical fluids. Bioresour Technol 288:121538. https://doi.org/10.1016/j.biortech.2019.121538

Markou G, Ilkiv B, Brulé M, Antonopoulos D, Chakalis L, Arapoglou D, Chatzipavlidis I (2022) Methane production through anaerobic digestion of residual microalgal biomass after the extraction of valuable compounds. Biomass Convers Biorefinery 12(2):419. https://doi.org/10.1007/s13399-020-00703-3

Mathimani T, Pugazhendhi A (2019) Utilization of algae for biofuel, bio-products and bioremediation. Biocatal Agric Biotechnol 17:326. https://doi.org/10.1016/j.bcab.2018.12.007

Mazepa E, Malburg BV, Mógor G, de Oliveira AC, Amatussi JO, Corrêa DO, Lemos JS, Ducatti DRB, Duarte MER, Mógor ÁF, Noseda MD (2021) Plant growth biostimulant activity of the green microalga Desmodesmus subspicatus. Algal Res 59:102434. https://doi.org/10.1016/j.algal.2021.102434

Mishra S, Roy M, Mohanty K (2019) Microalgal bioenergy production under zero-waste biorefinery approach: recent advances and future perspectives. Bioresour Technol 292:122008. https://doi.org/10.1016/j.biortech.2019.122008

Mobin SMA, Alam F (2018) A review of microalgal biofuels, challenges and future directions. In: Application of thermo-fluid processes in energy systems, Green energy and technology. Springer, Singapore. https://doi.org/10.1007/978-981-10-0697-5_4

Mohammed ASA, Naveed M, Jost N (2021) Polysaccharides; classification, chemical properties, and future perspective applications in fields of pharmacology and biological medicine (a review of current applications and upcoming potentialities). J Polym Environ 29(8):2359. https://doi.org/10.1007/s10924-021-02052-2

Montalvo GEB, Thomaz-Soccol V, Vandenberghe LPS, Carvalho JC, Faulds CB, Bertrand E, Prado MRM, Bonatto SJR, Soccol CR (2019) Arthrospira maxima OF15 biomass cultivation at laboratory and pilot scale from sugarcane vinasse for potential biological new peptides production. Bioresour Technol 273:103. https://doi.org/10.1016/j.biortech.2018.10.081

Moraes L, da Rosa GM, Cardias BB, dos Santos LO, Costa JAV (2016) Microalgal biotechnology for greenhouse gas control: carbon dioxide fixation by Spirulina sp. at different diffusers. Ecol Eng 91:426. https://doi.org/10.1016/j.ecoleng.2016.02.035

Moreno-Garcia L, Gariépy Y, Barnabé S, Raghavan V (2021) Factors affecting the fatty acid profile of wastewater-grown-algae oil as feedstock for biodiesel. Fuel 304:121367. https://doi.org/10.1016/j.fuel.2021.121367

Morris J, Hone D, Haigh M, Sokolov A, Paltsev S (2023) Future energy: in search of a scenario reflecting current and future pressures and trends. Environ Econ Policy Stud 25(1):31. https://doi.org/10.1007/s10018-021-00339-1

Mostafa SSM, El-Gendy NS (2017) Evaluation of fuel properties for microalgae Spirulina platensis bio-diesel and its blends with Egyptian petro-diesel. Arab J Chem 10:S2040. https://doi.org/10.1016/j.arabjc.2013.07.034

Mourelle ML, Gómez CP, Legido JL (2017) The potential use of marine microalgae and cyanobacteria in cosmetics and thalassotherapy. Cosmetics 4(4):46. https://doi.org/10.3390/cosmetics4040046

Mulders KJM, Weesepoel Y, Lamers PP, Vincken JP, Martens DE, Wijffels RH (2013) Growth and pigment accumulation in nutrient-depleted Isochrysis aff. galbana T-ISO. J Appl Phycol 25(5):1421. https://doi.org/10.1007/s10811-012-9954-6

Munaro D, Nunes A, Schmitz C, Bauer C, Coelho DS, Oliveira ER, Yunes RA, Moura S, Maraschin M (2021) Metabolites produced by macro- and microalgae as plant biostimulants. Stud Nat Prod Chem 71:87–120. https://doi.org/10.1016/B978-0-323-91095-8.00011-8

Mussgnug JH, Klassen V, Schlüter A, Kruse O (2010) Microalgae as substrates for fermentative biogas production in a combined biorefinery concept. J Biotechnol 150(1):51. https://doi.org/10.1016/j.jbiotec.2010.07.030

Naina Mohamed S, Ajit Hiraman P, Muthukumar K, Jayabalan T (2020) Bioelectricity production from kitchen wastewater using microbial fuel cell with photosynthetic algal cathode. Bioresour Technol 295:122226. https://doi.org/10.1016/j.biortech.2019.122226

Nayak M, Thirunavoukkarasu M, Mohanty RC (2016) Cultivation of freshwater microalga Scenedesmus sp. using a low-cost inorganic fertilizer for enhanced biomass and lipid yield. J Gen Appl Microbiol 62(1):7. https://doi.org/10.2323/jgam.62.7

Niphadkar S, Bagade P, Ahmed S (2018) Bioethanol production: insight into past, present and future perspectives. Biofuels 9(2):229. https://doi.org/10.1080/17597269.2017.1334338

Nookwam K, Cheirsilp B, Maneechote W, Boonsawang P, Sukkasem C (2022) Microbial fuel cells with photosynthetic-cathodic chamber in vertical cascade for integrated bioelectricity, biodiesel feedstock production and wastewater treatment. Bioresour Technol 346:126559. https://doi.org/10.1016/j.biortech.2021.126559

Nwoba EG, Parlevliet DA, Laird DW, Alameh K, Moheimani NR (2019) Light management technologies for increasing algal photobioreactor efficiency. Algal Res 39:101433. https://doi.org/10.1016/j.algal.2019.101433

Nwoba EG, Ogbonna CN, Ishika T, Vadiveloo A (2020) Microalgal pigments: a source of natural food colors. In: Microalgae biotechnology for food, health and high value products. Springer, Berlin. https://doi.org/10.1007/978-981-15-0169-2_3

Olabi AG, Shehata N, Sayed ET, Rodriguez C, Anyanwu RC, Russell C, Abdelkareem MA (2023) Role of microalgae in achieving sustainable development goals and circular economy. Sci Total Environ 854:158689. https://doi.org/10.1016/j.scitotenv.2022.158689

Osundeko O, Ansolia P, Gupta SK, Bag P, Bajhaiya AK (2019) Promises and challenges of growing microalgae in wastewater. In: Water conservation, recycling and reuse: issues and challenges. Springer, Berlin. https://doi.org/10.1007/978-981-13-3179-4_2

Pagels F, Pereira RN, Vicente AA, Guedes AC (2021) Extraction of pigments from microalgae and cyanobacteria-a review on current methodologies. Appl Sci 11(11):5187. https://doi.org/10.3390/app11115187

Paliwal C, Mitra M, Bhayani K, Bharadwaj SVV, Ghosh T, Dubey S, Mishra S (2017) Abiotic stresses as tools for metabolites in microalgae. Bioresour Technol 244:1216. https://doi.org/10.1016/j.biortech.2017.05.058

Pan M, Su Y, Zhu X, Pan G, Zhang Y, Angelidaki I (2021) Bioelectrochemically assisted sustainable conversion of industrial organic wastewater and clean production of microalgal protein. Resour Conserv Recycl 168:105441. https://doi.org/10.1016/j.resconrec.2021.105441

Parascanu MM, Sanchez N, Sandoval-Salas F, Carreto CM, Soreanu G, Sanchez-Silva L (2021) Environmental and economic analysis of bioethanol production from sugarcane molasses and agave juice. Environ Sci Pollut Res 28(45):64374. https://doi.org/10.1007/s11356-021-15471-4

Park YH, Han SI, Oh B, Kim HS, Jeon MS, Kim S, Choi YE (2022) Microalgal secondary metabolite productions as a component of biorefinery: a review. Bioresour Technol 344:126206. https://doi.org/10.1016/j.biortech.2021.126206

Parmar RS, Singh C (2018) A comprehensive study of eco-friendly natural pigment and its applications. Biochem Biophys Rep 13:22. https://doi.org/10.1016/j.bbrep.2017.11.002

Parmar P, Kumar R, Neha Y, Srivatsan V (2023) Microalgae as next generation plant growth additives: functions, applications, challenges and circular bioeconomy based solutions. Front Plant Sci 14:1073546. https://doi.org/10.3389/fpls.2023.1073546

Pauline, Spolaore Claire, Joannis-Cassan Elie, Duran Arsène, Isambert (2006) Commercial applications of microalgae Journal of Bioscience and Bioengineering 101(2):87–96. https://doi.org/10.1263/jbb.101.87

Perin G, Bellan A, Bernardi A, Bezzo F, Morosinotto T (2019) The potential of quantitative models to improve microalgae photosynthetic efficiency. Physiol Plant 166(1):380. https://doi.org/10.1111/ppl.12915

Plaza BM, Gómez-Serrano C, Acién-Fernández FG, Jimenez-Becker S (2018) Effect of microalgae hydrolysate foliar application (Arthrospira platensis and Scenedesmus sp.) on petunia x hybrida growth. J Appl Phycol 30(4):2359. https://doi.org/10.1007/s10811-018-1427-0

Rafa N, Ahmed SF, Badruddin IA, Mofijur M, Kamangar S (2021) Strategies to produce cost-effective third-generation biofuel from microalgae. Front Energy Res 9:1–11. https://doi.org/10.3389/fenrg.2021.749968

Rahimnejad M, Asghary M, Fallah M (2020) Microbial fuel cell (MFC): an innovative technology for wastewater treatment and power generation. In: Bioremediation of industrial waste for environmental safety. Springer, Berlin. https://doi.org/10.1007/978-981-13-3426-9_9

Rempel A, Gutkoski JP, Nazari MT, Biolchi GN, Biduski B, Treichel H, Colla LM (2022) Microalgae growth with a high concentration of emerging pollutants and phytotoxicity evaluation of cultivation wastewater. J Water Process Eng 46:102616. https://doi.org/10.1016/j.jwpe.2022.102616

Renuka N, Guldhe A, Prasanna R, Singh P, Bux F (2018) Microalgae as multi-functional options in modern agriculture: current trends, prospects and challenges. Biotechnol Adv 36(4):1255–1273. https://doi.org/10.1016/j.biotechadv.2018.04.004

Rodionova MV, Bozieva AM, Zharmukhamedov SK, Leong YK, Chi-Wei Lan J, Veziroglu A, Veziroglu TN, Tomo T, Chang JS, Allakhverdiev SI (2022) A comprehensive review on lignocellulosic biomass biorefinery for sustainable biofuel production. Int J Hydrog Energy 47(3):1481. https://doi.org/10.1016/j.ijhydene.2021.10.122

Rojas-Flores S, De La Cruz Noriega M, Benites SM, Aguirre Gonzales G, Salvador Salinas A, Silva Palacios F (2020) Generation of bioelectricity from fruit waste. Energy Rep 6:37. https://doi.org/10.1016/j.egyr.2020.10.025

Ronga D, Biazzi E, Parati K, Carminati D, Carminati E, Tava A (2019) Microalgal biostimulants and biofertilisers in crop productions. Agronomy 9(4):192. https://doi.org/10.3390/agronomy9040192

Rozina, Ahmad M, Elnaggar AY, Keat Teong L, Sultana S, Zafar M, Munir M, Hussein EE, Zain Ul Abidin S (2022) Sustainable and eco-friendly synthesis of biodiesel from novel and non-edible seed oil of Monotheca buxifolia using green nano-catalyst of calcium oxide. Energy Convers Manage X 13:100142. https://doi.org/10.1016/j.ecmx.2021.100142

Sadeghi S, Jalili H, Ranaei Siadat SO, Sedighi M (2018) Anticancer and antibacterial properties in peptide fractions from hydrolyzed spirulina protein. J Agric Sci Technol 20(4):673–683

Sahin D, Tas E, Altindag UH (2018) Enhancement of docosahexaenoic acid (DHA) production from Schizochytrium sp. S31 using different growth medium conditions. AMB Express 8(1):7. https://doi.org/10.1186/s13568-018-0540-4

Saide A, Martínez KA, Ianora A, Lauritano C (2021) Unlocking the health potential of microalgae as sustainable sources of bioactive compounds. Int J Mol Sci 22(9):4383. https://doi.org/10.3390/ijms22094383

Saini RK, Keum YS (2018) Omega-3 and omega-6 polyunsaturated fatty acids: dietary sources, metabolism, and significance—a review. Life Sci 203:255. https://doi.org/10.1016/j.lfs.2018.04.049

Salinas-Salazar C, Garcia-Perez JS, Chandra R, Castillo-Zacarias C, Iqbal HMN, Parra-Saldívar R (2019) Methods for extraction of valuable products from microalgae biomass. In: Microalgae biotechnology for development of biofuel and wastewater treatment. Springer, Berlin. https://doi.org/10.1007/978-981-13-2264-8_11

Santini G, Biondi N, Rodolfi L, Tredici MR (2021) Plant biostimulants from cyanobacteria: an emerging strategy to improve yields and sustainability in agriculture. Plants 10(4):643. https://doi.org/10.3390/plants10040643

Scaife MA, Merkx-Jacques A, Woodhall DL, Armenta RE (2015) Algal biofuels in Canada: status and potential. Renew Sust Energ Rev 44:620. https://doi.org/10.1016/j.rser.2014.12.024

Selvakumar P, Kavitha S, Sivashanmugam P (2019) Optimization of process parameters for efficient bioconversion of Thermo-chemo pretreated Manihot esculenta Crantz YTP1 stem to ethanol. Waste Biomass Valoriz 10:2177. https://doi.org/10.1007/s12649-018-0244-7

Sengupta B, Priyadarshinee R, Roy A, Banerjee A, Malaviya A, Singha S, Mandal T, Kumar A (2020) Toward sustainable and eco-friendly production of coffee: abatement of wastewater and evaluation of its potential valorization. In: Clean technologies and environmental policy, vol 22(5). Springer, Berlin. https://doi.org/10.1007/s10098-020-01841-y

Shuba ES, Kifle D (2018) Microalgae to biofuels: 'promising' alternative and renewable energy, review. Renew Sust Energ Rev 81:743. https://doi.org/10.1016/j.rser.2017.08.042

Siqueira SF, Queiroz MI, Zepka LQ, Jacob-Lopes E (2018) Introductory chapter: microalgae biotechnology—a brief introduction. In: Microalgal biotechnology. Springer, Berlin. https://doi.org/10.5772/intechopen.73250

Sivaramakrishnan R, Suresh S, Kanwal S, Ramadoss G, Ramprakash B, Incharoensakdi A (2022) Microalgal biorefinery concepts' developments for biofuel and bioproducts: current perspective and bottlenecks. Int J Mol Sci 23(5):2623. https://doi.org/10.3390/ijms23052623

Sosa-Hernández JE, Romero-Castillo KD, Parra-Arroyo L, Aguilar-Aguila-Isaías MA, García-Reyes IE, Ahmed I, Parra-Saldivar R, Bilal M, Iqbal HMN (2019) Mexican microalgae biodiversity and state-of-the-art extraction strategies to meet sustainable circular economy challenges: high-value compounds and their applied perspectives. Mar Drugs 17(3):174. https://doi.org/10.3390/md17030174

Su Y, Song K, Zhang P, Su Y, Cheng J, Chen X (2017) Progress of microalgae biofuel's commercialization. Renew Sust Energy Rev 74:402. https://doi.org/10.1016/j.rser.2016.12.078

Suganya T, Varman M, Masjuki HH, Renganathan S (2016) Macroalgae and microalgae as a potential source for commercial applications along with biofuels production: a biorefinery approach. Renew Sust Energ Rev 55:909. https://doi.org/10.1016/j.rser.2015.11.026

Sun H, Zhao W, Mao X, Li Y, Wu T, Chen F (2018) High-value biomass from microalgae production platforms: strategies and progress based on carbon metabolism and energy conversion. Biotechnol Biofuels 11:227. https://doi.org/10.1186/s13068-018-1225-6

Sun H, Wang Y, He Y, Liu B, Mou H, Chen F, Yang S (2023) Microalgae-derived pigments for the food industry. Mar Drugs 21(2):82. https://doi.org/10.3390/md21020082

Sundaram T, Rajendran S, Gnanasekaran L, Rachmadona N, Jiang JJ, Khoo KS, Show PL (2023) Bioengineering strategies of microalgae biomass for biofuel production: recent advancement and insight. Bioengineered 14(1):2252228. https://doi.org/10.1080/21655979.2023.2252228

Suparmaniam U, Lam MK, Uemura Y, Lim JW, Lee KT, Shuit SH (2019) Insights into the microalgae cultivation technology and harvesting process for biofuel production: a review. Renew Sust Energ Rev 115:109361. https://doi.org/10.1016/j.rser.2019.109361

Supraja KV, Behera B, Balasubramanian P (2020) Efficacy of microalgal extracts as biostimulants through seed treatment and foliar spray for tomato cultivation. Ind Crop Prod 151:112453. https://doi.org/10.1016/j.indcrop.2020.112453

Suriya S, Boomibalagan P, Eswaran S, Sivasangari Ramya S, Vijayanand N, Uthirapandi V, Kathiresan D (2018) Bio-fertilizer potential of seaweed liquid extracts of marine macro algae on growth and biochemical parameters of Ocimum sanctum. J Pharmacogn Phytochem 7(3):3528–3532

Tan CY, Dodd IC, Chen JE, Phang SM, Chin CF, Yow YY, Ratnayeke S (2021) Regulation of algal and cyanobacterial auxin production, physiology, and application in agriculture: an overview. J Appl Phycol 33:2995. https://doi.org/10.1007/s10811-021-02475-3

Telussa I, Fransina EG, Singerin J, Taipabu MI (2023) Bioethanol production from tropical marine microalgae Ambon Bay Navicula sp. of the inner Ambon Bay strain. Indo J Chem Res 10(3):136. https://doi.org/10.30598/ijcr.2023.10-ivo

Udayan A, Sirohi R, Sreekumar N, Sang BI, Sim SJ (2022) Mass cultivation and harvesting of microalgal biomass: current trends and future perspectives. Bioresour Technol 344:126406. https://doi.org/10.1016/j.biortech.2021.126406

Uribe-Wandurraga ZN, Igual M, Reino-Moyón J, García-Segovia P, Martínez-Monzó J (2021) Effect of microalgae (Arthrospira platensis and Chlorella vulgaris) addition on 3D printed cookies. Food Biophys 16(1):27. https://doi.org/10.1007/s11483-020-09642-y

Venkata Mohan S, Nikhil GN, Chiranjeevi P, Nagendranatha Reddy C, Rohit MV, Kumar AN, Sarkar O (2016) Waste biorefinery models towards sustainable circular bioeconomy: critical review and future perspectives. Bioresour Technol 215:2. https://doi.org/10.1016/j.biortech.2016.03.130

Ventura SPM, Nobre BP, Ertekin F, Hayes M, Garciá-Vaquero M, Vieira F, Koc M, Gouveia L, Aires-Barros MR, Palavra AMF (2017) Extraction of value-added compounds from microalgae. In: Microalgae-based biofuels and bioproducts: from feedstock cultivation to end-products. Springer, Berlin. https://doi.org/10.1016/B978-0-08-101023-5.00019-4

Voloshin RA, Rodionova MV, Zharmukhamedov SK, Nejat Veziroglu T, Allakhverdiev SI (2016) Review: biofuel production from plant and algal biomass. Int J Hydrog Energy 41(39):17257. https://doi.org/10.1016/j.ijhydene.2016.07.084

Wang J, Peng X, Chen X, Ma X (2019) Co-liquefaction of low-lipid microalgae and starch-rich biomass waste: the interaction effect on product distribution and composition. J Anal Appl Pyrolysis 139:250. https://doi.org/10.1016/j.jaap.2019.02.013

Weliwaththage SR, Yildirim M (2020) The review of innovation in renewable energy sector in the world. J Res Technol Eng 1(4):117–147

Wicker RJ, Kumar G, Khan E, Bhatnagar A (2021) Emergent green technologies for cost-effective valorization of microalgal biomass to renewable fuel products under a biorefinery scheme. Chem Eng J 415:128932. https://doi.org/10.1016/j.cej.2021.128932

Wu J, Gu X, Yang D, Xu S, Wang S, Chen X, Wang Z (2021) Bioactive substances and potentiality of marine microalgae. Food Sci Nutr 9:5279. https://doi.org/10.1002/fsn3.2471

Wusnah, Supardan MD, Haryani S, Yunardi (2021) Future production of bioethanol from microalgae as a renewable source of energy. IOP Conf Ser Earth Environ Sci 922(1):012010. https://doi.org/10.1088/1755-1315/922/1/012010

Yaakob Z, Ali E, Zainal A, Mohamad M, Takriff MS (2014) An overview: biomolecules from microalgae for animal feed and aquaculture. J Biol Res (Greece) 21(1):6. https://doi.org/10.1186/2241-5793-21-6

Yang Z, Guo R, Xu X, Fan X, Luo S (2011) Hydrogen and methane production from lipid-extracted microalgal biomass residues. Int J Hydrog Energy 36(5):3465. https://doi.org/10.1016/j.ijhydene.2010.12.018

Yen HW, Hu IC, Chen CY, Ho SH, Lee DJ, Chang JS (2013) Microalgae-based biorefinery—from biofuels to natural products. Bioresour Technol 135:166. https://doi.org/10.1016/j.biortech.2012.10.099

Yuan JP, Peng J, Yin K, Wang JH (2011) Potential health-promoting effects of astaxanthin: a high-value carotenoid mostly from microalgae. Mol Nutr Food Res 55(1):150–165. https://doi.org/10.1002/mnfr.201000414

Zhang P, Wang F, Yu M, Zhuang X, Feng X (2018) Two-dimensional materials for miniaturized energy storage devices: from individual devices to smart integrated systems. Chem Soc Rev 47(19):7426–7451. https://doi.org/10.1039/c8cs00561c

Microalgae: A Sustainable Bio-Factory for High-Value-Added Product

Arindam Adhikary

1 Introduction

An ever-increasing worldwide requirement for energy and overexploitation of fossil fuels incites the necessity for a better alternative economically and environmentally viable energy resource. Bioenergy from microalgae is one of the few renewable energy options that have gained attention in recent years as they can also be exploited for other valuable products like proteins, lipids and carbohydrates (Agrawal et al. 2023). The worldwide bioenergy market was estimated at $108.71 billion in 2021 and is projected to reach $206.24 billion by 2022–2030 ending with an estimated compound annual growth rate (CAGR) of 7.4%. The biorefinery market is anticipated to expand by 68.5 billion in 2027 with an expected CAGR of 8.2% (Chanana et al. 2023). However, commercialisation of microalgal bioenergy production is a challenge because of high product costs due to expensive equipment used in production, planning process layout, several trials to achieve high-value bioenergy products, equipment maintenance, process, etc. The overall cost for microalgal bioenergy production is considerably higher than energy products from fossil fuels (Goswami et al. 2022a). Utilisation of microalgal biomass as raw materials for the manufacture of high-value-added goods, like nutraceuticals/medicines, food, bioenergy, biofuels and fertiliser, make the method more economically sustainable and viable (Chandra et al. 2019; Mishra et al. 2019; Wijffels and Barbosa 2010; Goswami et al. 2023; Agrawal et al. 2020, 2023; Agrawal and Verma 2022; Bhardwaj et al. 2020). Different microalgal species contain different chemical compositions, and based on the chemical composition, the biomass from different algal species is further processed into different products, such as biodiesel, bioethanol, syngas, etc.

A. Adhikary (✉)
Department of Medical Microbiology, University School of Allied Health Sciences, Rayat Bahra University, Mohali, Punjab, India

K. Agrawal, P. Verma (eds.), *Biotechnological Advances in Biorefinery*, Interdisciplinary Biotechnological Advances,
https://doi.org/10.1007/978-981-97-5544-8_8

(Goswami et al. 2022a). Microalgae can be grown in large quantities for biomass in a range of natural environments that comprises various water bodies (Okeke et al. 2022; Rashid et al. 2019; Sharma et al. 2022). Usually, open raceway ponds along with photobioreactors are generally used for the mass production of microalgae. Recent advancements in microalgae cultivation, such as the use of both organic as well as inorganic sources of carbon in the availability of light by using several algal strains, reuse of waste medium from the previous batch or sewage sludge or feedstock from waste material as a nutrient source, also have a few major drawbacks that include product purification, eutrophication, chances of contamination from other microorganisms and high cost for temperature maintenance (Chanana et al. 2023; Molinuevo-Salces et al. 2019). Presently, various biotechnological approaches are used to genetically modify the genes to enhance the production of value-added products, stress tolerance and other characteristics in microalgae (Wijffels and Barbosa 2010). In this article, various approaches for the manufacture of high-value products and challenges confronted in the process were discussed that include the incorporation of different approaches towards augmenting microalgae into economically and environmentally viable high-value-added products.

2 Approaches Adapted for the Transformation of Microalgal Biomass to Biofuel

The transformation of the biomass from microalgae to biofuels can be done via three approaches: (1) biochemical approach, (2) thermochemical approach and (3) chemical approach.

2.1 Biochemical Approach

The biochemical approach for the transformation of biomass from microalgae to biofuels includes (1) photobiological hydrogen production, (2) alcoholic fermentation and (3) anaerobic digestion (Fig. 1). Photobiological hydrogen production is a natural process in which microalgae produce hydrogen from the photolysis of water (Eroglu and Melis 2011). Alcoholic fermentation is an energy efficient, economically viable process for bioethanol production from the carbohydrates present in microalgae (Abou El-Souod et al. 2021). Alcoholic fermentation is a yeast fermentation process, whereby monosaccharide such as glucose is fermented to ethanol. In bioethanol production, the microalgal complex carbohydrate mainly consists of starch, which is hydrolysed either chemically (H_2SO_4, HCl, NaOH, KOH) or enzymatically (pectinase, cellulase, amylase) into simple monosaccharide (glucose) before proceeding to the fermentation process (Lam and Lee 2015). Anaerobic digestion is another biochemical approach that utilises microalgal biomass to yield

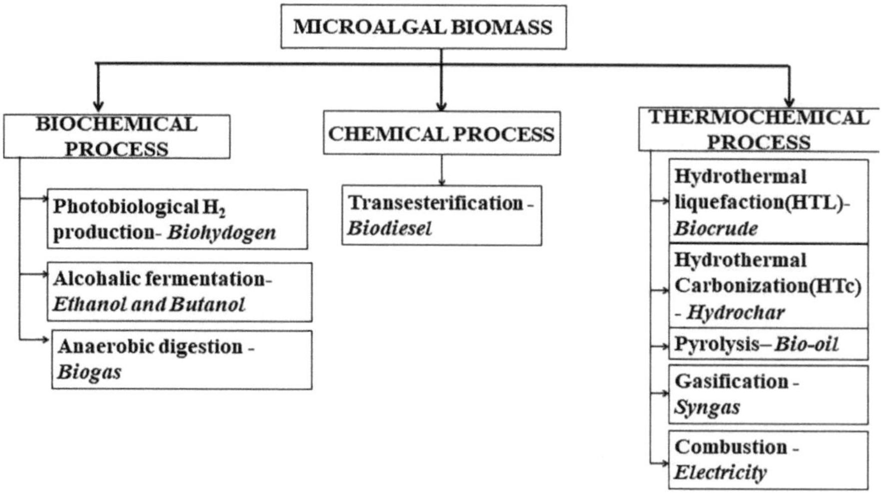

Fig. 1 Schematic representation of various processes for the manufacture of high-value-added products from microalgae

biogas. Anaerobic digestion of microalgal biomass contains multiple steps. In the first step, polysaccharides, lipids and proteins from microalgae are hydrolysed to yield soluble organic compounds such as amino acids, sugars and alcohols. In the second step, the soluble organic compounds so obtained are degraded by acidogenic bacteria, such as *Clostridium, Pseudomonas, Bacillus* and *Escherichia* to give rise to volatile organic compounds, ammonia and carbon dioxide—the process is known as acidogenesis. In the next step, the products from acidogenesis are utilised by acetogens such as *Acetobacterium, Butyribacterium* and *Eubacterium* to produce acetic acid, carbon dioxide and hydrogen (Cavinato et al. 2017). In the last step, methanogenesis occurs. In this step, products from the acetogenesis step are converted into methane, carbon dioxide and water by methanogens (De Bhowmick et al. 2019). The main limitation of anaerobic digestion of microalgal biomass is the rigidity of the cell wall that leads to low digestibility. To overcome this limitation, various pretreatment methods such as chemical (acid or alkali treatment), mechanical (high-pressure, high-speed homogenisation), thermal (low temperature, hydrothermal pretreatment) as well as biological (Enzymatic pretreatment) methods are being used (Cavinato et al. 2017).

2.2 *Thermochemical Approach*

Thermochemical approach includes Hydrothermal Liquefaction (HTL), Hydrothermal Carbonisation, Pyrolysis, Gasification and Combustion (Fig. 1). In HTL process wet microalgal biomass is transformed into liquid biofuel at around supercritical state. Supercritical state is a condition where water functions both as an acid as well as base catalyst. This initiates isomerisation, polymerisation causing conversion of biomolecules into biofuel. Hydrothermal carbonisation is the transformation of dry microalgal mass into biochar under high-temperature conditions. The biochar formation majorly consists of hydrolysis, condensation, decarboxylation and turning biomass into biochar (Wang et al. 2018). Biochar can also be utilised in the treatment of wastewater and soil. Pyrolysis is another process of thermochemical transformation of microalgae biomass. The process is executed at a very high temperature (400 °C–600 °C under 1 atmospheric pressure/800 °C for microwave pyrolysis). At this condition, metabolites in the microalgae crack and depolarise producing bio-oil, biochar and various gasses. Like HTL, pyrolysis is also influenced by multiple factors like composition of the biomass, heating temperature, particle size, reactor type, residence time and gas flow rate. Various pyrolytic conditions support the production of different types of products, such as very high temperature that increase gas production and fast hydrolysis with shorter residence time that yields bio-oil (Chen et al. 2015b). Gasification is a process by which the conversion of carbon content of microalgae into syngas (hydrogen, carbon monoxide, methane and other combustible gasses) is carried out. Gasification can be done by two methods; conventional and supercritical water gasification. In conventional gasification methods, the temperature range is quite high (800–1000 °C) at 1–10 bar pressure, whereas in supercritical water gasification, the temperature range is low (400–500 °C) at 24–36 MPa pressure. In this process, supercritical conditions decompose microalgal biomolecules into smaller gaseous molecules. Syngas is used in engines, turbines run by gaseous fuel and in manufacturing of methanol and other organic solvents (Raheem et al. 2018). Combustion is another process where dry microalgal biomass is burned at about 800 °C in boilers and steam engines to produce heat energy. For an economically sustainable approach, other feedstocks such as coal can be burned with microalgae (Mishra et al. 2019).

2.3 Chemical Approach

Transesterification is a chemical process by which microalgal biomass can be transformed into biodiesel (Fig. 1). The first step for biodiesel production is the extraction of oil from the microalgal biomass followed by transesterification. In the transesterification process, microalgal triglycerides are transformed into long-chain fatty acid alkyl esters (FAAE). These long-chain FAAEs are known as biodiesel. The transesterification process can be carried out in various ways such as enzymatic, microwave-assisted and supercritical transesterification (Chauhan et al. 2020; Mishra et al. 2019). In enzymatic transesterification reaction, lipase enzyme is usually used to convert microalgal triglycerides into fatty acids and glycerol. The main limitations of enzymatic transesterification reaction are slower rate of reaction, chances of enzyme deactivation and high cost (Jegannathan et al. 2009). In a microwave-assisted transesterification reaction, microalgae are mixed with methanol followed by exposure to microwave irradiation. After the reaction, phase separation is carried out using Buchner funnel under vacuum condition. The liquid phase so obtained is further separated using a decantation funnel to obtain biodiesel (Cancela et al. 2012; Koech et al. 2020). Supercritical transesterification process can be utilised for fast and considerable production of biodiesel (Chauhan et al. 2020; Mishra et al. 2019). Supercritical transesterification of microalgae can be done by two different processes. In the first process, the microalgae are dewatered followed by lipid extraction and supercritical transesterification which is very energy-intensive and costly. The second process is the direct transesterification of wet algae to yield biodiesel. In this process, water content of the microalgae serves as co-solvent along with supercritical alcohol for the production of biodiesel from microalgal lipid. Supercritical transesterification is a promising process for the production of biodiesel from microalgae as it takes less time, no catalyst as compared to conventional transesterification processes (Deshpande et al. 2017).

3 High-Value-Added Product Obtained from Microalgal Biorefinery

Microalgal biorefineries are effective tools for the manufacture of several high-value-added products (Mehariya et al. 2021) useful for industrial, agricultural and household uses. Microalgae convert CO_2 into various value-added products, viz. bioethanol, biogas, biofertilisers, etc. (Guo et al. 2018; Okeke et al. 2022). For the zero-waste strategy, complete microalgal biomass is utilised for the manufacture of various high-value-added products.

3.1 *Biodiesel*

Microalgae are autotrophic organisms and utilise CO_2 during photosynthesis and transform it into usable compounds for the production of biodiesel. The CO_2 produced during the burning of fossil fuels and animal respirations gets utilised, leading to environmentally sustainable fuel (Yaşar 2020). The high lipid contents (20–63%, comprising of myristic acid, palmitic acid, oleic acid, linoleic acid and linolenic acid) of microalgae make them a promising candidate in biodiesel production (Table 1) (Raslavičius et al. 2018). *Chlorella vulgaris* and *Chlorella protothecoides* are extensively studied for biodiesel production owing to their exceptional biomass and comparatively high PUFA content. Under stress conditions, such as N_2 deficiency, salt stress and light stress, microalga like *Chlorella vulgaris* showed increased triacylglycerol (TAG) and oleic acids content (Poh et al. 2020). Varied sources of nitrogen induce greater removal of nitrogen, improvement in biomass and higher lipid content and in turn greater biodiesel production in microalgae as compared to a single source of nitrogen (Zhu et al. 2019). Utilisation of wastewater

Table 1 Value-added product obtained from microalgae

Product	Microalgae	Reference
Biodiesel	*Aurantiochytrium* sp.	Kim et al. (2016b)
	Botryococcus braunii	Hidalgo et al. (2015)
	Chlamydomonas sp.	Chen et al. (2015a)
	Chlorella pyrenoidosa	Cheng et al. (2013)
	Chlorella vulgaris	Levine et al. (2010)
	Nannochloropsis oculata	Jafari et al. (2021)
Biohydrogen	*Chlamydomonas reinhardtii*	Mussgnug et al. (2010)
	Spirogyra sp.	Nobre et al. (2013)
	Acutodesmus obliquus	Nurdiawati et al. (2019)
	Cyanobacterium anabaena variablilis	Mahmudul et al. (2021)
Bioethanol	*Dunaliella tertiolecta*	Lee et al. (2013)
	Arthrospira platensis	Ismail et al. (2020)
	Porphyridium cruentum	Kim et al. (2017)
Biogas	*Dunaliella salina*	Sialve et al. (2009)

for biomass growth offers diverse fatty acid composition for better biodiesel yield (Sun et al. 2018; Yadav et al. 2022; Goswami et al. 2022b). Transesterification reaction is predominantly used for the conversion of algal oil to biodiesel. In this process, hydrochloric acid, sulphuric acid or potassium hydroxide and sodium hydroxide act as catalyst for reaction between the microalgal triacylglycerides (TAGs) and alcohol to yield fatty acid alkyl esters (FAAE). Enzymatic catalysts such as lipase can also be used for this process, but the efficiency is more in case of base catalyst as compared to acid and enzyme catalyst (Santos-Sánchez et al. 2016; Tang et al. 2020). Fatty acid alkyl ester profiling of the microalgae-based biodiesel matches the international standard for biodiesel for the use in vehicles (Rizwan et al. 2018). These standards cogitate about several features such as fuel characteristics, types of emission emitted by the vehicle and effect of oil content on the engine (Mofijur et al. 2022; Yadav et al. 2022).

3.2 *Bioethanol*

Bioethanol from microalgae can be produced from the alcoholic fermentation of feedstocks rich in carbohydrates, such as grains, potatoes, lignocellulolytic substrate, etc. (Tse et al. 2021). The lignocellulolytic substrate can be obtained from microalgal biomass. *Chlorococcum, Chlamydomonas* and *Chlorella* are generally used for producing bioethanol for their better transformation rate than other microalgal species (Table 1). The most difficult stage in producing bioethanol from microalgae is the breakdown of the recalcitrant biomass into readily fermentable sugars (de Farias Silva and Bertucco 2016; Lam and Lee 2015). A recent technology has emerged, where biotechnological approaches are used for the direct manufacture of ethanol by the microalgae utilising sunrays and nutrients from the media. Wong et al. (2022) reported that a cyanobacteria is bioengineered for direct production of bioethanol as a part of its constitutive fermentation process. Xu et al. (2019) in another study reported a combined process for the manufacture of biodiesel and bioethanol with higher efficiency. There are various components that affect bioethanol production, viz. sugar content, pH of the medium and the amount of the inoculum used. Brown algae are good candidates for bioethanol production because of their higher sugar content and ease of cultivation. *Chlorococcum* sp. contains 27% starch and can be utilised under dark conditions in 24 h at 25 °C to yield ethanol (Banerjee et al. 2022).

3.3 Biogas and Biomethane

Biogas is an inexhaustible fuel mainly composed of methane, CO_2, H_2 and volatile organic compounds. Microalgae with excessive lipid and protein content are not appropriate for biogas production through anaerobic digestion. In such cases, defatted microalgae biomass is used for the production of biogas. Though biogas can be exploited to produce heat energy through combustion and other forms of energy through gas turbines, its various components can be separated or refined for more interesting uses (Singh and Dhar 2019). Biogas production and its commercialisation are generally affected by the type of microalgal strain, the culture conditions and the pretreatment approaches (Jankowska et al. 2019). The production of only biogas from microalgae is not economically viable, but using a biorefinery approach, where other valuable products are also harvested, makes the overall production of biogas economically feasible. In addition, the process may be improved by integrating the usage of CO_2 from combustion engines, biomethane extracted from biogas, and volatile fatty acids obtained from acidogenic fermentation (Chen et al. 2018).

Extraction of biomethane from biogas mainly involves separating the methane part of the biogas from other components, such as CO_2, ammonia, carbon monoxide, hydrocarbons, hydrogen sulphide, nitrogen, oxygen, siloxanes and water vapour (Okeke et al. 2022). Various methods are being used for the removal of these components. Water vapour can be removed by using cyclone separators, demisters, moisture traps and various adsorption/absorption based drying methods, while H_2S can be removed from the biogas via adsorption onto iron chloride and air dosing (Adnan et al. 2019; Awe et al. 2017). CO_2 from biogas can be separated from biomethane by various approaches, such as physical/chemical absorption and membrane separation. Some CO_2 can be used for microalgae biomass production (Okeke et al. 2022).

3.4 Biohydrogen

Biohydrogen from algae can be used as a clean fuel for industries and automobiles (Table 1). With the recent increase in demand for a better and cleaner fuel, microalgae-based biohydrogen production (Goswami et al. 2021) gained a lot of industrial attention, and it can be produced by three different approaches: (1) direct biophotolysis, (2) indirect biophotolysis and (3) two-stage dark/photo fermentation (Eroglu and Melis 2011; González-Fernández et al. 2012). In the first process, molecular hydrogen is produced by photosystem II by splitting water under anaerobic condition. Algae play a pivotal part in the biohydrogen production because of its natural ability to do photolysis of water via hydrogenase enzyme activity (González-Fernández et al. 2012). The key limitation of this process is the oxygen sensitivity of the hydrogenase enzymes. In the second process, the microalgae are at first cultivated under normal conditions. During this step, the microalgae grow and store carbohydrates. In the next step, the nitrogenase enzyme under anaerobic condition utilises the carbohydrates for the production of biohydrogen. In the third process, at first, direct biophotolysis by the microalgae is carried out to produce hydrogen, in the subsequent step; the algal biomass is used to produce hydrogen through dark fermentation (Eroglu and Melis 2011).

4 Value-Added Product Obtained from Microalgal Biorefinery with Potential Therapeutic Value

Microalgae yield a diverse variety of value-added compounds with therapeutic value such as proteins, lipids, polysaccharides, vitamins, carotenoids, sterols, etc. These compounds have potential antioxidative activity, anti-inflammatory activity and anticancer activity and have significant application in pharmaceutical industry (Fu et al. 2019; Rizwan et al. 2018).

4.1 Proteins

Microalgae are a very good source of amino acids and proteins. These proteins are a rich source of bioactive peptides that have plenty of health benefits (Table 2) (Amorim et al. 2020). *Spirulina* contains around 60% protein of their biomass with other value-added products such as vitamins, minerals and other bioactive substances, which can be subsequently used as food supplements in many industries, viz. dairy, poultry, aquaculture, etc. (Bhattacharjee 2016).

Table 2 Value-added product with therapeutic value obtained from microalgae

Microalgae	Value-added product	Reference
Haematococcus pluvialis	Astaxanthin	Kim et al. (2016a)
Anabaena NCCU-9	Phycocyanin	Hemlata et al. (2011)
Spirulina platensis	Phycocyanin	Chen et al. (2013)
Chlorella saccharophila	B-carotene	Singh et al. (2013)
Chlorella saccharophila	Zeaxanthin	Singh et al. (2013)
Chlorella sp.	Lipids	Vanthoor-Koopmans et al. (2013)
Chlorococcum infusionum	Carbohydrates	Karemore and Sen (2016)
Scenedesmus almeriensis	Lipids	Hernández et al. (2014)
Nannochloropsis oculata	Lipids	Hernández et al. (2014)

4.2 Lipids

Microalgae are a very good source of lipids (Table 2). Under specific circumstances, microalgae can produce around 50% triglycerides (TAG), of which around 70% are poly-unsaturated fatty acids (PUFA). These lipid contents are used for biodiesel production. PUFA produced by microalgae shows various characteristics, such as antioxidant activity, antibacterial activity, antiviral activity, detoxification, improving brain activity and enhancing immune-stimulatory effects (de Morais et al. 2015; Lupette and Benning 2020). These microalgal triglycerides are an important source of omega-3-fatty acid.

4.3 Vitamins and Carotenoids

Vitamins are vital for good health. Microalgae are excellent source vitamins. Microalgae contains retinoids (vitamin A); alpha-tocopherol (vitamin E); vitamins B1, B2, B3, B6, B7, B9 and B12; and ascorbic acid (vitamin C) (Pistelli et al. 2021). Microalgae such as *Porphyridium cruentum* and *Tetraselmis suecica* yield vitamin E, while microalgae such as *Dunaliella salina* produce vitamin C in substantial quantities (Santiago-Morales et al. 2018). Microalgae also produce carotenoids (Table 2). Mehariya et al. (2021) reported that *Dunaliella salina* produces β-carotene. β-carotene is a precursor for vitamin A production in humans. Microalgae extracted β-carotene showed more antioxidant activity as compared to synthetic β-carotene (Yadav et al. 2022). Carotenoids show anti-cancer activity, anti-inflammatory activity, antibacterial activity, antidiabetic properties and neuroprotective properties (Nabi et al. 2020).

5 Sustainable Zero-Waste Approach

The waste generated from a microalgal biorefinery can be largely classified as liquid waste and solid waste. For a sustainable energy-efficient biorefinery, it is necessary to reduce, reuse and recycle these wastes.

5.1 Management of Liquid Waste

Cultivation of microalgae and its downstream processing to produce biofuel, a significant amount of liquid waste is produced. Utilisation of the liquid waste for the production of economically and environmentally viable energy requires recycling and digesting these liquid wastes. Various approaches viz. flocculation, filtration and centrifugation carry an important part in the processing of the effluent and subsequently valorisation of the waste.

5.1.1 Liquid Waste from Microalgal Culture

Cultivation of microalgae in photobioreactors as well as in open raceway ponds requires a large amount of culture medium. To fulfil the need of nutrients and water for the cultivation of microalgal, wastewater is used in place of the freshwater and commercial media. The main drawback of the use of waste water is that it has an elevated quantity of heavy metals and disposal of this heavy metal containing medium causes potential environmental contamination. Therefore, to avoid heavy metal contamination of soil and water as well as to fulfil the need of nutrient and water, the harvested medium (wastewater) is recycled as a substantial quantity of nutrient exists in the harvested medium. The harvesting method also determines the attributes of the wastewater medium. The use of chemicals as flocculants changes the chemical composition of the medium, in turn reduces the quality of the medium (Kim et al. 2011). To avoid the use of chemicals, bioflocculation technique is used for safe recycling of microalgal harvested medium. Other than the harvesting methods, other factors such as pathogenic microbes, extracellular metabolites and accumulation of toxic compounds limit the reusability of the culture medium. To further utilise the medium, it should be pretreated before reusing it for the cultivation of microalgal. There are various techniques that are employed for the pretreatment process, such as auto-sedimentation, centrifugation and membrane technology for purification of the harvested media. In the auto-sedimentation process, the microalgal culture is allowed to settle for overnight, and in the next, day the flocculated cells are removed/harvested. The leftover medium is further recycled and reused. In the case of centrifugation, it removes the microbes and particulate matter from the medium (Farooq et al. 2015; Mishra et al. 2019).

5.1.2 Liquid Waste from Biochemical Process

Production of biohydrogen via dark fermentation majorly produces organic acids as final products. These organic acids can be further utilised as the sources of carbon for supporting microalgal growth. Ren et al. (2014) described that utilisation of organic acids such as acetic acid and butyric acid improved microalgal biomass and lipid production. The fermentation media from bioethanol production produces liquid residue. This liquid residue contains high amount of carbon and other nutrients that can be further exploited as a low-cost substrate for biohydrogen and biomethane production (De Paoli et al. 2011; Mishra et al. 2019).

5.1.3 Liquid Waste from Thermochemical Process

Hydrothermal conversion (HTC) is a type of thermochemical conversion process that moves 20% to 50% of the organic compounds into the aqueous phase during conversion. During HTC of microalgal biomass, the coproduct produced is rich in organic compounds that include complex carbohydrates, proteins and organic acids. Recovery of resources from aqueous solution is necessary because if these organic molecules are not properly processed, they may cause damage to the environment (Chen et al. 2017). Phenol and carboxylic acid present in the HTC aqueous phase are utilised in pharmaceutical and chemical industries, while monosaccharide present in the HTC aqueous phase can be exploited as substrate for the production of bioethanol, and N, P and K present in the HTC aqueous phase can be exploited as nutrients for the cultivation of microalgae. *Chlorella vulgaris* can utilise 45.5%–59.9% of total N_2, 85.8%–94.6% of total P, and remove 50.0%–60.9% of chemical oxygen demand (COD) on diluted HTC aqueous phases (Mishra et al. 2019).

5.1.4 Liquid Waste from Transesterification

In transesterification reaction, the main by product formed during the biodiesel production is glycerol. Crude glycerol majorly consists of catalysts, methanol, soap, salts, organic matter and other impurities. Therefore, the discarding the crude glycerol into the environment can lead to severe ecological issues. Crude glycerol can be utilised as feedstock for animal and production of citric acid (Mishra et al. 2019). Crude glycerol can also be utilised for the manufacture of hydrogen via photofermentation.

5.2 Management of Solid Waste Residue

Biofuel production via the use of microalgae generates huge quantities of solid waste, such as fermentation residues, residues from anaerobic digestion, biochars and de oiled biomass. In order to produce value-added products for viable microalgal biorefinery, these solid residues can be processed further (Mishra et al. 2019).

5.2.1 Solid Waste from Biochemical Process

The fermentation process leaves undigested carbohydrates, proteins, other biomolecules and residual compounds. In order to convert complex undigested biomolecules into simpler utilisable compounds, the residual solid biomass is pretreated. This simpler compound can also be utilised for the process of fermentation and anaerobic digestion that yields biodiesel, bioethanol, biogas or any form of liquid biofuel. Lunprom et al. (2019) in a study reported that *Chlorella* sp. present in the anaerobic solid-state fermentation (ASSF) residue contains around 30% volatile solid that can be utilised as substrate for the production of bio-hythane. In another study, biohydrogen production by *Chlamydomonas reinhardtii* via anaerobic fermentation considerably enhanced biogas production in subsequent anaerobic digestion step with a greater quantity of methane (Mussgnug et al. 2010). Generally, the residue from anaerobic digestion can be utilised as an organic fertiliser to improve the nutritional value of the soil. However, advances in pretreatment technology helps the breakdown of biomolecules into sugar components and short-chain fatty acids that can be utilised as precursor materials to produce a variety of products, from bioenergy to biofuels. Other than biochemical transformation, thermochemical transformation of solid waste residue can also be utilised for economically and environmentally viable production of biofuel. A recent approach was proposed to integrate anaerobic digestion with pyrolysis for simultaneous yield of biogas and transformation of solid waste into biofuel (Fabbri and Torri 2016; Mishra et al. 2019).

5.2.2 Solid Waste from Thermochemical Process

The solid waste from the hydrothermal liquefaction is very limited. However, it contains many mineral elements, rich in ash content and low hydrogen and nitrogen content (Sharifzadeh et al. 2019). Due to the presence of mineral nutrients, the solid residue is used as fertiliser to improve the soil quality. Biochar is formed from lignocellulosic part of the biomass and is widely utilised as a soil conditioner, construction material and hydrogen storage material (Sharifzadeh et al. 2019).The major applications of solid waste residues are mostly in the agricultural sector, soil augmentation. In a recent study, Nayak et al. (2015) reported fly ash supplement at low concentration has significantly improved crop yield and soil microbiota. There have been numerous uses of fly ash that focused on lignocellulosic biomass.

Therefore, more research on the systematic characterisation of fly ash from microalgal biomass will provide insight into how to use it and generate value-added products for an economically and environmentally viable biorefinery.

6 Conclusion

Microalgal biomass is a pivotal substrate for bioenergy production that can produce a series of biofuels and other value-added products. Among several factors, choosing wet biomass of microalgae as a bioenergy feed can make the approach energy efficient and economically and environmentally viable. In addition to raw biomass, different processes produce different by-products such as liquid wastewater and solid residues from various up- and downstream processes that can be further utilised for the manufacture of various high-value-added products. Therefore, additional studies are necessary at larger scale to authenticate the viability of zero-waste microalgal biorefinery approach.

References

Abou El-Souod GW, Morsy EM, Hassan LH, El-Sheekh MM (2021) Efficient Saccharification of the microalga Chlorella vulgaris and its conversion into ethanol by fermentation. Iran J Sci Technol Trans A Sci 45:767–774

Adnan AI, Ong MY, Nomanbhay S, Chew KW, Show PL (2019) Technologies for biogas upgrading to biomethane: a review. Bioengineering 6(4):92

Agrawal K, Verma P (2022) An overview of various algal biomolecules and its applications. In: An integration of phycoremediation processes in wastewater treatment, pp 249–270

Agrawal K, Bhatt A, Bhardwaj N, Kumar B, Verma P (2020) Algal biomass: potential renewable feedstock for biofuels production–part I. In: Biofuel production technologies: critical analysis for sustainability, pp 203–237

Agrawal K, Ruhil T, Verma P (2023) Microalgae: a way toward sustainable development of a society. In: Basic research advancement for algal biofuels production. Singapore, Springer Nature, pp 259–277

Amorim ML, Soares J, Vieira BB, Batista-Silva W, Martins MA (2020) Extraction of proteins from the microalga Scenedesmus obliquus BR003 followed by lipid extraction of the wet deproteinized biomass using hexane and ethyl acetate. Bioresour Technol 307:123190

Awe OW, Zhao Y, Nzihou A, Minh DP, Lyczko N (2017) A review of biogas utilisation, purification and upgrading technologies. Waste Biomass Valor 8:267–283

Banerjee S, Das D, Ghosh AK (2022) Production of bioethanol from microalgal feedstock: a circular biorefinery approach. In: Potential and challenges of low carbon fuels for sustainable transport, pp 33–65

Bhardwaj N, Agrawal K, Verma P (2020) Algal biofuels: an economic and effective alternative of fossil fuels. In: Microbial strategies for techno-economic biofuel production, pp 207–227

Bhattacharjee M (2016) Pharmaceutically valuable bioactive compounds of algae. Asian J Pharm Clin Res 9(6):43

Cancela A, Maceiras R, Urrejola S, Sanchez A (2012) Microwave-assisted transesterification of macroalgae. Energies 5(4):862–871

Cavinato C, Ugurlu A, de Godos I, Kendir E, Gonzalez-Fernandez C (2017) Biogas production from microalgae. In: Microalgae-based biofuels and bioproducts. Woodhead Publishing, Sawston, pp 155–182
Chanana I, Kaur P, Kumar L, Kumar P, Kulshreshtha S (2023) Advancements in microalgal biorefinery technologies and their economic analysis and positioning in energy resource market. Fermentation 9(3):202
Chandra R, Iqbal HM, Vishal G, Lee H-S, Nagra S (2019) Algal biorefinery: a sustainable approach to valorize algal-based biomass towards multiple product recovery. Bioresour Technol 278:346–359
Chauhan DS, Goswami G, Dineshbabu G, Palabhanvi B, Das D (2020) Evaluation and optimization of feedstock quality for direct conversion of microalga chlorella sp. FC2 IITG into biodiesel via supercritical methanol transesterification. Biomass Convers Biorefinery 10:339–349
Chen CL, Huang CC, Ho KC, Hsiao PX, Wu MS, Chang JS (2015a) Biodiesel production from wet microalgae feedstock using sequential wet extraction/transesterification and direct transesterification processes. Bioresour Technol 194:179–186
Chen CY, Kao PC, Tsai CJ, Lee DJ, Chang JS (2013) Engineering strategies for simultaneous enhancement of C-phycocyanin production and CO2 fixation with Spirulina platensis. Bioresource Technology 145:307–312
Chen W-H, Lin B-J, Huang M-Y, Chang J-S (2015b) Thermochemical conversion of microalgal biomass into biofuels: a review. Bioresour Technol 184:314–327
Chen L, Zhu T, Fernandez JSM, Chen S, Li D (2017) Recycling nutrients from a sequential hydrothermal liquefaction process for microalgae culture. Algal Res 27:311–317
Chen H, Wang J, Zheng Y, Zhan J, He C, Wang Q (2018) Algal biofuel production coupled bioremediation of biomass power plant wastes based on chlorella sp. C2 cultivation. Appl Energy 211:296–305
Cheng J, Yu T, Li T, Zhou J, Cen K (2013) Using wet microalgae for direct biodiesel production via microwave irradiation. Bioresour Technol 131:531–535
De Bhowmick G, Sarmah AK, Sen R (2019) Zero-waste algal biorefinery for bioenergy and biochar: a green leap towards achieving energy and environmental sustainability. Sci Total Environ 650:2467–2482
de Farias Silva CE, Bertucco A (2016) Bioethanol from microalgae and cyanobacteria: a review and technological outlook. Process Biochem 51(11):1833–1842
de Morais MG, Vaz BDS, de Morais EG, Costa JAV (2015) Biologically active metabolites synthesized by microalgae. Biomed Res Int 2015:835761
De Paoli F, Bauer A, Leonhartsberger C, Amon B, Amon T (2011) Utilization of by-products from ethanol production as substrate for biogas production. Bioresour Technol 102(11):6621–6624
Deshpande SR, Sunol AK, Philippidis G (2017) Status and prospects of supercritical alcohol transesterification for biodiesel production. WIREs Energy Environ 6(5):e252
Eroglu E, Melis A (2011) Photobiological hydrogen production: recent advances and state of the art. Bioresour Technol 102(18):8403–8413
Fabbri D, Torri C (2016) Linking pyrolysis and anaerobic digestion (Py-AD) for the conversion of lignocellulosic biomass. Curr Opin Biotechnol 38:167–173
Farooq W, Suh WI, Park MS, Yang J-W (2015) Water use and its recycling in microalgae cultivation for biofuel application. Bioresour Technol 184:73–81
Fu W, Nelson DR, Mystikou A, Daakour S, Salehi-Ashtiani K (2019) Advances in microalgal research and engineering development. Curr Opin Biotechnol 59:157–164
González-Fernández C, Sialve B, Bernet N, Steyer JP (2012) Impact of microalgae characteristics on their conversion to biofuel. Part II: focus on biomethane production. Biofuels Bioprod Biorefin 6(2):205–218
Goswami RK, Mehariya S, Obulisamy PK, Verma P (2021) Advanced microalgae-based renewable biohydrogen production systems: a review. Bioresour Technol 320:124301
Goswami RK, Agrawal K, Upadhyaya HM, Gupta VK, Verma P (2022a) Microalgae conversion to alternative energy, operating environment and economic footprint: an influential approach towards energy conversion, and management. Energy Convers Manag 269:116118

Goswami RK, Agrawal K, Verma P (2022b) An exploration of natural synergy using microalgae for the remediation of pharmaceuticals and xenobiotics in wastewater. Algal Res 64:102703

Goswami RK, Mehariya S, Karthikeyan OP, Verma P (2023) Valorisation of organic carbons and organo-solvents by mixotrophic cultivation of methylotrophic Tetraselmis indica for enhanced biomolecules production. Sustain Chem Pharm 36:101281

Guo P, Zhang Y, Zhao Y (2018) Biocapture of CO2 by different microalgal-based technologies for biogas upgrading and simultaneous biogas slurry purification under various light intensities and photoperiods. Int J Environ Res Public Health 15(3):528

Hemlata, Pandey G, Bano F, Fatma T (2011) Studies on anabaena sp. NCCU-9 with special reference to phycocyanin. J Algal Biomass Utln 2(1):30–51

Hernández D, Solana M, Riaño B, García-González MC, Bertucco A (2014) Biofuels from microalgae: lipid extraction and methane production from the residual biomass in a biorefinery approach. Bioresour Technol 170:370–378

Hidalgo P, Ciudad G, Schober S, Mittelbach M, Navia R (2015) Biodiesel synthesis by direct transesterification of microalga Botryococcus braunii with continuous methanol reflux. Bioresour Technol 181:32–39

Ismail MM, Ismail GA, El-Sheekh MM (2020) Potential assessment of some micro-and macroalgal species for bioethanol and biodiesel production. Energy Sources Pt A Recov Util Environ Effects 46:1–17

Jafari A, Esmaeilzadeh F, Mowla D, Sadatshojaei E, Heidari S, Wood DA (2021) New insights to direct conversion of wet microalgae impregnated with ethanol to biodiesel exploiting extraction with supercritical carbon dioxide. Fuel 285:119199

Jankowska E, Zieliński M, Dębowski M, Oleśkowicz-Popiel P (2019) Anaerobic digestion of microalgae for biomethane production, second and third generation of feedstocks. Elsevier, Amsterdam, pp 405–436

Jegannathan KR, Jun-Yee L, Chan ES, Ravindra P (2009) Design an immobilized lipase enzyme for biodiesel production. J Renew Sustain Energy 1(6):063101

Karemore A, Sen R (2016) Downstream processing of microalgal feedstock for lipid and carbohydrate in a biorefinery concept: a holistic approach for biofuel applications. RSC Adv 6(35):29486–29496

Kim D-G, La H-J, Ahn C-Y, Park Y-H, Oh H-M (2011) Harvest of Scenedesmus sp. with bioflocculant and reuse of culture medium for subsequent high-density cultures. Bioresour Technol 102(3):3163–3168

Kim D-Y, Vijayan D, Praveenkumar R, Han J-I, Lee K, Park J-Y, Chang W-S, Lee J-S, Oh Y-K (2016a) Cell-wall disruption and lipid/astaxanthin extraction from microalgae: chlorella and Haematococcus. Bioresour Technol 199:300–310

Kim KH, Lee OK, Kim CH, Seo JW, Oh BR, Lee EY (2016b) Lipase-catalyzed in-situ biosynthesis of glycerol-free biodiesel from heterotrophic microalgae, Aurantiochytrium sp. KRS101 biomass. Bioresour Technol 211:472–477

Kim HM, Oh CH, Bae HJ (2017) Comparison of red microalgae (Porphyridium cruentum) culture conditions for bioethanol production. Bioresour Technol 233:44–50

Koech AK, Kumar A, Siagi ZO (2020) In situ transesterification of spirulina microalgae to produce biodiesel using microwave irradiation. J Energy 2020:1–10

Lam MK, Lee KT (2015) Bioethanol production from microalgae. In: Handbook of marine microalgae, pp 197–208

Lee OK, Kim AL, Seong DH, Lee CG, Jung YT, Lee JW, Lee EY (2013) Chemo-enzymatic saccharification and bioethanol fermentation of lipid-extracted residual biomass of the microalga, Dunaliella tertiolecta. Bioresour Technol 132:197–201

Levine RB, Pinnarat T, Savage PE (2010) Biodiesel production from wet algal biomass through in situ lipid hydrolysis and supercritical transesterification. Energy Fuel 24(9):5235–5243

Lunprom S, Phanduang O, Salakkam A, Liao Q, Imai T, Reungsang A (2019) Bio-hythane production from residual biomass of chlorella sp. biomass through a two-stage anaerobic digestion. Int J Hydrogen Energy 44(6):3339–3346

Lupette J, Benning C (2020) Human health benefits of very-long-chain polyunsaturated fatty acids from microalgae. Biochimie 178:15–25

Mahmudul HM, Rasul MG, Akbar D, Narayanan R, Mofijur M (2021) A comprehensive review of the recent development and challenges of a solar-assisted biodigester system. Sci Total Environ 753:141920

Mehariya S, Goswami RK, Karthikeysan OP, Verma P (2021) Microalgae for high-value products: a way towards green nutraceutical and pharmaceutical compounds. Chemosphere 280:130553

Mishra S, Roy M, Mohanty K (2019) Microalgal bioenergy production under zero-waste biorefinery approach: recent advances and future perspectives. Bioresour Technol 292:122008

Mofijur M, Rahman SA, Nguyen LN, Mahlia T, Nghiem L (2022) Selection of microalgae strains for sustainable production of aviation biofuel. Bioresour Technol 345:126408

Molinuevo-Salces B, Riaño B, Hernández D, Cruz García-González M (2019) Microalgae and wastewater treatment: advantages and disadvantages. In: Microalgae biotechnology for development of biofuel and wastewater treatment, pp 505–533

Mussgnug JH, Klassen V, Schlüter A, Kruse O (2010) Microalgae as substrates for fermentative biogas production in a combined biorefinery concept. J Biotechnol 150(1):51–56

Nabi F, Arain MA, Rajput N, Alagawany M, Soomro J, Umer M, Soomro F, Wang Z, Ye R, Liu J (2020) Health benefits of carotenoids and potential application in poultry industry: a review. J Anim Physiol Anim Nutr 104(6):1809–1818

Nayak A, Raja R, Rao K, Shukla A, Mohanty S, Shahid M, Tripathi R, Panda B, Bhattacharyya P, Kumar A (2015) Effect of fly ash application on soil microbial response and heavy metal accumulation in soil and rice plant. Ecotoxicol Environ Saf 114:257–262

Nobre BP, Villalobos F, Barragan BE, Oliveira AC, Batista AP, Marques PASS, Mendes RL, Sovová H, Palavra AF, Gouveia L (2013) A biorefinery from Nannochloropsis sp. microalga–extraction of oils and pigments. Production of biohydrogen from the leftover biomass. Bioresour Technol 135:128–136

Nurdiawati A, Zaini IN, Irhamna AR, Sasongko D, Aziz M (2019) Novel configuration of supercritical water gasification and chemical looping for highly-efficient hydrogen production from microalgae. Renew Sustain Energy Rev 112:369–381

Okeke ES, Ejeromedoghene O, Okoye CO, Ezeorba TPC, Nyaruaba R, Ikechukwu CK, Oladipo A, Orege JI (2022) Microalgae biorefinery: an integrated route for the sustainable production of high-value-added products. Energy Convers Manag 16:100323

Pistelli L, Mondo AD, Smerilli A, Corato F, Piscitelli C, Pellone P, Carbone DA, Sansone C, Brunet C (2021) Microalgal co-cultivation prospecting to modulate vitamin and bioactive compounds production. Antioxidants 10(9):1360

Poh ZL, Kadir WNA, Lam MK, Uemura Y, Suparmaniam U, Lim JW, Show PL, Lee KT (2020) The effect of stress environment towards lipid accumulation in microalgae after harvesting. Renew Energy 154:1083–1091

Raheem A, Ji G, Memon A, Sivasangar S, Wang W, Zhao M, Taufiq-Yap YH (2018) Catalytic gasification of algal biomass for hydrogen-rich gas production: parametric optimization via central composite design. Energy Convers Manag 158:235–245

Rashid N, Ryu AJ, Jeong KJ, Lee B, Chang Y-K (2019) Co-cultivation of two freshwater microalgae species to improve biomass productivity and biodiesel production. Energy Convers Manag 196:640–648

Raslavičius L, Striūgas N, Felneris M (2018) New insights into algae factories of the future. Renew Sustain Energy Rev 81:643–654

Ren H-Y, Liu B-F, Kong F, Zhao L, Xing D, Ren N-Q (2014) Enhanced energy conversion efficiency from high strength synthetic organic wastewater by sequential dark fermentative hydrogen production and algal lipid accumulation. Bioresour Technol 157:355–359

Rizwan M, Mujtaba G, Memon SA, Lee K, Rashid N (2018) Exploring the potential of microalgae for new biotechnology applications and beyond: a review. Renew Sustain Energy Rev 92:394–404

Santiago-Morales IS, Trujillo-Valle L, Márquez-Rocha FJ, Hernández JFL (2018) Tocopherols, phycocyanin and superoxide dismutase from microalgae: as potential food antioxidants. Appl Food Biotechnol 5(1):19–27

Santos-Sánchez N, Valadez-Blanco R, Hernández-Carlos B, Torres-Ariño A, Guadarrama-Mendoza P, Salas-Coronado R (2016) Lipids rich in ω-3 polyunsaturated fatty acids from microalgae. Appl Microbiol Biotechnol 100:8667–8684

Sharifzadeh M, Sadeqzadeh M, Guo M, Borhani TN, Konda NM, Garcia MC, Wang L, Hallett J, Shah N (2019) The multi-scale challenges of biomass fast pyrolysis and bio-oil upgrading: review of the state of art and future research directions. Prog Energy Combust Sci 71:1–80

Sharma P, Gujjala LKS, Varjani S, Kumar S (2022) Emerging microalgae-based technologies in biorefinery and risk assessment issues: bioeconomy for sustainable development. Sci Total Environ 813:152417

Sialve B, Bernet N, Bernard O (2009) Anaerobic digestion of microalgae as a necessary step to make microalgal biodiesel sustainable. Biotechnol Adv 27(4):409–416

Singh J, Dhar DW (2019) Overview of carbon capture technology: microalgal biorefinery concept and state-of-the-art. Front Mar Sci 6:29

Singh D, Puri M, Wilkens S, Mathur AS, Tuli DK, Barrow CJ (2013) Characterization of a new zeaxanthin producing strain of Chlorella saccharophila isolated from New Zealand marine waters. Bioresour Technol 143:308–314

Sun X-M, Ren L-J, Zhao Q-Y, Ji X-J, Huang H (2018) Microalgae for the production of lipid and carotenoids: a review with focus on stress regulation and adaptation. Biotechnol Biofuels 11:1–16

Tang DYY, Khoo KS, Chew KW, Tao Y, Ho S-H, Show PL (2020) Potential utilization of bio-products from microalgae for the quality enhancement of natural products. Bioresour Technol 304:122997

Tse TJ, Wiens DJ, Reaney MJ (2021) Production of bioethanol—a review of factors affecting ethanol yield. Fermentation 7(4):268

Vanthoor-Koopmans M, Wijffels RH, Barbosa MJ, Eppink MHM (2013) Biorefinery of microalgae for food and fuel. Bioresour Technol 135:142–149

Wang T, Zhai Y, Zhu Y, Li C, Zeng G (2018) A review of the hydrothermal carbonization of biomass waste for hydrochar formation: process conditions, fundamentals, and physicochemical properties. Renew Sustain Energy Rev 90:223–247

Wijffels RH, Barbosa MJ (2010) An outlook on microalgal biofuels. Science 329(5993):796–799

Wong KH, Tan IS, Foo HCY, Chin LM, Cheah JRN, Sia JK, Tong KTX, Lam MK (2022) Third-generation bioethanol and L-lactic acid production from red macroalgae cellulosic residue: prospects of industry 5.0 algae. Energy Convers Manag 253:115155

Xu S, Elsayed M, Ismail GA, Li C, Wang S, Abomohra AE-F (2019) Evaluation of bioethanol and biodiesel production from Scenedesmus obliquus grown in biodiesel waste glycerol: a sequential integrated route for enhanced energy recovery. Energy Convers Manag 197:111907

Yadav K, Vasistha S, Nawkarkar P, Kumar S, Rai MP (2022) Algal biorefinery culminating multiple value-added products: recent advances, emerging trends, opportunities, and challenges. 3 Biotech 12(10):244

Yaşar F (2020) Comparison of fuel properties of biodiesel fuels produced from different oils to determine the most suitable feedstock type. Fuel 264:116817

Zhu L, Li S, Hu T, Nugroho YK, Yin Z, Hu D, Chu R, Mo F, Liu C, Hiltunen E (2019) Effects of nitrogen source heterogeneity on nutrient removal and biodiesel production of mono-and mix-cultured microalgae. Energy Convers Manag 201:112144

Feedstock to Fortune: Microalgae Bioconversion to Value-Added Products Via Circular Bioeconomy

Sankari Mohan, G. Sai Sri Rekha, M. Reshma Anjum, and Narasimha Golla

1 Introduction

According to the United Nations (UN), the current global population is around 7.8×10^9, and it is expected to reach 8.5×10^9 and 9.7×10^9 by the year 2030 and 2050, respectively (Orejuela-Escobar et al. 2021a, b). A special report issued by the International Energy Agency stated that the fuel consumption of economically developing nations expanded rapidly during the past few decades (Dineshkumar and Sen 2020). Fossil fuels are nonrenewable resources as they are the remnants of living creatures from the geological period, and it would take millions of years to produce them. It has been estimated that the majority of our energy consumption needs are accomplished by coal, natural gas, and petroleum with less than 10% leftover renewable energy sources, and it has been very clear that there will be no more fossil fuel resources in the upcoming decades (Yen et al. 2013). A study conducted by International Energy reported nearly 10 kg of carbon dioxide (CO_2) is released per gallon of diesel fuel consumed (Dineshkumar and Sen 2020) that strongly triggers global warming by increasing the temperature of the earth's atmosphere (Leong et al. 2021). Based on the current consumption rate, the depletion of fossil fuels and negative impacts on climate are the most challenging concerns that will be faced by the human race in future generation.

The transition from coal-based resources to sustainable and renewable sources is a golden thread to an environment that positively promotes the meteorological conditions (Rodrigues Dias et al. 2023) Clearing lands to utilize fossil fuels has been a

S. Mohan · G. S. S. Rekha · M. R. Anjum
Sri Padmavathi Mahila Visvavidyalayam (Women's University), Tirupati, Andhra Pradesh, India

N. Golla (✉)
Department of Virology, Sri Venkateswara University, Tirupati, Andhra Pradesh, India

K. Agrawal, P. Verma (eds.), *Biotechnological Advances in Biorefinery*, Interdisciplinary Biotechnological Advances,
https://doi.org/10.1007/978-981-97-5544-8_9

major concerning problem for the past few decades. The progression from a fossil-based economy to a circular bio-based economy is a worldwide strategy currently followed by emergent countries (Silva et al. 2022; Tsakona et al. 2019). Globally, biofuels will play a pivotal role in fulfilling the current societal demands for fossil fuels. Various potential feedstocks including plants, cyanobacteria, bacteria, and microalgae have been used for the synthesis of biofuels by replacing conventional fossil fuels (Kumar et al. 2020). However, plant-derived biofuels require large agricultural land to produce huge yields. Moreover, it competes with food production and thereby adversely affects both the bio-economy and sustainability of the country (Masri et al. 2018). Among various other renewable alternatives, microalgae have gained more attention in bioenergy production (Mishra et al. 2019).

Microalgae are phototrophic organisms, capable of utilizing natural resources such as water, sunlight, and atmospheric CO_2 to generate diverse products, including biofuel, biochemical, nutraceuticals, feed, bio-fertilizer, and other high-value-added products (Sarma et al. 2021; Rehmanji et al. 2021). They are a key source of bioactive compounds, such as proteins, polysaccharides, lipids, polyunsaturated fatty acids, vitamins, pigments, and enzymes that possess a wide range of biomedical and nutritional properties, such as antioxidant, antibacterial, antiviral, antitumor, regenerative, antihypertensive, neuroprotective, and immunostimulant activity in the body (Kapoor et al. 2023; Goswami et al. 2022a; Mehariya et al. 2021a, b; Dolganyuk et al. 2020). Microalgae have outstanding efficiency for CO_2 fixation representing their promising role in eliminating toxic substances from the environment (Koller et al. 2014). A study reported that 100 t of algal biomass can fix approximately 183 t of CO_2 by producing biofuels from microalgae, thereby promoting circular bio-economy (De Bhowmick et al. 2019). The concept of circular bio-economy is a win-win scenario to establish economic and eco-friendly biofuels by maximizing the consumption and recirculation of renewable resources and by minimizing waste production (Solis et al. 2020).

The generation of a single yield one at a time cannot maintain the economy of the production process (Kumar et al. 2020). Therefore, a multiproduct microalgae biorefinery process combined with the zero-waste discharge has emerged as a promising approach to produce multivariate products, including bioethanol, biodiesel, biohydrogen, biochar, biomethane, etc., from a single feedstock (Goswami et al. 2021a, b; Okeke et al. 2022). This waste-to-wealth algal biorefinery cost-effective method aids in the reduction of greenhouse gases by maintaining a low carbon footprint, thus promoting a sustainable and greener world (Leong et al. 2021). This study describes the conventional harvesting methods used for microalgae cultivation and highlights the potential role of the microalgae biorefinery process in terms of sustainability followed by challenges and future perspectives related to this process.

2 Biofuel Generations

Currently, depending on their feedstock, numerous techniques have been adopted to fulfill the global energy requirements (Li et al. 2015). The first-generation biofuels are derived from sugar-rich plant-based biomass, such as corn, soybean, sugarcane, and oil palm, as their feedstock (Kumar et al. 2020). For instance, first-generation ethanol was produced by the conversion of sugar from sugarcane and starch from corn (Orejuela-Escobar et al. 2021a, b). This production process utilizes food crops as their feedstock, thereby raising food prices, thus creating a food vs. fuel debate, which is a prime bottleneck in the development of first-generation biofuel technology (Kumar et al. 2020). Considering the limitations, the first-generation biofuels are replaced by second-generation biofuels, which are produced from bio-waste including nonedible parts of the plant, municipal sludge, industrial, food, fishery, and waste oils, thereby promoting low carbon footprint and recycling of waste (Kumar et al. 2020). In Brazil, bioethanol was produced from sugars isolated and fermented from sugarcane bagasse processing (Orejuela-Escobar et al. 2021a, b). However, they are not economically feasible as they require sophisticated and expensive technologies. While the second generation of biofuels experienced lukewarm attention due to their cost-effective methods, there is a need for the new generation to fulfill the requirements of bioenergy production around the world (Catone et al. 2021). The third generation utilizes aqueous phototrophic organisms (microalgae and macroalgae) as a precursor for the production of biofuels (Bochenski 2017). This transition from food feedstock to nonfood feedstock has gained much attention as they pose numerous benefits, such as minimal production period, low water requirements for their metabolism, high productivity rate, and low negative impacts on the environment (De Bhowmick et al. 2019). Apart from the wider applications of microalgae as potential feedstock, third-generation biofuels require high capital investments for the production and accumulation of lipids, contributing to a negative energy balance (Kumar et al. 2020). Nevertheless, these hindrances were resolved by fourth-generation biofuels, which use genetically modifying microalgae cell factories as their feedstock for the production of biofuels (Okeke et al. 2022).

3 Microalgae: A Renewable Feedstock

Microalgae are non-taxonomic single-cellular organisms and are considered a promising sustainable feedstock due to their high photosynthetic efficiency (Bochenski 2017). It has been proved that microalgae can produce a high yield by converting 10–20% more sunlight than traditional crops. Although the biochemistry of each microalgal strain varies, the majority of them consist of 6–70% of proteins, 4–64% of carbohydrates, 2–50% of lipids, and 1–6% of nucleic acid, depending upon their mode of cultivation and habitat (De Bhowmick et al. 2019). These microscopic solar energy-driven cell factories are receiving significant attention due to their exceptional characteristics, such as high growth rate, less cultivation time, minimal requirement of agricultural land for cultivation, production of multi products from a single feedstock, and restoration of carbon balance, thus controlling the emission of greenhouse gases and global warming and resolving food vs. fuel debates (Sarma et al. 2021). Apart from biofuels, microalgae have wider applications in pharmaceutical and nutraceutical industries. For instance, *Chlorella* has been used as a dietary food source of protein (Kumar et al. 2020). According to the WHO, *Arthrospira* is referred to as a superfood as it is rich in protein, iron, carbohydrates, vitamins, and minerals and can be safely administrated to children as well (Mitra and Mishra 2019). Moreover, microalgae can be used as a chief food supplement that plays a key role in the treatment of various diseases (Yen et al. 2013). These features make microalgae a potential candidate for the production of biofuels and economically valued products (Rizwan et al. 2015) (Table 1).

4 Microalgae Biorefinery

The renewable conversion of biomass into a variety of marketable goods (food, feed, materials, chemicals), as well as energy (fuels, power, and heat), is known as biorefinery (Olguín et al. 2022). Similar to petroleum refineries, biorefineries typically generate two types of products: high volumes of low-value products, like biofuels, and low-volume high-value products, like nutraceuticals. The aim behind integrating such a notion is to generate fuels in large quantities to meet energy demands, which will boost the profitability of high-value products by lowering energy costs and, crucially, reducing greenhouse gas emissions (De Bhowmick et al. 2019). Since microalgae may produce a variety of products, they are categorized as efficient sources in biorefinery processes. They are regarded as a sustainable source of biomass, which has advantages over other biomass sources like fast growth and no risk of fuel vs. food debate (Chew et al. 2017). A microalgae biorefinery offers several benefits over traditional biorefineries, including the ability to target several products, such as energy and bioactive chemicals, utilization of atmospheric working conditions, and renewable resources, like sunlight, air, and wastewater for their development and production. Microalgae biomass is an excellent precursor for the biorefinery approach to generate a variety of components suited for a range of industries, including the food, energy, pharmaceutical, and nutraceutical sectors. This versatility allows it to adapt to evolving market demands (Sarma et al. 2021). A zero-waste microalgal biorefinery that is both sustainable and energy-efficient will result from the reduction, recycling, and reuse of waste generated by the algal bioenergy conversion process. A biorefinery strategy involves the production of fuels, food, medicines, and other value-added goods on a single platform from a possible biomass feedstock (Agrawal et al. 2020, 2023; Goswami et al. 2022b; Mishra et al. 2019). The use of these biorefineries will contribute to reducing the financial disparity that currently exists between the expenses of fossil fuels and the cultivation and extraction of microalgae (Li et al. 2015).

Table 1 Various applications of microalgae as value-added products

Microalgal species	Bioactive compounds	Applications	References
Porphyridium purpureum	Phycoerythrin, phycocyanin, arachidonic acid, polysaccharides, superoxide dismutase, and GABA	Diagnostic tool, nutritional supplement, food additive, antioxidant, and inhibitory neurotransmitter in CNS	De Jesus Raposo et al. (2013), Chu (2012) and Olguín et al. (2022)
Nannochloropsis gaditana	Lutein, chlorophyll, astaxanthin, and β-carotene	Aqua culture feed, and nutritional supplements	Chu (2012)
Hematococcus pluvialis	Astaxanthin	Pigment for salmon	Chu (2012)
Arthrospira platensis	Phycocyanin, phycoerythrin, linolenic acid, algae protein, and vitamin B_{12}	Cosmetics, antioxidant, infant formula and health supplement	Chu (2012) and Olguín et al. (2022)
Botryococcus braunii	Chlorophyll and lipids—Triglycerides	Biofuels	Chu (2012)
Nephroselmis	β-Carotene and violaxanthin	Food colorant	Olguín et al. (2022)
Chlorella sorokiniana	Lutein, chlorophyll, astaxanthin, and β-carotene	Food supplement	Olguín et al. (2022)
Desmodesmus sp.	β-Carotene and violaxanthin	Cancer prevention	Olguín et al. (2022)
Aphanizomenon sp.	Mycosporine-like amino acids	Sunscreen	Chu (2012)
Chlorella protothecoides	Lipids-triglycerides	Biofuels	Chu (2012)
Nannochloropsis oculata	Omega-3 PUFA	Omega-3 supplements	Balasubramaniam et al. (2021)
Crypthecodinium cohnii	Docosahexaenoic acid (DHA)	Infant formula and bio supplement	Balasubramaniam et al. (2021)
Gambierdiscus	Okadaic acid	Antifungal activity	de Jesus Raposo et al. (2013)
Schizochytrium	Docosahexaenoic acid (DHA)	Aqua culture feed	Balasubramaniam et al. (2021)
Porphyridium cruentum	Phycoerythrin and polysaccharides	Anti viral agent and tool for medicinal studies	Chu (2012)
Euglena gracilis	Organic biomass	Dietary supplements	Balasubramaniam et al. (2021)
Amphidinium	Phycotoxins	To study neurodegenerative disorders	Chu (2012)
Ulkenia	Docosahexaenoic acid (DHA)	Food supplements	Balasubramaniam et al. (2021)

(continued)

Table 1 (continued)

Microalgal species	Bioactive compounds	Applications	References
Pavlova	Eicosapentaenoic acid (EPA)	Hypocholesterolemic	De Jesus Raposo et al. (2013)
Odontella aurita	Fatty acids	Nutrient supplements	Balasubramaniam et al. (2021)
Tetraselmis	Alpha-tocopherol	Antioxidant	De Jesus Raposo et al. (2013)
Phaeodactylum	Eicosapentaenoic acid (EPA)	Anti-inflammatory	De Jesus Raposo et al. (2013) and Chu (2012)
Dunaliella salina	β-Carotene and violaxanthin	Bio-colorant	Balasubramaniam et al. (2021) and Chu (2012)

5 Cultivation

Depending on the mode of nutrition, algal biomass can be cultivated by three approaches: photoautotrophic, heterotrophic, and mixotrophic processes (Kumar et al. 2020). Photoautotrophic cultivation is the commonly adopted mode, as the majority of the microalgae are phototropic organisms (Yin et al. 2020). In this process, the microalgae directly convert inorganic carbon into organic matter by utilizing light and CO_2 as energy and inorganic carbon sources through photosynthesis (Alavianghavanini et al. 2023). Thus, the productivity of this model will highly rely on lighting conditions (Moreno-Garcia et al. 2017). The photoautotrophic microalgae can grow by adsorbing industrial effluents as their precursor, which aids in maintaining environmental pollution. Hence, this mode is referred to as the most sustainable and cost-effective method (Yin et al. 2020; Kadimpati et al. 2020a).

In contrast to the photoautotrophic method, the heterotrophic mode of cultivation utilizes organic compounds as carbon sources for their growth and metabolism (Yin et al. 2020). They do not require an external solar energy source; hence, they can be cultivated in large fermenters under aseptic conditions, thereby minimizing microbial contaminations unlike the traditional autotrophic cultures (Lopes da Silva et al. 2019). They can be cultivated in dual methods: fed-batch system or heterotrophy-photoinduction system, thus contributing to higher yield (Moreno-Garcia et al. 2017). However, this mode of cultivation has also a few hindrances such as follows: (1) it requires sophisticated equipment, (2) the process is not eco-friendly and economically feasible, (3) and not all microalgae strains are adapted to this mode (Kumar et al. 2020).

The mixotrophic mode of cultivation is the combination of both photoautotrophic and heterotrophic methods, in which both CO_2 and organic carbon are assimilated simultaneously by the photosynthesis process (Yin et al. 2020). This is the most viable process employed in wastewater treatment to produce high biomass with large lipid content (Fal et al. 2023). It was reported that the efficiency of the mixotrophic mode is higher than the phototrophic and slightly lower than the heterotrophic mode of cultivation (Kumar et al. 2020). This was observed by (Ummalyma et al. 2021) in their study, which reported a sevenfold rise in the production of lipid content through the mixotrophic cultivation of *Chlorella pyrenoidosa* than with phototrophic mode. Moreover, this method can treat both organic and inorganic waste, thus contributing to a sustainable environment (Patel et al. 2020).

A well-organized and commercially viable cultivation system is essential to produce highly valued economical multivariate products from a single feedstock, thereby maintaining a circular economy (Yen et al. 2013). The cultivation of microalgal systems is broadly classified into three approaches: open, closed, and hybrid systems (Fig. 1) (Fal et al. 2023).

5.1 Open Pond System

The transition from conventional systems of cultivation to advanced systems is initiated by the Germans during World War II (Koller et al. 2014). Cultivation in open ponds is the most fundamental system for algal growth commonly constructed with concrete or plastic. They are large open-air reactors that usually extend up to 0.5 m in depth (Tomaz et al. 2017). They show high efficiency in the presence of sunlight, which facilitates the production of high-volume low-value products, including biofuels and health food (Siddiki et al. 2022). Relatively 80% of global algal biomass was generated in open ponds as they are easy to maintain and require low investment (Yin et al. 2020; Moreno-Garcia et al. 2017). *Scenedesmus, Spirulina*, and *Chlorella* are the most commonly cultivated species in open pond systems (Siddiki et al. 2022). Despite these advantages, it has many limitations, such as large arable land requirement, high degree of contamination, and poor regulation of growth parameters, like assimilation of CO_2 and evaporation of water (Mehariya et al. 2021a).

The most widely used design of an open pond is the raceway pond model proposed by Oswald in the mid-twentieth century (Mehariya et al. 2021a). They are closed oval-shaped structures extended with paddling to ensure efficient mixing of nutrients, thus enhancing the light absorption rate. These paddlewheels regulate the sedimentation of algal biomass to avoid contamination (Sarma et al. 2021). The most common strains that are cultivated in the raceway pond model are *Chlorella, Dunaliella, Spirulina*, and *Hematococcus* (Mehariya et al. 2021a). Comparatively, circular ponds are large containers equipped with lengthy rotary arms present at the center that are similar to paddle wheels in function. The diameter of these ponds varies from 40 to 50 m, while the depth ranges from 20 to 30 cm. It is the first artificial pond system used for the large-scale production of microalgae (Siddiki et al. 2022).

An inverted pyramid-shaped pond, also referred as V-shaped pond, is another alternative to raceway pond system, which provides maximum surface area to volume ratio that ensures high photosynthesis. It is considered as most cost-effective

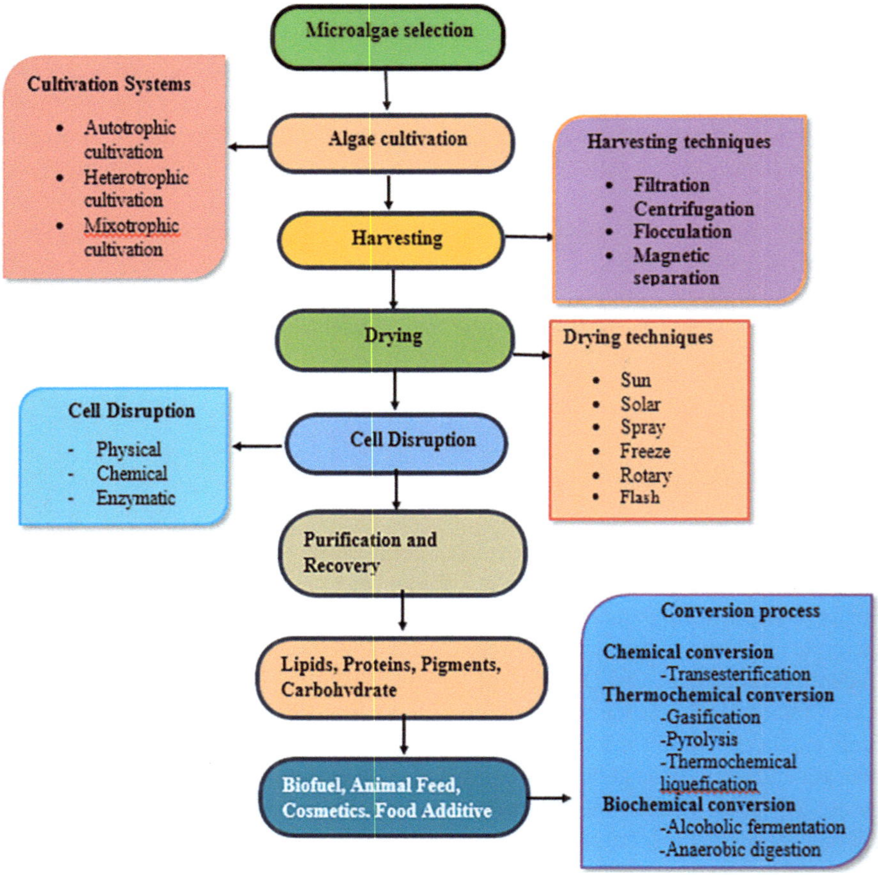

Fig. 1 Various steps involved in the cultivation of microalgae

system due to their high-capacity range of around 3000 L and low operational cost (Sarma et al. 2021). The unstirred open pond is a simple system with basic equipment required for the growth of microalgae. They do not contain a stirrer unit, leads to poor mixing, which is the major hindering element for production of algal biomass. There is a high chance of contamination due to their easy and plain mode of construction system (Mehariya et al. 2021a).

5.2 *Closed Systems*

Closed photobioreactors emerged as the best alternative to open pond systems to resolve the contamination problem observed in the latter (Mehariya et al. 2021a). The prime objective of these reactors is the large-scale production of pure culture of microalgae strains in aseptic conditions (Siddiki et al. 2022). Hence, it is the most accepted model for the cultivation of algal species to produce germ-free commercially valued compounds (Sarma et al. 2021). Moreover, it (1) requires minimal capital costs and (2) low power consumption; (3) it can have accurate monitoring of operational conditions, regular stirring, and biomass transfer; (4) it can be cleaned and sterilized easily (Kumar et al. 2020). Despite these features, it has various critical concerns, such as a high chance of excessive heating and accumulation of oxygen (O_2), thereby successive inhibition of growth (Deprá et al. 2018).

Tubular photobioreactors (PBR) are economically viable transparent cultivation systems, usually made up of polyvinyl chloride or polypropylene acrylic for the bulk production of high-priced compounds. The agitation and stirring process was carried out by an air pump equipped with the system. Based on the columns, they are widely classified into two categories: vertical and horizontal tubular photobioreactors (Mehariya et al. 2021a). Vertical column photobioreactors is a mono-cylindrical column formed with plastic. It comprises of air sparger for efficient mixing and assimilation of CO_2 for the systematic growth and metabolism of the culture. Due to their poor exposure to sunlight, they are the least preferred cultivation system. Bubble column and flat plate PBRs are the two forms of vertical column reactors (Sarma et al. 2021). Bubble columns are cylindrical with a 10 cm radius, and the height is approximately twice the radius. This accounts for their high surface-to-volume ratio prone to accumulation of O_2, and the formation of biofilms on the walls of the container leads to photoinhibition (Paladino and Neviani 2024). Flat-plate PBRs are characterized by rectangular-shaped boxes made up of clear substances, such as glass, plexiglass, and polycarbonate. The air is passed either from one side or supplied through a stirring mechanism by a motor. The cuboidal shape of flat-plate PBRs is responsible for a minimal light path that enables efficient mixing and thereby high production rate (Siddiki et al. 2022).

Horizontal tubular reactors are a group of tubes aligned horizontally toward light for maximum exposure of biomass to solar energy resulting in a higher light transition rate. They are placed in different possible orientations, including helicoidal, spiral, inclined, etc., that work with uniform principles. They have a high air retention period that increases dissolved CO_2 concentration in the culture medium. Nevertheless, this system has been reported to have a minimum photosynthetic rate due to the piling up of O_2 resulting in excessive energy consumption (Moreno-Garcia et al. 2017).

5.3 Hybrid System

Hybrid mode is a breakthrough in microalgae cultivation systems, which is formed by the unification of both open and closed systems (Mehariya et al. 2021a). It is a two-stage process, initiated in a closed system followed by open mode (Sarma et al. 2021). The first stage takes place in tubular photobioreactors, which promote steady growth of the algae, while the open raceway system ensures lipid buildup under various limiting factors (Sarma et al. 2021). For instance, it was proved in a study that the hybrid mode of cultivation produces 40% more biomass and 62% higher lipid content than produced in an open cultivation system (Mehariya et al. 2021a). The integration of both systems promotes a contamination-free medium resulting in commercial-scale productivity of biomass (Sarma et al. 2021). This hybrid system was constructed by combining the advantages of both systems while mastering the limitations of one another (Sarma et al. 2021).

Down streaming is a natural step followed soon after the microalgae cultivation process due to their elevated algal-to-water ratio and compact cytoplasmic lipid globules (De Bhowmick et al. 2019). Down streaming is a multistage procedure resulting in purification and product recovery. The primary goal of downstream processing is to fully value the leftover biomass to optimize product recovery and reduce costs (Malik et al. 2022). The conversion of algal biomass to biofuel is hindered by the high operational costs of downstream processing (De Bhowmick et al. 2019). For instance, the anticipated cost of the downstream processes in a multiproduct biorefinery ranges from 50% to 60% to produce algal biomass, which is significantly more than the 20%– 40% cost of conventional downstream processes in large-scale industries. Complete recovery of biorefinery multiproduct can be achieved by an efficient downstream approach, categorized into a series of steps that include harvesting, drying, cell disruption, extraction, product purification, and recovery (Mitra and Mishra 2019).

6 Harvesting

The crucial step in the microalgal biorefinery process is harvesting, which accounts for 20–30% of the overall expense, and it might differ based on the strain and strategy employed in the process (Sivaramakrishnan et al. 2022). The significant features of a harvesting technique itself determine the best harvesting approach. The best harvesting techniques should provide a germ-free environment throughout the process. Furthermore, the materials used in the harvesting process should prevent secondary pollution and biomass settling and produce biomass with high efficiency (Yin et al. 2020), and it should cut down the cost of down streaming process to a minimal range (Sivaramakrishnan et al. 2022). The harvested biomass can be categorized into two portions: the commercially valued and leftover products. The former is treated and purified under aseptic conditions to maintain and produce globally standard products, while the latter is employed in large-scale production of biofuels. The settled residues with high dietary values could be exploited as animal and aquaculture feed (Li et al. 2015). Centrifugation, flocculation, filtration, gravity sedimentation, and flotation are the most widely accepted techniques employed for microalgae harvesting (Deprá et al. 2018).

Centrifugation is a promising and expeditious harvesting technique used for successive extraction of water from biomass in a short period (Sarma et al. 2021). The most widely accepted and employed techniques of centrifugation by bulk industries for high-scale production of biomass are hydro-cyclone, solid bowl decanter, nozzle type, and solid ejecting disc. Among them nozzle approach is the most versatile technique, even though it requires high investment and energy consumption, which is the most concerning issue (Yin et al. 2020). Flocculation is a highly productive, cost-effective approach with minimum energy demands (Delrue et al. 2016). Various flocculation agents, such as inorganic flocculants, cationic polymers, flocculants, and recyclable flocculants, have been exploited for the harvesting process (Vermuë et al. 2018). The commonly used chemical flocculants are iron and aluminum salts that are capable of eliminating the surface charge of cells, resulting in the aggregation of biomass. But these are harmful and may cause pollution, which can be resolved by eco-friendly organic flocculants, such as chitosan and acrylic acid (Sarma et al. 2021). A nother alternative to this are inorganic flocculants that include the following: $FeCl_3$, $Al_2(SO_4)_3$, $AlCl_3$, $Fe_2(SO4)_3$, $CaCl_2$, $ZnCl_2$, $CaSO_4$, $CaCl_2$, $ZnSO_4$, $MgSO_4$, $MgCl_2$, and NH_4C (Yin et al. 2020). Moreover, biopolymers, such as chitosan, cationic starch, cassia gum, and guar gum, are also widely accepted flocculants (Dineshkumar and Sen 2020).

Filtration is the most manageable form of harvesting method that requires a basic selectively permeable membrane for its function (Sarma et al. 2021). The medium, along with algal biomass, passed over a porous membrane using electromotive force (Yin et al. 2020). Even though the mode of operation is very simple compared to other techniques with zero power consumption, it is not preferred, mostly because of its high risk of contamination and clogging problems (Malik et al. 2022). Different forms of filtration strategies have been encouraged in microalgae harvesting like micro-filtration, macro-filtration, ultra-filtration, vacuum filtration, pressure filtration, and tangential flow filtration (TFF) (Yin et al. 2020).

7 Drying

The algal effluent should be treated immediately after harvesting because of its fragile and fouling nature. Drying or dehydration is the popularly employed method to prolong the viability of microalgae for subsequent synthesis of highly prized products (Deprá et al. 2018). This method helps in the formation of undiluted biomass, thereby producing 90–95% of heavy biomass. In contrast, it requires a high-power supply for its efficient function, which is not an economically feasible approach. The generally used drying methods for this purpose are spray drying, freeze drying, drum drying, and sun drying. Among them, sun drying is the primary method with low-cost requirements, but it also has various limitations, like high chance of material damage, extended drying period, and need for wide drying surfaces. Spray drying is usually employed to obtain commercially valued products from various algal strains. Spray drying is a typical method used to dry several microalga species. This process can maintain the product's homogeneity throughout the procedure. But the main disadvantage is the increased risk of algal pigments fouling during the process (Siddiki et al. 2022).

8 Cell Disruption

Cell disruption is an essential step in the whole biorefinery process; it removes the cell wall barrier and aids in the collection of intracellular biomolecules and secondary products of our interest. It should perform strictly under moderate conditions to protect the native cell components of algal cells (Vermuë et al. 2018). Cell disruption is commonly used to extract cytoplasmic lipid content of the algal biomass that is used to generate high-standard products. The three important strategies used in microalgal cell disruption are (1) physical or mechanical, such as ultrasonication, high-pressure homogenization, bead-beating, and microwave radiation; (2) chemical, which includes cationic detergents and alkali; (3) and biological methods like the use of enzymes. Among these techniques, the physical method has a relatively higher capability of disruption than other processes (Dineshkumar and Sen 2020). They did not require specific parameters, like high temperature and optimum pH for their high efficiency; hence, it can be easily carried out in mild conditions (Vermuë et al. 2018). Mechanical methods like bead mill and ball mill involve the formation of ice molecules that eventually cleave the cell wall of algae to reveal the intracellular contents. Further, the microwave is an advanced process with around 80% efficiency rate (Mohan et al. 2020). Organic solvent extraction is the most popular chemical method used to extract microalgae lipids from ruptured cells. The frequently used organic solvents, like chloroform, methane, and benzene, have high toxic levels and volatile properties that eventually hamper the environmental conditions. Hence, they are replaced by highly preferred alternatives, such as bio-based solvents, ionic liquids and convertible solvents, supercritical fluids, subcritical water, and pressurized solvents, which are relatively more sustainable than organic solvents. The enzymatic or biological method is the most versatile method of cell disruption that utilizes cellulose, hemicellulose, and pectinase enzymes as precursors to open the cell. They do not require high energy nor do they release harmful volatile substances unlike the previous method (Siddiki et al. 2022).

9 Extraction

Suitable extraction techniques are essential for the generation of commercially prized products like biofuels from potential algal biomass. There are various extraction techniques exploited in microalgae biorefinery methods; hence, it is very important to select the most promising method for the production of maximum yield and minimum residue (Kumar et al. 2020). The key factors in choosing the most convenient approach are the perquisite solvent system, short extraction period, and optimum temperature. Moreover, the appropriate system has been employed based on the by-products produced in the subsequent steps. For example, an aqueous system is preferred for the synthesis of hydrophilic pigment proteins, while an organic solvent system is considered for the production of hydrophobic pigments, such as carotenoids and chlorophylls. However, the toxic native elements of these solvent systems will show a negative impact on the ecosystem. Hence, these conventional systems are merged with green solvents that eventually reduce the hazardous nature of organic solvents. The commonly used green solvents in the extraction process are supercritical fluids, switchable solvents, deep eutectic solvents, pressurized liquids, and ionic liquids, which will positively hamper the sustainability of the environment. A study proved that coproducts such as carotenoids, tocopherols, and fatty acids are extracted using a green solvent, called supercritical CO_2 fluid (Mitra and Mishra 2019).

10 Purification and Recovery

Within a biorefinery system, the extracted microbial elements are the primary criteria for the selection of appropriate purification methods, thus promoting a circular economy. The 50% ammonium sulfate precipitation is the most often used method to retrieve pure C-PC. Despite its high operation time, chitosan/activated charcoal-based C-PC purification technology is also a viable method for the synthesis of high-purity compounds. One-stage chromatographic strategy is a progressive method employed in the bulk purification of C-PC. Moreover, ultrafiltration and packed bed chromatographic approaches are suitable for the purification of essential algal pigment C-phycocyanin. In addition, membrane-dependent purification technologies also have a significant role in the recovery of high-end products. Various purification methods have been introduced to recover glacial acetic acid, including fractional distillation, argentite silica gel column chromatography, preparative-high-performance liquid chromatography, urea complexation, winterization, and argentated silica gel column chromatography (Mitra and Mishra 2019).

11 Conversion of Biomass to Biofuels

The transition from algal biomass to bioenergy or biofuels is commonly processed by three chief approaches: biochemical, thermochemical, and chemical conversion (Mishra et al. 2019).

11.1 Biochemical Conversion

The biochemical conversion is associated with the biological transition of algal biomass into biofuels for efficient energy transformation (Chew et al. 2017). The primary processes involved in this conversion are photobiological hydrogen production, alcoholic fermentation, anaerobic digestion, and microalgae-based microbial fuel cells (mMFCs) (Mishra et al. 2019).

11.1.1 Photobiological Hydrogen Production

Photobiological hydrogen production illustrates the production of hydrogen and oxygen gases by microscopic algae using water as a substrate (Chew et al. 2017). Algal species, namely, *Scenedesmus obliquus*, *A. variabilis*, *Chlorella fusca*, *Chlorococcum littorale*, *Platymonas subcordiformis*, and *Chlamydomonas reinhardtii*, are considered the potential candidates for the production of biohydrogen due to presence of hydrogenase enzyme encrypted in their DNA. Biohydrogen from algal biomass can be generated by using three mechanisms: (1) direct biophotolysis, (2) indirect photobiolysis, (3) and two-step dark or photo-fermentation. In direct biophotolysis, H_2 is produced in oxygen-free conditions by the hydrolytic activity of photosystem II via hydrogenase enzyme. Indirect photobiolysis is a two-stage process in which the microalgae are initially grown naturally in the phototropic environment under optimal parameters followed by a subsequent shift to anaerobic conditions to induce the production of molecular hydrogen gas. In the second phase, both the H_2 and O_2 gases are produced simultaneously and separated immediately (Mishra et al. 2019).

11.1.2 Alcoholic Fermentation

Alcoholic fermentation is the bioconversion of algal polysaccharides, such as starch, glycogen, and cellulose, into bioethanol in anaerobic conditions. The microalgal cell walls are rigid and intact; hence, they should be pretreated, which aids in the release of cytoplasmic saccharides, which are hydrolyzed into simple sugar through saccharification (Deprá et al. 2018). Acetone-butanol-ethanol (ABE) fermentation is a widely used technique for the production of biobutanol, which is a great alternative to bioethanol. The major product produced in this method is utilized for the production of biofuels, while the leftover portion is used as animal feed, thus promoting both sustainability and bioeconomy (Mishra et al. 2019).

11.1.3 Anaerobic Digestion

Anaerobic digestion is an economically viable process that involves the biological metabolism of organic biomass to biogas through a cascade of reactions under oxygen-deprived conditions (Tsegaye et al. 2021). This transition occurs in fourfold stages including (1) hydrolysis, (2) acidogenesis, (3) acetogenesis, and (4) methanogenesis (Sivaramakrishnan et al. 2022). In the initial step, the insoluble biochemical compounds, such as carbohydrates, lipids, proteins, and nucleic acids, are hydrolyzed into soluble inorganic substances. As the name suggests, the hydrolyzed substances are disrupted by acidogenic organisms, resulting in the formation of volatile fatty acids, CO_2, and NH_3. These are treated by acetogenins, which are degraded into carbon dioxide, CH_3COOH, and H_2 (Mishra et al. 2019). Finally, the methanogenic bacteria act upon the products formed in the previous stage, resulting in the production of 55–75% methane and 25–45% CO_2 and a small portion of nourishing liquid residue known as digestate (Farooq 2021).

11.1.4 Microalgae-Based Microbial Fuel Cells

The microalgae-based microbial fuel cell technology is a bioelectrochemical approach that has immense potential for the production of bioelectricity at the commercial level. The integration of this model into microalgae harvesting is not a zero-sum game, as it eliminates waste from the environment and simultaneously induces the production of electricity (Chew et al. 2017). This biochemical transition occurs in two phases: (a) oxidation and (b) reduction. In the first step, the algal biomass is oxidized at anode chamber to produce a mixture of protons and electrons. The synthesized electrons are directed toward the cathode chamber by a terminal electron acceptor through a superficial circuit, while the protons are driven through an ion exchange filter. In the next phase, the collected oxygen is reduced at the cathode chamber which reacts with the proton to form a water molecule. Apart from bioelectricity synthesis, this method also aids in the assimilation of nitrogen, carbon dioxide, and phosphorous (Mishra et al. 2019). Hence, they are considered a novel system for the generation of mass electricity by biodegradation of organic biomass, which promotes sustainability (Bhattacharya and Goswami 2020).

11.2 Thermochemical Conversion

The thermochemical approach illustrates the thermal deterioration of microalgal organic biomass into economically valued biofuels, such as bio-oil, syngas bio-char, etc. (Chew et al. 2017). This strategy is preferred over chemical and biochemical because of its complete conversion of biomass into final products with no production of waste (Catone et al. 2021). Depending on parameters, like temperature, pressure, and heating time, they are further classified into three methods, namely, hydrothermal liquefaction, gasification, and pyrolysis (Siddiki et al. 2022).

11.2.1 Hydrothermal Liquefaction

Hydrothermal liquefaction is a low-budget strategy widely employed for the synthesis of bio-crude. It is a novel approach that converts aqueous algal organic matter into liquid fuel. This method is based upon the key principle that supercritical condition water can behave as both an acidic and alkali catalyst, which induces a cascade of reactions, including isomerization, depolymerization, and repolymerization, that ensures the transition of organic compounds of algae into biofuels (Mishra et al. 2019). The prerequisite parameters, such as high-pressure range of up to 20 MPa while low temperature varies from 300 to 350 °C, are essential for the production of high-standard biofuels and valued added compounds (Deprá et al. 2018). This method can directly convert wet biomass into bio-crude, thereby reducing the energy consumption required for the drying process. These features made hydrothermal liquefaction a potentially accepted technique in microalgal biorefinery (Zaky 2021).

11.2.2 Gasification

Gasification is associated with the conversion of microalgae carbonic compounds into a mixture of volatile gases, such as syngas, methane (CH_4), hydrogen (H_2), carbon dioxide (CO_2), nitrogen (N_2), and carbon monoxide (CO). The production of syngas can be carried out by two approaches, namely, (1) traditional gasification and (2) supercritical water gasification. Traditional gasification requires high temperatures ranging from 800 °C to 1000 °C, and low pressure varies from 1 to 10 bars. In contrast, supercritical water gasification requires a relatively low-temperature range of 400 °C to 500 °C and high pressure ranging from 24 to 36 MPa. Rehmanji et al. (2021) reported a study that illustrates the conversion of organic biomass of *C. vulgaris* and *Spirulina* into combustible syngas under non-equivalence conditions through gasification. Syngas have a wide application associated with biofuels; it can also act as a chemical intermediate or can be directly used as fuel.

11.2.3 Pyrolysis

The thermochemical process of pyrolysis is carried out in inert and oxygen-deprived conditions. In this process, biomolecules, such as carbohydrates, proteins, and lipid, present in the algae are thermally decomposed into bio-oil and biochar. The main mechanism behind this process is the cracking and depolymerization of microalgal biomass due to excessive heating. An optimum temperature of 400–600 °C is most preferred for this process. The temperature may vary based upon the type of pyrolysis used, for example, in the case of microwave pyrolysis, the temperature increases to 800 °C, whereas in catalytic pyrolysis, it was carried out at 300 °C. In this process, temperature is the key parameter that eventually decides the final product. For instance, biochar is produced at high temperature, while low temperature favors the production of bio-oil. Hence, the efficiency of the yield will greatly depend upon the reaction temperature of pyrolysis (Mishra et al. 2019).

11.3 Chemical Conversion

Chemical conversion is an easy and commercially reliable method for the conversion of algae organic matter into biofuels like biodiesel. The most employed process in chemical conversion is transesterification (Chew et al. 2017).

11.3.1 Transesterification

Transesterification involves the conversion of triglycerides of algal biomass into fatty acid alkyl esters (FAAEs) by the action of a suitable catalyst. It occurs in a series of reactions initiated by lipid extraction from algal biomass, and they are further processed to generate FAAE and glycerol. FAAEs also referred to as biodiesel can be obtained by using three methods: (a) enzymatic transesterification, (b) microwave-assisted transesterification, and (c) supercritical transesterification (Mishra et al. 2019). Novozym 435 and Lipozyme TL IM are the most frequently used enzymes in the transesterification process (Haddadi et al. 2018). However, enzymatic transesterification is a time-consuming process compared to other techniques. Microwave-assisted transesterification and supercritical transesterification can utilize aqueous algae as their feedstock, thus reducing the operational cost of the drying procedure (Mishra et al. 2019). The molecular composition of long-chain fatty acids has a significant role in maintaining the quality of biodiesel. Hence, 5:4:1 ratio of C_{16}:1, C_{18}:1, and C_{14}:0 fatty acids are considered the most accepted composition for the production of high-standard biodiesel (Moreira et al. 2023). Alcohol is also a limiting factor as it determines the quantity of biodiesel produced in the final product (Deprá et al. 2018).

12 Microalga Biorefinery: For High-Valued Products

Microalgae is a nutrient-rich factory of bioactive compounds, such as lipids, carbohydrates, and proteins, thus making them a potential feedstock for the extraction of numerous commercially valued products, including biofuels, pigments, vitamins, pharmaceuticals, and nutraceuticals that help to encounter global malnutrition problem (Table 1) (Orejuela-Escobar et al. 2021a, b). They also have a broad-spectrum role in antioxidant, immune response, anticancer, anti-inflammatory, antiviral, antifungal, neuroprotective, and antibiotic activity, which leads them as a potential agent in biotechnological applications (Koller et al. 2014). Residual algal biomass can also be used as animal and aqua feed, and they can be employed as biofertilizer, and chemical feedstock (Rizwan et al. 2015), thus maintaining an integrated biorefinery approach for circular bio-economy and sustainable environment (Fig. 2) (Caetano et al. 2020).

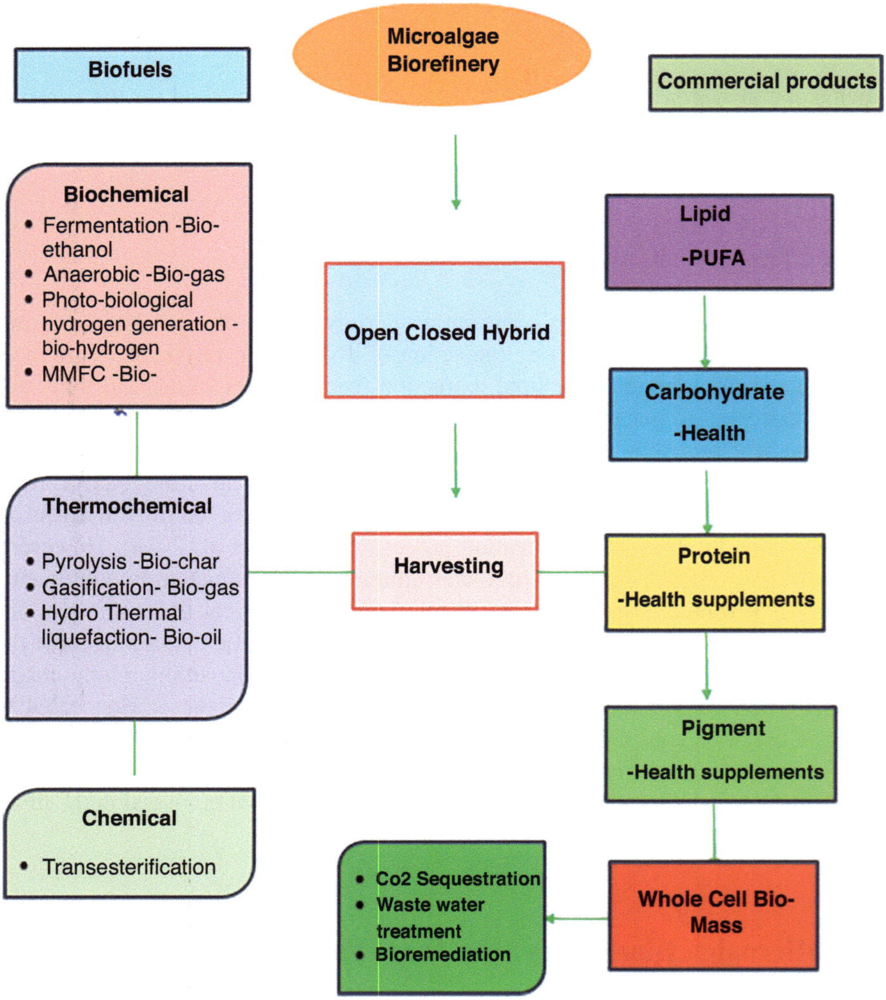

Fig. 2 Different methods involved in microalgal biorefinery

12.1 Biofuels

The four-generation biofuels are promising alternatives to nonrenewable fossil fuels that greatly influence the ecosystem across the world (Li et al. 2015). The biofuels extracted from microalgae are low toxic, have high biodegradability properties, and are devoid of sulfur; hence, they are mostly preferred in bulk industries for mass production (Yen et al. 2013). Various energy-rich biofuels have been produced from microalgae, such as biodiesel, bioethanol, biogas, biomethane, biohydrogen bio-char, etc. Biodiesel and bioethanol are the chief products in biorefinery systems, while others are considered secondary products (Kumar et al. 2020; Kadimpati et al. 2020b, c).

12.1.1 Bioethanol

Microalgae are the cell factories of polysaccharides, such as starch, cellulose, and glycogen, which are the key components for the production of bioethanol (Moreno-Garcia et al. 2017). The low lignin content of microalgal carbohydrates facilitates them as a suitable feedstock for the production of bioethanol, thus easing the purification process (Catone et al. 2021). The main mechanism employed for the production of bioethanol is the scarification or fermentation of sugars (Kumar et al. 2020). Microalgae strains, such as *Dunaliella*, *Chlamydomonas*, *Scenedesmus*, *Spirulina*, *S. cerevisiae*, and *Z. mobilis*, are the most widely used microalgal strains for the mass production of bioethanol (De Bhowmick et al. 2019). Moreover, in China, *Mychonastes* is also considered a potential candidate for the generation of bioethanol (Li et al. 2015). A study documented by Xiu et al. (2017) showed that *Chlorella vulgaris* is a promising feedstock for the production of bioethanol. Despite numerous benefits, this process shows a negative impact on climate, thus influencing the yield of bioethanol (Olguín et al. 2022). The production of bioethanol can be summarized in the following reaction (Sivaramakrishnan et al. 2022; Kadimpati et al. 2020a)

$$C_6H_{12}O_6 \rightarrow 2C_2H_5OH + 2CO_2 + \text{heat}$$

12.1.2 Biodiesel

Microalgal-based biodiesel is commonly produced by the transesterification of algal lipids, like triglycerides, monoglycerides, diglycerides, and trace portion of free fatty acids (De Bhowmick et al. 2019). Microalgae are considered a potential feedstock for the production of biodiesel due to their huge cytoplasmic lipid content (16%–77%) and rapid metabolism rate. Oleic, palmitic, and stearic acids are the standard fatty acids present in the microalgae used for the production of ideal biodiesel (De Bhowmick et al. 2019). Different studies reported that microalgal species, like *Arthrospira* (3), *N. oleoabundans* (Malik et al. 2022), *Chlorella vulgaris*, *Chlorella salina* (Okeke et al. 2022), *P. tricornutum* (48), *C. pyrenoidosa*, and *S. mangroves* (Sharma et al. 2020), and marine algal strains, such as *Nannochloropsis oceanica*, *N. gaditana*, *Dunaliella salina*, *Tetraselmis suecica*, and *N. salina* (Zaky 2021), are used as a promising feedstock for the production of biodiesel. Relatively, microalgal extracted biodiesel requires smaller arable land for their cultivation than plant cultivation (Yen et al. 2013), and it greatly minimizes greenhouse gas emissions and thereby reduces global warming; hence, it is referred to as green fuel (Sarma et al. 2021). The most addressed issue of algal-derived biodiesel is the requirement of high investment, and it can be encountered by the biorefinery model (Rizwan et al. 2015).

12.1.3 Biogas

Biogas is a group of gases mainly composed of methane, carbon dioxide, hydrogen, and other volatile organic compounds under oxygen-deprived conditions. Delipidified spent microalgae biomass is considered a suitable feedstock for the mass production of biogas. The major constituent of biogas is methane, which covers almost 55–70% of the total biogas produced during the process (Catone et al. 2021). B iomethane is the methane portion of biogas, while biohydrogen is the part of hydrogen extracted from biogas, whereas biohythane refers to the combination of methane and hydrogen gas produced from biogas (Sharma et al. 2020). Algal-derived biomethane is a good alternative to conventional sources, but it is not a viable option due to its high operation cost and low sustainability (Paul et al. 2023). Therefore, the integrated biorefinery model aids in the production of biomethane as a coproduct and biodiesel or bioethanol as a chief product from a single feedstock. The common species used for the production of biogas are *Melosira*, *Oscillatoria*, *Spirulina*, *Scenedesmus*, *Chlorella vulgaris*, and *Euglena* (Sivaramakrishnan et al. 2022). Currently, biogas production is considered a nonprofitable process; hence, zero-waste approaches like waste effluent treatment and a decrease in eutrophication are employed to improve bioeconomy (Mehariya et al. 2021a).

12.1.4 Biohydrogen

Hydrogen is a promising geothermal source with high energy production. Hydrogen is not easily found in nature, but it can be created using primary energy sources (Deprá et al. 2018). Due to its carbon-free nature, high energy production, and ability to produce water upon combustion, hydrogen is being utilized worldwide as an alternative to fossil fuels. Hydrogen gas created through biological processes is known as biohydrogen (Sharma et al. 2020). The future-generation fuel, biohydrogen ($BioH_2$), has many advantages over existing liquid or gaseous fuels. Through photolysis, microalgae can naturally produce $bioH_2$, and their biomass can also be utilized as feedstock for fermentative biohydrogen. It has been utilized to produce power using internal combustion engines or fuel cells over the past decades. The creation of hydrogen ($bioH_2$) using biological means is a more energy-efficient and ecologically friendly process than traditional thermochemical and electrochemical methods, including water electrolysis and gasification (Catone et al. 2021). Because of their comparatively greater potential to produce hydrogen than other green algae, *Chlamydomonas reinhardtii* and *A. variabilis* (De Bhowmick et al. 2019), Scenedesmus (Sarma et al. 2021), *Acutodesmus obliquus*, *Galdieria sulphuraria* (Siddiki et al. 2022), *Chlorella vulgaris, Scenedesmus obliquus*, and *natural algal Consortium* (Catone et al. 2021) have drawn more attention than other species. Even though microalgae are used to produce biohydrogen products, they have two drawbacks: low capacity for producing hydrogen and no waste utilization (Haddadi et al. 2018).

12.1.5 Biochar

Biochar is a carbonaceous compound generated by thermal degradation (hydrothermal liquefaction and pyrolysis) of organic biomass under the subcritical range of oxygen and elevated temperature (Mehariya et al. 2021a). Certain functional groups in biochar improve the ability to remove hazardous substances from soil and wastewater (Kumar et al. 2020). It has been shown that biochar has the potential to be used as a fertilizer for the soil, for pollutant cleanup, in a rapid aging process, and in the formation of biocomposite materials (De Bhowmick et al. 2019). Nonetheless, new research indicates that it can serve as a thermal coal or fuel source for the production of power (Mehariya et al. 2021a). Because of its 1–100 nm porous structure, biochar has the potential to improve soil's ability to retain water. Its porous structure can also absorb nutrients and serve as a home for microorganisms, which is good for plant growth. The high carbon content and strong chemical stability of biochar combine its application with the soil into a concept of sequestering carbon. Additionally, biochar may be used as a bio-adsorbent for water and air treatment processes to eliminate certain harmful organic and inorganic contaminants. It was demonstrated that the *Tetraselmis chui pyrolyzed* biochar exhibited a high electrical conductivity (Saiyud et al. 2022). Moreover, biochar produced from *Scenedesmus* is also a potential source for bioelectricity production (Catone et al. 2021).

12.2 Bioactive Compounds

Microalgae are abundant in biochemical compounds; hence, they are regarded as renewable resources for the next generation (Vermuë et al. 2018). The important components that inhabit algal biomass, include lipids, proteins, carbohydrates, and pigments (Malik et al. 2022). The new search for natural sources of energy, food, and high-value commercial products like food supplements and cosmetics was perfectly fulfilled by the bioactive compounds of microalgae, which could significantly reduce the expenses associated with biomass production, harvesting, and processing (Caetano et al. 2020).

12.2.1 Polyunsaturated Lipids

Long-chain fatty acids possessing more than 20 double-bonded carbons in their structure are known as polyunsaturated fatty acids (PUFAs). Since the long-chain PUFAs found in fish come from digested microalgae in oceanic habitats, it makes sense to think of microalgae as a direct possible source of this compound (Yen et al. 2013). Among various strains, *Crypthecodinium cohnii* (Chalima et al. 2019), *Chlorella vulgaris, Auxenochlorella protothecoides* (Chalima et al. 2017), and *Ulkenia* (Lopes da Silva et al. 2019) are widely employed for the production of PUFAs. Apart from diverse bio-based products, PUFAs are the most significant multifunctional compounds derived from algal biomass. It has been found that two well-researched PUFAs, namely, eicosapentaenoic acid (EPA) and docosahexaenoic acid (DHA), are extracted from microalgae. While these two molecules are essential for good health and are not naturally synthesized by the human race, they possess immense value in the food and dietary supplement industries. PUFAs are extensively utilized for promoting the growth and metabolism of brain and vascular systems, particularly among young children. Moreover, they are widely accepted as vital nutritional components that can assist in preventing some heart conditions, and they also have significance in human health and well-being (Kumar et al. 2020).

12.2.2 Carbohydrates

Microalgae are a good source of carbohydrates that can readily store sugars and have a relatively high photosynthesis rate, hence commonly possessing an extremely high level of carbohydrates that comprises around 50% higher than their dry weight. The primary components of algal carbohydrates are glucose, starch, cellulose, and a mixture of polysaccharides. Among these, the generation of biofuels, like bioethanol and hydrogen, often uses algae glucose or starch. Microalgal polysaccharides are excellent sources of metabolically active compounds for use as dietary supplements, cosmetic compounds, and biological medicinal agents, because they can regulate the body's defenses and inflammatory responses (Chew et al. 2017). Recently, industries primarily concentrate more on exopolysaccharides, which are firmly linked to the cell wall surface, compared to polysaccharides found inside the cell (Haddadi et al. 2018). They are valuable substances derived from the secretions of algae that have extraordinary medicinal characteristics corresponding to their molecular weight and shape (Dolganyuk et al. 2020). It has recently been discovered that they have biological activities, such as anti-allergic, anti-ulcer, tumor-fighting, anticoagulant, antimutagenic, antimicrobial, and radioprotective (Rehmanji et al. 2021). In addition, fucose and rhamnose derived from *Porphyridium* are used as thickening agents, food additives, and stabilizers (Orejuela-Escobar et al. 2021a, b).

12.2.3 Proteins

Approximately 50–70% of microalgae is composed of proteins, making them its key components. One of the essential compounds of microalgae biorefineries is protein, which is valuable nutritional for both humans and animals (Chew et al. 2017). The total number of necessary amino acids that are needed for various kinds of activities determines the nutritional value of algal protein (Orejuela-Escobar et al. 2021a, b). For instance, the amino acid profile of diatom microalgae includes serine, alanine, arginine, leucine, glycine, aspartate, and threonine, whereas the microalgal strains *Porphyridium cruentum, Haematococcus pluvialis* and *Nannochloropsis gaditana* have aspartic acid, threonine, serine, glutamic acid, glycine, alanine, cysteine, and valine as their core amino acids (Dolganyuk et al. 2022). The most researched strains are those that yield the largest quantity of proteins, such as *Arthrospira maxima, Spirulina platensis, Scenedesmus obliquus, Chlorella vulgaris, Dunaliella salina,* and *Synechococcus*. Huge quantities of *Chlorella vulgaris* and *Arthrospira platensis*, often known as *spirulina,* are manufactured and marketed as healthful products for nutrition and food supplements. Industrial enzymes and a wide variety of high-value therapeutic proteins, such as human growth hormone, erythropoietin, and functional antibodies, have all been produced using *Dunaliella tertiolecta* (Kumar et al. 2020). Additionally, the microalgal peptides have medicinal applications, for example, *Tetraselmis suecica* produced antimicrobial peptides that have bactericidal properties against both Gram-positive and Gram-negative bacteria (Okeke et al. 2022).

12.2.4 Pigments

Microalgae have a colorful appearance owing to pigments that can absorb light and trigger photosynthetic activities (Yen et al. 2013). The best-analyzed example of small-quantity, high-value components derived from microalgae is pigments. Pigments are valuable assets that are widely used in a broad range of industries, including paint, ink, food, textiles, personal-care products, and numerous other fields. These pigments have been used as nutritional coloring agents, biomaterials, supplements, cosmetics, therapeutic, and animal feed precursors for vitamins (Chew et al. 2017). Algae have three main kinds of photosynthetic pigments: phycobilins, carotenoids, and chlorophylls (Yen et al. 2013). *Dunaliella* and *Haematococcus* have been emphasized as potent pigment producers (Koller et al. 2014). Subsequently, chlorophyll a is present in nearly every type of algae and captures the greatest percentage of sunlight; hence, it plays a significant role in the process of photosynthesis. On the other hand, only green algae contain chlorophyll b, brown algae and diatoms include chlorophyll c, and red algae contain chlorophyll d. In addition, cyanobacteria also contain chlorophyll c, d, e, and f. The food, paper, and textile sectors make use of the important natural colorants obtained from chlorophyll derivatives (Deprá et al. 2018). Carotenoids, which include xanthophylls and carotenes, exhibit secondary light photon-capturing pigment that aids in the absorption of light energy by chlorophyll, the primary pigment (K). It is especially well-known that carotenoids have antioxidant qualities. Astaxanthin, a carotenoid enhanced by β-carotene oxygenation, has received an extremely high score (Li et al. 2015). It is used for medicinal purposes to deal with a variety of illnesses, such as cancer and diabetes as well as inflammatory, ocular, skin, liver, and cardiovascular problems. Cyanobacteria and red algae are the only organisms that have phycobilins, which include phycocyanins, phycoerythrins, and allophycocyanins. *Spirulina* and *Arthrospira* are widely recognized for the production of phycobilin pigments, such as phycocyanin, which can be employed as an antiseptic drug. C-phycocyanin is directly used as a nutritional food produced by *Spirulina* (Kumar et al. 2020).

12.3 Commercial Products

Microalgae are solar energy-driven cell factories that produce extremely valuable bioactive chemicals with health significance that include proteins, pigments, polyphenols, antioxidants, polyunsaturated fatty acids, vitamins, minerals, sterols, and polysaccharides. These organic compounds are known to provide a variety of health benefits, including hepatoprotective, neuroprotective, antiviral, anticancer, antioxidant, hypoglycemic, antiseptic, and antifungal effects. The discussion below provides an overview of current functional research on extraordinary biological activities of microalgae-bioactive substances that promote health (Olguín et al. 2022).

12.3.1 Pharmaceuticals

The need for natural pharmaceutical goods is steadily increasing worldwide, and algae are predicted to be able to supply a portion of this demand. A variety of medications can be made using primary and secondary metabolites of algal biomass, and the most commonly employed species for this purpose are *Arthrospira*, *Chlorella*, *Dunaliella*, and *Haematococcus* (Orejuela-Escobar et al. 2021a, b). In the early research on microalgae as potential sources for pharmaceutical uses, *Chlorella vulgaris* was found to create a chemical that suppresses its growth in antibiotic-related experiments (Rehmanji et al. 2021). It is anticipated that several bioactive molecules derived from microalgae will be employed as effective drugs to regulate the body's immune responses and to develop oral vaccines against the influenza A and HIV viruses. These molecules also act as histamine inhibitors and can be used for antiallergenic therapies. Additionally, the neuroprotective compounds that microalgae create help to reduce the progression of disorders that impact the neurological system. For example, *Spirulina platensis* produces neuroprotective compounds that have demonstrated efficaciousness against Parkinson's and Alzheimer's illnesses (Rehmanji et al. 2021). Chlorella growth factor (CGF), a nucleotide peptide complex, comprised of nucleic acid derivatives discovered in only *Chlorella* sp. In addition to protecting the cells from physiological impacts, this promotes tissue repair (Koller et al. 2014).

12.3.2 Functional Foods

Algae have been considered a vital source of nourishment for thousands of years. A healthy lifestyle demands a balanced nutritional diet, adding functional foods from microalgae to the diet will be an appropriate solution (Rehmanji et al. 2021). Microalgal cells are regarded as ideal candidates for the popular "functional food" (Koller et al. 2014). The foods made from microalgae are currently being sold as dietary supplements. They come in the form of liquids, capsules, tablets, powders, and gums. They may also be found as components in chocolates, gums, refreshments, noodles, and pasta, as well as morning grains, wine, and other drinks (Rehmanji et al. 2021). Human consumption of microalgae is restricted to a few numbers of species, including *Dunaliella, Aphanizomenon, Nostoc, Chlorella vulgaris*, and *Arthrospira* (Dolganyuk et al. 2020). Several popular dishes, including pasta, noodles, desserts, ice cream, bean curd sweets, and baked goods, are enhanced nutritionally and healthily by adding microalgal species, like *Spirulina, Chlorella, and Dunaliella.* Moreover, various kinds of food, such as baby meals, formula, eggs, dairy, pastries, and nonalcoholic drinks, contain microalgae-derived DHA as a key ingredient (Siddiki et al. 2022).

12.3.3 Cosmetics

The pharmacological effects of sunscreens and photoprotector medications are identical to properties like photoprotection, antiaging, and free radical scavenging of algal-based cosmetics. Furthermore, the moisture-retention qualities of microalgal proteins and their derivatives give skin and hair prerequisite texture. Owing to their exceptional qualities, HVABs derived from microalgae play a vital role in the cosmetic industry, acting as renewable components and mitigating the negative effects of the hazardous chemicals used in cosmetic products. The cosmetics industry uses a wide range of extracts from microalgae organisms, including *Padina, Polysiphonia, Spirulina, Colpomenia*, and *Halymenia* as antiaging agents, *Ulva lactuca* as moisturizing and skin-softening agents, *Kappaphycus alvarezii* as wound-healing agents, and *Ecklonia cava* as hair growth promoters (Siddiki et al. 2022). A few microalgal products used in cosmetics include Givaudan Active Beauty's proprietary polluting, an antiaging cream made of a combination of organic components. A natural astaxanthin extract from *H. pluvialis* is used by *Solasta astaxanthin* to minimize wrinkles and enhance skin elasticity and hydration. Based on these investigations, bioactive substances obtained from marine algae show promise for use as skin-whitening agents (Rehmanji et al. 2021).

12.4 Environmental Protection

The various aspects of environmental protection have been discussed as follows:

12.4.1 Carbon Sequestration

Over the past few decades, human activity has caused the level of carbon dioxide in the Earth's atmosphere to rise to a dangerous extent. This has led to global environmental change, leading to the greenhouse effect, which has negative effects on the globe (Gupta et al. 2022). The atmosphere now contains 411 parts per million of CO_2, up from 313 parts per million in 1960. Due to the rise in atmospheric CO_2 levels, ocean water becomes more acidic, which has a major impact on marine life. As a result, we must immediately create a suitable plan to lower or stabilize the amount of CO_2 in the atmosphere. Compared to vascular plants, microalgae can eliminate CO_2, 10 to 50 times more successfully without affecting people or other animals (Okeke et al. 2022). Considering a primary focus on the CO_2 concentration mechanism (CCM), which comprises carbonic anhydrase and rubisco as the primary enzymes for CO_2 fixation, microalgae have a great potential for CO_2 biosequestration and can assimilate carbon dioxide in their cells (Fal et al. 2023). Furthermore, while microalgae grow in wastewater, their photosynthesis sequesters carbon dioxide and builds oxygen, which lowers greenhouse gas emissions and minimizes climate change (Alavianghavanini et al. 2023). Microalgae's capacity to absorb nutrients and CO_2 sequestration is receiving more attention in the development of circular economy models, with self-suitability as an ultimate goal (Mohan et al. 2020).

12.4.2 Wastewater Treatment

Municipalities and some industrial sectors, including aquaculture, dairy, food, pharmaceuticals, swine, and textiles, produce wastewater that is high in nutrients. The organic and inorganic nutrients in the wastewater from these sectors increase eutrophication, a serious hazard to ecosystems (Mehariya et al. 2021a). Globally, 600 million m^3 of wastewater is estimated to be produced annually by agricultural industries, including dairy farms, piggeries, poultry businesses, and meat processing facilities (Alavianghavanini et al. 2023). Elimination of organic matter and nutrients like phosphorus and nitrogen in industrial and urban wastewaters requires the use of effective remediation strategies. Choosing microalgae has become a viable method that could be able to eradicate harmful substances (Okeke et al. 2022). Because of their rapid development and strong barrier to harmful substances, *Chlorella* and *Scenedesmus* have been determined to be the most ideal microalgae for growing on wastewater. Nevertheless, wastewater can also support the growth of other freshwater microalgae like *Desmodesmus and Cyanobacteria* as well as marine microalgae like *Nannochloris and Tetraselmis* (Catone et al. 2021). Microalgae release a mixture of bioactive compounds like lipids, carbohydrates, and other substances during their growth in wastewater. These can be exploited to manufacture biofuel. Additionally, the filtered water can be utilized for irrigation in farming. These technologies can therefore combine the treatment of effluent with the synthesis of biofuel and the recycling of water for crop cultivation.

12.4.3 Bioremediation

Environmental degradation is currently rising exponentially due to the substantial rapid increase in the human population. It is widely recognized that cutting down the human-caused pollution from air, water, soil, and food is a significant problem and has led to the emergence of ecological and health hazards (Olguín et al. 2022). Phytoremediation is an eco-friendly method of removing heavy metals from polluted waste water (Okeke et al. 2022). More than 50 years ago, Oswald and Golueke in the United States developed the first method of growing microalgae in wastewater with the purpose of producing biomass and bioremediating the effluent (Alavianghavanini et al. 2023). By serving as bioremediation agents for the chemical components, these microorganisms encourage the reuse and treatment of water as well as the generation of biomass with a high market value (Moreira et al. 2023). Thanks to carbon capture and utilization (CCU) technology, algal bioremediation is becoming interesting, since it can clean up trash and create biomass for biofuels while reducing carbon emissions (Pekkoh et al. 2024). As a result, the use of microalgae biomass from biorefineries for wastewater bioremediation as well as for biofertilizers and biostimulants is emerging as a new trend that can improve soil quality and increase food production in an economically viable, environmentally sustainable, and beneficial manner (Olguín et al. 2022).

13 Challenges and Future Perspectives

Although the generation of microalgal bioenergy is a renewable and environmentally friendly process, several significant obstacles prevent its commercialization including choosing a viable strain, the amount of water needed, collecting biomass, the availability of feedstock during the winter and wet seasons, and choosing an appropriate method for converting biomass into fuel. (De Bhowmick et al. 2019). Because of the increased expenses involved in the downstream processing of the biomass, algal biorefineries are not profitable. Moreover, the separation of various metabolites individually without impacting other chemicals presents another difficulty for the biorefinery concept (Malik et al. 2022). Research on this cost-cutting trend, which will result in a circular economy and a financially sustainable biorefinery system, will increase in the upcoming years (Moreira et al. 2023). In this case, even if there are obvious bottlenecks in the algae-based methods, the obstacles have been reduced, since a foundation for the development, processing, and industrial utilization of microalgae has been created through intensive study utilizing recently developed new technologies. Microalgae-based processes have the potential to significantly contribute to world development in this way (Deprá et al. 2018).

14 Conclusion

Algal biomass has demonstrated enormous promise for producing biofuels and natural products in more sustainable, environmentally friendly ways. The subject of microalgae-based biorefinery research is still in its infancy and has not yet developed to the point, where it can generate a consistent and trustworthy dataset comparing the many technological options at each stage of the processing of microalgae. Further research is still extremely important to reveal the progressive biorefinery concept as financially viable under realistic conditions. In this situation, it is necessary to use creative biorefinery systems to produce high-value goods and biofuels. It is also possible to manufacture other bioproducts with microalgal origins, like biofertilizers. Furthermore, circular economy-based integrated processes in microalgae biorefineries boost resource recovery while also improving process profit ability and efficiency.

References

Agrawal K, Bhatt A, Bhardwaj N, Kumar B, Verma P (2020) Algal biomass: potential renewable feedstock for biofuels production–part I. In: Biofuel production technologies: critical analysis for sustainability, pp 203–237

Agrawal K, Ruhil T, Verma P (2023) Microalgae: a way toward sustainable development of a society. In: Srivastava N, Mishra P (eds) Basic research advancement for algal biofuels production. Springer, Singapore, pp 259–277

Alavianghavanini A, Shayesteh H, Bahri PA, Vadiveloo A, Moheimani NR (2023) Microalgae cultivation for treating agricultural effluent and producing value-added products. Sci Total Environ 912:169369

Balasubramaniam V, Gunasegavan RDN, Mustar S, Lee JC, Mohd Noh MF (2021) Isolation of industrial important bioactive compounds from microalgae. Molecules 26(4):943

Bhattacharya M, Goswami S (2020) Microalgae–a green multi-product biorefinery for future industrial prospects. Biocatal Agric Biotechnol 25:101580

Bochenski T (2017) Techno-economic analysis of microalgae biorefinery in arid environments for co-production of biofuels and value-added compounds. Unpublished doctoral thesis, Masdar Institute of Science and Technology, Abu Dhabi, United Arab Emirates

Caetano N, Melo AR, Gorgich M, Branco-Vieira M, Martins AA, Mata TM (2020) Influence of cultivation conditions on the bioenergy potential and bio-compounds of Chlorella vulgaris. Energy Rep 6:378–384

Catone CM, Ripa M, Geremia E, Ulgiati S (2021) Bio-products from algae-based biorefinery on wastewater: a review. J Environ Manage 293:112792

Chalima A, Oliver L, Fernández de Castro L, Karnaouri A, Dietrich T, Topakas E (2017) Utilization of volatile fatty acids from microalgae for the production of high added value compounds. Fermentation 3(4):54

Chalima A, Hatzidaki A, Karnaouri A, Topakas E (2019) Integration of a dark fermentation effluent in a microalgal-based biorefinery for the production of high-added value omega-3 fatty acids. Appl Energy 241:130–138

Chew KW, Yap JY, Show PL, Suan NH, Juan JC, Ling TC et al (2017) Microalgae biorefinery: high value products perspectives. Bioresour Technol 229:53–62

Chu WL (2012) Biotechnological applications of microalgae. IeJSME 6(1):S24–S37

De Bhowmick G, Sarmah AK, Sen R (2019) Zero-waste algal biorefinery for bioenergy and biochar: a green leap towards achieving energy and environmental sustainability. Sci Total Environ 650:2467–2482

de Jesus Raposo MF, de Morais RMSC, de Morais AMMB (2013) Health applications of bioactive compounds from marine microalgae. Life Sci 93(15):479–486

Delrue F, Álvarez-Díaz PD, Fon-Sing S, Fleury G, Sassi JF (2016) The environmental biorefinery: using microalgae to remediate wastewater, a win-win paradigm. Energies 9(3):132

Deprá MC, dos Santos AM, Severo IA, Santos AB, Zepka LQ, Jacob-Lopes E (2018) Microalgal biorefineries for bioenergy production: can we move from concept to industrial reality? Bioenergy Res 11:727–747

Dineshkumar R, Sen R (2020) A sustainable perspective of microalgal biorefinery for co-production and recovery of high-value carotenoid and biofuel with CO2 valorization. Biofuels Bioprod Biorefin 14(4):879–897

Dolganyuk V, Belova D, Babich O, Prosekov A, Ivanova S, Katserov D et al (2020) Microalgae: a promising source of valuable bioproducts. Biomol Ther 10(8):1153

Dolganyuk V, Andreeva A, Sukhikh S, Kashirskikh E, Prosekov A, Ivanova S, Michaud P, Babich O (2022) Study of the physicochemical and biological properties of the lipid complex of marine microalgae isolated from the coastal areas of the eastern water area of the Baltic Sea. Molecules 27(18):5871

Fal S, Smouni A, El Arroussi H (2023) Integrated microalgae-based biorefinery for wastewater treatment, industrial CO2 sequestration and microalgal biomass valorization: a circular bioeconomy approach. Environ Adv 12:100365

Farooq W (2021) Sustainable production of microalgae biomass for biofuel and chemicals through recycling of water and nutrient within the biorefinery context: a review. GCB Bioenergy 13(6):914–940

Goswami RK, Agrawal K, Verma P (2021a) Microalgae-based biofuel-integrated biorefinery approach as sustainable feedstock for resolving energy crisis. In: Bioenergy research: commercial opportunities & challenges, pp 267–293

Goswami RK, Mehariya S, Obulisamy PK, Verma P (2021b) Advanced microalgae-based renewable biohydrogen production systems: a review. Bioresour Technol 320:124301

Goswami RK, Agrawal K, Verma P (2022a) Microalgae Dunaliella as biofuel feedstock and β-carotene production: an influential step towards environmental sustainability. Energ Conver Manage 13:100154

Goswami RK, Agrawal K, Upadhyaya HM, Gupta VK, Verma P (2022b) Microalgae conversion to alternative energy, operating environment and economic footprint: an influential approach towards energy conversion, and management. Energ Conver Manage 269:116118

Gupta R, Mishra A, Thirupathaiah Y, Chandel AK (2022) Biochemical conversion of CO2 in fuels and chemicals: status, innovation, and industrial aspects. Biomass Convers Biorefin 14:1–24

Haddadi MH, Aiyelabegan HT, Negahdari B (2018) Advanced biotechnology in biorefinery: a new insight into municipal waste management to the production of high-value products. Int J Environ Sci Technol 15:675–686

Kadimpati KK, Sanneboina S, Golla N, Kumpati R, Skarka W (2020a) Cultivation of microalgae: effects of nutrient focus on biofuels. In: Microbial strategies for techno-economic biofuel production, pp 85–127

Kadimpati KK, Sanneboina S, Golla N, Ayla S, Kumpati R, Skarka W (2020b) Microalgae as an efficient feedstock biomass for biofuel production. In: Microbial strategies for techno-economic biofuel production, pp 129–169

Kadimpati KK, Sanneboina S, Golla N, Ayla S, Skarka W, Mitoma Y (2020c) Microalgae potential feedstock for the production of biohydrogen and bioactive compounds. In: Microbial strategies for techno-economic biofuel production, pp 171–206

Kapoor S, Giri A, Bhardwaj P, Singh M, Chavali M, Verma P (2023) Strategies for improved production of microalgae-derived carotenoids and pigments. In: Verma P (ed) Industrial microbiology and biotechnology: emerging concepts in microbial technology. Springer Nature, Singapore, pp 661–693

Koller M, Muhr A, Braunegg G (2014) Microalgae as versatile cellular factories for valued products. Algal Res 6:52–63

Kumar M, Sun Y, Rathour R, Pandey A, Thakur IS, Tsang DC (2020) Algae as potential feedstock for the production of biofuels and value-added products: opportunities and challenges. Sci Total Environ 716:137116

Leong HY, Chang CK, Khoo KS, Chew KW, Chia SR, Lim JW et al (2021) Waste biorefinery towards a sustainable circular bioeconomy: a solution to global issues. Biotechnol Biofuels 14(1):1–15

Li J, Liu Y, Cheng JJ, Mos M, Daroch M (2015) Biological potential of microalgae in China for biorefinery-based production of biofuels and high value compounds. N Biotechnol 32(6):588–596

Lopes da Silva T, Moniz P, Silva C, Reis A (2019) The dark side of microalgae biotechnology: a heterotrophic biorefinery platform directed to ω-3 rich lipid production. Microorganisms 7(12):670

Malik S, Shahid A, Haider MN, Amin M, Betenbaugh MJ, Mehmood MA et al (2022) Prospects of multiproduct algal biorefineries involving cascading processing of the biomass employing a zero-waste approach. Curr Pollut Rep 8(2):147–158

Masri MA, Jurkowski W, Shaigani P, Haack M, Mehlmer N, Brück T (2018) A waste-free, microbial oil centered cyclic bio-refinery approach based on flexible macroalgae biomass. Appl Energy 224:1–12

Mehariya S, Goswami RK, Verma P, Lavecchia R, Zuorro A (2021a) Integrated approach for wastewater treatment and biofuel production in microalgae biorefineries. Energies 14(8):2282

Mehariya S, Goswami RK, Karthikeysan OP, Verma P (2021b) Microalgae for high-value products: a way towards green nutraceutical and pharmaceutical compounds. Chemosphere 280:130553

Mishra S, Roy M, Mohanty K (2019) Microalgal bioenergy production under zero-waste biorefinery approach: recent advances and future perspectives. Bioresour Technol 292:122008

Mitra M, Mishra S (2019) Multiproduct biorefinery from Arthrospira spp. towards zero waste: current status and future trends. Bioresour Technol 291:121928

Mohan SV, Hemalatha M, Chakraborty D, Chatterjee S, Ranadheer P, Kona R (2020) Algal biorefinery models with self-sustainable closed loop approach: trends and prospective for blue-bioeconomy. Bioresour Technol 295:122128

Moreira JB, Santos TD, Duarte JH, Bezerra PQM, de Morais MG, Costa JAV (2023) Role of microalgae in circular bioeconomy: from waste treatment to biofuel production. Clean Techn Environ Policy 25(2):427–437

Moreno-Garcia L, Adjallé K, Barnabé S, Raghavan GSV (2017) Microalgae biomass production for a biorefinery system: recent advances and the way towards sustainability. Renew Sustain Energy Rev 76:493–506

Okeke ES, Ejeromedoghene O, Okoye CO, Ezeorba TPC, Nyaruaba R, Ikechukwu CK, Orege JI (2022) Microalgae biorefinery: an integrated route for the sustainable production of high-value-added products. Energ Conver Manage 16:100323

Olguín EJ, Sánchez-Galván G, Arias-Olguín II, Melo FJ, González-Portela RE, Cruz L, Adessi A (2022) Microalgae-based biorefineries: challenges and future trends to produce carbohydrate enriched biomass, high-added value products and bioactive compounds. Biology 11(8):1146

Orejuela-Escobar LM, Landázuri AC, Goodell B (2021a) Second generation biorefining in Ecuador: circular bioeconomy, zero waste technology, environment and sustainable development: the nexus. J Bioresour Bioprod 6(2):83–107

Orejuela-Escobar L, Gualle A, Ochoa-Herrera V, Philippidis GP (2021b) Prospects of microalgae for biomaterial production and environmental applications at biorefineries. Sustainability 13(6):3063

Paladino O, Neviani M (2024) Interchangeable modular design and operation of photo-bioreactors for Chlorella vulgaris cultivation towards a zero-waste biorefinery. Enzyme Microb Technol 173:110371

Patel AK, Choi YY, Sim SJ (2020) Emerging prospects of mixotrophic microalgae: way forward to sustainable bioprocess for environmental remediation and cost-effective biofuels. Bioresour Technol 300:122741

Paul R, Silkina A, Melville L, Suhartini S, Sulu M (2023) Optimization of ultrasound pretreatment of microalgal biomass for effective biogas production through anaerobic digestion process. Energies 16(1):553

Pekkoh J, Lomakool S, Chankham J, Duangjan K, Thurakit T, Phinyo K (2024) Maximizing biomass productivity of cyanobacterium Nostoc sp. through high-throughput bioprocess optimization and application in multiproduct biorefinery towards a holistic zero waste. Biomass Convers Biorefin 14(1):327–347

Rehmanji M, Suresh S, Nesamma AA, Jutur PP (2021) Microalgal cell factories, a platform for high-value-added biorenewables to improve the economics of the biorefinery. In: Microbial and natural macromolecules. Academic Press, San Diego, CA, pp 689–731

Rizwan M, Lee JH, Gani R (2015) Optimal design of microalgae-based biorefinery: economics, opportunities and challenges. Appl Energy 150:69–79

Rodrigues Dias R, Deprá MC, Ragagnin de Menezes C, Queiroz Zepka L, Jacob-Lopes E (2023) The high-value product, bio-waste, and eco-friendly energy as the tripod of the microalgae biorefinery: connecting the dots. Sustainability 15(15):11494

Saiyud N, Deethayat T, Asanakham A, Duongbia N, Kamopas W, Kiatsiriroat T (2022) Biochar production from co-pyrolysis of coffee ground and native microalgae consortium. Biomass Convers Biorefin 14:1–9

Sarma S, Sharma S, Rudakiya D, Upadhyay J, Rathod V, Patel A, Narra M (2021) Valorization of microalgae biomass into bioproducts promoting circular bioeconomy: a holistic approach of bioremediation and biorefinery. 3 Biotech 11:1–29

Sharma P, Gaur VK, Kim SH, Pandey A (2020) Microbial strategies for bio-transforming food waste into resources. Bioresour Technol 299:122580

Siddiki SYA, Mofijur M, Kumar PS, Ahmed SF, Inayat A, Kusumo F et al (2022) Microalgae biomass as a sustainable source for biofuel, biochemical and biobased value-added products: an integrated biorefinery concept. Fuel 307:121782

Silva C, Moniz P, Oliveira AC, Vercelli S, Reis A, da Silva TL (2022) Cascading *Crypthecodinium cohnii* biorefinery: global warming potential and techno-economic assessment. Energies 15(10):3784

Sivaramakrishnan R, Suresh S, Kanwal S, Ramadoss G, Ramprakash B, Incharoensakdi A (2022) Microalgal biorefinery concepts' developments for biofuel and bioproducts: current perspective and bottlenecks. Int J Mol Sci 23(5):2623

Solis CA, Mayol AP, San Juan JG, Ubando AT, Culaba AB (2020) Multi-objective optimal synthesis of algal biorefineries toward a sustainable circular bioeconomy. IOP Conf Ser Earth Environ Sci 463:012051

Tsakona S, Papadaki A, Kopsahelis N, Kachrimanidou V, Papanikolaou S, Koutinas A (2019) Development of a circular oriented bioprocess for microbial oil production using diversified mixed confectionery side-streams. Food Secur 8(8):300

Tsegaye B, Jaiswal S, Jaiswal AK (2021) Food waste biorefinery: pathway towards circular bioeconomy. Food Secur 10(6):1174

Ummalyma SB, Sahoo D, Pandey A (2021) Resource recovery through bioremediation of wastewaters and waste carbon by microalgae: a circular bioeconomy approach. Environ Sci Pollut Res 28(42):58837–58856

Vermuë MH, Eppink MHM, Wijffels RH, Van Den Berg C (2018) Multi-product microalgae biorefineries: from concept towards reality. Trends Biotechnol 36(2):216–227

Xiu S, Zhang B, Boakye-Boaten NA, Shahbazi A (2017) Green biorefinery of giant miscanthus for growing microalgae and biofuel production. Fermentation 3(4):66

Yen HW, Hu IC, Chen CY, Ho SH, Lee DJ, Chang JS (2013) Microalgae-based biorefinery–from biofuels to natural products. Bioresour Technol 135:166–174

Yin Z, Zhu L, Li S, Hu T, Chu R, Mo F et al (2020) A comprehensive review on cultivation and harvesting of microalgae for biodiesel production: environmental pollution control and future directions. Bioresour Technol 301:122804

Zaky AS (2021) Introducing a marine biorefinery system for the integrated production of biofuels, high-value-chemicals, and co-products: a path forward to a sustainable future. PRO 9(10):1841

Development and Scale-Up of the Bioreactor System in Biorefinery: A Significant Step Toward a Green and Bio-Based Economy

Swethaa Venkatraman, Devi Sri Rajendran, and Vinoth Kumar Vaidyanathan

1 Introduction

Globally, the indiscriminate use of fossil fuels has been prevalent in obtaining various forms of affordable energy and fuels. This extensive reliance has led not only to the depletion of fossil fuel reserves but also to the generation of immense quantities of waste. Simultaneously, the world is grappling with the pressing challenges of climate change (Xiong et al. 2023). These intertwined issues have led to a significant interest in developing renewable, environmentally friendly, and carbon-neutral fuels to cater to the rising energy demands. Consequently, there is a growing focus on harnessing the potential of waste as a valuable resource for producing beneficial forms of bioenergy (Rama Mohan 2016). To counter the strain on finite resources and ensure sustained availability, countries worldwide are strategically focusing on advancing the bioeconomy (Usmani et al. 2021). Within this framework, biorefining pertains to the sustainable transformation of biomass into a wide array of commercially viable goods and sources of energy.

Various attempts have been undertaken to classify biorefineries based on the utilized biomass and technologies. Biorefinery types concerning the substrates employed encompass lignocellulosic feedstock biorefineries, agricultural residues biorefineries, forestry biorefineries, algal biorefineries, and food waste biorefineries (Kumar and Verma 2021). Within this scenario, the integration of all waste treatment components to produce bio-based products and recover energy within a single, self-sufficient, and highly sustainable waste biorefinery proves significantly more

S. Venkatraman · D. S. Rajendran · V. K. Vaidyanathan (✉)
Integrated Bioprocessing Laboratory, Department of Biotechnology, School of Bioengineering, Faculty of Engineering and Technology, SRM Institute of Science and Technology (SRM IST), Chennai, Tamil Nadu, India
e-mail: vinothkv@srmist.edu.in

K. Agrawal, P. Verma (eds.), *Biotechnological Advances in Biorefinery*, Interdisciplinary Biotechnological Advances,
https://doi.org/10.1007/978-981-97-5544-8_10

advantageous than pursuing a unified approach (Rama Mohan 2016). However, prevalent obstacles to industrial biorefining processes include high initial and operational expenses, technical infancy, and challenges in scaling up (Mahendrasinh Kosamia et al. 2022).

Utilizing resources derived from biological sources offers a clear environmental benefit, given their lack of contribution to the buildup of carbon dioxide, unlike their fossil-based counterparts. Nevertheless, industries may be reluctant to adopt products yielding lower financial returns unless they incorporate the environmental benefits into their monetary assessments or encompass them within the overall cost analysis (Louw et al. 2023). Nevertheless, enhancing productivity and yield from microorganism-based biorefineries through microbial fermentation is vital for cost-effective commercialization. Scaling up in a biorefinery involves transitioning from laboratory or pilot-scale experiments to a larger, commercial scale, and this process presents several challenges. It begins with a feasibility study to assess the technical, economic, and environmental viability, followed by process optimization at the pilot scale to ensure reproducibility and consistency.

Effective management of manufacturing processes, including utilities, sterility, unit operations, regulatory compliance, and emissions, is imperative for successful biorefinery commercialization. Challenges faced in industrializing renewable feedstock biorefineries include supply chain feasibility, transitioning experiments to commercially viable processes, and technological maturity. The book chapter emphasizes the utilization of renewable feedstock in bioreactor (pilot, large and industrial scale), promoting a circular economy and sustainable energy and bio-based chemicals production from eco-friendly resources. Also, it accentuates current advancements in overcoming these challenges, providing a road map for renewable feedstock biorefineries future by aligning technical progress with market demands and obstacles through development and scale up of bioreactors.

2 Renewable Feedstock

2.1 Agricultural Residues

Agricultural residues primarily comprise carbon-based materials. These residues can be categorized into two groups: primary residues, consisting of stalks, leaves, and straw remaining in the field postharvest, and secondary residues, which remain post-processing, presenting a valuable resource. Materials within the scope of process residues include husks, molasses, seeds, rhizomes, and bagasse (Akshaya et al. 2023). Agricultural residues find utility as fertilizers, sources of energy, animal feedstock, enhancers for soil health, and in various manufacturing processes. On a global scale, the production of agricultural residues, in terms of dry matter, was approximated to reach 3.8 billion metric tons. In India, the annual generation of crop residue is estimated to vary between 500 and 550 million tons. Estimates indicate that, globally, approximately 377 million tons of corn stover are available for utilization (Bhuyan et al. 2020). Additionally, it's been estimated that there's an availability of 658 million tons of rice straw worldwide. Moreover, the production of cereal straw, totaling about 1580 million tons, comprises crops such as barley and oat in Europe, corn and sorghum in the United States, and rice and wheat in China (Ginni et al. 2021).

Agricultural residues, as naturally occurring lignocellulosic biomass with a high carbon content, face imminent disposal to prevent pollution. However, an alternative approach involves harnessing these residues to produce bioenergy and bio-based products (Jain et al. 2022). The agricultural industry produces organic and biodegradable waste comprising minerals, proteins, fibers, and vitamins. This waste primarily consists of lignocellulosic materials, such as straw (dry crop stalks), molasses, bagasse, husks, spent grains (from rice, wheat, and maize), shells (from walnut, groundnut, and coconut), and various plant residues annually on a global scale (Singh et al. 2021). In 2013, according to the FAO, 250 million tons of plant waste were generated through diverse crop processing methods. Nevertheless, agricultural biowaste holds promise as a valuable resource for producing a diverse range of materials, including bioplastics and bioethanol, by utilizing various parts of plants (Kassim et al. 2022).

2.2 *Algae*

Algae, as efficient cellular factories powered by sunlight, effectively convert carbon dioxide into a range of biomolecules, including lipids, carbohydrates, and proteins. The composition of these biomolecules varies significantly across different species of algae (Muthuraj et al. 2013). Algae are broadly divided into two categories: microalgae and macroalgae. Microalgae are unicellular and smaller, typically measuring up to a few millimeters, while macroalgae are larger, reaching sizes of a few meters and are multicellular. In nature, there exist over 50,000 species of microalgae, of which 4000 have been identified. Remarkably, only a limited number, fewer than 50 species, have been commercially exploited. Consequently, the biomass derived from microalgae holds vast untapped and undiscovered potential for various applications (Siddiki et al. 2022). Microalgae serve as microbial factories capable of producing a wide array of compounds beyond lipids suitable for biodiesel. They consist of lipids (7–23%), carbohydrates (5–23%), and proteins (6–52%), enabling the potential production of diverse compounds. Microalgae represent an ideal feedstock to commercial the value-added products widely utilized in food, nutraceuticals, cosmetics, and pharmaceuticals due to their ability to yield various compounds with significant commercial value (Chandra et al. 2019).

Currently, both microalgae and macroalgae serve as essential raw materials for producing an extensive array of bioproducts, categorized into fuel and nonfuel products. Fuel-based products include bioethanol, biogas, biodiesel, and biohydrogen, while nonfuel products include carbohydrates, pigments (such as lutein and astaxanthin), recombinant proteins, biomaterials, and other bioproducts (Javed et al. 2022). Factors such as wastewater treatment, carbon neutrality, and biodiesel production drive research in algal biofuels. Value-added products derived from microalgae are primarily limited to a few species that are easily cultured, have established market demand, and hold a recognized market value (Olguín 2012). Commercially viable products derived from algae primarily originate from marine algae, commonly known as seaweed, which is primarily cultivated for food purposes. These products are frequently obtained through harvesting wild populations rather than cultivated methods. Nonfuel products derived from algae enjoy accessible markets, offering higher profit margins (Chandra et al. 2019).

2.3 *Forestry and Woody Biomass*

The forestry, timber, pulp, and paper (FTPP) industry consists of two primary sectors: the timber industry, which includes lumber, particle board, plywood, and fiberboard, and the pulp and paper industry. This industry faces significant challenges, such as declining worldwide demand for pulp, timber, and paper products, water scarcity concerns, rising energy costs, and increasing regulatory pressure favoring cleaner technologies (Andrew et al. 2020). Additionally, environmental concerns regarding waste management and pollution are also impacting the industry. The FTPP industry serves as a promising platform for the integration of biorefinery technologies owing to its established infrastructure capable of handling and processing substantial amounts of biomass. Moreover, there are significant prospects to utilize forestry biomass for the creation of value-added products (Rathour et al. 2023). In paper-making, approximately 50% of the pulpwood is comprised of residues and by-products due to the removal of lignin, hemicellulose, and extractives during pulping and bleaching processes. Similarly, in timber production, up to 60% of the sawlogs consist of residues and waste materials, such as sawdust, bark, branches, and other trimmings (Henrich et al. 2015).

2.4 *Municipal Solid Waste*

Urban and rural communities discard a varied mixture of solid waste daily, known as municipal solid waste, comprising garbage, debris, and refuse. Presently, the global generation of municipal solid waste totals approximately two billion tons, with nearly 33% of this waste uncollected by municipal authorities (Karak et al. 2012). Municipal solid waste emerges from collection of solid waste from residences, offices, small-scale institutions, and commercial enterprises. The composition and categorization of waste vary significantly among municipalities worldwide, encompassing both biodegradable and nonbiodegradable elements from organic and inorganic sources. Municipal solid waste typically includes kitchen and yard waste, cardboard, rubber, plastic, electronic waste, metal, glass, paper, inert materials, and miscellaneous debris (Nanda and Berruti 2021). The quality of municipal solid waste compost depends on numerous variables, encompassing the design of the composting facility, the source and proportions of feedstock employed, the composting methodology, and the duration of maturation. Moreover, when municipal solid waste compost is applied to diverse field soils, variations in plant response further accentuate differences between compost types (Hargreaves et al. 2008). A range of waste-to-energy technologies, including thermochemical and biological conversion processes, provide opportunities to convert municipal solid waste into solid, liquid, and gaseous fuels. These technologies help meet the increasing demands for energy resources (Nanda and Berruti 2021). Composting municipal solid waste is regarded as a means of diverting organic waste from landfills while generating a cost-effective product suitable for agricultural use. This inclination is influenced by various economic and environmental factors, including municipal landfill capacity constraints, expenses related to landfilling and material transportation, legislative measures aimed at environmental preservation, declining reliance on commercial fertilizers, expanding household waste recycling capabilities, and the enhanced quality of compost products (Hargreaves et al. 2008).

2.5 Food Waste

Food waste constitutes a significant portion of bio-waste globally, with approximately 1.3 billion tons, equivalent to one-third of the total global food production, being wasted annually. Food waste primarily consists of organic components, including carbohydrates, proteins, lipids, and small amounts of inorganic elements (Sharma et al. 2020). These components can undergo further digestion, breaking down into simpler organic compounds such as glucose, amino acids, and fatty acids. However, the specific composition of food waste may vary depending on its source. Food waste generation arises from multiple stages: production, handling, storage, distribution, processing, and consumption (Gustafsson et al. 2013). Before disposal, food waste can be categorized into two fractions: avoidable (edible) and unavoidable (nonedible). Implementing mitigation measures along the production-consumption chain can effectively reduce avoidable food waste. As for unavoidable food waste, effective waste management strategies and policies are imperative for appropriate handling, reuse, and disposal (Garcia-Garcia et al. 2015). Food waste can undergo bioconversion processes to yield a range of bio-commodity chemicals and bioenergy. This bioprocessing allows for the transformation of food waste into sustainable platform chemicals, presenting alternative resource recovery solutions that significantly reduce reliance on fossil-based chemicals (Tapia et al. 2019).

3 Development of Biorefinery Product Spectrum by Renewable Feedstock

Biorefineries are poised to play a pivotal role in advancing the development of a bioeconomy. The concept of a biorefinery is similar to traditional petrochemical refineries but focuses on the sustainable utilization of biomass (Sadhukhan et al. 2018). The International Energy Agency Bioenergy Task 42 defines a biorefinery as the "sustainable processing of biomass into a spectrum of marketable bio-based products (chemicals, materials) and bioenergy (fuels, power, and heat)." This transformative approach aligns with the broader objectives of the bioeconomy, emphasizing sustainability, resource efficiency, and bio-based alternatives to traditional petrochemical products (Wenger and Stern 2019). The process encompasses the transformation of the above renewable feedstocks, including agricultural residues, algae, forestry by-products, municipal solid waste, and industrial by-products, into a spectrum of valuable and eco-friendly products (Fig. 1). The development of a biorefinery product spectrum, such as biofuels, biochemicals, organic acid, enzyme, biosurfactant, and antibiotics, through the utilization of renewable feedstocks involves a multifaceted approach to harness the diverse array of compounds present in these sustainable resources. This development is driven by the principles of the bioeconomy and the circular economy, aiming to extract maximum value from available resources while minimizing waste and environmental impact (Mohan et al. 2016).

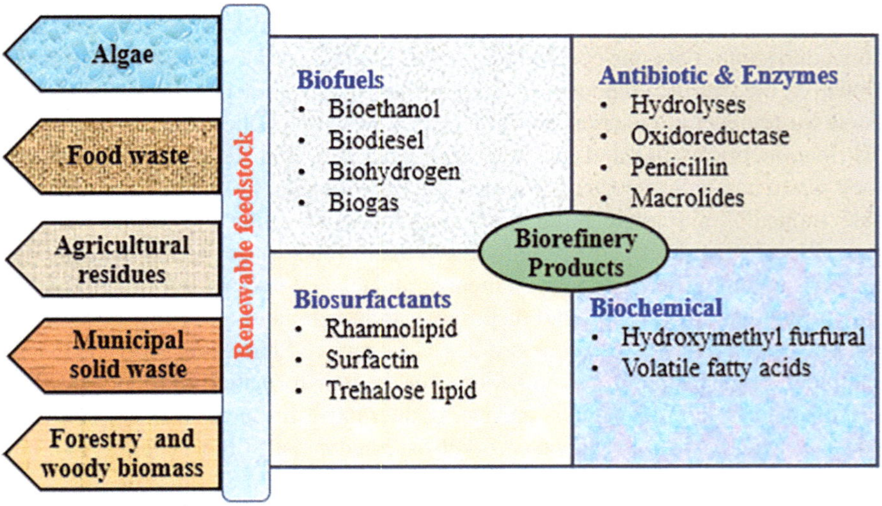

Fig. 1 Various biorefinery production from the different waste renewable products for achieving sustainability

3.1 Biofuels: Shaping Sustainable Energy

Biofuel production can be achieved through an integrated process that maximizes by-product utilization and adds overall value. Certain microorganisms exhibit preferences for specific substrate constituents in specific biofuel production processes. For instance, in the bioethanol production process, hemicellulose materials are more readily degraded than cellulose (Srivastava et al. 2021). The selective degradation leaves behind agro-residue rich in lignin, which is subsequently employed in thermochemical conversion processes for the generation of both electric and heat energy. The efficiency of biomass in yielding bioethanol is primarily influenced by the composition of cellulose and hemicellulose, with lignin content posing a notable impediment to hydrolysis and the biological conversion into ethanol (Limayem and Ricke 2012). Apart from the lignocellulosic composition, additional factors such as increased ash and silica content in rice straw, as well as notable alkali and protein content in wheat straw and corn stover, pose challenges in the thermochemical conversion of biomass into ethanol (Binod et al. 2010).

A significant ethanol yield of 61.3% containing 29.1 g/L was achieved from rice straw for bioethanol production. This was accomplished through a combination of microwave and alkali pretreatment, followed by hydrolysis and fermentation using cellulase from *T. reesei* and *S. cerevisiae* in simultaneous saccharification and fermentation (SSF). In comparison, pretreatments involving only alkali (yielding 40% to 74% of ethanol and acid treatment yielding 30% resulted in lower ethanol yields (Pattanaik et al. 2019). Microalgae have the potential to serve as a source of renewable energy, particularly to produce biofuels. Utilizing microalgae for biodiesel production offers the advantage of not compromising the production of food, fodder, and other products derived from traditional crops. Notable microalgae species, such as *Ankistrodesmus fusiformis*, *Kirchneriella lunaris*, *Ankistrodesmus falcatus*, and *Chlamydocapsa bacillus,* are particularly well-suited to produce biodiesel due to their high content of polyunsaturated fatty acid methyl esters (Ahmad et al. 2011).

The sustainable food waste valorization development is crucial in addressing waste disposal and environmental concerns. The starch residue obtained from instant noodle waste has been effectively converted into bioethanol using *S. cerevisiae* K35 through the simultaneous saccharification and fermentation process. The fermentation process was optimized, resulting in an impressive ethanol conversion rate of 96.8% achieved within 36 h (Yang et al. 2014). Researchers made biodiesel using fat that was removed from food waste. Using KOH-catalyzed transesterification with a 1:10 M ratio (lipid to methanol) at 60 °C, a 100% biodiesel yield was obtained in 2 h. On the other hand, Novozyme-435 produced 90% biodiesel in 24 h at 40 °C and a 1:5 M ratio (lipid to methanol) (Karmee et al. 2015).

The results of previous research have provided important new information to further the development of bacterial consortia that break down cellulosic materials in termite guts. Consequently, there is great potential to improve the biological pretreatment of wood in lignocellulosic anaerobic digestion biorefineries. The anaerobic digester M-2, comprising a blend of pretreated wood sawdust (WSD) and cattle dung in a 50/50 ratio, exhibited the highest biogas production at 66.1 NL/kg VS, with methane constituting 66% of the generated biogas (Ali et al. 2023).

3.2 Biochemicals: Building Blocks for Industry

Biochemicals are substances derived from biological sources, such as plants, microorganisms, or biomass, and are utilized for the production of wide range of materials essential for industry. In this context, biochemicals serve as the building blocks or raw materials for the manufacturing of diverse products across different sectors (Ahorsu et al. 2018). They are employed in the synthesis of chemicals, materials, and fuels, providing a sustainable and often environmentally friendly alternative to the traditional petrochemical-based processes. These bio-based building blocks contribute to the development of a more sustainable and circular economy by reducing the dependence on fossil fuels as well as minimize environmental impact (Dahiya et al. 2020).

Biochemicals, also known as value-added goods, are significant industrial components that include proteins, lipids, carbohydrates, polyunsaturated fatty acids, starch, cellulose, and vitamins. For instance, the efficiency of producing polyhydroxyalkanoates from food wastes was increased by combining anaerobic (hydrogen) fermentation with aerobic conversion of volatile fatty acids by Ralstonia eutropha (Patel et al. 2021). Mango industry residues have been investigated by researchers as a possible source of sugars for the production of 5-hydroxymethylfurfural (HMF), an important platform chemical. The feedstock for the hydrothermal, microwave-assisted synthesis of HMF was overripe mangoes. HPLC-DAD was utilized to quantify the product. A maximum HMF production of 21.2 mol% was achieved under ideal experimental circumstances in a 20-min reaction period at 150 °C and a 1:19 dilution ratio (Liu 2017). The generation of volatile fatty acids (VFAs) via the anaerobic digestion of agro-industrial residues represents a valuable production process and offers an alternative and sustainable method for supplementing ruminant feed. During batch assays, the co-digestion yielded the highest concentration of VFAs (14.2 g/L) at pH 10. This result suggests a yield of 0.85 g CODVFAs/g volatile solids added (Parchami et al. 2023).

The valuable algal-derived products find applications in various industries, such as food, organic fertilizer, cosmetics, livestock feed, pesticides, natural dyes, animal feed, pigments, nutraceuticals, and even in combustion-based electricity generation (Chew et al. 2017). Certain microalgae species boast a protein content of 70%, providing all essential amino acids crucial for human health (Wells et al. 2017). Similarly, microalgae are rich in long-chain PUFAs that play an important role in neurological and cardiac functions. These PUFAs contribute to the body's defense against conditions such as heart disease, hypertension, cancer, and cholesterol issues (López et al. 2019). Examples of these beneficial long-chain PUFAs include eicosapentaenoic acid, docosahexaenoic acid, linolenic acid, α-linolenic acid, and arachidonic acid.

3.3 *Organic Acid: Diverse Applications Across Industries*

A biorefinery dedicated to organic acid production operates by employing renewable biomass sources like agricultural residues or food waste to extract, process, and refine organic acids through biological processes. These refineries focus on utilizing the components of raw materials, often sugars or other organic compounds, which microbial fermentation or enzymatic reactions convert into organic acids like lactic acid, citric acid, or succinic acid (Menon and Rao 2012). Through downstream processing methods involving purification and separation, these organic acids are isolated from the fermentation broth, rendering them suitable for various industries. These acids find applications across sectors such as food, pharmaceuticals, agriculture, and chemical manufacturing due to their versatile properties (Alexandri et al. 2019). Importantly, such biorefineries significantly contribute to sustainability efforts, leveraging renewable resources and reducing reliance on nonrenewable fossil fuels, thus minimizing environmental impact while offering economically viable and eco-friendly alternatives to traditional chemical production methods.

Food waste, typically abundant in carbohydrates, has been extensively utilized for producing hydrogen and biochemicals. This type of waste predominantly consists of carbon chains and holds a substantial amount of energy (Fatima et al. 2020). In a mixed sample containing meat, fruits, and vegetables, the average energy content measured was 14.31 MJ/kg (Dangol et al. 2022). Annually, about a third of food intended for human consumption ends up as waste, a trend expected to escalate alongside the growth of the global population. Several types of microorganisms possess the ability to accumulate substantial quantities of citric acid in culture media. Citric acid production is demonstrated by fungi, such as *Aspergillus niger*, *Trichoderma viride*, *Penicillium lactum*, *and Mucor pyriformis* (Merrylin et al. 2020).

3.4 *Antibiotics and Enzymes: Harnessing Microbial Capabilities*

Microbes, including bacteria and fungi, generate diverse array of enzymes with applications in different industries. The concept of harnessing microbial capabilities involves tapping into the extensive genetic and metabolic diversity of microorganisms. Biotechnological advances allow scientists to manipulate microbial genomes and optimize conditions for the enhancement of antibiotics and enzyme production (Baweja et al. 2016). Fermentation processes, often driven by microbial activity, are employed on a large scale to produce these compounds. This utilization of microbial potential has wide-ranging applications, from medicine, where antibiotics are crucial in combating infections, to industry, where enzymes play key roles in various processes such as food production and biofuel manufacturing. The phrase emphasizes the symbiotic relationship between scientific understanding, biotechnological innovation, and the vast capabilities inherent in microbial life (Yu et al. 2016).

Current research focused on enzyme production utilizing agro-industry wastes has predominantly examined the impact of pretreatment procedures on enhancing yield. Pretreatments represent techniques commonly applied in production processes using lignocellulose-based raw materials. These methods involve the use of chemical, physical, or a combination of both approaches to disrupt the intricate plant structure, aiming to improve digestibility. A captivating investigation by Salim et al. (2017) involved the utilization of a novel *Bacillus* strain to explore to produce four enzymes—α-amylase, protease, cellulase, and pectinase (Salim et al. 2017). The study employed various agricultural residues as growth substrates, including sunflower meal, soybean meal, maize pericarp, maize bran, wheat bran and olive oil cake (Ravindran et al. 2018).

Galactan hydrolases, which exhibit specificity for insoluble and polyanionic polysaccharides, are predominantly categorized as carrageenases and agarases. The extraction and production of these hydrolytic enzymes from red seaweed using various technologies pave the way for innovative applications within the biorefinery framework (Siller-Sánchez et al. 2019). *P. sumatrense* AQ67100, demonstrated the ability to degrade heat-killed *Chlorella vulgaris* cells using an enzyme repertoire that included β-1,3-glucanases, glycosidases, and dipeptidyl- and amino-peptidases. β-glucuronidase, α- and β-glucosidases, α-mannosidases, and β-galactosidases were among the glycosidases. Chlorophylls and lipids were released from the algal cells in a significant manner when *P. sumatrense* AQ67100 filtrate was applied to *C. vulgaris*; the increases were 42.6% and 48.9%, respectively. (Giovannoni et al. 2021).

3.5 Biosurfactants

Microorganisms produce surface-active chemicals called biosurfactants, which have unique abilities to lower surface tension and emulsify substances. In the biosurfactant biorefinery, microorganisms are cultivated to produce biosurfactants through fermentation or other biotechnological processes (Venkataraman et al. 2022). The biorefinery utilizes renewable biomass feedstock, including agricultural residues or waste materials, as the raw materials for biosurfactant production. The biosurfactants produced can serve as effective and environmentally friendly alternatives to synthetic surfactants in various applications (Mohanty et al. 2021). This integrated approach not only focuses on biosurfactant production but also aligns with the broader biorefinery concept. The facility explores the use of biomass for the production of various bio-based products, including biofuels, bio-based chemicals, and other value-added compounds. The biosurfactant biorefinery emphasizes sustainability by utilizing renewable resources, reducing dependence on fossil fuels, and promoting the environmentally friendly characteristics of biosurfactants. As a result, this innovative system opens new avenues to develop the bio-based products with applications in diverse industries, contributing to more sustainable and eco-friendly practices in the production of surfactants and other valuable compounds (Satpute et al. 2017).

Hazra et al. (2011) conducted an assessment for the production of rhamnolipid by *Pseudomonas aeruginosa* AB4 utilizing de-oiled cakes derived from mahua (*Madhuca indica*), karanj (*Pongamia pinnata*), neem (*Azadirachta indica*), and jatropha (*Jatropha curcas*). After optimizing the production conditions to pH 8.5, 40 °C, and a 10:0.6 carbon-to-nitrogen (C:N) ratio, the scientists recorded a peak production of 40 g/L of rhamnolipids. Additionally, when the strain was grown in a basal mineral medium supplemented with 50 g/L of mahua oil waste, an interfacial tension of 40 mN/m was attained (Hazra et al. 2011).

The biosurfactant produced from defatted *Chlorella* biomass by *Bacillus subtilis* C9 in a conventional culture medium was assessed. Despite the expectation that increased acid catalyst concentration would enhance dilute acid hydrolysis, optimal results were observed with 0.25 N HCl treatment for 15 min at 121 °C, yielding the highest fermentable monosaccharides from DCB. These findings suggest the potential integration of biosurfactant fermentation into algal biorefinery processes for valorizing defatted algal biomass. Further optimization efforts are warranted for industrial-scale algal biorefineries (Yun et al. 2020).

3.6 Meeting Market Demands in the Bioeconomy Era

In the bioeconomy era, meeting market demands is increasingly achieved through the implementation of biorefinery processes. Biorefineries serve as versatile platforms, integrating various technologies to convert biomass into a spectrum of valuable products, ranging from biofuels and biochemicals to materials and energy (Conteratto et al. 2021). By efficiently utilizing renewable resources, such as agricultural residues, forest biomass, and algae, biorefineries contribute to sustainability and reduce dependence on fossil fuels. This holistic approach aligns with the global emphasis on environmentally friendly practices and the circular economy. Biorefineries play a pivotal role in transforming diverse feedstocks into a multitude of products, addressing market demands while promoting a more sustainable and resource-efficient future (Sillanpää et al. 2017).

4 Bioreactor System in the Production of Biorefineries

Bioreactors in biorefineries come in various types, each tailored to specific production needs and optimized for the characteristics of the biological processes involved. Common bioreactor systems include stirred-tank bioreactors, and photobioreactors, designed for photosynthetic microorganisms like algae or cyanobacteria. Additionally, membrane bioreactors (MBRs) are frequently employed to facilitate efficient separation and concentration processes (Fig. 2).

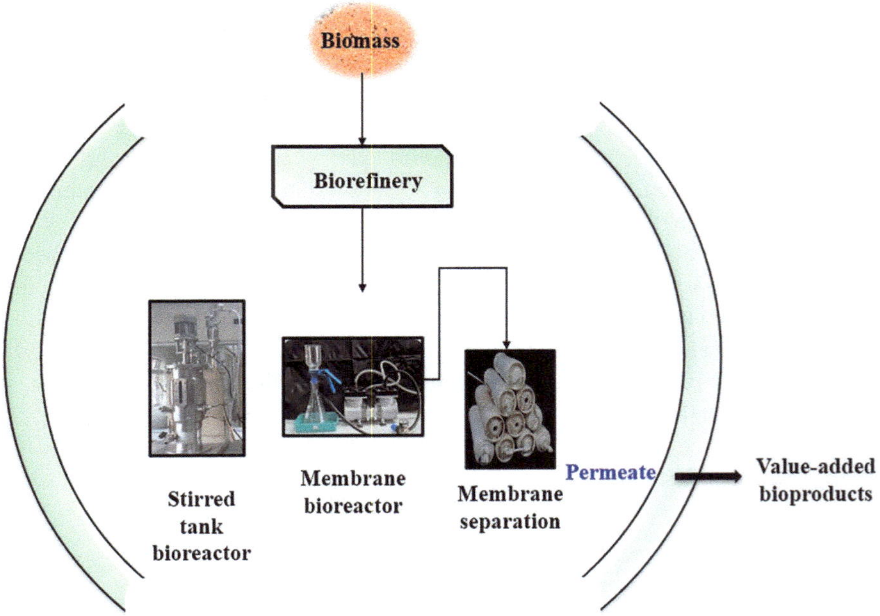

Fig. 2 Different types of bioreactor for the production of value-added products from renewable feedstocks

4.1 *Stirred-Tank Bioreactors*

Stirred-tank bioreactors play a pivotal role in the modern biorefinery landscape, serving as indispensable vessels for the cultivation of microorganisms and the production of bio-based products. These bioreactors are designed to maintain well-mixed and controlled environment, providing optimal conditions for microbial growth and the synthesis of valuable compounds (Awasthi et al. 2019). The critical aim is to guarantee the ideal environment for the biological entity, facilitating a high yield in the bioprocess. In the context of biorefineries, stirred-tank bioreactors are instrumental in converting diverse biomass feedstocks, including food waste, industrial waste, and agricultural residues, into a spectrum of products such as biofuels, biochemicals, and other high-value compounds. Their flexibility allows for the adaptation to various raw materials, and they enable precise control over critical parameters such as temperature, pH, and nutrient concentrations (Jha and Siddiqui 2023). With the capability for easy scale-up, stirred-tank bioreactors are well-suited for industrial-scale biorefinery operations, supporting continuous processing and the production of a wide range of bio-based products. As the bioeconomy continues to evolve, stirred-tank bioreactors remain indispensable tools in meeting market demands through efficient and sustainable biorefinery processes (Humbird et al. 2017).

Bioethanol production using a stirred tank reactor involved the utilization of compatible mixed cultures, including *Phanerochaete chrysosporium*, *Trichoderma harzianum*, *Mucor hiemalis*, and *Saccharomyces cerevisiae*. Among these, the mixed cultures tested, the simultaneous inoculation of *S. cerevisiae* and *T. harzianum* was identified as the most effective, yielding the maximum ethanol production at 4% v/v or 31.6 g/L (Alam et al. 2009). The optimal condition for biomass production with elevated protein content and a diverse amino acid profile was identified to be a blend of glucose and xylose. *P. ostreatus* LGAM 1123 was grown in a 4 L stirred-tank bioreactor with aspen hydrolysate, producing 25.0 ± 3.4 g/L of biomass, 1.8 ± 0.4/day of specific growth rate, and 54.5 ± 0.5% (g/100 g sugars) of protein (Bakratsas et al. 2023). Rice straw has emerged as a promising and economically viable feedstock for the production of biobutanol through fermentation. However, to optimize the fermentation process, the medium derived from rice straw hydrolysate requires supplementation with essential nutritional elements. Tests conducted in a 2 L bioreactor utilizing this optimal fermentation medium demonstrated nearly exhaustive consumption of soluble sugars, resulting 8.7 g/L of total solvents, including butanol (6 g/L) (Ranjan et al. 2013).

4.2 Photobioreactors

Various photobioreactor designs, such as flat plate, tubular, column, and biofilm photobioreactors, are employed for microalgae cultivation. Among these, tubular photobioreactors are the most prevalent. Various configurations, including airlift, bubble column, flat inclined, solar penthouse-roof, column aeration, and multistage continuous flow, are viable options for these photobioreactors (Singh and Sharma 2012). There are two major classifications of photobioreactor, namely, closed photobioreactor and open photobioreactor. As the name implies, closed algae cultivation involves growing algae in a closed system. While a closed photobioreactor (PBR) has the capacity to yield higher biomass compared to other closed systems, its cost-effectiveness is a limiting factor. The predominant technology for closed cultivation is the photobioreactor, with tubular and flat bioreactors being commonly employed. These closed systems are constructed using materials such as glass or plastic, with glass PBR being a prevalent choice for large-scale production (Brennan and Owende 2010). In closed systems, glass tubes are arranged vertically, helically, or horizontally, with mechanical pumps facilitating CO_2 and O_2 exchange. Closed PBR offers advantages over open systems by efficiently harnessing more sunlight, thereby enhancing productivity (20 to 40 g/m^2/d) within a short timeframe. However, its complexity in structure contributes to higher costs (Buhrer 2000). Open cultivation systems have the advantage of being cost-effective and easy to maintain. However, they come with several disadvantages. The biomass production in open systems tends to be low, and their light use efficiency is also limited. Additionally, there is a high risk of contamination by other microorganisms, making open systems less suitable for cultivating sensitive microalgae species (Kunjapur and Eldridge 2010).

These are the operating parameters for a series of AL-PBRs (airlift photobioreactors) that are used to grow *Chlorella vulgaris*. These reactors are part of a pilot-scale biowaste-to-biofuels manufacturing facility, where wood chips are gasified to make syngas and waste frying oils are transesterified to produce biodiesel. Outdoor experiments have substantiated that a set of 12–15 airlift photobioreactors (AL-PBRs), with a cumulative volume ranging from 125 to 155 L, effectively repurpose 28.8 L/d of wastewater produced by the biodiesel units. The amount of biodiesel produced by this technique is about 228 L/d. Additionally, they process approximately 90 m^3/h of flue gas that is produced by the syngas combustion unit, producing 9–9.5 kW of power. Every cycle, the battery's beginning concentration is roughly 0.5 g/L, and it reaches a concentration of 1.05 g/L upon extraction (Paladino and Neviani 2021).

The production of biofuels by pilot scale derived from solid organic wastes, waste frying oil, and algal biomass. The study aims to establish a closed-loop process that effectively utilizes organic wastes and by-products. The biodiesel's cetane number ranged between 47.7 and 58.4, with LHVs spanning in the range of 36,080 kJ/kg–36,992 kJ/kg. The introduction of glycerol led to an improvement in syngas quality, and the gas composition of flue was deemed suitable for partially fueling the airlift reactors (Paladino and Neviani 2018).

4.3 Membrane Bioreactors (MBR)

Membrane bioreactors (MBRs) are reactors that combine a method of separation with a biological response. These systems make use of specialized membranes that act as physical barriers, allowing certain species to pass through while impeding the mobilization of the entire mass. On a commercial scale, the innovative Aerobic Membrane Bioreactor (AeMBR) has proven effective in treating municipal and industrial wastewaters. An offshoot of AeMBR, the Anaerobic Membrane Bioreactor (AnMBR), has emerged as a novel research focus. AnMBR demonstrates effective retention of crucial microorganisms, such as methanogens, within the reactor through direct membrane interception, thereby achieving the complete separation of sludge retention time (SRT) and hydraulic retention time (HRT) (Zhen et al. 2019).

Since the year 2000, Kubota Corp. has successfully established approximately 26 full-scale AnMBR installations globally, designed to treating various biowastes (Hu and Stuckey 2006). Two AnMBRs were built in Kobe, Japan, as part of a specialized project, and each reactor was outfitted with 150 Kubota flat-sheet membranes (each measuring 0.8 m^2). With an effective volume of 100 m^3 and a total treatment capacity of 30 t/d, each AnMBR was built with the purpose of digesting effluent from dairy production. The system produced an impressive 880 m^3/d of biogas (Kanai et al. 2010). In a study where CO_2 served as the exclusive carbon source, researcher reported a production of methane rate of 75.8 mL/L/d at a cathodic potential of −0.9 V Ag/AgCl (Zhen et al. 2018). The study compared the effectiveness of anaerobic (AnMBR) and aerobic (AeMBR) membrane bioreactors as viable options for fertilizer, energy, and water recovery in the aim of vinasse recycling. Concerning organic matter removal efficiency, the AnMBR exhibited superior performance (RAnMBR: 97%) compared to the AeMBR (RAeMBR: 92%), showcasing significant differences in their capabilities (Moreira et al. 2022).

Specialized membranes for cellulose saccharification facilitate enzyme immobilization by trapping them within the membrane because of their impermeable character. The primary advantage of such bioreactors lies in their capacity to recover and reuse the biocatalyst. Additionally, MBRs exhibit minimal shear stress forces, minimizing the loss of biocatalyst activity. Membrane separation techniques have been noted for their cost-effectiveness and ease of scalability compared to chromatography processes. Therefore, the adoption of membrane efforts to increase cellulose conversion rates and improve enzyme efficiency have led to the consideration of bioreactors (MBR) as a potentially effective substitute for feedback inhibition. In this arrangement, glucose passes through the membrane due to its lower molecular weight, but enzymes are retained due to the membrane's impermeability to polysaccharides, or substrate. The cellulose saccharification processes are more efficient overall as a result of this selective separation technique (Pino et al. 2018).

4.4 Factors Influencing Bioreactor Selection

Selecting a bioreactor for biorefinery processes at the reactor level involves thorough consideration of multiple factors to ensure optimal performance. Reactor-level factors include accommodating specific reaction kinetics and residence times, providing efficient mixing and mass transfer capabilities, addressing oxygen transfer and aeration requirements, and maintaining precise control over temperature and pH. Compatibility with the substrate, scalability considerations, and downstream processing compatibility are crucial for successful biorefinery operations. Additionally, the bioreactor must possess reliable sterilization methods and aseptic design features to prevent contamination. Additionally, advanced monitoring and control systems, along with energy efficiency considerations, contribute to the overall efficiency and sustainability of the biorefinery process. Ensuring regulatory compliance at the reactor level is essential for meeting industry standards and guidelines, ultimately facilitating the production of biorefinery products with high efficiency and quality (Drewnowski et al. 2019).

5 Limitation of Scaling up of the Biorefinery Product

Scaling up a biorefinery to meet growing demand for bio-based products presents several challenges. One critical limitation involves the consistency and availability of feedstock. The quality and composition of biomass can vary, impacting the efficiency of the biorefinery process. Logistical hurdles, such as transporting and storing larger quantities of diverse feedstock, add complexity. Technologically, scaling up may reveal inefficiencies not evident at smaller scales, necessitating intricate optimization efforts and substantial research and development investments. Environmental concerns arise as well, with larger biorefineries requiring significant land use and potentially posing ecological impacts. Regulatory compliance becomes more intricate, demanding careful attention to permits and adherence to complex environmental and safety regulations. The economic viability of scaling up is contingent on accurately estimating market demand and navigating potential price fluctuations of both feedstock and bio-based products. Additionally, addressing increased energy and water requirements becomes imperative, as does ensuring sustainable sourcing and utilization is crucial. As biorefineries expand, they must navigate with scale-dependent issues, competitive dynamics, and the evolving market landscape to achieve successful and sustainable growth.

Stirred tank reactor: While stirred-tank bioreactors are widely employed in various bioprocesses, including those utilizing biomass as a substrate, they do come with certain disadvantages. One significant drawback is the potential for high energy consumption. The constant agitation and mixing required to maintain a homogeneous environment for microbial activity demand substantial energy input, contributing to operational costs. Additionally, shear forces generated during stirring may negatively impact delicate microorganisms or cells, leading to reduced cell viability and product yield. The complexity of the stirred-tank bioreactor design, along with the need for regular maintenance and sterilization, can also increase the overall operational complexity and cost. Moreover, the scalability of stirred-tank bioreactors to larger volumes may encounter limitations, particularly in achieving efficient mass transfer and maintaining uniform conditions throughout the entire bioreactor (Berovic and Zhong 2023).

Photobioreactor: Closed cultivation systems offer distinct advantages. They boast high biomass yield and sunlight use efficiency. Closed systems require less space, making them more versatile and suitable for a variety of environments. Moreover, they are particularly well-suited for monoculture and sensitive microalgae species. Despite these benefits, the construction and maintenance of closed systems, including the cleaning of reactors, incur high costs. This financial aspect can be a significant drawback for those considering closed cultivation methods (Shahar et al. 2023).

Membrane bioreactor: Although this technology has made significant advancements, there are still many obstacles to overcome, especially with regard to severe membrane fouling, biogas upgrading, effluent ammonia emission, high level of dissolved methane, phosphate loss, and other issues. To tackle these issues, a cutting-edge process termed "Integrated Multistage Bio-Process (IMBP)" comprises Bioelectrochemical Systems (BES), Anaerobic Membrane Bioreactor (AnMBR), PN/A (possibly referring to phosphorus/nitrogen removal/attenuation), anaerobic oxidation of methane, and biological or chemical precipitation units. This unique integration improves biogas, controls in situ fouling, eliminates ammonia more successfully, and speeds up the conversion of biopolymeric components and CO2 electromethanogenesis through bioelectrochemical regulation (Zhen et al. 2019).

6 Circular Economical Perspective of Production of Bio-Based Products Toward Sustainability

In order to solve the economic, environmental, and social difficulties connected with the existing linear resource consumption, the idea of a circular economy is utilized to transform the value chain from a linear to a closed-loop model, improving the efficiency of resource utilization. The bioeconomy is based on the transformation of organic waste and agricultural biomass, which are renewable carbon reservoirs, into a variety of materials and final products. These include feedstock, food, fuels, biopolymers, bio-based synthetic chemicals, and bioenergy (Zeller et al. 2019). Many opportunities have been explored for the bio-based economy to support sustainability. For instance, the outcomes of the 2015 Global Bioeconomy Summit show how at least 11 of the 17 Sustainable Development Goals might be accomplished with the help of an industrial bio-based economy. The concept that the bio-based economy could effectively respond to various societal challenges is, in part, based on its cross-sectoral reach. This reach is seen as a distinctive opportunity to holistically tackle interconnected societal issues such as food security, scarcity of natural resources, dependence on fossil resources, and climate change, all while fostering sustainable economic growth (Bennich and Belyazid 2017). The fundamental principles of the circular economy align closely with those of the bioeconomy, aiming to promote the recycling and reutilization of materials in the development of integrated sustainable approaches. Transitioning toward a circular bioeconomy is anticipated to provide a robust framework for fostering positive economic development and generating employment through organizational, social, and technological advancements (Yaashikaa et al. 2020). Bio-based technologies find significant applicability in optimizing wastewater treatment and managing organic waste and sewage sludges.

This chapter includes noteworthy achievements that have been reported in the literature for both large-scale implementations and pilot- or laboratory-scale experiments. Beyond the typical biological treatment with activated sludge, wastewaters can be directly biomethanized or used as a source for several kinds of bioproducts. Biorefineries can also be successfully fed organic residues, such as sludge from municipal sewage systems and specially grown biomass crops. Nevertheless, anaerobic digestion is the prevailing technology for processing organic waste. This expanding technology not only addresses waste treatment but also facilitates biogas production. As waste-utilizing biogas plants operate based on the circulation of organic matter, they emerge as crucial components capable of closing product/material loops in a modern bioeconomy (Kaszycki et al. 2021). Similarly, proposing a circular economy framework harmonized with environmental waste management, this model is introduced to facilitate a closed-cascade flow of ethanol off-gases and spent streams. The implementation of such a closed-loop approach is critically urgent in the contemporary world, aligning with the evolving need for a sustainable livelihood and the ongoing transition toward the fourth industrial revolution, commonly known as Industry 4.0 (I4.0). The synergy of I4.0 with a circular economy, empowered by biodegradable municipal solid waste (BMSW) and food waste (FW), emerges as a transformative paradigm for the sustainable production of bioethanol and other valuable biochemicals (Althuri and Venkata Mohan 2022). Capitalizing on renewable feedstocks to fuel biorefineries, capable of generating a diverse range of valuable outputs such as biofuels, biochemicals, bioplastics, antibiotics, and enzymes, presents a strategic approach. This not only adds value to these feedstock residues but also aligns with the growing demands of an expanding bioeconomy market emphasizing renewable products. Utilizing advanced technologies and microorganisms, biorefineries extract valuable products, energy, and biochemicals from a variety of feedstocks, leading to a substantial reduction in waste and environmental impacts and concurrently diminishing dependence on fossil fuels through a circular economy (Fig. 3).

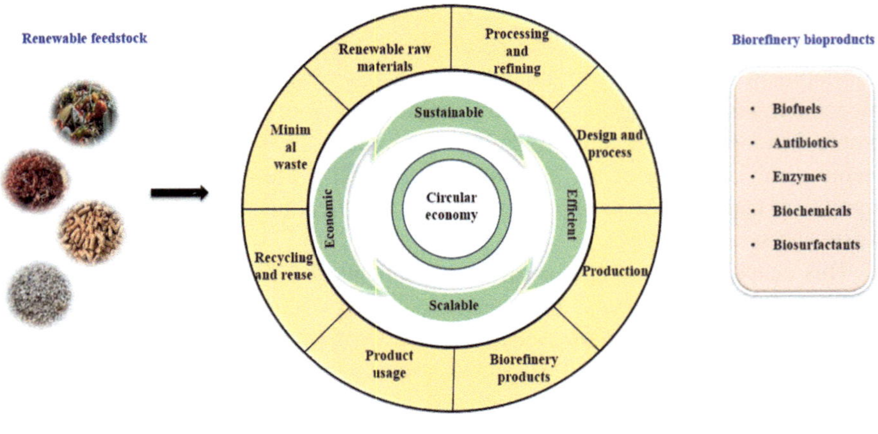

Fig. 3 Circular economical perspective of development of biorefinery from renewable feedstock

7 Future Perspective and Conclusion

The utilization of materials once considered as by-products, such as lignocellulose, holds immense significance for the global economy and environmental preservation. This approach aims to optimize the use of natural resources and promote the principles of a circular economy. Currently, there is a substantial surge in investment toward the development of products and technologies related to lignocellulose and algal biomass. However, more substantial investments are required to advance the maturation and commercialization of existing processing and product development approaches. The valorization processes and technologies involved must be economically efficient to ensure that the products derived from biomass can compete effectively in the market. Scaling up reactor operations within an integrated biorefinery approach becomes crucial for the entire value chain of biomass. Key aspects of scaling up include the implementation of cost-effective pretreatment methods, the advancement of efficient enzymes, the complete conversion of polysaccharides into sugars, and seamless process integration.

In conclusion, biorefineries stand as versatile facilities capable of integrating diverse biomass feedstocks, ranging from agricultural residues to industrial by-products, and harnessing the potential of renewable resources to produce a wide array of valuable outputs. This chapter underscores the significance of renewable feedstocks in biorefineries within the broader contexts of the bioeconomy and circular economy with various bioreactor types, including stirred-tank bioreactors, membrane bioreactor, and photobioreactors. It evaluates the potential for generating value-added bioproducts from a variety of residue sources while also addressing the associated advantages and challenges, offering insights into the sustainable and innovative future of biorefinery processes. Furthermore, scaling up a biorefinery involves navigating complex technical, economic, and environmental challenges, but with careful planning and optimization, it can lead to efficient and sustainable large-scale operations.

Declaration of Competing Interest **The authors declare that they have no known competing financial interests or personal relationships that could have appeared to influence the work reported in this paper.**

Funding **This research did not receive any specific grant from funding agencies in the public, commercial, or not-for-profit sectors.**

References

Ahmad AL, Yasin NHM, Derek CJC, Lim JK (2011) Microalgae as a sustainable energy source for biodiesel production: a review. Renew Sustain Energy Rev 15(1):584–593

Ahorsu R, Medina F, Constantí M (2018) Significance and challenges of biomass as a suitable feedstock for bioenergy and biochemical production: a review. Energies 11(12):3366

Akshaya V, Akila I, Murali R, Raajasubramanian D, Kuppan N, Srinivasan S (2023) Rice straw biomass and agricultural residues as strategic bioenergy: effects on the environment and economy path with new directions. In: Bioenergy: impacts on environment and economy. Springer, Berlin, pp 139–164

Alam MZ, Kabbashi NA, Hussin SNIS (2009) Production of bioethanol by direct bioconversion of oil-palm industrial effluent in a stirred-tank bioreactor. J Ind Microbiol Biotechnol 36(6):801. https://doi.org/10.1007/s10295-009-0554-7

Alexandri M, Vlysidis A, Papapostolou H, Tverezovskaya O, Tverezovskiy V, Kookos IK, Koutinas A (2019) Downstream separation and purification of succinic acid from fermentation broths using spent sulphite liquor as feedstock. Sep Purif Technol 209:666–675

Ali SS, Jiao H, El-Sapagh S, Sun J (2023) Biodegradation of willow sawdust by novel cellulase-producing bacterial consortium from wood-feeding termites for enhancing methane production. Bioresour Technol 383:129232. https://doi.org/10.1016/j.biortech.2023.129232

Althuri A, Venkata Mohan S (2022) Emerging innovations for sustainable production of bioethanol and other mercantile products from circular economy perspective. Bioresour Technol 363:128013. https://doi.org/10.1016/j.biortech.2022.128013

Andrew J, Masetlwa J, Tesfaye T, Sithole B (2020) Beneficiation of eucalyptus tree barks in the context of an integrated biorefinery–optimisation of accelerated solvent extraction (ASE) of polyphenolic compounds using response surface methodology. Sustain Chem Pharm 18:100327

Awasthi MK, Sarsaiya S, Wainaina S, Rajendran K, Kumar S, Quan W, Duan Y, Awasthi SK, Chen H, Pandey A (2019) A critical review of organic manure biorefinery models toward sustainable circular bioeconomy: technological challenges, advancements, innovations, and future perspectives. Renew Sustain Energy Rev 111:115–131

Bakratsas G, Polydera A, Nilson O, Chatzikonstantinou AV, Xiros C, Katapodis P, Stamatis H (2023) Mycoprotein production by submerged fermentation of the edible mushroom Pleurotus ostreatus in a batch stirred tank bioreactor using agro-industrial hydrolysate. Food Secur 12(12):2295

Baweja M, Nain L, Kawarabayasi Y, Shukla P (2016) Current technological improvements in enzymes toward their biotechnological applications. Front Microbiol 7:965

Bennich T, Belyazid S (2017) The route to sustainability—prospects and challenges of the bio-based economy. Sustainability 9(6):887

Berovic M, Zhong J-J (2023) Advances in pilot-scale stirred bioreactors in solid-state and submerged cultivations of medicinal mushrooms. In: Biochemical engineering and biotechnology of medicinal mushrooms, pp 163–185

Bhuyan N, Narzari R, Gogoi L, Bordoloi N, Hiloidhari M, Palsaniya DR, Deb U, Gogoi N, Kataki R (2020) Chapter 2. Valorization of agricultural wastes for multidimensional use. In: Kataki R, Pandey A, Khanal SK, Pant D (eds) Current developments in biotechnology and bioengineering. Elsevier, Amsterdam, pp 41–78. https://doi.org/10.1016/B978-0-444-64309-4.00002-7

Binod P, Sindhu R, Singhania RR, Vikram S, Devi L, Nagalakshmi S, Kurien N, Sukumaran RK, Pandey A (2010) Bioethanol production from rice straw: an overview. Bioresour Technol 101(13):4767–4774

Brennan L, Owende P (2010) Biofuels from microalgae—a review of technologies for production, processing, and extractions of biofuels and co-products. Renew Sustain Energy Rev 14(2):557–577

Buhrer H (2000) Light within algal cultures; implications from light intensities within a lens. Aquat Sci 62(1):91–103

Chandra R, Iqbal HMN, Vishal G, Lee H-S, Nagra S (2019) Algal biorefinery: a sustainable approach to valorize algal-based biomass towards multiple product recovery. Bioresour Technol 278:346–359. https://doi.org/10.1016/j.biortech.2019.01.104

Chew KW, Yap JY, Show PL, Suan NH, Juan JC, Ling TC, Lee D-J, Chang J-S (2017) Microalgae biorefinery: high value products perspectives. Bioresour Technol 229:53–62

Conteratto C, Artuzo FD, Santos OIB, Talamini E (2021) Biorefinery: a comprehensive concept for the sociotechnical transition toward bioeconomy. Renew Sustain Energy Rev 151:111527

Dahiya S, Katakojwala R, Ramakrishna S, Mohan SV (2020) Biobased products and life cycle assessment in the context of circular economy and sustainability. Mater Circ Econ 2:1–28

Dangol S, Ghimire A, Tuladhar S, Khadka A, Thapa B, Sapkota L (2022) Biohythane and organic acid production from food waste by two-stage anaerobic digestion: a review within biorefinery framework. Int J Environ Sci Technol 19(12):12791–12824

Drewnowski J, Remiszewska-Skwarek A, Duda S, Łagód G (2019) Aeration process in bioreactors as the main energy consumer in a wastewater treatment plant. Review of solutions and methods of process optimization. Processes 7(5):311

Fatima A, Basak B, Ganguly A, Chatterjee PK, Dey A (2020) Biohydrogen production through dark fermentation of food wastes by anaerobic digester sludge mixed microbial consortium. In: Recent developments in waste management: select proceedings of recycle, pp 57–70

Garcia-Garcia G, Woolley E, Rahimifard S (2015) A framework for a more efficient approach to food waste management. Int J Food Eng 1(1):65–72

Ginni G, Kavitha S, Yukesh Kannah R, Bhatia SK, Adish Kumar S, Rajkumar M, Kumar G, Pugazhendhi A, Chi NTL, Rajesh Banu J (2021) Valorization of agricultural residues: different biorefinery routes. J Environ Chem Eng 9(4):105435. https://doi.org/10.1016/j.jece.2021.105435

Giovannoni M, Larini I, Scafati V, Scortica A, Compri M, Pontiggia D, Zapparoli G, Vitulo N, Benedetti M, Mattei B (2021) A novel Penicillium sumatraense isolate reveals an arsenal of degrading enzymes exploitable in algal bio-refinery processes. Biotechnol Biofuels 14(1):180. https://doi.org/10.1186/s13068-021-02030-9

Gustafsson J, Cederberg C, Sonesson U, Emanuelsson A (2013) The methodology of the FAO study: global food losses and food waste-extent, causes and prevention—FAO, 2011. SIK Institutet för livsmedel och bioteknik, Gothenburg

Hargreaves JC, Adl MS, Warman PR (2008) A review of the use of composted municipal solid waste in agriculture. Agric Ecosyst Environ 123(1):1–14. https://doi.org/10.1016/j.agee.2007.07.004

Hazra C, Kundu D, Ghosh P, Joshi S, Dandi N, Chaudhari A (2011) Screening and identification of Pseudomonas aeruginosa AB4 for improved production, characterization and application of a glycolipid biosurfactant using low-cost agro-based raw materials. J Chem Technol Biotechnol 86(2):185–198

Henrich E, Dahmen N, Dinjus E, Sauer J (2015) The role of biomass in a future world without fossil fuels. Chem Ing Tech 87(12):1667–1685

Hu AY, Stuckey DC (2006) Treatment of dilute wastewaters using a novel submerged anaerobic membrane bioreactor. J Environ Eng 132(2):190–198

Humbird D, Davis R, McMillan JD (2017) Aeration costs in stirred-tank and bubble column bioreactors. Biochem Eng J 127:161–166

Jain A, Sarsaiya S, Awasthi MK, Singh R, Rajput R, Mishra UC, Chen J, Shi J (2022) Bioenergy and bio-products from bio-waste and its associated modern circular economy: current research trends, challenges, and future outlooks. Fuel 307:121859

Javed MU, Mukhtar H, Hayat MT, Rashid U, Mumtaz MW, Ngamcharussrivichai C (2022) Sustainable processing of algal biomass for a comprehensive biorefinery. J Biotechnol 352:47–58

Jha S, Siddiqui N (2023) Municipal wastewater as potential bio-refinery. In: Biorefinery for water and wastewater treatment. Springer, pp 89–108

Kanai M, Ferre V, Wakahara S, Yamamoto T, Moro M (2010) A novel combination of methane fermentation and MBR—Kubota submerged anaerobic membrane bioreactor process. Desalination 250(3):964–967

Karak T, Bhagat RM, Bhattacharyya P (2012) Municipal solid waste generation, composition, and management: the world scenario. Crit Rev Environ Sci Technol 42(15):1509–1630

Karmee SK, Linardi D, Lee J, Lin CSK (2015) Conversion of lipid from food waste to biodiesel. Waste Manag 41:169–173

Kassim FO, Thomas CLP, Afolabi OOD (2022) Integrated conversion technologies for sustainable Agri-food waste valorization: a critical review. Biomass Bioenergy 156:106314

Kaszycki P, Głodniok M, Petryszak P (2021) Towards a bio-based circular economy in organic waste management and wastewater treatment—the polish perspective. N Biotechnol 61:80–89. https://doi.org/10.1016/j.nbt.2020.11.005
Kumar B, Verma P (2021) Biomass-based biorefineries: an important architype towards a circular economy. Fuel 288:119622. https://doi.org/10.1016/j.fuel.2020.119622
Kunjapur AM, Eldridge RB (2010) Photobioreactor design for commercial biofuel production from microalgae. Ind Eng Chem Res 49(8):3516–3526. https://doi.org/10.1021/ie901459u
Limayem A, Ricke SC (2012) Lignocellulosic biomass for bioethanol production: current perspectives, potential issues and future prospects. Prog Energy Combust Sci 38(4):449–467
Liu L (2017) Catalytic conversion of biomass sugars to value-added platform chemicals. North Carolina State University, Raleigh, NC
López G, Yate C, Ramos FA, Cala MP, Restrepo S, Baena S (2019) Production of polyunsaturated fatty acids and lipids from autotrophic, mixotrophic and heterotrophic cultivation of Galdieria sp. strain USBA-GBX-832. Sci Rep 9(1):10791
Louw J, Dogbe ES, Yang B, Görgens JF (2023) Prioritisation of biomass-derived products for biorefineries based on economic feasibility: a review on the comparability of techno-economic assessment results. Renew Sustain Energy Rev 188:113840. https://doi.org/10.1016/j.rser.2023.113840
Mahendrasinh Kosamia N, Samavi M, Piok K, Kumar Rakshit S (2022) Perspectives for scale up of biorefineries using biochemical conversion pathways: technology status, techno-economic, and sustainable approaches. Fuel 324:124532. https://doi.org/10.1016/j.fuel.2022.124532
Menon V, Rao M (2012) Trends in bioconversion of lignocellulose: biofuels, platform chemicals & biorefinery concept. Prog Energy Combust Sci 38(4):522–550
Merrylin J, Kannah RY, Banu JR, Yeom IT (2020) Production of organic acids and enzymes/biocatalysts from food waste. In: Food waste to valuable resources. Elsevier, Amsterdam, pp 119–141
Mohan SV, Nikhil GN, Chiranjeevi P, Reddy CN, Rohit MV, Kumar AN, Sarkar O (2016) Waste biorefinery models towards sustainable circular bioeconomy: critical review and future perspectives. Bioresour Technol 215:2–12
Mohanty SS, Koul Y, Varjani S, Pandey A, Ngo HH, Chang J-S, Wong JWC, Bui X-T (2021) A critical review on various feedstocks as sustainable substrates for biosurfactants production: a way towards cleaner production. Microb Cell Fact 20(1):1–13
Moreira VR, Carpanez TG, dos Santos FS, Santos LS, dos Santos Fernandes D, França-Neta LS, Lange LC, Amaral MCS (2022) Circular economy in biorefineries: scale-up of anaerobic/aerobic membrane bioreactors for vinasse recycling. J Clean Prod 377:134448. https://doi.org/10.1016/j.jclepro.2022.134448
Muthuraj M, Palabhanvi B, Misra S, Kumar V, Sivalingavasu K, Das D (2013) Flux balance analysis of chlorella sp. FC2 IITG under photoautotrophic and heterotrophic growth conditions. Photosynth Res 118:167–179
Nanda S, Berruti F (2021) Municipal solid waste management and landfilling technologies: a review. Environ Chem Lett 19(2):1433–1456. https://doi.org/10.1007/s10311-020-01100-y
Olguín EJ (2012) Dual purpose microalgae–bacteria-based systems that treat wastewater and produce biodiesel and chemical products within a biorefinery. Biotechnol Adv 30(5):1031–1046
Paladino O, Neviani M (2018) A closed loop biowaste to biofuel integrated process fed with waste frying oil, organic waste and algal biomass: feasibility at pilot scale. Renew Energy 124:61–74. https://doi.org/10.1016/j.renene.2017.08.027
Paladino O, Neviani M (2021) Airlift photo-bioreactors for Chlorella vulgaris cultivation in closed-loop zero waste biorefineries. Biomass Bioenergy 144:105926. https://doi.org/10.1016/j.biombioe.2020.105926
Parchami M, Uwineza C, Ibeabuchi OH, Rustas B-O, Taherzadeh MJ, Mahboubi A (2023) Membrane bioreactor assisted volatile fatty acids production from agro-industrial residues for ruminant feed application. Waste Manag 170:62–74. https://doi.org/10.1016/j.wasman.2023.07.032

Patel A, Sarkar O, Rova U, Christakopoulos P, Matsakas L (2021) Valorization of volatile fatty acids derived from low-cost organic waste for lipogenesis in oleaginous microorganisms—a review. Bioresour Technol 321:124457

Pattanaik L, Pattnaik F, Saxena DK, Naik SN (2019) Chapter 5. Biofuels from agricultural wastes. In: Basile A, Dalena F (eds) Second and third generation of feedstocks. Elsevier, pp 103–142. https://doi.org/10.1016/B978-0-12-815162-4.00005-7

Pino MS, Rodríguez-Jasso RM, Michelin M, Flores-Gallegos AC, Morales-Rodriguez R, Teixeira JA, Ruiz HA (2018) Bioreactor design for enzymatic hydrolysis of biomass under the biorefinery concept. Chem Eng J 347:119–136

Rama Mohan S (2016) Strategy and design of innovation policy road mapping for a waste biorefinery. Bioresour Technol 215:76–83. https://doi.org/10.1016/j.biortech.2016.03.090

Ranjan A, Mayank R, Moholkar VS (2013) Development of semi-defined rice straw-based medium for butanol production and its kinetic study. 3 Biotech 3(5):353–364. https://doi.org/10.1007/s13205-013-0120-x

Rathour RK, Behl M, Dhashmana K, Sakhuja D, Ghai H, Sharma N, Meena KR, Bhatt AK, Bhatia RK (2023) Non-food crops derived lignocellulose biorefinery for sustainable production of biomaterials, biochemicals and bioenergy: a review on trends and techniques. Ind Crop Prod 204:117220

Ravindran R, Hassan SS, Williams GA, Jaiswal AK (2018) A review on bioconversion of agro-industrial wastes to industrially important enzymes. Bioengineering 5(4):93

Sadhukhan J, Martinez-Hernandez E, Murphy RJ, Ng DKS, Hassim MH, Ng KS, Kin WY, Jaye IFM, Hang MYLP, Andiappan V (2018) Role of bioenergy, biorefinery and bioeconomy in sustainable development: strategic pathways for Malaysia. Renew Sustain Energy Rev 81:1966–1987

Salim AA, Grbavčić S, Šekuljica N, Stefanović A, Tansaković SJ, Luković N, Knežević-Jugović Z (2017) Production of enzymes by a newly isolated bacillus sp. TMF-1 in solid state fermentation on agricultural by-products: the evaluation of substrate pretreatment methods. Bioresour Technol 228:193–200

Satpute SK, Płaza GA, Banpurkar AG (2017) Biosurfactants' production from renewable natural resources: example of innovative and smart technology in circular bioeconomy. Manag Syst Prod Eng 25(1):46–54

Shahar B, Haim E, Kuc ME, Azerrad SP, Dudai N, Kurzbaum E (2023) Simplified and cost-effective modulatory photobioreactor setup for upscaling microalgal culture for research and semi-industrial purposes. Algal Res 74:103200

Sharma P, Gaur VK, Kim S-H, Pandey A (2020) Microbial strategies for bio-transforming food waste into resources. Bioresour Technol 299:122580

Siddiki SYA, Mofijur M, Kumar PS, Ahmed SF, Inayat A, Kusumo F, Badruddin IA, Khan TMY, Nghiem LD, Ong HC (2022) Microalgae biomass as a sustainable source for biofuel, biochemical and biobased value-added products: an integrated biorefinery concept. Fuel 307:121782

Sillanpää M, Ncibi C, Sillanpää M, Ncibi C (2017) Biorefineries: industrial-scale production paving the way for bioeconomy. In: A sustainable bioeconomy: the Green Industrial Revolution, pp 233–270

Siller-Sánchez A, Ruiz HA, Aguilar CN, Rodríguez-Jasso RM (2019) Biorefinery approach for red seaweeds biomass as source for enzymes production: food and biofuels industry. In: Parameswaran B, Varjani S, Raveendran S (eds) Green bio-processes: enzymes in industrial food processing. Springer, Singapore, pp 413–446. https://doi.org/10.1007/978-981-13-3263-0_21

Singh RN, Sharma S (2012) Development of suitable photobioreactor for algae production—a review. Renew Sustain Energy Rev 16(4):2347–2353

Singh R, Das R, Sangwan S, Rohatgi B, Khanam R, Peera SKPG, Das S, Lyngdoh YA, Langyan S, Shukla A (2021) Utilisation of agro-industrial waste for sustainable green production: a review. Environ Sustain 4(4):619–636

Srivastava RK, Shetti NP, Reddy KR, Kwon EE, Nadagouda MN, Aminabhavi TM (2021) Biomass utilization and production of biofuels from carbon neutral materials. Environ Pollut 276:116731

Tapia JFD, Samsatli S, Doliente SS, Martinez-Hernandez E, Ab Karim WABW, Lim KL, Shafri HZM, Shaharum NSNB (2019) Design of biomass value chains that are synergistic with the food–energy–water nexus: strategies and opportunities. Food Bioprod Process 116:170–185

Usmani Z, Sharma M, Awasthi AK, Lukk T, Tuohy MG, Gong L, Nguyen-Tri P, Goddard AD, Bill RM, Nayak SC, Gupta VK (2021) Lignocellulosic biorefineries: the current state of challenges and strategies for efficient commercialization. Renew Sustain Energy Rev 148:111258. https://doi.org/10.1016/j.rser.2021.111258

Venkataraman S, Rajendran DS, Kumar PS, Vo D-VN, Vaidyanathan VK (2022) Extraction, purification and applications of biosurfactants based on microbial-derived glycolipids and lipopeptides: a review. Environ Chem Lett 20:1–22

Wells ML, Potin P, Craigie JS, Raven JA, Merchant SS, Helliwell KE, Smith AG, Camire ME, Brawley SH (2017) Algae as nutritional and functional food sources: revisiting our understanding. J Appl Phycol 29:949–982

Wenger J, Stern T (2019) Reflection on the research on and implementation of biorefinery systems—a systematic literature review with a focus on feedstock. Biofuels Bioprod Biorefin 13(5):1347–1364. https://doi.org/10.1002/bbb.2021

Xiong Y, Guo H, Nor DDMM, Song A, Dai L (2023) Mineral resources depletion, environmental degradation, and exploitation of natural resources: COVID-19 aftereffects. Resour Policy 85:103907

Yaashikaa PR, Kumar PS, Saravanan A, Varjani S, Ramamurthy R (2020) Bioconversion of municipal solid waste into bio-based products: a review on valorisation and sustainable approach for circular bioeconomy. Sci Total Environ 748:141312. https://doi.org/10.1016/j.scitotenv.2020.141312

Yang X, Lee SJ, Yoo HY, Choi HS, Park C, Kim SW (2014) Biorefinery of instant noodle waste to biofuels. Bioresour Technol 159:17–23

Yu A-Q, Pratomo N, Ng T-K, Ling H, Cho H-S, Leong SSJ, Chang MW (2016) Genetic engineering of an unconventional yeast for renewable biofuel and biochemical production. J Vis Exp 115:e54371

Yun J-H, Cho D-H, Lee B, Lee YJ, Choi D-Y, Kim H-S, Chang YK (2020) Utilization of the acid hydrolysate of defatted chlorella biomass as a sole fermentation substrate for the production of biosurfactant from Bacillus subtilis C9. Algal Res 47:101868

Zeller V, Towa E, Degrez M, Achten WMJ (2019) Urban waste flows and their potential for a circular economy model at city-region level. Waste Manag 83:83–94

Zhen G, Zheng S, Lu X, Zhu X, Mei J, Kobayashi T, Xu K, Li Y-Y, Zhao Y (2018) A comprehensive comparison of five different carbon-based cathode materials in CO2 electromethanogenesis: long-term performance, cell-electrode contact behaviors and extracellular electron transfer pathways. Bioresour Technol 266:382–388

Zhen G, Pan Y, Lu X, Li Y-Y, Zhang Z, Niu C, Kumar G, Kobayashi T, Zhao Y, Xu K (2019) Anaerobic membrane bioreactor towards biowaste biorefinery and chemical energy harvest: recent progress, membrane fouling and future perspectives. Renew Sustain Energy Rev 115:109392. https://doi.org/10.1016/j.rser.2019.109392

Scaling Up Bioreactor Systems for Sustainable Biorefinery: A Crucial Step in Advancing the Green Economy

Shivani Uniyal and Pavan Kumar Bijalwan

1 Introduction

Due to the continuous rise in energy needs and gradual depletion of fossil resources, the world is currently experiencing a catastrophic energy crisis (Maity 2015). There is substantial scientific evidence that the usage of fossil fuels had a negative impact on environmental health because of the increase in greenhouse gas emissions (Shahid et al. 2021). Renewable energy currently has a high potential to meet the difficulties of energy supply and demand (Ravindran and Jaiswal 2016), as well as the depletion of fossil fuels (Sudiyani et al. 2013). Biorefinery systems have emerged as an appealing route for chemical production, complementing biofuel generation, with tremendous promise for diminished environmental impact (Kamm and Kamm 2005; Cherubini et al. 2009). Biorefineries convert lignin, cellulose, and hemicellulose into valuable products including biofuels and other platform chemicals (Usmani et al. 2020). Commercial-scale conversions of raw materials into useful products typically necessitate the use of bioreactors, which in turn offer the ideal environmental conditions for the production of the desired product. Nonetheless, there are several restrictions related to the bioprocesses. Enzymatic hydrolysis, for example, is a limiting step in the synthesis of biofuels due to the high cost of enzymes and the development of inhibitors of enzymatic activity (Pino et al. 2018). The application of an appropriate bioreactor gets around a number of the drawbacks pertaining to the bioprocesses.

The production of high-value materials in biological processes relies heavily on bioreactors (Taya and Kino-oka 2011). Bioreactors are typically used for the

S. Uniyal
Department of Botany, H.N.B. Garhwal University, Pauri, Uttarakhand, India

P. K. Bijalwan (✉)
Department of Physics, R.C.U. Gov. P. G. College, Uttarkashi, Uttarakhand, India

K. Agrawal, P. Verma (eds.), *Biotechnological Advances in Biorefinery*, Interdisciplinary Biotechnological Advances,
https://doi.org/10.1007/978-981-97-5544-8_11

large-scale manufacturing of biorefinery products (Chanana et al. 2023). An extensive study of the biological system, entailing growth, and metabolism is requisite to apprehend the needs of the cells in order to design an appropriate bioreactor for a given bioprocess (Zhong 2011). Furthermore, scaling up bioreactors in the biotech sector is essential to commercializing breakthroughs in bioprocess and bioproduct (Delvigne et al. 2017). Reactor scale-up involves raising the intended product's production amount in a very complex way that depends on the interplay of hydrodynamics, kinetics, and mass transfer (Bastidas-Oyandeel and Schmidt 2019). Ensuring the project's economic and technological feasibility while ensuring the right circumstances for microbial development and growth is crucial throughout scale-up (de Mello et al. 2023). Inevitably, there are a number of obstacles that come with the scale-up of the bioreactor in biorefinery operations, including the adverse consequences of fluctuating bioreactor conditions on microorganisms, genotypic drift over extended cultivation, inadequate mixing, etc. (Bailey and Ollis 1986; Chubukov et al. 2016; Rugbjerg et al. 2018). Therefore, improvements in bioreactor design hold considerable potential for ensuring their large-scale application and profitable biofuel harvesting. The present chapter aims to provide an overview of the biorefinery concept, the latest studies, and technological advancements related to the various kinds of bioreactors utilized in biorefinery processes, as well as the obstacles that need to be overcome to get this technology implemented in practical applications. Subsequently, prospects for the growth and expansion of bioreactors in biorefineries are also emphasized.

2 Biorefinery Concept

Biorefineries are indispensable for lowering global emissions and decarbonizing our energy systems (Dhamodharan et al. 2020). A biorefinery is a framework or structure that aims to be environmentally friendly and self-sustaining and maximizes the use of biomass to produce a variety of products (Saral et. al. 2022). Using various raw materials from virgin and residual sources is a key feature of biorefineries (Kumar et al. 2020). The framework of the burgeoning bioeconomy revolves around biorefining, which is a key component since the broad spectrum of biomass resources presents excellent prospects for a varied product portfolio to meet societal requirements (Hingsamer and Jungmeier 2019). The majority of them are classified according to their feedstock, such as corn-based biorefinery, wood-based biorefinery, forest-based biorefinery, palm-based biorefinery, algae-based biorefinery, and so on (Ng et al. 2017). However, from a biorefinery perspective, feedstocks can be classified into three main generations: the first, second, and third generations belonging to use crops, residues, and algae as feedstocks, respectively (Cherubini 2010; Cherubini and Ulgiati 2010; Jonker and Faaij 2013). However, some workers have identified nonedible crops as a fourth generation) (Cherubini and Ulgiati 2010).

Integrated biorefinery, which focuses on the conversion of biomass feedstocks into a wide range of value-added products, including biofuels and specialty chemicals, has successfully extended the biorefinery concept (Tay et al. 2012; Mongkhonsiri et al. 2021). These days, sustainable production and the economy require a biorefinery that is affordable (Chanana et al. 2023).

Nonetheless, biorefinery concepts still require significant work to meet all requirements for producing high-quality biofuels, chemicals with added value, or other products. These requirements primarily relate to optimizing and upgrading current conversion processes, developing new, cost-justifiable processes and products, and scaling up existing ideas for industrial use (Bušić et al. 2018). Advances in biomass pretreatment technologies, metabolic engineering, synthetic biology tools, high-performance strain production techniques, fermentation processes, and downstream technologies have largely guided the design of the biorefinery process (Andhalkar et al. 2023). By utilizing several modern forms of bioreactors, it is possible to successfully transform the significant productivity of biorefinery processes into a commercial scale with economic feasibility.

3 Development and Scale-Up of Bioreactor

Since their inception, bioreactors have facilitated the expansion of lab-scale processes into industrial settings. They have been a crucial component of the downstream processes. Aside from product development, they are also engaged in sustainable environmental practices such as waste reduction, waste product treatment, wastewater treatment, and detoxification. The primary functions of a bioreactor include maintaining optimal conditions for the growth and metabolism of cultured organisms. This entails controlling parameters including agitation, aeration, pH, temperature, and nutrition supply. Depending on the particular use and the size of the manufacturing process, bioreactors can have a variety of sizes and designs. (Verdú-Navarro et al. 2023). There are several varieties of bioreactors used in biorefineries, and the choice of particular bioreactors depends on the nature of the processes and the requirements of the bioconversion or bioprocessing involved (Jaibiba et al. 2020). Different types of bioreactors have been for subsequent use in biorefinery (Fig. 1). Here are some common types of bioreactors commonly used in biorefineries:

3.1 Continuous Stirred Tank Bioreactors (CSTBRs)

Continuous stirred tank bioreactors are the most popular and traditional types of bioreactors. The CSTBRs are made up of a large tank or vessel, typically made of stainless steel, furnished with a stirring mechanism, which keeps the reactants moving and well-mixed (Butler et al. 2011; Saran et al. 2019). Application of stirring

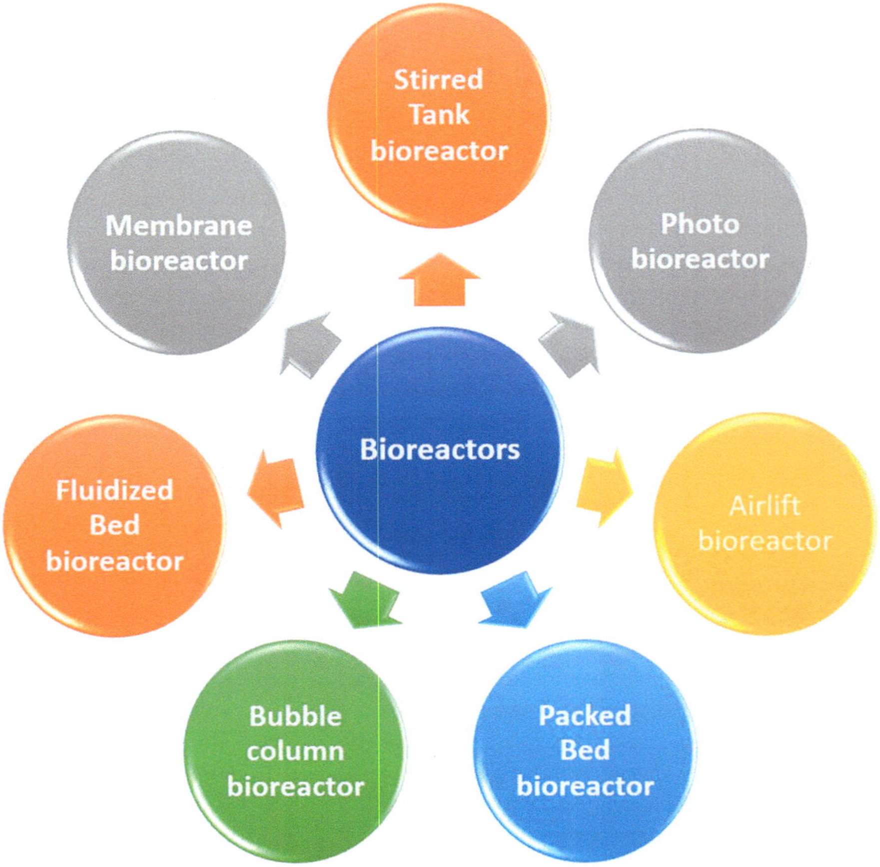

Fig. 1 Different types of bioreactors used in biorefinery studies

mechanism ensures the process to restraint in terms of reaction parameters (Yang 2007).

CSTBRs are especially well-suited for microbial cultivation and reactions of intermediate product formation (Cherkasov et al. 2023). In general, CSTBRs are a versatile and trustworthy solution to industrial operations that demand continuous, high-volume biological, or other product manufacturing.

3.2 *Bubble Column Bioreactors*

A bubble column bioreactor is a cylindrical container designed to cultivate and nurture microbial cultures for bioprocessing. It works on the concept of mass transfer via a continuous flow of gas bubbles, often air or a gas mixture, rising through a liquid medium containing the desired microorganisms (Heijnen and Van't Riet

1984; Kantarci et al. 2005). The bubbles in the culture supply oxygen and agitation, boosting microbial growth and increasing the synthesis of target compounds, such as biofuels, medicines, enzymes, and organic acids.

Bubble column bioreactors are often used for aerobic fermentations. They are prominent because of their modest structure, easy maintenance, and low operating costs. Bubble column bioreactors entail much lower maintenance and operating expenses compared to other reactor types due to the absence of any moving parts (Ribeiro 2008). They offer a wide range of applications in biochemical, chemical, and petrochemical processes and thus are suitable to be used in industries (Kantarci et al. 2005).

3.3 Solid-State Fermentation Reactor

The creation of bioprocesses and products through microbial cultivation may benefit from the three-phase heterogeneous process known as solid-state fermentation (SSF), which consists of solid, liquid, and gaseous phases (Costa et al. 2018). Most of the unfavorable characteristics in different SSF bioreactors can be mitigated by choosing the right design and operational variables (Krishania et al. 2018). There are several types of SSF bioreactors, including rotating chamber reactors (Stuart et al. 1999), packed-bed bioreactors, fluidized-bed bioreactors, chamber trays (Fujian et al. 2002; Zeng and Chen 2009), and PlaFractorTM reactors (Pallín et al. 2022). Submerged column designs are described in the ORSTOM patent (Raimbault and Germon 1976). These reactors are suitable for laboratory scales. The two most popular product categories from SSF are industrial enzymes and organic acids (Wang and Yang 2007). Among various types of SSF reactors, packed-bed bioreactors and fluidized-bed bioreactors are generally utilized in large-scale biorefinery processes.

3.4 Packed-Bed Bioreactors

Packed-bed bioreactors also known as fixed-bed bioreactors is a type of bioreactor that comprises of a solids bed with a biocatalyst on or inside a solid's matrix. It has two modes of operation: submersible, with or without aeration, and drip flow (Sen et al. 2017). Packed-bed reactors are extensively employed in chemical processes, such as absorption, separation, catalytic reactions, nutraceutical synthesis, and waste treatment (Roy et al. 2023; Mitchell et al. 2023). Because reactions in packed-bed bioreactors are heterogeneous, the reactor-to-catalyst-particle-diameter ratio and length-to-diameter ratio may play a significant role in determining the design of bioreactor (Sen et al. 2017).

3.5 Fluidized-Bed Bioreactors

This reactor is made up of packed beds (solid-liquid material) in which solid particles are held while liquid particles flow out. Similar to fixed-bed bioreactors, fluidized-bed bioreactors use a solid support matrix for microbial immobilization. However, in fluidized beds, the solid support particles are kept in constant motion by the upward flow of liquid or gas, providing better mixing and mass transfer. Geometrically, the fluidized-bed bioreactor is similar to a bubble column bioreactor except that the cross-section is prolonged near the top to shrink the superficial velocity of the fluidizing liquid to a value beneath that required to keep the solids in suspension. Fluidized-bed bioreactors are widely used in the environmental, chemical, and process industries (de Mello et al. 2023).

3.6 Membrane Bioreactors (MBRs)

Membrane bioreactors (MBRs) comprise of a biological reactor with suspended biomass and solids removal by ultra- and microfiltration membranes. MBRs incorporate biological treatment with membrane filtration. They syndicate microbial processes with a membrane separation unit, permitting for better control over the bioprocess and separation of biomass from the liquid phase (Enrico and Giorno 2010). The use of MBR offers certain advantages like small-space requirement, owing to its ability to retain a large-size microbial population.

3.7 Airlift Bioreactors

It is a type of bioreactor that is also known as a tower reactor and is used for gas-liquid or gas-liquid-solid processes. Within the reactor, it uses a two-zone approach to enhance circulation and oxygen transfer (Poltronieri and D'Urso 2016). It is similar to a bubble column bioreactor with a draft tube that improves circulation and air transfer. Furthermore, a draught tube enhances shear forces in a bioreactor. The pumping of air, along with the circulation of liquid, determines the efficacy of air-lifted bioreactors. It is different from a stirred bioreactor, which requires the use of a heating cover around the tank for establishing an insulated bioreactor. Clearly, the airlift bioreactor removes heat more effectively than a stirred tank. Because of its basic design, it does not require any agitators, making airlift bioreactors economical, low-energy, and easy to clean (AL-Mashhadani et al. 2015).

3.8 Photobioreactors

In the context of bioreactors, photobioreactors are a relatively recent technology. These bioreactors are a novel reactor, which uses photosynthetic organism to converts natural solar light into useful energy, which is then captured and used to power a range of various applications. Photobioreactors, as opposed to standard bioreactors, rely exclusively on the abundantly and freely available sunlight to produce energy, making them extremely eco-friendly and efficient (Singh and Sharma 2012; Huang et al. 2017).

The extensive application of photobioreactors is to cultivate microalgae in wastewater treatment plants. The photobioreactor generally has a light source to permit necessary and sufficient light for photosynthesis and a culture media. These bioreactors are usually made up of glass or any other transparent polymers. The array made up of these glass tubes or transparent materials constitute light receive systems. It is important that the cells are in uninterrupted circulation without developing sediments, which is usually maintained through the application of centrifugal pumps or airlift pumps. Moreover, more light exposure is needed and the tubes must be cooled to prevent them from overheating.

Photobioreactor is a type of bioreactor that is used to grow photosynthesizing microorganisms such as algae, cyanobacteria, etc. The purpose of a photobioreactor is to create controlled conditions for the growth of photosynthesized microorganisms (Znad 2020). These conditions include exposure to light, temperature, nutrient supply, and the ability to maximize biomass or biofuel production. A photobioreactor can be designed in many different ways, depending on the type of microorganism to be grown and the end product to be produced. For example, a photobioreactor may use mechanical stirrers or bubbling mechanisms to mix the culture medium, allowing air to pass through. During the application process, the temperature and pH must be carefully regulated to ensure the growth of the microorganisms (Delavar and Wang 2022; de Mello et al. 2023).

Generally, there are four basic types of photobioreactors used in biorefinery (Branyikova and Lucakova 2020). These are open raceway pond photobioreactor, flat-plate photobioreactor, inclined tubular photobioreactor, and horizontal tubular photobioreactor.

The selection of bioreactor type depends on factors such as the type of microorganisms being used, the specific biochemical processes involved, the scale of production, and the desired outcome of the biorefinery process. Diverse bioreactors may be used at various stages of the process to achieve efficient and scalable production in biorefineries (Mitra and Murthy 2021).

Since their invention, bioreactors have made lab-scale operations more feasible at the industrial scale (Prakash et al. 2022). The parameters that influence the particular process behavior are typically taken into consideration when choosing the scaling parameter (Najafpour 2007; Burke 2008). A variety of bioreactor systems, including stirred tank reactors, membrane bioreactors (MBRs), bubble columns, and others, have been employed in biorefineries. But when it comes to working, conventional designs have some drawbacks, including low productivity and economic constraints (Gomes et al. 2023) (Table 1). Furthermore, there has been little research done on how different physiological parameters affect scale-up performance (Chubukov et al. 2016; Rugbjerg et al. 2018). As a result, the focus of current research has shifted to encompass the development and scalability of bioreactors using various approaches. Process modification and optimization are currently the main areas of research for the bioreactor system's scale-up (Table 2).

4 Process Optimization

A number of researchers optimized bioreactor operating conditions in an effort to boost biorefinery process efficiency. Ferreira-Torres et al. (2005), for example, established the viability of microscale experimentation for the design of pilot and process-scale operating parameters. The findings indicated that in order to maintain optimal circumstances for high enzyme activity and efficient protein production, fermentation parameters need to be optimized (Zare et al. 2019).

The optimization studies have also made use of statistical designs. For example, a work by Dhaver et al. (2023) successfully optimized the expression of the recombinant XT6 xylanase utilizing response surface methodology (RSM) and Plackett-Burman design (PBD). The results indicated that the specific activity produced from the scaled-up manufacturing is 5.23 times higher than commercial XT6 xylanase from Megazyme. Fenila and Shastri (2019) investigated the optimization of lignocellulose enzymatic hydrolysis in a nonideally mixed reactor in a different study, improving the operation strategies for a batch and fed-batch reactor to carry out the enzymatic hydrolysis of biomass while taking nonideal mixing into account. The findings showed that, when nonideal mixing was taken into account, the glucose concentration and profit increased by 3.19 and 6.63%, respectively. In a similar investigation, Caspeta et al. (2014) employed factorial designs to maximize the output of ethanol by optimizing the temperature, residence duration, and acid and ethanol concentrations.

4.1 Process Modification

Process modification can also improve the productivity of biorefinery processes and thus aid in scale-up of the bioreactor system in biorefinery.

Table 1 Types of bioreactors commonly used in biorefineries studies, their advantages, disadvantages, and applications

S. no	Types of bioreactor	Advantages	Drawbacks	Applications
1	Continuous stirred tank bioreactors	Simple design, flexibility, continuous operation, less labor and capital intensive, excellent control on temperature and other ambient conditions, easy to clean and operate	High power consumption, moving internal parts, high-shear stress	Wastewater treatment, maintenance of animal cell culture
2	Bubble column bioreactor	Easily scalable for both small- and large-scale operations	Limited mixing, foaming, mass transfer limitation, temperature control limitation	In petrochemical and biochemical process, biofuel production, enzyme production, cultivation of algae bioreactor
3	Packed-bed bioreactor	Low operational expenses, no moving parts to wear and tear, effective and efficient at high temperature and pressure	Inadequate temperature regulation, difficult cleaning process, undesired heat gradient	Mammalian cell culture, production of hydrogen and organic acids
4	Fluidized-bed bioreactor	Uniform proper mixing, uniform temperature gradient	Wear and tear of pipe and vessel walls due to particle collisions	Petroleum processing, wastewater treatment
5	Membrane bioreactor	Low downstream contamination	Requirement of periodic cleaning due to biofilm overgrowth, higher power consumption, high foaming propensity, membrane maintenance, and tedious cleaning process	Wastewater treatment, water quality management, remediation of contaminated soil
6	Airlift bioreactor	Simple design with no moving parts, lower shear stress, well-control flow and efficient mixing, easy to scale up	Nonuniform nutrient supply, insufficient mixing	Commonly used for aerobic bioprocessing technology, wastewater treatment, methanol production, single-cell protein production

(continued)

Table 1 (continued)

S. no	Types of bioreactor	Advantages	Drawbacks	Applications
7	Photobioreactors	Reduced risk of contamination, smaller area requirement	Requirement of maximum light exposure, regular cleaning due to light exposure	Production of biofuels, high-value chemicals, and wastewater treatment

Table 2 Process optimization and modification employed in bioreactors used in the scale-up of bioreactor studies

S. no	Type of bioreactor	Variable studied	Application	References
1	Stirred tank bioreactors	Agitation systems, elephant ear impellers	Enzymatic hydrolysis of sugarcane bagasse	Correâ et al. (2016)
2	Helical stirring bioreactor	Improvement in bioreactor design with novel helical impeller	Production of bioethanol	Zhang et al. (2010)
3	Stirred in-house-made mini-reactor	Optimization of environmental condition using factorial design	Enzymatic bioconversion of cellulose to glucose	Caspeta et al. (2014)
4	Stirred tank bioreactors	Power input	Enzymatic hydrolysis of lignocellulosic materials	Palmqvist and Lidén (2012)
5	Stirred tank bioreactors	Improvement in the recombinant enzyme production by optimization of experimental conditions	Enhanced production of a recombinant xylanase (XT6)	Dhaver et al. (2023)
6	Stirred tank reactor	Process optimization and employment of a fed-batch strategy	Increased glucose concentration	Battista et al. (2018)
7	Miniaturized stirred tank bioreactor	HTP, fed-batch strategy	Fermentation	Ali et al. (2012)
8	Membrane bioreactors	Higher-solid levels	Enzymatic liquefaction and saccharification	Du et al. (2014)
9	Solid-state fermentation reactor	Biomass pretreatment, wheat straw digestibility	Bioethanol production	Pallín et al. (2022)
10	Bubble column photobioreactor	Technical and economic feasibility of biodiesel production	Biodiesel production	Branco-Vieira et al. (2020)

4.1.1 Pre-Saccharification and High-Solid Operation

A parameter that prevents mass transfer phenomena while scaling up a bioreactor is higher solid levels because they raise the viscosity of the slurry, which turns it into a non-Newtonian fluid and reduces the saccharification efficiency (Du et al. 2014;

Hodge et al. 2008). To prevent resistance brought on by lignocellulose slurry during stirring, a horizontal rotating bioreactor (HRR) was employed in a study to apply high-solid concentrations of liquefaction and saccharification to the pretreated maize stover. The mixing technique used in this study worked well for saccharification and liquefaction of corn stover at high-solid concentrations (Du et al. 2014). Pre-saccharification can also help with the process of producing biofuel. According to a study, bioethanol production from *Eucalyptus globulus* bark can be enhanced by simultaneous saccharification and fermentation using pre-saccharification and high-solid operation (Amândio et al. 2023). The findings showed that the SSF setup for simultaneous saccharification and fermentation, especially under the optimum operating circumstances, can be effectively used to produce bioethanol from the waste material eucalyptus bark.

4.1.2 Biomass Deconstruction

Enhancing the outcomes of scale-up research can also be facilitated by an affordable biomass deconstruction method. In comparison to the bench scale, at 10% solid loading, Li et al. (2017) successfully scaled up the demonstration of acid-assisted ionic liquid (IL) deconstruction on feedstock blends by a factor of 30. Their study's findings indicated that the technology for acid-assisted IL deconstruction might be successfully expanded to larger operations.

4.1.3 Biomass Pretreatment

Owing to complex lignocellulosic structure of the feedstock and the resultant requirement of their breakdown into their simpler monomeric units, a pretreatment step is indispensable for their transformation into useful product (Mosier et al. 2005). Several researches used different pretreatment techniques to assess how they affected product recovery. The white-rot fungus *Irpex lacteus* was used in a study to assess the feasibility of a fungal pretreatment of wheat straw at a demonstration size in order to increase the digestibility of the straw. After 21 days of pretreatment, the wheat straw digestibility (60.6%) produced in the SSF bioreactor was comparable to that attained on a small scale (57.9%) (Pallín et al. 2022).

4.1.4 High-Throughput Technology (HTP)

One of the main elements that makes HTP culture both practically and economically feasible is the miniaturization of bioreactors. Recently, a variety of stirred tank-type microscale bioreactors have been developed (Klein et al. 2013). Miniaturized stirred tank bioreactors (MSBRs) were described by Ali et al. (2012), who also successfully applied a fed-batch technique in the MSBR. The outcomes showed that MSBR can perform HTP in-parallel fermentation and offer an economical and an affordable way to build bioprocesses.

4.1.5 Advancement in Bioreactor Design

Innovative bioreactor designs can be employed in the bioconversion process to get around the drawbacks of traditional designs. In a recent study, a microalgal sp. *Phaeodactylum tricornutum* was employed for the production of biodiesel using a bubble column photobioreactor. Results indicated that biofuel with a concentration of about 171,705 L per year can be synthesized using tons of microalgal biomass (Branco-Vieira et al. 2020). Additionally, research was conducted to create a hybrid type of photobioreactor system, since closed and open systems have greater construction and maintenance costs and higher evaporation rates, respectively (Chanana et al. 2023). According to a recent study, cultivating in a hybrid system yields a maximum biomass production of 0.74 g L^{-1} d^{-1}, along with a 46.9% carbon bioconversion efficiency (Song et al. 2020). The use of helical impellers in helical stirring bioreactors can further modify bioreactor designs and enhance the reactor's mixing performance (Zhang et al. 2010).

4.1.6 Use of High-Performance Strains

Recombinant DNA technology can be used to create high-performance strains that are then utilized in a variety of operations. Solvent-tolerant strains have been obtained using a variety of mechanisms, including chaperones and transcriptional factors (Mukhopadhyay 2015).

4.1.7 Modification of the Feeding Strategy

Due to their poor mixing performance, bioreactors with conventional designs, such as the horizontal rotating reactor (HRR) structure, were not suitable for large-scale applications, like industrial ones (Bailey and Ollis 1986). In order to guarantee multi-direction mixing and transport, which is crucial for the dissemination of enzymes and to maximize product synthesis, numerous attempts were made to improve the bioreactor's rheology (Sherritt et al. 2003; Hehl et al. 1978). Mondebach and Nokel (2013) shown in a study that the substrate feeding approach might reduce the diffusion and mixing restrictions by enhancing the bioreactor's rheology. In a different investigation, a specific fed-batch technique made it possible to gradually increase the reactor's dry matter content while maintaining a low viscosity in the reaction media. The results revealed an efficient synthesis of glucose with a concentration of about 85 g L^{-1} (Battista et al. 2018).

5 Conclusion

A highly efficient and economical biorefinery system is necessary to meet the complex and enormous challenges of adhering to decreasing greenhouse gas emissions and moving toward an environmentally and economically sustainable development (Calvo-Flores and Martin-Martinez 2022). Furthermore, efficiency of the process can be enhanced by adoption of integration of bioconversion approach, which could help in maintaining sustainability in terms of resources utilization (Ahmed et al. 2023). These changes can take place in a bioreactor under carefully monitored environmental conditions. The most common bioreactors used in biorefineries to improve the environmentally friendly process are continuous-type stirring, plug flow, packed-bed bioreactors, bubble column, fluidized-bed, membrane, airlift, and photobioreactor. The two major approaches, utilized to scale up the bioreactor system in the biorefinery and so increase the productivity and efficiency of bioconversion in these bioreactors, are process optimization and process modification.

6 Futures Outlook

Considering the relevance of the biorefinery concept, we advocate the following areas for deeper investigation for future research and development.

1. Screening and employment of feedstock is one of the most important parameters, during scale-up operations, as it might get affected due to the seasonal availability of biomass variations (Pyrgakis and Kokossis 2018). Therefore, more advancements in the areas of process integration and intensification in biorefineries are needed in order to compete with fossil fuels (Saral et al. 2022).
2. Large-scale production economics can be enhanced via metabolic engineering to lower an organism's oxygen demand (Humbird and Fei 2016).
3. A novel biomass pretreatment process needs to be explored in order to ensure the advancement in the industrial lignocellulose applications and enhanced chemical output (Chen et al. 2017).
4. While the development of bioprocesses has accelerated due in large part to the miniaturization of bioreactors, there is still much room for advancement in the current HTP platforms as emphasized in recent studies (Chen et al. 2009; Huber et al. 2009, 2011).

References

Ahmed SF, Kabir M, Mehjabin A, Oishi FTZ, Ahmed S, Mannan S, Mofijur M, Almomani F, Badruddin IA, Kamangar S (2023) Waste biorefinery to produce renewable energy: bioconversion process and circular bioeconomy. Energy Rep 10:3073–3091

Ali S, Perez-Pardo MA, Aucamp JP, Craig A, Bracewell DG, Baganz F (2012) Characterization and feasibility of a miniaturized stirred tank bioreactor to perform E. coli high cell density fed-batch fermentations. Biotechnol Prog 28(1):66–75

AL-Mashhadani MKH, Wilkinson SJ, Zimmerman WB (2015) Airlift bioreactor for biological applications with microbubble mediated transport processes. Chem Eng Sci 137:243–253. https://doi.org/10.1016/j.ces.2015.06.032

Amândio MST, Rocha JMS, Xavier AMRB (2023) Improving simultaneous saccharification and fermentation by pre-saccharification and high solids operation for bioethanol production from Eucalyptus globulus bark. J Environ Chem Eng 11(5):110763

Andhalkar VV, Foong SY, Kee SH, Lam SS, Chan YH, Djellabi R, Bhubalan K, Medina F, Constantí M (2023) Macromol. Integrated biorefinery design with techno-economic and life cycle assessment tools in polyhydroxyalkanoates processing. J Mol Eng Mater 308:1–20

Bailey J, Ollis D (1986) Biochemical engineering fundamentals, vol 348, 2nd edn. McGraw-Hill, New York, p 349

Bastidas-Oyandeel JR, Schmidt JE (2019) Biorefinery. Integrated sustainable processes for biomass conversion to biomaterials, biofuels, and fertilisers. Springer, Cham. https://doi.org/10.1007/978-3-030-10961-5

Battista F, Almendros MG, Rousset R, Bouillon P (2018) Enzymatic hydrolysis at high lignocellulosic content: optimization of the mixing system geometry and of a fed-batch strategy to increase glucose concentration. Renew Energy 131:152–158

Branco-Vieira M, Mata TM, Martins AA, Freitas MAV, Caetano NS (2020) Economic analysis of microalgae biodiesel production in a small-scale facility. Energy Rep 6(8):325–332

Branyikova I, Lucakova S (2020) Technical and physiological aspects of microalgae cultivation and productivity—spirulina as a promising and feasible choice. Org Agric 11(2):269–276. https://doi.org/10.1007/s13165-020-00323-1

Burke F (2008) Scale up and scale down of fermentation processes. In: McNeil B, Harvey LM (eds) Practical fermentation technology. Wiley, Chichester

Bušić A, Mardetko N, Kundas S, Morzak G, Belskaya H, Ivančić Šantek M, Komes D, Novak S, Šantek B (2018) Bioethanol production from renewable raw materials and its separation and purification: a review. Food Technol Biotechnol 56(3):289–311

Butler M, Moo-Young M, Al E (2011) Comprehensive biotechnology. Elsevier, Amsterdam

Calvo-Flores FG, Martin-Martinez FJ (2022) Biorefineries: achievements and challenges for a bio-based economy. Front Chem 10(10):973417

Caspeta L, Caro-Bermúdez MA, Ponce-Noyola T, Martinez A (2014) Enzymatic hydrolysis at high-solids loadings for the conversion of agave bagasse to fuel ethanol. Appl Energy 113:277–286

Chanana I, Kaur P, Kumar L, Kumar P, Kulshreshtha S (2023) Advancements in microalgal biorefinery technologies and their economic analysis and positioning in energy resource market. Fermentation 9:202

Chen A, Chitta R, Chang D, Amanullah A (2009) Twenty-four well plate miniature bioreactor system as a scale-down model for cell culture process development. Biotechnol Bioeng 102(1):148–160

Chen H, Liu J, Chang X, Chen D, Xue Y, Liu P, Lin H, Han S (2017) A review on the pretreatment of lignocellulose for high-value chemicals. Fuel Process Technol 160:196–206

Cherkasov N, Adams SJ, Bainbridge EGA, Thornton JAM (2023) Continuous stirred tank reactors in fine chemical synthesis for efficient mixing, solids-handling, and rapid scale-up. React Chem Eng 8(2):266–277. https://doi.org/10.1039/D2RE00232A

Cherubini F (2010) The biorefinery concept: using biomass instead of oil for producing energy and chemicals. Energ Conver Manage 51(7):1412–1421

Cherubini F, Ulgiati S (2010) Crop residues as raw materials for biorefinery systems—a LCA case study. Appl Energy 87(1):47–57

Cherubini F, Jungmeier G, Wellisch M, Willke T, Skiadas I, Van Ree R, Jong E (2009) Toward a common classification approach for biorefinery systems. Biofuels Bioprod Biorefin 3(534):546

Chubukov V et al (2016) Synthetic and systems biology for microbial production of commodity chemicals. NPJ Syst Biol Appl 2:16009

Correâ LJ, Badino AC, Cruz AJG (2016) Mixing design for enzymatic hydrolysis of sugarcane bagasse: methodology for selection of impeller configuration. Bioprocess Biosyst Eng 39:285–294

Costa JAV, Treichel H, Kumar V, Pandey A (2018) Advances in solid-state fermentation. In: Pandey A, Larroche C, Soccol CR (eds) Current developments in biotechnology and bioengineering. Elsevier, Amsterdam, pp 1–17. https://doi.org/10.1016/B978-0-444-63990-5.00001-3

de Mello AFM, de Souza Vandenberghe LP, Herrmann LW et al (2023) Strategies and engineering aspects on the scale-up of bioreactors for different bioprocesses. Syst Microbiol Biomanuf 4:365. https://doi.org/10.1007/s43393-023-00205-z

Delavar MA, Wang J (2022) Advanced methods and mathematical modeling of biofilm: applications in health care, medicine, food, aquaculture, environment, and industry. Elsevier Science & Technology, Amsterdam

Delvigne F, Takors R, Mudde R, Gulik WV, Noorman H (2017) Bioprocess scale-up/down as integrative enabling technology: from fluid mechanics to systems biology and beyond. J Microbial Biotechnol 10(5):1267–1274

Dhamodharan K, Ahlawat S, Kaushal M, Rajendran K (2020) Economics and cost analysis of waste biorefineries. In: Kumar RP, Gnansounou E, Raman JK, Baskar G (eds) Refining biomass residues for sustainable energy and bioproducts. Academic Press, San Diego, CA, pp 545–565. ISBN 9780128189962

Dhaver P, Sithole T, Pletschke B, Sithole B, Govinden R (2023) Enhanced production of a recombinant xylanase (XT6): optimization of production and purification, and scaled-up batch fermentation in a stirred tank bioreactor. Sci Rep 13:20895

Du J, Zhang F, Li Y, Zhang H, Liang J, Zheng H, Huang H (2014) Enzymatic liquefaction and saccharification of pretreated corn Stover at high-solids concentrations in a horizontal rotating bioreactor. Bioprocess Biosyst Eng 37:173–181

Enrico D, Giorno L (2010) Comprehensive membrane science and engineering: basic aspects of membrane science and engineering, vol 1. Elsevier, Amsterdam

Fenila F, Shastri Y (2019) Optimization of cellulose hydrolysis in a nonideally mixed reactors. Comput Chem Eng 128:340–351

Ferreira-Torres C, Micheletti M, Lye GJ (2005) Microscale process evaluation of recombinant biocatalyst libraries: application to Baeyer-Villiger monooxygenase catalysed lactone synthesis. Bioprocess Biosyst Eng 28:83–93

Fujian X, Hongzhang C, Zuohu L (2002) Effect of periodically dynamic changes of air on cellulase production in solid-state fermentation. Enzyme Microb Technol 30:45–48. https://doi.org/10.1016/S0141-0229(01)00454-9

Gomes DG, Coelho E, Silva R, Domingues L, Teixeira JA (2023) Bioreactors and engineering of filamentous fungi cultivation. In: Taherzadeh MJ, Ferreira JA, Pandey A (eds) Current developments in biotechnology and bioengineering. Elsevier, Amsterdam, pp 219–250., ISBN 9780323918725. https://doi.org/10.1016/B978-0-323-91872-5.00018-1

Hehl M, Kroger H, Helmrich H, Schugerl K (1978) Longitudinal mixing in horizontal rotary drum reactors. Powder Technol 20:29–37

Heijnen JJ, Van't Riet K (1984) Mass transfer, mixing and heat transfer phenomena in low viscosity bubble column reactors. Chem Eng J 28(2):B21–B42. https://doi.org/10.1016/0300-9467(84)85025-x

Hingsamer M, Jungmeier G (2019) Biorefineries. In: Lago C, Caldés N, Lechón Y (eds) The role of bioenergy in the bioeconomy. Academic Press, San Diego, CA, pp 179–222.,ISBN 9780128130568. https://doi.org/10.1016/B978-0-12-813056-8.00005-4

Hodge DB, Karim MN, Schell DJ, McMillan JD (2008) Model-based fed-batch for high-solids enzymatic cellulose hydrolysis. Appl Biochem Biotechnol 152:88

Huang Q, Jiang F, Wang L, Yang C (2017) Design of photobioreactors for mass cultivation of photosynthetic organisms. Engineering 3(3):318–329. https://doi.org/10.1016/j.eng.2017.03.020

Huber R, Ritter D, Hering T, Hillmer AK, Kensy F, Muller C, Buchs J (2009) Robo-lector—a novel platform for automated high-throughput cultivations in microtiter plates with high information content. Microb Cell Fact 8(1):42
Huber R, Roth S, Rahmen N, Buchs J (2011) Utilizing high-throughput experimentation to enhance specific productivity of an *E. Coli* T7 expression system by phosphate limitation. BMC Biotechnol 11:22
Humbird D, Fei Q (2016) Scale-up considerations for biofuels. In: Eckert CA, Trinh CT (eds) Biotechnology for biofuel production and optimization. Elsevier, Amsterdam, pp 513–537., ISBN 9780444634757. https://doi.org/10.1016/B978-0-444-63475-7.00020-0
Jaibiba P, Naga Vignesh S, Hariharan S (2020) Working principle of typical bioreactors. In: Bioreactors, pp 145–173. https://doi.org/10.1016/b978-0-12-821264-6.00010-3
Jonker JGG, Faaij APC (2013) Techno-economic assessment of micro-algae as feedstock for renewable bio-energy production. Appl Energy 102:461–475
Kamm B, Kamm M (2005) Principles of biorefineries. Appl Microbiol Biotechnol 64:137–145
Kantarci N, Borak F, Ulgen KO (2005) Bubble column reactors. Process Biochem 40(7):2263–2283. https://doi.org/10.1016/j.procbio.2004.10.004
Klein T, Schneider K, Heinzle E (2013) A system of miniaturized stirred bioreactors for parallel continuous cultivation of yeast with online measurement of dissolved oxygen and off-gas. Biotechnol Bioeng 110(2):535–542
Krishania M, Sindhu R, Binod P, Ahluwalia V, Kumar V, Sangwan RS, Pandey A (2018) Design of bioreactors in solid-state fermentation. In: Pandey A, Larroche C, Soccol CR (eds) Current developments in biotechnology and bioengineering. Elsevier, Amsterdam, pp 83–96
Kumar PS, Saravanan A, Jayasree R, Jeevanantham S (2020) Food industry waste biorefineries. In: Kumar RP, Gnansounou E, Raman JK, Baskar G (eds) Refining biomass residues for sustainable energy and bioproducts. Academic Press, San Diego, CA, pp 407–426. ISBN 9780128189962
Li C, Liang L, Sun N et al (2017) Scale-up and process integration of sugar production by acidolysis of municipal solid waste/corn Stover blends in ionic liquids. Biotechnol Biofuels 10:13
Maity SK (2015) Opportunities, recent trends and challenges of integrated biorefinery: part II. Renew Sustain Energy Rev 43:1446–1466
Mitchell DA, Ruiz HA, Krieger N (2023) A critical evaluation of recent studies on packed-bed bioreactors for solid-state fermentation. 11(3):872–872. https://doi.org/10.3390/pr11030872
Mitra S, Murthy GS (2021) Bioreactor control systems in the biopharmaceutical industry: a critical perspective. Syst Microbiol Biomanuf 2(1):91. https://doi.org/10.1007/s43393-021-00048-6
Mondebach AA, Nokel SE (2013) Enzymatic hydrolysis of biomass at high-solids loadings—a review. Biomass Bioenergy 56:526–544
Mongkhonsiri G, Anantpinijwatna A, Charoensuppanimit P, Arpornwichanop A, Gani R, Assabumrungrat S (2021) Development of sustainable integrated biorefinery networks in pulp and paper industries. In: Türkay M, Gani R (eds) Computer aided chemical engineering, vol 50. Elsevier, Amsterdam, pp 1517–1522., ISSN 1570–7946, ISBN 9780323885065. https://doi.org/10.1016/B978-0-323-88506-5.50234-5
Mosier N, Wyman C, Dale B et al (2005) Features of promising technologies for pretreatment of lignocellulosic biomass. Bioresour Technol 96:673–686
Mukhopadhyay A (2015) Tolerance engineering in bacteria for the production of advanced biofuels and chemicals. Trends Microbiol 23(8):498–508
Najafpour GD (2007) Biochemical engineering and biotechnology. Elsevier Science, Amsterdam
Ng DKS, Ng KS, Ng RTL (2017) Integrated biorefineries. In: Abraham MA (ed) Encyclopedia of sustainable technologies. Elsevier, Amsterdam, pp 299–314., ISBN 9780128047927. https://doi.org/10.1016/B978-0-12-409548-9.10138-1
Pallín MÁ, González-Rodríguez S, Eibes G et al (2022) Towards industrial application of fungal pretreatment in 2G biorefinery: scale-up of solid-state fermentation of wheat straw. Biomass Conv Bioref 14:593. https://doi.org/10.1007/s13399-022-02319-1

Palmqvist B, Lidén G (2012) Torque measurements reveal large process differences between materials during high solid enzymatic hydrolysis of pretreated lignocellulose. Biotechnol Biofuels 5:57

Pino MS, Rodríguez-Jasso RM, Michelin M, Flores-Gallegos AC, Morales-Rodriguez R, Teixeira JA, Ruiz HA (2018) Bioreactor design for enzymatic hydrolysis of biomass under the biorefinery concept. Chem Eng J 347:119–136

Poltronieri P, D'Urso OF (2016) Biotransformation of agricultural waste and by-products. Elsevier, Amsterdam

Prakash P, Pandey S, Jha SK, Singh HR (2022) Development of bioreactors: current scenario and future challenges. In: Mishra P, Singh L, Ghosh P (eds) Techno-economics and life cycle assessment of bioreactors. Elsevier, Amsterdam, pp 37–54., ISBN 9780323898485. https://doi.org/10.1016/B978-0-323-89848-5.00012-3

Pyrgakis KA, Kokossis AC (2018) An engineering tool to screen and integrate biomass valorization paths in multiple-feedstock biorefineries. In: Friedl A, Klemeš JJ, Radl S, Varbanov PS, Wallek T (eds) Computer aided chemical engineering, vol 43. Elsevier, Amsterdam, pp 573–578

Raimbault M, Germon JC (1976) Procédé d'enrichissement en protéines de produits comestibles solides. French Patent no. 76-06-677

Ravindran R, Jaiswal AK (2016) A comprehensive review on pre-treatment strategy for lignocellulosic food industry waste: challenges and opportunities. Bioresour Technol 199:92–102

Ribeiro CP (2008) On the estimation of the regime transition point in bubble columns. Chem Eng J 140(1–3):473–482. https://doi.org/10.1016/j.cej.2007.11.029

Roy C, Klausner JF, Bénard A (2023) Numerical investigation of the blockage phenomenon in packed columns. Chem Eng Res Des 190:508–519. https://doi.org/10.1016/j.cherd.2022.11.026

Rugbjerg P et al (2018) Diverse genetic error modes constrain large-scale bio-based production. Nat Commun 9:787

Saral JS, Ajmal RS, Ranganathan P (2022) Bioeconomy of hydrocarbon biorefinery processes, Hydrocarbon Biorefinery, Elsevier, 355-385, https://doi.org/10.1016/B978-0-12-823306-1.00011-X

Saran S, Malaviya A, Chaubey A (2019) Introduction, scope and significance of fermentation technology. Wiley & Sons, EBooks, Hoboken, NJ, pp 1–25. https://doi.org/10.1002/9781119460053.ch1

Sen P, Nath A, Bhattacharjee C (2017) Packed-bed bioreactor and its application in dairy, food, and beverage industry. In: Current developments in biotechnology and bioengineering, pp 235–277. https://doi.org/10.1016/b978-0-444-63663-8.00009-4

Shahid MK, Batool A, Kashif A, Nawaz MH, Aslam M, Iqbal N, Choi Y (2021) Biofuels and biorefineries: development, application and future perspectives emphasizing the environmental and economic aspects. J Environ Manage 297:113268

Sherritt RG, Chaouki J, Mehrotra AK, Behie LA (2003) Axial dispersion in the three- dimensional mixing of particles in a rotating drum reactor. Chem Eng Sci 58:401–415

Singh RN, Sharma S (2012) Development of suitable photobioreactor for algae production—a review. Renew Sustain Energy Rev 16(4):2347–2353. https://doi.org/10.1016/j.rser.2012.01.026

Song C, Han X, Qiu Y, Liu Z, Li S, Kitamura Y (2020) Microalgae carbon fixation integrated with organic matters recycling from soybean wastewater: effect of pH on the performance of hybrid system. Chemosphere 248:126094

Stuart DM, Mitchell DA, Johns MR, Litster JD (1999) Solid-state fermentation in rotating drum bioreactors: operating variables affect performance through their effects on transport phenomena. Biotechnol Bioeng 63:383–391

Sudiyani Y, Styarini D, Triwahyuni E, Sudiyarmanto SKC, Aristiawan Y, Abimanyu H, Han MH (2013) Utilization of biomass waste empty fruit bunch fiber of palm oil for bioethanol production using pilot–scale unit. Energy Procedia 32:31–38

Tay DHS, Ng RTL, Ng DKS (2012) Modular optimization approach for process synthesis and integration of an integrated biorefinery. In: Karimi IA, Srinivasan R (eds) Computer aided chemical engineering, vol 31. Elsevier, Amsterdam, pp 1045–1049., ISSN 1570-7946, ISBN 9780444595058. https://doi.org/10.1016/B978-0-444-59506-5.50040-7

Taya M, Kino-oka M (2011) Bioreactors for animal cell cultures. In: Moo-Young M (ed) Comprehensive biotechnology, 2nd edn. Academic Press, New York, pp 373–382., ISBN 9780080885049. https://doi.org/10.1016/B978-0-08-088504-9.000969

Usmani Z, Sharma M, Karpichev Y, Pandey A, Kuhad RC, Bhat R, Punia R, Aghbashlo M, Tabatabaei M, Gupta VK (2020) Advancement in valorization technologies to improve utilization of bio-based waste in bioeconomy context. Renew Sustain Energy Rev 131:109965

Verdú-Navarro F, Moreno-Cid JA, Weiss J, Egea-Cortines M (2023) The advent of plant cells in bioreactors. Front Plant Sci 14:1310405. https://doi.org/10.3389/fpls.2023.1310405

Wang L, Yang ST (2007) Solid state fermentation and its applications. In: Yang S-T (ed) Bioprocessing for value-added products from renewable resources. Elsevier, Amsterdam, pp 465–489

Yang ST (2007) Bioprocessing for value-added products from renewable resources: new technologies and applications. Elsevier, Amsterdam

Zare H, Mir Mohammad Sadeghi H, Akbari V (2019) Optimization of fermentation conditions for reteplase expression by Escherichia coli using response surface methodology. AJMB 11(2):162–168

Zeng W, Chen HZ (2009) Air pressure pulsation solid state fermentation of feruloyl esterase by *Aspergillus niger*. Bioresour Technol 100:1371–1375

Zhang J, Chu D, Huang J, Yu Z, Dai G, Bao J (2010) Simultaneous saccharification and ethanol fermentation at high corn Stover solids loading in a helical stirring bioreactor. Biotechnol Bioeng 105:718–728

Zhong JJ (2011) Bioreactor engineering. In: Moo-Young M (ed) Comprehensive biotechnology, 2nd edn. Academic Press, New York, pp 165–177., ISBN 9780080885049. https://doi.org/10.1016/B978-0-08-088504-9.00097-0

Znad H (2020) Microalgae culture technology for carbon dioxide biomitigation. In: Handbook of algal science, technology and medicine, pp 303–316. https://doi.org/10.1016/b978-0-12-818305-2.00019

Bioconversion of Waste to Bioethanol Production

G. P. Dhanvanti, J. Tharunkumar, R. Bobita, and S. Rakesh

1 Introduction

In this current generation, the rise in population has led toward deep concern about the high dependency on fossil fuels for transportation. The emission of CO_2 is more importantly impacted as a negative consequence due to the production of more greenhouse gases. At the same time, these have mainly led to global warming and climate change, which has boosted interest in alternative energy sources such as biofuels (Ray et al. 2023). Biofuels have emerged as a sustainable energy source; bioethanol is an attractive substitute for gasoline (Broda et al. 2022; Kumar and Verma 2020; Kumar et al. 2020; Chaturvedi and Verma 2013). Such biofuels represent significantly renewable energy that will help sustain our ever-demanding civilization shortly. Many researchers in their study cracked their heads for biofuel production. Another great result of the challenge was the dumping of waste in an open place, which caused problems for the dumpsite's surrounding environment (Moneruzzaman Khandaker et al. 2021). In addition, the solution for this crisis can be worked on by applying the concept of producing energy by utilizing this waste, representing a cost-effective method and sustainable impact with no harm.

In this continuous evolution, biofuel technology moved forward with several generations of significant advancements. Among them, bioethanol is an attractive substitute for gasoline (Broda et al. 2022). First-generation bioethanol was typically produced from sugar, starch, or vegetable oil-based feedstock that competed directly with food production. Hence, the requirement of soil, water, and fertilization became a barrier to bioethanol yield. On the other hand, second-generation biofuels were

G. P. Dhanvanti · J. Tharunkumar · R. Bobita · S. Rakesh (✉)
Biofuel Research Laboratory, Department of Microbiology, School of Life Sciences, Central University of Tamil Nadu, Thiruvarur, Tamil Nadu, India
e-mail: suchitrar@cutn.ac.in

K. Agrawal, P. Verma (eds.), *Biotechnological Advances in Biorefinery*, Interdisciplinary Biotechnological Advances,
https://doi.org/10.1007/978-981-97-5544-8_12

derived from nonedible lignocellulosic residues of plants or crops that consisted of 70% polymerized sugar (Isikgor and Becer 2015). Biomass accumulated from vegetation or related wastages such as grass, cornstocks, sugarcane, sorghum, etc. consists of cellulosic properties that can be utilized for bioethanol production (Bhardwaj et al. 2020b). Such research on using lignocellulose wastes to produce clean energy sources reduced the burden of greenhouse gases. With a cost-effective approach, the production of bioethanol from lignocellulosic wastes is expected to increase significantly, potentially reaching approximately 134.5 billion liters and 39 billion liters in the coming future (Aulitto et al. 2019; Prasad et al. 2019). In countries such as the United States and Brazil, in 2021, bioethanol production was primarily derived from corn and sugarcane, producing 82% of global production. However, the United States was considered the world's most significant ethanol producer. In 2019, bioethanol production rose to 15.7 billion gallons in developed countries such as the United States (US Department of Energy 2020).

Many plant materials, mainly sugarcane, pineapple, and potato, are considered beneficial for bioethanol production due to a significant amount of sugar. Recent research stated that pineapple and oranges, having high glucose content, are excellent substrates for bioethanol production (Khandaker et al. 2018). On the other hand, the microbial digestion of vegetable waste can be another source for bioethanol production with high humus content. The chemical potentials in vegetable waste contain a high amount of saccharide, leading to lignocellulose for bioethanol conversion. Through properly controlled fermentation conditions, a high yield can be expected. However, vegetable waste biomass is rich in carbohydrates that can be used to generate potent renewable energy (Supriyanto et al. 2019). Basically, three types of processes are dependent upon in the study of fermentation for bioethanol production. These processes are simultaneous saccharification and fermentation, simultaneous saccharification and co-fermentation, and lastly, separate hydrolysis and fermentation (Fig. 1). Among these processes, simultaneous saccharification and co-fermentation and simultaneous saccharification and fermentation resulted in efficient outcomes compared to the separate hydrolysis and fermentation processes. Overall, it thus depends on the feedstock concentration as prolonged exposure to higher concentrations generally leads to less bioethanol yield (Triwahyuni et al. 2015). This research regarding lignocellulose waste abundantly resulted in greenhouse gas reduction and clean energy source production.

2 Sources of Waste for Bioethanol Production

The predominant sources used for bioethanol production, according to various authors and industry practices, include the following (Table 1):

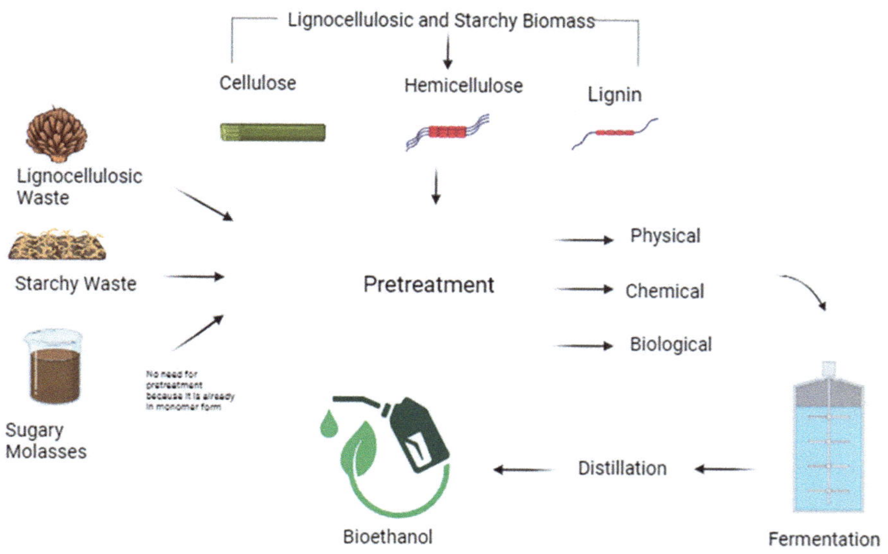

Fig. 1 Strategy design for bioethanol production from various sources

2.1 High Lignocellulosic Content Waste Sources

2.1.1 Paddy Straw

Paddy straw has several characteristics that make it a suitable feedstock for ethanol production. It has a high proportion of cellulose and hemicellulose, making it readily hydrolyzed into fermentable sugar. The paddy straw mainly comprises cellulose (38.7%), hemicellulose xylose (22.9%), and lignin (22.13%) (Jia Wu1 2018). Rice straw is a promising candidate due to its high cellulose and hemicellulose content and its large annual global production of 370–520 million tons yearly (Valles et al. 2021). The Indian government's targets for blending ethanol at 10% by 2022 and 20% by 2030, as outlined in the 2018 National Biofuel Policy, are part of an effort to reduce the country's dependence on imported fossil fuels (Sharma et al. 2021).

2.1.2 Wheat Straw

Wheat straw, as an abundant agricultural waste, has the potential to be utilized for various applications, providing sustainable alternatives to traditional products and contributing to the transition to more circular and bio-based chemicals and materials (Zemam et al. 2019). The wheat straw's cellulose, hemicellulose, and lignin contents are 42.8%, 23.8%, and 15.1%, respectively (Karagöz and Özkan 2014). Organosolvent pre-extraction and alkaline ethanol concentration at 65 g/L have resulted in a fermentation yield of 97%. This indicates that the process has

Table 1 Ethanol production from high lignocellulose and starch waste sources

Waste source	Various pretreatments used for waste material	Hydrolysis method	Microbes used in fermentation	Bioethanol yield	References
Rice stalk	4% NaOH alkali soaked for 4 h and autoclaved for 30 min	Enzymatic hydrolysis with cellulase, α-amylase, and amyloglucosidase	*Saccharomyces cerevisiae*	21.45%	Ningthoujam et al. (2023)
Wheat straw	Acidic pretreatment with 1:4 v/v of nitric acid-ethanol mixture, boiled for 1 h	Enzymatic hydrolysis using Cellic CTec3 enzyme cocktail	*Saccharomyces cerevisiae* J672	23.85 gL^{-1}	Brandenburg et al. (2018)
Corn straw	Alkaline pretreatment with calcium hydroxide under 0.15 MPa pressure and thermal treatment at 135 °C for 30 min	Enzymatic hydrolysis with cellulase, arabanase, β-glucanase, hemicellulase, and xylanase carried out at 50 °C for 4 h	*Saccharomyces cerevisiae* (D-2)	16.98 gL^{-1}	Kotarska et al. (2019)
Sugarcane bagasse	Alkaline pretreatment with 1.84% w/v calcium hydroxide solution, incubated at 100 °C for 4 h	Enzymatic hydrolysis with cellulase enzyme (10% v/v) at 50 °C for 48 h	*Candida tropicalis* TISTR 5306	16.6 ± 0.26 gL^{-1}	Nunta et al. (2023)
Potato peel waste	Hydrothermal pretreatment	Enzymatic hydrolysis with cellulase, hemicellulase, and amylase at 50 °C for 48 h	*Saccharomyces cerevisiae*	20.08 ± 0.36 gL^{-1}	Ben Taher et al. (2017)
Taro waste	Simultaneous saccharification and fermentation (SSF) with addition of 30 μL amyloglucosidase, agitated at 120 rpm in a shaker water bath for 3 h at 60 °C	Enzymatic hydrolysis with α-amylase, agitated at 120 rpm in a shaker water bath for 5 h at 79.2 °C	*Kluyveromyces marxianus* K21	48.98 gL^{-1}	Wu et al. (2016)

(continued)

Table 1 (continued)

Waste source	Various pretreatments used for waste material	Hydrolysis method	Microbes used in fermentation	Bioethanol yield	References
Sweet potato peel waste	Simultaneous saccharification and fermentation (SSF)	Enzymatic hydrolysis—α--amylase and amyloglucosidase	*Saccharomyces cerevisiae* (Fleischmann)	38–45 gL^{-1}	Lareo et al. (2013)

successfully extracted a high percentage of fermentable sugars from the biomass and converted them into ethanol.

2.1.3 Corn Straw

The quantities of agricultural waste (in million tons) for cornstarch reported to be available for bioethanol production are as follows: Asia, 33.90 million tons; Europe, 28.6 million tons; and America, 140.86 million tons (Ronzon and Sanjuán 2020). The degradation of cornstalks by different cultures showed that coculture achieved a degradation rate of 65.55%, while monoculture achieved a degradation rate of 40.79%. This indicates that coculture is more effective in degrading cornstalks (Wei et al. 2016). *L. botulinus* can potentially be used for bioethanol production at 0.22 g and 0.16 g of ethanol per gram of substrates (Kyung Hoan et al. 2016).

2.1.4 Sugarcane Bagasse

The reported quantities of agricultural waste (in million tons) available for bagasse in ethanol production, according to the provided data, are as follows: Asia, 74.88 million tons; America, 87.62 million tons; and Africa, 11.73 million tons (Sarkar et al. 2012). These figures represent the potential availability of bagasse, a by-product of sugarcane processing, for ethanol production in the respective regions. Utilizing bagasse for ethanol production is essential to sustainable and renewable energy initiatives. The agricultural waste contains approximately 65% of the total carbohydrate. Hemicellulose contains 18% of the lignin content of agricultural waste, which is the same as the hemicellulose content (Sarkar et al. 2012).

2.2 *High Sugar Content Waste Sources*

2.2.1 Sugarcane Molasses

The by-product of sugar production, i.e., molasses, is abundantly available due to the global sugar industry. Sugarcane molasses contains high concentrations of fermentable sugars, such as sucrose, glucose, and fructose, which microorganisms can metabolize directly during fermentation (Parascanu et al. 2021). Sugarcane molasses is the cheapest raw material for bioethanol production due to easier processing and higher productivity (Pavlečić et al. 2017).

2.3 *High Starch Content Waste Sources*

2.3.1 Potato Peel Waste

Among several varieties of processed foods, potato products are in high demand. Potato contains a high amount of carbohydrates ranging from 13 to 30% with a little protein (0.7–4.6%) (Puttongsiri et al. 2012). Potato industries release vast amounts of waste that can serve as a source for ethanol production. This will be eco-friendly and help in waste management (Chintagunta et al. 2016). The absence of furfural, formic acid, and acetic acid in the samples from the degradation of potato peel waste was expected, as the production of these compounds is typically associated with the degradation of xylose or the solubilization of hemicellulose due to bond cleavage (Antonopoulou et al. 2015).

3 Pretreatment of Biowaste

In biorefinery based on lignocellulosic materials, the feedstock's structure needs to be broken down to obtain sugars from cellulose and hemicellulose. The pretreatment process is known to have a significant impact on the disintegration of the cross-linked fraction of lignocellulosic material, which ultimately enhances the accessibility and biodegradability of cellulose and hemicellulose (Soltanian et al. 2020) (Table 2).

3.1 *Biological Pretreatment*

Biological pretreatment of lignocellulosic biomass involves using whole-cell bacteria or enzymes to break down and remove lignin from the biomass (Behera et al. 2014). Whole-cell microorganisms used in biological pretreatment of

Table 2 Pros and conditions required for different biomass pretreatments

Pretreatment	Pros	Conditions required	References
Mechanical pretreatment	Combining these methods, the overall yield of sugar can be significantly increased	Temperature required above 300 °C at 25–250 rpm	Kumar and Sharma (2017)
Microwave-assisted pretreatment	Less production of inhibitor compound Required short processing time	Irradiation time needed 5–30 min Power requirements in the range of 550–900 watts	Boonmanumsin et al. (2012), Zhu et al. (2021)
Physiochemical pretreatment: Steam explosion	Reduced production of inhibitory compounds and enhanced the efficiency of enzymatic hydrolysis Minimum chemical usage Less energy required	Provided temperature 160–260 °C, pressure 0.7–4.8 MPa	Wang et al. (2018)
Physiochemical pretreatment: Liquid hot water	Possible to minimize the formation of inhibitors at high temperatures Need for low temperature	Temperature needed 170–230 °C, pressure Up to 5 MPa	Hongdan et al. (2013)
Biological pretreatment	Safe and eco-friendly Less energy required	Incubation time provided 4–8 weeks	Baramee et al. (2020), Heap et al. (2014)
Advanced pretreatment—Sc-CO2	Cost for CO_2 is less Required low temperature Toxin formation null	Temperature required, 25–200 °C; pressure required, 1000–4000 psi; residence time, 5–72 h	Cha et al. (2014), Alayoubi et al. (2020)

lignocellulosic biomass are typically lignocellulose-degrading bacteria or fungi that produce enzymes such as ligninase, manganese peroxidase, laccase, and other lignocellulolytic enzymes (Janusz et al. 2017). Soft-rot, white, and brown fungi are also used for breaking down lignin and hemicellulose in the biological pretreatment of lignocellulosic biomass (Agrawal et al. 2019; Agrawal and Verma 2020). Among these, white fungi are particularly efficient for delignification due to their lignin degradation mechanism (Ren et al. 2016).

3.2 Chemical Pretreatment

3.2.1 Dilute Acid Hydrolysis

Acid pretreatment is commonly used in the biorefinery industry to solubilize hemicellulose and disrupt the lignin structure in lignocellulosic biomass. This process enhances the accessibility of cellulose chains, making them more amenable to subsequent enzymatic hydrolysis (Wirawan et al. 2020). The study by Rojas-Chamorro et al. (2020) focused on optimizing the acid pretreatment of brewers' spent grain. They identified the optimal pretreatment conditions as 130 °C temperature, 1% w/v H_2SO_4 concentration, and 26 min' reaction time. Under these conditions, the researchers recovered 94% of sugars from the hemicellulose fraction in the liquor and 90% of cellulose in the pretreated solid (Chen et al. 2018a, b). Acid treatment effectively breaks down the complex structure of biomass, including disrupting the lignin structure and removing inhibitory substances such as furfural, acetic acid, and formic acid (Lopez-Hidalgo et al. 2017). High-temperature conditions (>150 °C) in combination with dilute acid treatment are indeed effective in the removal of hemicellulose from lignocellulosic biomass (Kundu et al. 2021).

3.2.2 Alkaline Extraction

Various alkali solutions, such as sodium, calcium, and ammonium hydroxide, have been found to be effective in treating lignocellulosic biomass by removing undesirable components, particularly lignin. Among the alkali solutions, sodium hydroxide is the most commonly used for treating lignocellulosic biomass (Kim et al. 2016). These alkaline solutions, including sodium hydroxide, readily dissolve and remove lignin from the lignocellulosic biomass. Additionally, they can facilitate cellulose decrystallization by degrading the biomass's esters and glycolic side chains (Loow et al. 2016). The pretreatment of hemp biomass with NaOH successfully split lignin and removed it in the 58.6–75.3% range, leaving the glucan content at 66.9–75%. This allowed glucose concentration from 59.1 to 88.9% and ethanol yield from 95.8 to 96.7% (Zhao et al. 2020a, b). The NaOH-catalyzed organic solvent pretreatment mentioned, with 10% loading of NaOH and a mixture of ethanol and water in a 3:2 (%v/v) ratio for 30 min at 180 °C, leads to solid recovery of 81.2% and a lignin removal of 40.7% from SCB biomass. This indicates a highly efficient pretreatment process for biomass (Zhang et al. 2020).

3.3 Physical Pretreatment

Physical pretreatment of lignocellulosic biomass focuses on disrupting the complex bonds in the lignin and increasing the accessibility of sugars for biofuel conversion (Da Silva et al. 2018).

3.3.1 Mechanical Extrusion

Mechanical extrusion has a crucial effect on disrupting cellulose and hemicellulose fractions in lignocellulosic biomass. Mechanical extrusion processes involve subjecting the biomass to mechanical forces, often through pressure, shear, and friction, which can result in physical disruption and breakdown of the biomass structure (Kumar and Sharma 2017). Simple process, low energy consumption, short residence time, moderate temperature and pressure, and ease of commercial usability are all desirable for an efficient and cost-effective pretreatment method, and a high amount of compressive force is applied in the physical pretreatment (Cha et al. 2016; Ahmed et al. 2017).

3.3.2 Microwave-Assisted

Microwave-assisted pretreatment has been widely used in biomass conversion processes due to several benefits, such as less energy requirement and minimum production of toxic/inhibitor compounds (Bhardwaj et al. 2020a; Jasmine et al. 2023). The process of microwave-assisted acid hydrolysis of sago pith waste using dilute sulfuric acid yields glucose at a rate of 0.67 g/g of biomass and an ethanol yield of 0.31 g/g, with a fermentation efficiency of 91% (Thangavelu et al. 2019). The study used a factorial design to predict bioethanol yield based on different combinations of fermentation duration, temperature, and inoculum concentration. The highest bioethanol yield was achieved using a fermentation duration of 36 hours, a temperature of 30 °C, and an inoculum concentration of 1.5 g/L (Shenbagamuthuraman and Kasianantham 2023).

4 Fermentation

4.1 Microorganism Based Treatment

The commercial viability of bioethanol production and other value-added products from biomass depends on developing microorganisms that can effectively handle inhibitory compounds in the feedstock (Chandel et al. 2013). Two specific strains of yeast, *S. cerevisiae* ATCC 26602 and *S. stipitis* DSM 365, are commonly used in ethanol production. Parameters related to the process include the use of wheat straw, alginate beads, temperature control, and flow rate, resulting in an ethanol productivity of 7.3 g/L (Karagöz and Özkan 2014). *S. cerevisiae* BY4741 was used after pretreatment with 0.2% w/v H_2SO_4 at 170 °C for 10 min. The fermentation process was conducted with a solid loading of 7% w/v at 50 °C, pH 5.0, and 200 rpm. The fermentation occurs at 30 °C with an inoculum size of OD600 10, and the duration is 118 h with the productivity of ethanol at 7.3 g/L (Ko et al. 2018). Bagasse was pretreated with phosphoric acid (H_3PO_4), and the enzyme Accellerase® 1500 was used along with *Zymomonas mobilis* ATCC 29191 for separate hydrolysis and fermentation (SHF), resulting in an ethanol yield of 5.52 g/L (Wirawan et al. 2012). The HybriScan D Yeast kit is a powerful tool for detecting various yeast species, including those relevant to the fuel ethanol industry. It can identify almost 30 species from genera such as *Saccharomyces*, *Zygosaccharomyces*, *Pichia*, *Candida*, *Torulaspora*, *Brettanomyces*, *Debaryomyces*, and *Hanseniaspora* (Sigma-Aldrich 2019). This provides comprehensive coverage, making valuable resources for quality control and monitoring in ethanol production.

4.2 Types of Fermenters Used in Bioethanol Production

4.2.1 Stirred Tank Fermenter

Stirred tank bioreactors (STR) are widely used in ethanol production. This technology has been extensively utilized in laboratory research and industrial-scale ethanol production due to its scalability (Antunes et al. 2018). However, achieving perfect uniformity in a STR can be challenging, especially for large-scale industrial reactors (Pino et al. 2018). The process involved enzymatic hydrolysis with specific conditions, resulting in a hydrolysate with 66 g/L of glucose after hydrolysis in both Erlenmeyer flasks and a 1 L stirred tank reactor. Additionally, a gain of 11.2% in glucose concentration (reaching 72.2 g/L) was achieved in the stirred tank. A new isolate of *Candida pelliculosa* CCT 7734 was used in a separate hydrolysis and fermentation (SHF) process, resulting in bioethanol production of 16.6 g/L in batch operation and 23.0 g/L in fed-batch operation, respectively. The process involves a separate hydrolysis and fermentation (SHF) process, which resulted in 16.6 g/L and 23.0 g/L of bioethanol through batch and fed-batch operation, respectively (Campioni et al. 2020).

4.2.2 Airlift Fermenters

Airlift bioreactors are a suitable option for bioethanol production from various biomass sources, as they have advantages over mechanically stirred bioreactors, such as lower energy consumption, higher oxygen transfer, and better suitability for shear stress-sensitive microorganisms. Airlift bioreactors offer an attractive renewable resource for producing biofuels (particularly bioethanol), the most widely used biofuel (Chávez-Camarillo et al. 2022). This type of bioreactor utilizes a loop of gas and liquid to create a continuous flow, allowing for efficient mixing and mass transfer. The design also allows for easy scale-up and control of operating parameters, making it suitable for various applications like biotechnology and pharmaceutical industries. Additionally, the ALR has a smaller footprint compared to traditional bioreactors, making it a cost-effective and space-saving option for industrial production (Wang et al. 2011).

5 Life Cycle Assessment

More specifically, these studies indicate several factors which might impact the conclusions: feedstock choices, processing technology, and sectors of deployment, even within the same generation of biofuel, can make a significant difference in terms of whether or not a specific biofuel is environmentally beneficial (Acheampong et al. (2017). This analysis should consider the specific feedstock used, the conversion process employed, and the various production, distribution, and use stages. By considering the unique characteristics of each bioethanol production method, a more accurate assessment of its environmental sustainability can be achieved (Costa et al. 2018). By conducting an LCA that accounts explicitly for biomass transportation, researchers and industry professionals can gain insights into the environmental impact of this stage of the bioenergy life cycle (Beagle and Belmont 2019). For bioenergy production, conducting a sensitivity analysis on the estimation of variation in power plant efficiency and biomass heating values is crucial to understand the impact of various variables on overall performances (Kumar and Verma 2021). Environmental parameters such as soil, air, and water quality, availability, overall greenhouse gas (GHG) profile, and carbon sequestration via agricultural approaches, as well as eutrophication and atmospheric acidification, are critical factors to consider when evaluating the sustainability and environmental impact of bioenergy production, particularly from biomass sources (Caffrey and Veal 2013). Various studies have been conducted to estimate the environmental and technical aspects of different bioenergy production processes, such as direct combustion of biomass and gasification of biomass for power generation (Chen et al. 2020). A tailored life cycle analysis is essential for each bioethanol conversion process to ensure a thorough understanding of its environmental implications.

6 Perspective and Future Directions

Biomass-based fuel can be produced from various biomass sources. When used as a transportation fuel, bioethanol has the potential to reduce overall carbon emissions and decrease the reliance on nonrenewable resources. Pretreatment is an important step in bioethanol production from lignocellulosic feedstocks. Lignocellulosic biomasses like agricultural and forestry residues, as well as organic domestic wastes, contain complex structures of cellulose, hemicellulose, and lignin (Jury et al. 2022). Different methods for biomass pretreatment can have varying environmental impacts, including acidification potential, particulate matter emissions, eutrophication potential, and energy demand (Borrion et al. 2012). The low substrate conversion and ethanol production yield, often resulting from incomplete breakdown of the complex biomass structure during pretreatment and enzymatic hydrolysis, can limit the overall efficiency and cost-effectiveness of the process (Rijn et al. 2018). Innovative conversion processes and environmentally benign, cost-effective pretreatment approaches are essential for advancing the industrial production of bioethanol (Costa et al. 2018). Crucial aspects of sustainable biofuel production and utilization are as follows:

1. Reutilization of raw substances for biofuels with increased value.
2. Developing liquid markets for biofuel products enhancing mutual economic growth by creating opportunities for trade and exchange among users.
3. Integrating bioprocesses with zero waste creation which is a crucial goal for sustainable biofuel production.
4. The reusable potential of biofuel products for varied purposes (Nogué et al. 2018; Chandel et al. 2020).

7 Conclusion

This chapter comprises second-generation bioethanol production, crucial in enabling society toward sustainable development. Bioethanol is a renewable and sustainable source of energy that can be produced from various organic materials such as crops, agricultural residues, and waste materials. This includes carefully selecting raw materials for pretreatment, using the most effective enzymes for hydrolysis, and closely monitoring fermentation conditions. Microorganism engineering is a crucial area of focus for improving enzymatic hydrolysis and fermentation processes in the production of biofuels. The increasing demand for second-generation biofuels reflects a growing recognition of the need to transition toward more sustainable and environmentally friendly energy sources. As research and development efforts continue to advance, the production of bioethanol from nonfood sustainable biomass resources is expected to play a crucial role in meeting energy demands while mitigating environmental impacts associated with fossil fuel use.

References

Acheampong M, Ertem FC, Kappler B, Neubauer P (2017) In pursuit of sustainable development goal (SDG) number 7: will biofuels be reliable? Renew Sust Energ Rev 75:927–937

Agrawal K, Verma P (2020) Production optimization of yellow laccase from *Stropharia* sp. ITCC 8422 and enzyme-mediated depolymerization and hydrolysis of lignocellulosic biomass for biorefinery application. Biomass Convers Biorefinery 12:1–20. https://doi.org/10.1007/s13399-020-00869-w

Agrawal K, Bhardwaj N, Kumar B, Chaturvedi V, Verma P (2019) Process optimization, purification and characterization of alkaline stable white laccase from *Myrothecium verrucaria* ITCC-8447 and its application in delignification of agroresidues. Int J Biol Macromol 125:1042–1055. https://doi.org/10.1016/j.ijbiomac.2018.12.108

Ahmed MA, Rehman MSU, Terán-Hilares R, Khalid S, Han JI (2017) Optimization of twin gear-based pretreatment of rice straw for bioethanol production. Energy Convers Manag 141:120–125. https://doi.org/10.1016/j.enconman.2016.06.022

Alayoubi R, Mehmood N, Husson E et al (2020) Low temperature ionic liquid pretreatment of lignocellulosic biomass to enhance bioethanol yield. Renew Energy 145:1808–1816. https://doi.org/10.1016/j.renene.2019.07.091

Antonopoulou G, Dimitrellos G, Beobide AS, Vayenas D, Lyberatos G (2015) Chemical pretreatment of straw biomass: the effect on chemical composition and structural changes. Waste Biomass Valor 6:733–746. https://doi.org/10.1007/s12649-015-9388-x

Antunes FAF, Chandel AK, Brumano LP, Teran Hilares R, Peres GFD, Ayabe LES, Sorato VS, Santos JR, Santos JC, Da Silva SS (2018) A novel process intensification strategy for second-generation ethanol production from sugarcane bagasse in fluidized bed reactor. Renew Energy 124:189–196. https://doi.org/10.1016/j.renene.2017.06.004

Aulitto M, Fusco S, Nickel DB, Bartolucci S, Contursi P, Franzén CJ (2019) Seed culture pre-adaptation of Bacillus coagulans MA-13 improves lactic acid production in simultaneous saccharification and fermentation. Biotechnol Biofuels 12(1):1–11. https://doi.org/10.1186/s13068-019-1382-2

Baramee S, Siriatcharanon AK, Ketbot P et al (2020) Biological pretreatment of rice straw with cellulase-free xylanolytic enzyme producing Bacillus firmus K-1: structural modification and biomass digestibility. Renew Energy 160:555–563. https://doi.org/10.1016/j.renene.2020.06.061

Borrion AL, McManus MC, Hammond GP (2012) Environmental life cycle assessment of bioethanol production from wheat straw. Biomass and Bioenergy 47:9–19. https://doi.org/10.1016/j.biombioe.2012.10.017

Beagle E, Belmont E (2019) Comparative life cycle assessment of biomass utilization for electricity generation in the European Union and the United States. Ener Poll 28:267–275. https://doi.org/10.1016/j.enpol.2019.01.006

Behera S, Arora R, Nandhagopal N, Kumar S (2014) Importance of chemical pretreatment for bioconversion of lignocellulosic biomass. Renew Sust Energ Rev 36:91–106

Ben Taher I, Fickers P, Chniti S, Hassouna M (2017) Optimization of enzymatic hydrolysis and fermentation conditions for improved bioethanol production from potato peel residues. Biotechnol Prog 33:397–406. https://doi.org/10.1002/btpr.2427

Bhardwaj N, Kumar B, Verma P (2020a) Microwave-assisted pretreatment using alkali metal salt in combination with orthophosphoric acid for generation of enhanced sugar and bioethanol. Biomass Convers Biorefinery 12:1–8. https://doi.org/10.1007/s13399-020-00640-1

Bhardwaj N, Kumar B, Agrawal K, Verma P (2020b) Bioconversion of rice straw by synergistic effect of in-house produced ligno-hemicellulolytic enzymes for enhanced bioethanol production. Bioresour Technol Rep 10:100352. https://doi.org/10.1016/j.biteb.2019.100352

Boonmanumsin P, Treeboobpha S, Jeamjumnunja K, Luengnaruemitchai A, Chaisuwan T, Wongkasemjit S (2012) Release of monomeric sugars from Miscanthus sinensis by microwave-assisted ammonia and phosphoric acid treatments. Bioresour Technol 103(1):425–431

Brandenburg J, Poppele I, Blomqvist J, Puke M, Pickova J, Sandgren M, Rapoport A, Vedernikovs N, Passoth V (2018) Bioethanol and lipid production from the enzymatic hydrolysate of wheat straw after furfural extraction. Appl Microbiol Biotechnol 102:6269–6277. https://doi.org/10.1007/s00253-018-9081-7

Broda M, Yelle DJ, Serwańska K (2022) Bioethanol production from lignocellulosic biomass—challenges and solutions. Molecules 27:8717. https://doi.org/10.3390/molecules27248717

Caffrey KR, Veal MW (2013) Conducting an agricultural life cycle assessment: challenges and perspectives. Sci World J 2013:472431

Campioni TS, Soccol CR, Libardi Junior N, Rodrigues C, Woiciechowski AL, Letti LAJ, Vandenberghe LPS (2020) Sequential chemical and enzymatic pretreatment of palm empty fruit bunches for Candida pelliculosa bioethanol production. Biotechnol Appl Biochem 67(5):723–731. https://doi.org/10.1002/bab.1826

Cha YL, Yang J, Ahn JW et al (2014) The optimized CO_2-added ammonia explosion pretreatment for bioethanol production from rice straw. Bioprocess Biosyst Eng 37:1907–1915. https://doi.org/10.1007/s00449-014-1165-x

Cha YL, Yang J, Seo SI, An GH, Moon YH, You GD et al (2016) Alkaline twin-screw extrusion pretreatment of Miscanthus with recycled black liquor at the pilot scale. Fuel 164:322–328. https://doi.org/10.1016/j.fuel.2015.10.006

Chandel AK, da Silva SS, Singh OV (2013) Detoxification of lignocellulose hydrolysates: biochemical and metabolic engineering toward white biotechnology. Bioenergy Res 6:388–401. https://doi.org/10.1007/s12155-012-9241-z

Chandel AK, Garlapati VK, Kumar SPJ, Hans M, Singh AK, Kumar S (2020) The role of renewable chemicals and biofuels in building a bioeconomy. Biofuels Bioprod Biorefin 14:830–844. https://doi.org/10.1002/bbb.2104

Chaturvedi V, Verma P (2013) An overview of key pretreatment processes employed for bioconversion of lignocellulosic biomass into biofuels and value added products. 3 Biotech 3:415–431. https://doi.org/10.1007/s13205-013-0167-8

Chávez-Camarillo GM, Lopez-Nuñez PV, Jiménez-Nava RA, Aranda-García E, Cristiani-Urbina E (2022) Production of extracellular α-amylase by single-stage steady-state continuous cultures of Candida wangnamkhiaoensis in an airlift bioreactor. PLoS One 17(3):e0264734. https://doi.org/10.1371/journal.pone.0264734

Chen X, Zhai R, Shi K, Yuan Y, Dale BE, Gao Z, Jin M (2018a) Mixing alkali pretreated and acid pretreated biomass for cellulosic ethanol production featuring reduced chemical use and decreased inhibitory effect. Ind Crop Prod 124:719–725. https://doi.org/10.1016/j.indcrop.2018.08.056

Chen YA, Zhou Y, Liu D, Zhao X, Qin Y (2018b) Evaluation of the action of tween 20 non-ionic surfactant during enzymatic hydrolysis of lignocellulose: pretreatment, hydrolysis conditions and lignin structure. Bioresour Technol 269:329–338. https://doi.org/10.1016/j.biortech.2018.08.119

Chen S, Feng H, Zheng J, Ye J, Songm Y, Yang H, Zhou M (2020) Life cycle assessment and economic analysis of biomass energy technology in China: a brief review. PRO 8(9):1112. https://doi.org/10.3390/pr8091112

Chintagunta AD, Jacob S, Banerjee R (2016) Integrated bioethanol and biomanure production from potato waste. Waste Manag 49:320–325. https://doi.org/10.1016/j.wasman.2015.08.010

Costa D, Jesus J, Virgínio Silva J, Silveira M (2018) Life cycle assessment of bioethanol production from sweet potato (Ipomoea batatas L.) in an experimental plant. Bioenergy Res 11:715–725. https://doi.org/10.1007/s12155-018-9932-1

Da Silva ARG, Errico M, Rong BG (2018) Systematic procedure and framework for synthesis and evaluation of bioethanol production processes from lignocellulosic biomass. Bioresour Technol Rep 4:29–39. https://doi.org/10.1016/j.biteb.2018.08.015

Heap L, Green A, Brown D, van Dongen B, Turner N (2014) Role of laccase as an enzymatic pretreatment method to improve lignocellulosic saccharification. Cat Sci Technol 4(8):2251–2259

Hongdan Z, Shaohua X, Shubin W (2013) Enhancement of enzymatic saccharifcation of sugarcane baghigh solid content. Biomass Convers Biorefn 2:123–132. https://doi.org/10.1016/j.biortech.2011.09.136

Isikgor FH, Becer CR (2015) Lignocellulosic biomass: a sustainable platform for the production of bio-based chemicals and polymers. Polym Chem 6(25):4497–4559. https://doi.org/10.1039/c5py00263j

Janusz G, Pawlik A, Sulej J et al (2017) Lignin degradation: microorganisms, enzymes involved, genomes analysis and evolution. FEMS Microbiol Rev 41:941–962

Jasmine A, Rajendran M, Thirunavukkarasu K, Abinandan S, Vaidyanathan VK, Krishnamurthi T (2023) Microwave-assisted alkali pre-treatment medium for fractionation of rice straw and catalytic conversion to value-added 5-hydroxymethyl furfural and lignin production. Int J Biol Macromol 236:123999

Jury C, Thomas HL, Carrere H (2022) Life cycle assessment of two alkaline pretreatments of sorghum and miscanthus and of their batch co-digestion with cow manure. Bioenergy Res 15:1–24. https://doi.org/10.1007/s12155-021-10369-y

Karagöz P, Özkan M (2014) Ethanol production from wheat straw by Saccharomyces cerevisiae and Scheffersomyces stipitis co-culture in batch and continuous system. Bioresour Technol 158:286–293. https://doi.org/10.1016/j.biortech.2014.02.022

Khandaker MM, Qiamuddin KB, Majrashi A, Dalorima T, Sajili MH, Sharif Hossain ABM (2018) Bio-ethanol production from fruit and vegetable waste by using saccharomyces cerevisiae. Biosci Res 15(3):1703–1711

Kim JS, Lee YY, Kim TH (2016) A review on alkaline pretreatment technology for bioconversion of lignocellulosic biomass. Bioresour Technol 199:42–48. https://doi.org/10.1016/J.BIORTECH.2015.08.085

Ko JK, Jung JH, Altpeter F, Kannan B, Kim HE, Kim KH, Alper HS, Um Y, Lee SM (2018) Largely enhanced bioethanol production through the combined use of lignin-modified sugarcane and xylose fermenting yeast strain. Bioresour Technol 256:312–320. https://doi.org/10.1016/j.biortech.2018.01.123

Kotarska K, Dziemianowicz W, Świerczyńska A (2019) Study on the sequential combination of bioethanol and biogas production from corn straw. Mol Basel Switz 24:4558. https://doi.org/10.3390/molecules24244558

Kumar AK, Sharma S (2017) Recent updates on different methods of pretreatment of lignocellulosic feedstocks: a review. Bioresour Bioprocess 4:7

Kumar B, Verma P (2020) Application of hydrolytic enzymes in biorefinery and its future prospects. In: Srivastava N, Srivastava M, Mishra PK, Gupta VK (eds) Microbial strategies for techno-economic biofuel production, clean energy production technologies. Springer, Singapore, pp 59–83. https://doi.org/10.1007/978-981-15-7190-9_3

Kumar B, Verma P (2021) Life cycle assessment: blazing a trail for bioresources management. Energy Convers Manag X 10:100063. https://doi.org/10.1016/j.ecmx.2020.100063

Kumar B, Bhardwaj N, Agrawal K, Verma P (2020) In: Srivastava N, Srivastava M, Mishra P, Gupta V (eds) Bioethanol production: generation-based comparative status measurements. Biofuel production technologies: critical analysis for sustainability, clean energy production technologies. Springer, Singapore, pp 155–201. https://doi.org/10.1007/978-981-13-8637-4_7

Kundu C, Samudrala SP, Kibria MA, Bhattacharya S (2021) Onestep peracetic acid pretreatment of hardwood and softwood biomass for platform chemicals production. Sci Rep 11:1–11. https://doi.org/10.1038/s41598-021-90667-9

Kyung Hoan IM et al (2016) Ethanol production from various sugars and cellulosic biomass by white rot fungus Lenzites betulinus. Mycobiology 44(1):48–53. https://doi.org/10.5941/MYCO.2016.44.1.48

Lareo C, Ferrari MD, Guigou M et al (2013) Evaluation of sweet potato for fuel bioethanol production: hydrolysis and fermentation. Springer Plus 2:493. https://doi.org/10.1186/2193-1801-2-493

Loow YL, Wu TY, Jahim JM et al (2016) Typical conversion of lignocellulosic biomass into reducing sugars using dilute acid hydrolysis and alkaline pretreatment. Cellulose 23:1491–1520

Lopez-Hidalgo AM, Sánchez A, De León-Rodríguez A (2017) Simultaneous production of bioethanol and biohydrogen by Escherichia coli WDHL using wheat straw hydrolysate as substrate. Fuel 188:19–27

Moneruzzaman Khandaker M, Aliyu Abdullahi U, Dogara Abdulrahman M, Afiza Badaluddin N, Suryati Mohd K (2021) Bio-ethanol production from fruit and vegetable waste by using Saccharomyces cerevisiae. Bioethanol Technol 188:19–27. https://doi.org/10.1016/j.fuel.2016.10.022

Ningthoujam R, Jangid P, Yadav VK, Sahoo DK, Patel A, Dhingra HK (2023) Bioethanol production from alkali-pretreated rice straw: effects on fermentation yield, structural characterization, and ethanol analysis. Front Bioeng Biotechnol 11:1243856. https://doi.org/10.3389/fbioe.2023.1243856

Nogué VS, Black BA, Kruger JS, Singer CA, Ramirez KJ, Reed ML et al (2018) Integrated diesel production from lignocellulosic sugars via oleaginous yeast. Green Chem 20:4349–4365. https://doi.org/10.1039/C8GC01905C

Nunta R, Techapun C, Sommanee S, Mahakuntha C, Porninta K, Punyodom W, Phimolsiripol Y, Rachtanapun P, Wang W, Zhuang X, Qi W, Jantanasakulwong K, Reungsang A, Kumar A, Leksawasdi N (2023) Valorization of rice straw, sugarcane bagasse and sweet sorghum bagasse for the production of bioethanol and phenylacetylcarbinol. Sci Rep 13:727. https://doi.org/10.1038/s41598-023-27451-4

Parascanu MM, Sanchez N, Sandoval-Salas F, Carreto CM, Soreanu G, Sanchez-Silva L (2021) Environmental and economic analysis of bioethanol production from sugarcane molasses and agave juice. Environ Sci Pollut Res Int 28(45):64374–64393. https://doi.org/10.1007/s11356-021-15471-4

Pavlečić M, Rezić T, Šantek MI, Horvat P, Šantek B (2017) Bioethanol production from raw sugar beet cossettes in horizontal rotating tubular bioreactor. Bioprocess Biosyst Eng 40(11):1679–1688. https://doi.org/10.1007/s00449-017-1823-x

Pino MS, Rodríguez-Jasso RM, Michelin M, Flores-Gallegos AC, Morales Rodriguez R, Teixeira JA, Ruiz HA (2018) Bioreactor design for enzymatic hydrolysis of biomass under the biorefinery concept. Chem Eng J 347:119–136. https://doi.org/10.1016/j.cej.2018.04.057

Prasad RK, Chatterjee S, Mazumder PB, Gupta SK, Sharma S, Vairale MG, Datta S, Dwivedi SK, Gupta DK (2019) Bioethanol production from waste lignocelluloses: a review on microbial degradation potential. Chemosphere 231:588–606. https://doi.org/10.1016/j.chemosphere.2019.05.142

Puttongsiri T, Choosakul N, Sakulwilaingam D (2012) Moisture content and physical properties of instant mashed potato. In: International conference on nutrition and food sciences, vol 39. IACSIT Press, Singapore

Rojas-Chamorro JA, Romero-García JM, Cara C, Romero I, Castro E (2020) Improved ethanol production from the slurry of pretreated brewers' spent grain through different co-fermentation strategies. Bioresource technology 296:122367. https://doi.org/10.1016/j.biortech.2019.122367

Ray B, Jayakumar T, Rakesh S (2023) Comparative analysis of lipid content in mono and co-culture microalgae for potential biodiesel production using diverse cell disruption techniques. Biomass Conv Biorefinery. https://doi.org/10.1007/s13399-023-04482-5

Ren NQ, Zhao L, Chen C et al (2016) A review on bioconversion of lignocellulosic biomass to H2: key challenges and new insights. Bioresour Techno 215:92–99

Rijn RV, Nieves IU, Shanmugam KT et al (2018) Techno-economic evaluation of cellulosic ethanol production based on pilot biorefinery data: a case study of sweet sorghum bagasse processed via L+SScF. Bioenergy Res 11:414–425. https://doi.org/10.1007/s12155-018-9906-3

Ronzon T, Sanjuán AI (2020) Friends or foes? A compatibility assessment of bioeconomy-related sustainable development goals for European policy coherence. J Clean Prod 254:119832

Sarkar N, Ghosh SK, Bannerjee S, Aikat K (2012) Bioethanol production from agricultural wastes: an overview. Renew Energy 37:19–27

Sharma S, Swain MR, Mishra A, Mathur AS, Gupta RP, Puri SK, Ramakumar SS, Sharma AK (2021) High solid loading and multiple-fed simultaneous saccharifcation and co-fermentation (mf-SSCF) of rice straw for high titer ethanol production at low cost. Renew Energy 179:1915–1924. https://doi.org/10.1016/j.renene.2021.07.146

Shenbagamuthuraman V, Kasianantham N (2023) Microwave irradiation pretreated fermentation of bioethanol production from Chlorella vulgaris biomasses: comparative analysis of response surface methodology and artificial neural network techniques. Bioresour Technol 390:129867. https://doi.org/10.1016/j.biortech.2023.129867

Sigma-Aldrich (2019) HybriScan®. https://www.sigmaaldrich.com/analytical-chromatography/microbiology/identification-tests/hybriscan-rapid-test.html. Accessed Sept 2019

Soltanian S, Aghbashlo M, Almasi F, Hosseinzadeh-Bandbafha H, Nizami AS, Ok YS, Lam SS, Tabatabaei M (2020) A critical review of the effects of pretreatment methods on the exergetic aspects of lignocellulosic biofuels. Energy Conversion and Management 212:112792. https://doi.org/10.1016/j.enconman.2020.112792

Supriyanto A, Lestari I, Citrasari N, Putri SS (2019) The bioconversion of vegetable waste extract from Osowilangun Central Market Surabaya into bioethanol using Saccharomyces cerevisiae. In: IOP Conference Series: Earth and Environmental Science, vol 245, No. 1. IOP Publishing, p 012042. https://doi.org/10.1088/1755-1315/245/1/012042

Thangavelu SK, Rajkumar T, Pandi DK, Ahmed AS, Ani FN (2019) Microwave assisted acid hydrolysis for bioethanol fuel production from sago pith waste. Waste management 86:80–86. https://doi.org/10.1016/j.wasman.2019.01.035

Triwahyuni E, Sudiyani Y, Abimanyu H (2015) The effect of substrate loading on simultaneous saccharification and fermentation process for bioethanol production from oil palm empty fruit bunches. Energy Procedia 68:138–146. https://doi.org/10.1016/j.egypro.2015.03.242

Valles A, Capilla M, Álvarez-Hornos FJ, García-Puchol M, SanValero P, Gabaldón C (2021) Optimization of alkali pretreatment to enhance rice straw conversion to butanol. Biomass Bioenergy 150:106131. https://doi.org/10.1016/j.biombioe.2021.106131

Wang X, Jia XQ, Wen JP (2011) Transient CFD modeling of toluene waste gas biodegradation in a gas–liquid–solid three-phase airlift loop reactor by immobilized pseudomonas putida. Chem Eng J 172:735–745

Wang Z, Wu G, Jönsson LJ (2018) Efects of impregnation of softwood with sulfuric acid and sulfur dioxide on chemical and physical characteristics, enzymatic digestibility, andfermentability. Bioresour Technol 247:200–208. https://doi.org/10.1016/j.biortech.2017.09.081

Wei YQ, Long RJ, Yang H, Yang HJ, Shen XH, Shi RF, Wang ZY, Du JG, Qi XJ, Ye QH (2016) Fiber degradation potential of natural co-cultures of Neocallimastix frontalis and Methanobrevibacter ruminantium isolated from yaks (Bos grunniens) grazing on the Qinghai Tibetan Plateau. Anaerobe 39:158–164. https://doi.org/10.1016/j.anaerobe.2016.03.005

Wirawan F, Cheng CL, Kao WC, Lee DJ, Chang JS (2012) Cellulosic ethanol production performance with SSF and SHF processes using immobilized Zymomonas mobilis. Appl Energy 100:19–26. https://doi.org/10.1016/j.apenergy.2012.04.032

Wirawan F, Cheng C-L, Lo Y-C, Chen C-Y, Chang J-S, Leu S-Y, Lee D-J (2020) Continuous cellulosic bioethanol co-fermentation by immobilized Zymomonas mobilis and suspended Pichia stipitis in a two-stage process. Appl Energy 266:114871. https://doi.org/10.1016/j.apenergy.2020.114871

Wu W-H, Hung W-C, Lo K-Y, Chen Y-H, Wan H-P, Cheng K-C (2016) Bioethanol production from taro waste using thermo-tolerant yeast Kluyveromyces marxianus K21. Bioresour Technol 201:27–32

Zemam S, Resan SF, Abed MS (2019) Structural characteristics of developed sustainable lime-straw composite. Civil Eng J 5:2587–2597. https://doi.org/10.28991/cej2019-0309

Zhang H, Zhang J, Xie J, Qin Y (2020) Effects of NaOH-catalyzed organosolv pretreatment and surfactant on the sugar production from sugarcane bagasse. Bioresour Technol 312:123601. https://doi.org/10.1016/j.biortech.2020.123601

Zhao G, Kuang G, Wang Y, Yao Y, Zhang J, Pan Z-H (2020a) Effect of steam explosion on physicochemical properties and fermentation characteristics of sorghum (Sorghum bicolor (L.) Moench). LWT 129:109579. https://doi.org/10.1016/j.lwt.2020.109579

Zhao J, Xu Y, Wang W (2020b) Conversion of liquid hot water, acid and alkali pretreated industrial hemp biomasses to bioethanol. Bioresour Technol 309:123383. https://doi.org/10.1016/j.biortech.2020.123383

Zhu Z, Liu Y, Yang X et al (2021) Comparative evaluation of microwave-assisted acid, alkaline, and inorganic salt pretreatments of sugarcane bagasse for sugar recovery. Biomass Convers Biorefn 11:2681–2693. https://doi.org/10.1007/s13399-020-00680-7

The Development in Nanotechnology and Tailor-Made Enzymes as the Future of Biobased Economy

Vivek Chintada, K. Veeraiah, and Narasimha Golla

1 Introduction

Significant advancements have been achieved in the realm of biotechnology, nanotechnology, and enzyme development in recent years, driving scientific research forward and revolutionizing various industries (Kulkarni 2007; Betz et al. 2023; Sahu et al. 2023). This chapter aims to investigate the relationship between nanotechnology and enzymes, emphasizing their diverse applications and the potential to transform the biobased economy. Nanotechnology, renowned for its precise manipulation of matter at the atomic and molecular levels, has transformed material design and fabrication (Feynman 1959; Mansoori 2002; Mansoori and Soelaiman 2005; Bhagyaraj and Oluwafemi 2018). Particularly in biobased systems, nanotechnology plays a crucial role in enhancing enzymatic processes by regulating interactions between enzymes and nanoparticles, thereby facilitating advancements in industries such as healthcare and renewable energy production (Salata 2004; Roco 2011; Thanigaivel et al. 2022; Ahmed et al. 2022; Sheikh et al. 2023; Preethi et al. 2024).

Enzymes, the catalysts driving biological reactions, play a key role in biocatalysis, showcasing significant potential in the biobased economy (Otero et al. 2007; Agger and Zeuner 2022; Rodríguez-Núñez et al. 2023; Zhang et al. 2024). By employing advanced techniques like genome editing, synthetic biology, and bioinformatics, enzymes can be designed and optimized to exhibit specific functions tailored for a range of industrial applications (Bornscheuer and Schwaneberg 2019;

V. Chintada
Department of Zoology, Sri Venkateswara University, Tirupati, Andhra Pradesh, India

K. Veeraiah
Department of Zoology, Acharya Nagarjuna University, Guntur, Andhra Pradesh, India

N. Golla (✉)
Department of Virology, Sri Venkateswara University, Tirupati, Andhra Pradesh, India

K. Agrawal, P. Verma (eds.), *Biotechnological Advances in Biorefinery*, Interdisciplinary Biotechnological Advances,
https://doi.org/10.1007/978-981-97-5544-8_13

Biz et al. 2019; Lawson et al. 2021; Vasina et al. 2022; Gayathiri et al. 2023; Kharbikar et al. 2023). These customized enzymes show great promise in sectors such as sustainable manufacturing, waste management, and healthcare (Sanchez-Rexach et al. 2020; Clark and Ray 2023; Alzoubi et al. 2023; Xu et al. 2023; Bhatia et al. 2024).

The fusion of nanotechnology with tailored enzymes opens up a wealth of opportunities within the biobased economy (Thapa et al. 2019; Gkantzou et al. 2021). Through the integration of enzymes into nanoparticles or nanodevices using nanoscale manipulation, researchers can develop efficient enzymatic systems characterized by improved stability, selectivity, and reusability, paving the way for applications in targeted drug delivery, sustainable agriculture, biofuel production, and environmental remediation (Yin et al. 2016; Chimento et al. 2018; Zhang et al. 2020; Agrawal and Verma 2021, 2023; Chen et al. 2021; Zdarta et al. 2023; Manikandan et al. 2023; Gericke et al. 2024).

1.1 Key Concepts

1. Nanotechnology: the precise manipulation of materials at the atomic and molecular scale, leading to advancements in material design and production.
2. Enzyme development: the engineering of tailor-made enzymes using cutting-edge techniques like genome editing, synthetic biology, and bioinformatics to exhibit specific functionalities suitable for various industrial applications.
3. Biobased economy: The utilization of biological systems, including nanotechnology and custom-designed enzymes, to drive economic growth and sustainability across industries.
4. Industrial applications: The application of nanotechnology and enzymes in industries such as healthcare, renewable energy production, sustainable manufacturing, waste management, and healthcare to optimize processes and create innovative solutions.
5. Transformative technologies: the advancement of nanotechnology and tailor-made enzymes as fundamental pillars of the biobased economy, leading to significant shifts in harnessing and optimizing biological systems for sustainable development.

2 Applications in Biotechnology

2.1 *Targeted Drug Delivery*

Nanotechnology-enhanced enzymes offer a promising strategy for delivering therapeutic agents directly to specific cells or tissues, thereby improving treatment efficacy and minimizing side effects (Nel et al. 2009; Pal et al. 2021; Del Grosso et al. 2022). By creating nanoparticles that are packed with enzymes for the controlled release of therapeutic agents, the efficiency of drug delivery can be significantly enhanced (De La Rica et al. 2012; Shahriari et al. 2019; Zhang et al. 2017). Furthermore, the advancement of enzyme-nanoparticle systems that are capable of crossing the blood-brain barrier paves the way for innovative treatments for neurological disorders (Sonvico et al. 2009; Manu et al. 2021; Padma et al. 2023).

2.2 *Sustainable Agriculture*

The incorporation of nanotechnology-enhanced enzymes in agriculture holds the promise of improving crop productivity and mitigating environmental impact (Ditta 2012; Magnabosco et al. 2023; Fayaz et al. 2021). Nanoparticle-enzyme complexes have the potential to enhance nutrient absorption in plants, thereby stimulating growth and increasing yields (Khoshgoftarmanesh et al. 2019; Verma et al. 2019; Tang et al. 2022). Additionally, the precise delivery of enzymes to the soil can facilitate the breakdown of pollutants, offering an environmentally friendly solution for soil remediation (Kumar et al. 2018a, b; Li et al. 2020; Somu et al. 2022; Aparicio et al. 2022; Shahid et al. 2023).

2.3 *Biofuel Production*

Revolutionizing biofuel production, the integration of nanotechnology and customized enzymes has the potential to enhance the efficiency of enzymatic reactions by utilizing nanoparticles as carriers for enzymes involved in biomass conversion (Syed et al. 2023; Manikandan et al. 2023). This advancement not only decreases manufacturing expenses but also facilitates the production of biofuels from a diverse array of raw material origins (Pandey et al. 2018; Singh et al. 2023; Pfleger and Takors 2023). In conclusion, the advancement of nanotechnology and customized enzymes as key elements of the biobased economy signifies a significant shift in leveraging and optimizing biological processes (Zhong et al. 2023; Iñiguez-Moreno et al. 2023; Gericke et al. 2024). This chapter has highlighted the latest progress in enzyme engineering and has illuminated the abundant opportunities and challenges that lie ahead. By embracing these transformative technologies, we can pave the way for a sustainable future where biobased solutions drive economic growth while safeguarding the planet for future generations.

3 The Intersection of Nanotechnology and Enzymes

Nanotechnology and enzyme research represent two vibrant fields with immense potential for diverse applications in medicine, biotechnology, environmental science, and beyond (Chen et al. 2019; Kang et al. 2018; Villalba-Rodríguez et al. 2023; Haleem et al. 2023). Recently, experts have been acknowledging the substantial advantages of combining these two fields. The collaboration of nanotechnology and enzymes has opened up fresh avenues for enhancing incredibly efficient and accurate biochemical processes (Wang et al. 2023a, b; Paramasivam et al. 2024). This section explores the present status of this fusion, providing insights into its impacts and providing a preview of the promising prospects it brings.

Nanotechnology involves the manipulation and utilization of materials at the nanoscale, usually between 1 and 100 nm in size. At this scale, materials display unique properties distinct from their larger counterparts, enabling the development of innovative applications (Ramanathan et al. 2019; Nasrollahzadeh et al. 2019; Patel et al. 2021; Suhag et al. 2023). Nanomaterials, such as nanoparticles, nanowires, and nanotubes, boast high surface area-to-volume ratios, enhancing reactivity and selectivity in diverse chemical reactions (Ding et al. 2023; Harish et al. 2023).

Enzymes are natural catalysts that effectively catalyze organic reactions within living organisms. These proteins are vital in speeding up biochemical reactions by lowering activation energies. Enzymes demonstrate a remarkable specificity toward their substrates, ensuring accuracy and efficiency in catalysis (Sujitha and Shanthi 2023; Anwar et al. 2023; Issaka et al. 2023a, b). As biological catalysts, enzymes provide several advantages, such as functioning under gentle reaction conditions, displaying strong selectivity, and being compatible with biological systems (Kang et al. 2018; Villalba-Rodríguez et al. 2023).

The integration of nanotechnology with enzymes creates a symbiotic relationship that amplifies efficiency and widens the scope of applications. Scientists have explored different approaches to combine nanomaterials with enzymes to enhance efficiency and precision in biochemical reactions (Wang et al. 2023a, b; Issaka et al. 2023a, b; Bharti 2024). One approach involves anchoring enzymes onto nanostructured substrates such as carbon nanotubes or nanoparticles, establishing a robust and reusable platform for catalytic reactions (Tan et al. 2023; Jamuna et al. 2023; Liu and Dong 2020). These nanomaterials not only enhance the stability of enzymes but also elevate their efficacy and versatility, leading to improved overall performance (Wang et al. 2019; Sharifi et al. 2020; Anwar et al. 2023).

3.1 Applications in Biomedicine

The incorporation of nanotechnology with enzymes has demonstrated significant advancements in the field of biomedicine (Lu et al. 2023; Soozanipour et al. 2023). Nanocarriers loaded with enzymes have been developed to ferry therapeutic enzymes directly to specific sites in the body, bypassing degradation and enhancing efficacy (Dean et al. 2017; Cao et al. 2020; Tripathi et al. 2023). This method shows great potential in treating a range of illnesses, including cancer, metabolic disorders, and genetic conditions (Dalziel et al. 2014; Rani et al. 2016; Sharma et al. 2019). Moreover, nanomaterials have played a critical role in enhancing the activity of enzymes used in diagnostics, enabling more sensitive and accurate detection methods (Ju 2018; Das et al. 2021).

3.2 Environmental Remediation

The collaboration between nanotechnology and enzymes has brought about substantial advantages for environmental cleanup efforts. Enzymes immobilized on nanomaterials exhibit the capacity to degrade organic pollutants in polluted soil and water systems, offering a more effective and environmentally sustainable alternative to conventional remediation methods (Bakshi and Abhilash 2020; Hidangmayum et al. 2023). Furthermore, modified nanomaterials can selectively capture or break down particular pollutants, thus enhancing focused environmental remediation endeavors (Ramanathan et al. 2019; Chaudhary et al. 2019; Rafeeq et al. 2022).

3.3 Industrial Processes and Energy Production

The convergence of nanotechnology and enzymes holds significant implications for industrial processes and energy production. Enzyme-catalyzed reactions are instrumental in the production of pharmaceuticals, biofuels, and fine chemicals (Chaudhary et al. 2019). Leveraging nanomaterials can enhance enzyme stability, selectivity, and recyclability, thereby driving cost-effective and sustainable industrial processes (Brar et al. 2022; Bilal et al. 2023). Additionally, nanomaterials can facilitate the development of efficient enzyme-based energy conversion devices, such as biofuel cells and enzymatic batteries (Kang et al. 2018; Zhang et al. 2021; Serag et al. 2022; ul Haque et al. 2023).

3.4 Future Perspectives

The fusion of nanotechnology and enzymes represents a dynamic and rapidly advancing field of study (Navya and Daima 2016; Yu et al. 2022). Progress in material synthesis, characterization techniques, and enzyme engineering is opening up new avenues for functional nanomaterials and enzyme systems (Rinaldi et al. 2016; Reshmy et al. 2021; Ai et al. 2022). Future research should focus on advanced bio-interfacing techniques like bioconjugation and self-assembly to tailor nanomaterials with specific enzyme functionalities (Ali et al. 2021; Liang and Liang 2021). Overcoming challenges related to scalability, long-term stability, and regulatory considerations will be crucial for the practical application of nanotechnology-enzyme systems (Oke et al. 2023; Cheah et al. 2016).

The convergence of nanotechnology and enzymes offers significant potential for various uses in biomedicine, environmental science, and industrial operations (Shahcheraghi et al. 2022; Gholami-Shabani et al. 2016). By combining the distinct characteristics of nanomaterials with the effectiveness and specificity of enzymes, scientists can create exceptionally efficient and accurate catalysts. Continuous research is essential for further exploring the capabilities of this fusion and overcoming challenges to facilitate its practical integration (Nothling et al. 2018; Khanzada et al. 2023; Kong et al. 2023).

4 Nanotechnology: Revolutionizing Material Design and Fabrication

Nanotechnology, the manipulation of matter at the nanoscale, is a groundbreaking field that is shaping material design and production methods. This chapter explores the profound impact of nanotechnology on diverse industries and applications (Ahire et al. 2022; Behera and Behera 2022). Nanotechnology's ability to engineer materials at the molecular level has led to the development of advanced materials with exceptional properties (National Nanotechnology Initiative) (Roco 2011; Nasrollahzadeh et al. 2019; Tegart 2004). Through precise control of size, structure, and composition, scientists have created materials with enhanced characteristics such as increased strength, improved conductivity, and superior durability (Harish et al. 2023; Chafiq et al. 2023). These advancements have enabled the innovation of novel materials for a wide range of applications in fields like electronics, healthcare, energy, and environmental sustainability (Lim et al. 2020; Baxter et al. 2009). One of the key advantages of nanotechnology is its capability to improve manufacturing processes (Ravichandran 2010). By using nanomaterials, manufacturers can produce products that are lighter, stronger, and more efficient (Bawa et al. 2005; Adlakha-Hutcheon et al. 2009; Grewal et al. 2021). The precise control provided by nanotechnology in the fabrication process leads to products that demonstrate enhanced performance and reliability.

Moreover, nanotechnology is instrumental in addressing global challenges concerning pollution and resource scarcity (Mauter et al. 2018). Nanomaterials find applications in environmental remediation, water purification, and the development of renewable energy technologies, offering sustainable solutions to urgent issues (Khan 2020; Ahmed et al. 2022). In summary, nanotechnology signifies a paradigm shift in material design and fabrication, presenting boundless opportunities for innovation across different sectors. By harnessing the potential of nanotechnology, researchers and engineers can continue to explore new frontiers and contribute to building a more sustainable and technologically advanced future.

The field of nanotechnology has had a substantial influence on enzymatic processes by facilitating the creation and production of nanotechnology-enhanced enzymes that improve interactions and catalytic efficiency (Wei et al. 2010; Fu et al. 2021). The integration of nanomaterials with enzymes has unlocked new possibilities in industries such as healthcare, biotechnology, and environmental science by enhancing the efficiency and specificity of enzymatic reactions (Chapman et al. 2018). This synergy between nanotechnology and enzymes has not only improved the stability and activity of enzymes but also enabled the development of novel enzyme-based technologies with enhanced functionalities (Othman et al. 2016). Furthermore, the precise control afforded by nanotechnology in modifying enzyme surfaces and encapsulating enzymes within nanomaterials has paved the way for tailored enzymatic systems with improved substrate binding and reaction rates (Wei et al. 2020). Nanotechnology-enhanced enzymes show potential for use in targeted drug delivery, biosensors, biofuel production, and wastewater treatment, among other fields, emphasizing the importance of optimizing enzymatic interactions to achieve specific goals (Gupta and Roy 2020).

In summary, the combination of nanotechnology and enzymatic processes marks a new chapter of innovation and sustainability in diverse industries. Leveraging nanotechnology to enhance enzyme performance and functionality enables researchers and industry professionals to tackle present obstacles and create advanced solutions that enhance efficiency, specificity, and scalability in enzymatic processes.

5 Nanotechnology in Enzymatic Processes: Enhancing Interactions

The field of nanotechnology has had a profound impact on enzymatic processes, facilitating the development of nanotechnology-enhanced enzymes that optimize interactions and catalytic performance (Nile et al. 2020). Through the integration of nanomaterials with enzymes, new opportunities have emerged in industries such as healthcare, biotechnology, and environmental science (Alizadeh and Salimi 2021). This collaboration has not only bolstered the efficiency and specificity of enzymatic reactions but has also led to the creation of innovative enzyme-based technologies with expanded functionalities (Zheng et al. 2024; Bilal et al. 2020). The precise manipulation enabled by nanotechnology, whether in altering enzyme surfaces or encapsulating enzymes within nanomaterials, has paved the way for customized enzymatic systems that exhibit improved substrate binding and reaction rates (Wei et al. 2020; Gkantzou et al. 2021). Nanotechnology-enhanced enzymes show great promise for applications in targeted drug delivery, biosensors, biofuel production, and wastewater treatment, where optimizing enzymatic interactions is pivotal to achieving desired outcomes.

In conclusion, the synergy between nanotechnology and enzymatic processes represents a transformative phase of advancement and sustainability across diverse industries. By leveraging nanotechnology to enhance enzyme performance and functionality, researchers and industry practitioners can tackle existing challenges and pioneer cutting-edge solutions that offer enhanced efficiency, specificity, and scalability in enzymatic processes.

6 Tailor-Made Enzymes: Engineered for Industrial Use

Tailor-made enzymes are bespoke biological catalysts specifically designed for industrial applications. This unique process entails using advanced techniques and technologies to modify the genetic characteristics of enzymes, improving their effectiveness, specificity, and durability to meet the distinct requirements of different industrial operations (Adrio and Demain 2014). Genome editing plays a crucial role in achieving accuracy in enzyme engineering (Terns and Terns 2017). Through direct manipulation of the genetic material of organisms, scientists can fine-tune enzyme characteristics like substrate specificity and thermal stability to optimize performance in industrial contexts (Johannes and Zhao 2006; Heux et al. 2015). This degree of manipulation enables the creation of enzymes with customized characteristics that can greatly improve overall operational efficiency and the quality of the end products.

Additionally, synthetic biology offers an avenue for crafting enzymes with specific functionalities tailored for industrial use (Le Feuvre et al. 2018). By amalgamating principles from engineering, biology, and chemistry, researchers can fabricate novel enzyme variants with desired traits, such as heightened activity under specific conditions or unique substrate specificities (Zhuang et al. 2023). These custom-designed enzymes have the capacity to transform various sectors by facilitating the development of innovative and enhanced processes.

Moreover, bioinformatics plays a vital role in harnessing the power of enzyme engineering by scrutinizing and understanding extensive biological data sets to guide the creation and enhancement of enzymes (Borneman et al. 2006). Through computational modeling and predictive algorithms, bioinformatics aids researchers in identifying key regions in enzyme structure that can be modified to achieve desired functionalities (Honarparvar et al. 2014). This approach expedites the enzyme engineering process by furnishing valuable insights into the correlation between enzyme structure and function. To sum up, the combination of genome editing, synthetic biology, and bioinformatics in enzyme engineering empowers the development of tailor-made enzymes optimized for industrial applications (Banerjee et al. 2018). By harnessing the capabilities of these advanced technologies, scientists can create enzymes with enhanced properties that drive innovation and sustainability across various sectors.

7 The Synergy of Nanotechnology and Tailor-Made Enzymes

The integration of nanotechnology with tailor-made enzymes represents a powerful synergy that offers transformative solutions across various sectors. Efficient enzymatic systems are being developed for targeted drug delivery through the combination of nanotechnology and custom-designed enzymes (Nultsch and Germershaus 2018). By encapsulating enzymes within nanoparticles, researchers can enhance the stability and bioavailability of the enzymes, enabling precise delivery to specific target tissues or cells (Soppimath et al. 2001). This approach holds great promise for improving the efficacy and safety of therapeutic treatments by enabling controlled release and localized action of enzymes within the body.

In the field of sustainable agriculture, the synergy of nanotechnology and tailor-made enzymes is driving innovations to improve crop productivity. Nanoparticles loaded with engineered enzymes can be used to enhance nutrient uptake, pest resistance, and stress tolerance in plants (Sarkar et al. 2019). By delivering enzymes directly to plant systems, farmers can optimize agricultural processes while minimizing environmental impact, ultimately leading to higher yields and sustainable farming practices (Trivedi et al. 2021; Elnahal et al. 2022). Biofuel production is another area benefiting from the synergy of nanotechnology and tailor-made enzymes. Engineered enzymes immobilized on nanomaterials can significantly enhance the efficiency of biofuel conversion processes, such as cellulose hydrolysis and biodiesel production (Mishra et al. 2020). This approach not only improves the overall yield of biofuels but also reduces production costs and environmental footprint, making biofuels a more viable and sustainable alternative to fossil fuels.

Furthermore, the combination of nanotechnology and tailor-made enzymes is being leveraged for environmental remediation purposes, particularly in cleaning up contaminated sites. Nanoparticles functionalized with specialized enzymes offer efficient and targeted degradation of environmental pollutants, such as heavy metals and organic compounds (Cirino et al. 2013; Rao et al. 2014). This innovative approach holds great potential for rehabilitating polluted environments and restoring ecosystem health through tailored enzymatic processes (Oliveira et al. 2024). In summary, the combination of nanotechnology and custom-engineered enzymes offers a diverse strategy for tackling significant issues in targeted drug delivery, sustainable agriculture, biofuel production, and environmental cleanup. By leveraging the distinct strengths of nanotechnology and personalized enzymes, scientists are leading the path toward inventive solutions that promote human health, food security, energy sustainability, and environmental conservation (Table 1).

Table 1 Nanotechnology and tailor-made enzymes as the future of biobased economy

Aspect	Nanotechnology	Tailor-made enzymes
Definition	Utilizes materials at the nanoscale for various applications in biobased economy	Enzymes designed for specific functions tailored for specific processes in biobased economy
Advantages	Enhanced efficiency and productivity in biobased products	Enhanced specificity and performance in biobased processes
	Improved biocatalytic properties	Reduced energy consumption and resource wastage
	Increased surface area-to-volume ratio for better interactions	Customizable for specific tasks
	Unique physicochemical properties for novel applications	Faster reaction rates and higher
Applications	Drug delivery	Detergents and cleaning agents
	Tissue engineering	Food and beverage industries
	Environmental monitoring	Biofuels and bioenergy
	Smart agriculture	Pharmaceutical industry
Challenges	Safety concerns at the nanoscale	Cost of production and scaling
	Regulation and ethical issues	Stability and compatibility with existing industrial processes
	Environmental impact	

This table provides a comparison between the use of nanotechnology and tailor-made enzymes in the context of the biobased economy, outlining their respective definitions, advantages, applications, and challenges.

8 Challenges and Opportunities in Implementing Nanotechnology-Enhanced Enzymes

Introducing nanotechnology-enhanced enzymes poses challenges and opportunities that demand thorough evaluation and strategic preparation to maximize their effectiveness. One crucial obstacle in their implementation is scalability, moving from small-scale laboratory tests to large-scale industrial manufacturing (Porter et al. 2019). While advancements in nanotechnology have enabled precise control over enzyme properties, replicating these results on a larger scale poses logistical and technical hurdles (Kong et al. 2023). Efforts are needed to optimize manufacturing processes, scale up production, and ensure consistent quality and performance of nanotechnology-enhanced enzymes in industrial settings.

Cost-effectiveness is another critical factor that must be addressed when implementing nanotechnology-enhanced enzymes (Trivedi et al. 2023). Balancing the initial investments required for developing and incorporating nanotechnologies with the long-term benefits they offer is essential for sustainability and commercial viability (Kumar et al. 2018a, b). Strategic cost analysis, optimization of production processes, and exploring novel business models are necessary to maximize the economic efficiency of nanotechnology-enhanced enzymes and make them competitive in the market. Regulatory compliance presents a significant challenge in the implementation of nanotechnology-enhanced enzymes, particularly concerning safety and approval processes (Babu et al. 2016). Given the innovative characteristics of nanotechnologies and their possible effects on human health and the environment, navigating regulatory structures and securing essential approvals can be intricate and time-consuming (Yadav et al. 2023). Understanding and addressing regulatory requirements early in the development phase are crucial to ensure compliance and facilitate market acceptance of nanotechnology-enhanced enzymes.

Despite these obstacles, there are substantial prospects linked to the application of nanotechnology-enhanced enzymes (Sargazi et al. 2022; Del Grosso et al. 2022). Their distinct properties and improved functionalities offer vast potential for utilization across multiple sectors, such as healthcare, agriculture, energy, and environmental remediation (Garg et al. 2021). Through the utilization of nanotechnology, researchers and industries can craft pioneering solutions that address crucial societal needs and contribute to sustainable progress (Roco et al. 2010). In conclusion, while challenges exist in the implementation of nanotechnology-enhanced enzymes, such as scalability, cost considerations, and regulatory requirements, there are abundant opportunities to leverage the unique capabilities of nanotechnologies for fostering innovation and making positive impacts across diverse industries. In today's rapidly advancing world, interdisciplinary collaborations have become increasingly essential for bridging knowledge disparities and propelling innovation in various industries. By uniting experts from various fields, interdisciplinary collaborations furnish a distinctive platform for sharing ideas, expertise, and viewpoints to tackle intricate problems and generate imaginative solutions. This approach not only optimizes the potential for unveiling fresh insights but also cultivates a climate of innovation and originality (National Academy of Sciences 2015).

Preserving intellectual property is crucial for safeguarding the outcomes of interdisciplinary collaborations. Developing precise protocols and legal structures to uphold intellectual property rights guarantees that creators can fully capitalize on their innovations without concerns of infringement or unauthorized utilization (Menell and Scotchmer 2007). This safeguarding acts as a driver for fostering additional innovation and encouraging investments in research and development activities (World Intellectual Property Organization 2019). However, amidst the excitement of collaborative innovation and intellectual property protection, ethical concerns must also be carefully considered and addressed. Balancing the benefits of interdisciplinary collaborations with the potential risks of ethical misconduct, such as conflicts of interest, data privacy breaches, and misuse of technology, is essential for maintaining trust, transparency, and integrity within the collaborative ecosystem. Ethical guidelines and oversight mechanisms play a key role in ensuring that interdisciplinary collaborations adhere to the highest ethical standards and uphold the well-being of all stakeholders involved (National Institutes of Health 2020).

By embracing interdisciplinary collaborations, upholding intellectual property protection, and addressing ethical concerns, organizations can harness the full impending of collaborative innovation while mitigating risks and maximizing benefits for all stakeholders. This holistic approach not only accelerates the pace of innovation but also fosters a culture of responsible and sustainable development in an increasingly interconnected world.

9 Embracing Nanotechnology and Tailor-Made Enzymes for a Sustainable Future

Nanotechnology and tailor-made enzymes represent cutting-edge technologies that hold immense potential in shaping a sustainable future across various sectors (Harish et al. 2023). Nanotechnology involves the management of matter at the molecular and atomic scales to create materials, devices, and systems with unique properties and functionalities (Begley, 2019). This field offers innovative solutions for addressing environmental challenges, improving healthcare, enhancing energy efficiency, and revolutionizing various industries. Conversely, custom-designed enzymes are specialized proteins created to efficiently and selectively catalyze particular chemical reactions (Lim, 2010). Through the customization of enzyme structures and functionalities, scientists can establish sustainable methods for activities such as biofuel production, waste management, pharmaceutical synthesis, and other applications.

The integration of nanotechnology and tailor-made enzymes offers a synergistic approach to addressing complex sustainability issues (Ribeiro, 2019). For instance, nanomaterials can be used to immobilize enzymes, enhance their stability and activity, and enable enzyme recycling in industrial processes. This enables the development of efficient and eco-friendly biocatalytic systems for green chemistry applications (Manikandan et al. 2023). Moreover, nanotechnology can be utilized to produce nanoscale sensors and diagnostic tools that incorporate customized enzymes for swift and precise detection of environmental contaminants, pathogens, and biomarkers, thereby promoting personalized medicine and environmental monitoring (Chen et al. 2021).

In the pursuit of a sustainable future, it is vital to assess the environmental and societal implications of nanotechnology and personalized enzymes. Measures for sustainable innovation and risk evaluation must be enacted to ensure the safe and ethical utilization of these advancements (Li et al. 2023). Collaboration among researchers, industry players, policymakers, and the community is essential for navigating the potential advantages and obstacles linked with nanotechnology and personalized enzymes in pursuit of sustainable progress. By seizing the prospects afforded by nanotechnology and personalized enzymes, we can steer the transformation toward a more sustainable and resilient future. Through interdisciplinary partnerships, ethical deliberations, and continual innovation, these technologies have the capacity to transform various sectors and lay the groundwork for a more sustainable and prosperous society.

The advancement of nanotechnology and customized enzymes serves as a catalytic factor shaping the evolution of the biobased economy (Thanigaivel et al. 2022). These cutting-edge technologies offer sustainable solutions for addressing environmental challenges, improving healthcare, enhancing energy efficiency, and revolutionizing various industries (Bibri et al. 2024). By harnessing the power of nanotechnology and tailor-made enzymes, we can unlock unprecedented opportunities for sustainable development and innovation. Nanotechnology enables precise manipulation at the molecular level, leading to the creation of materials, devices, and systems with unique properties and functionalities (Nasrollahzadeh et al. 2019). When integrated with tailor-made enzymes, nanotechnology enhances the efficiency, stability, and recyclability of enzymes, paving the way for eco-friendly biocatalytic systems in green chemistry applications. Furthermore, nanotechnology-based sensors and diagnostic devices utilizing tailor-made enzymes enable rapid and accurate detection of environmental pollutants, pathogens, and biomarkers, advancing personalized medicine and environmental monitoring.

As we transition toward a biobased economy, the responsible deployment of nanotechnology and tailor-made enzymes is essential. Collaboration among researchers, industry stakeholders, policymakers, and the public is crucial in navigating the ethical, environmental, and societal implications of these technologies (Dwivedi et al. 2024; Polyportis and Pahos 2024). By leveraging interdisciplinary collaborations, ethical considerations, and continuous innovation, we can realize the full potential of nanotechnology and tailor-made enzymes in driving sustainable economic growth and societal well-being. The advancement of nanotechnology and custom-engineered enzymes offers the potential to redefine the biobased economy by providing sustainable solutions across various applications. By prioritizing responsible innovation, conducting risk assessments, and engaging with stakeholders, we can leverage the transformative capabilities of these technologies to establish a more sustainable and prosperous future for future generations. Embracing the opportunities that nanotechnology and tailor-made enzymes present is vital for the future of the biobased economy, where innovation and sustainability converge to catalyze positive global transformations.

10 Conclusion

In conclusion, the advancements in biotechnology, nanotechnology, and enzyme development present a transformative potential for the biobased economy. The precise manipulation enabled by nanotechnology enhances enzymatic processes, leading to advancements in material design and production. Tailor-made enzymes, engineered using cutting-edge techniques, offer extensive functionalities suitable for various industrial applications and revolutionize sectors such as sustainable manufacturing, waste management, and healthcare. The integration of nanotechnology and custom-designed enzymes opens up numerous opportunities in fields like targeted drug delivery, sustainable agriculture, biofuel production, and environmental cleanup. However, practical implementation requires considerations such as scalability, cost-effectiveness, and regulatory compliance. Interdisciplinary collaborations, intellectual property protection, and ethical considerations are crucial for successful technology adoption in this space.

Embracing these transformative technologies as fundamental pillars of the biobased economy paves the way for a sustainable future where biobased solutions drive economic growth and preserve the planet for future generations. The dynamic intersection of nanotechnology and enzymes holds promise for a wide range of industries, offering innovative solutions to pressing challenges and shaping a more sustainable and prosperous world.

References

Adlakha-Hutcheon G, Khaydarov R, Korenstein R, Varma R, Vaseashta A, Stamm H, Abdel-Mottaleb M (2009) Nanomaterials, nanotechnology: applications, consumer products, and benefits. In: Nanomaterials: risks and benefits. Springer, Netherlands, pp 195–207

Adrio JL, Demain AL (2014) Microbial enzymes: tools for biotechnological processes. Biomol Ther 4(1):117–139

Agger JW, Zeuner B (2022) Bio-based surfactants: enzymatic functionalization and production from renewable resources. Curr Opin Biotechnol 78:102842

Agrawal K, Verma P (2021) The interest in nanotechnology: A step towards bioremediation. In: Shah MP (ed) Removal of emerging contaminants through microbial processes. Springer, Singapore, pp 265–282

Agrawal K, Verma P (2023) Nanofiltration: unravelling the potential of the future. In: Shah MP (ed) Bio-nano filtration in industrial effluent treatment. CRC Press, Boca Raton United States, pp 35–48

Ahire SA, Bachhav AA, Pawar TB, Jagdale BS, Patil AV, Koli PB (2022) The augmentation of nanotechnology era: A concise review on fundamental concepts of nanotechnology and applications in material science and technology. Results Chemistry 4:100633

Ahmed SF, Mofijur M, Rafa N, Chowdhury AT, Chowdhury S, Nahrin M et al (2022) Green approaches in synthesising nanomaterials for environmental nanobioremediation: technological advancements, applications, benefits and challenges. Environ Res 204:111967

Ai Y, Hu ZN, Liang X, Sun HB, Xin H, Liang Q (2022) Recent advances in nanozymes: from matters to bioapplications. Adv Funct Mater 32(14):2110432

Ali ES, Sharker SM, Islam MT, Khan IN, Shaw S, Rahman MA et al (2021) Targeting cancer cells with nanotherapeutics and nanodiagnostics: current status and future perspectives. In: Seminars in cancer biology, vol 69. Academic Press, pp 52–68

Alizadeh N, Salimi A (2021) Multienzymes activity of metals and metal oxide nanomaterials: applications from biotechnology to medicine and environmental engineering. J Nanobiotechnol 19(1):1–31

Alzoubi L, Aljabali AA, Tambuwala MM (2023) Empowering precision medicine: the impact of 3d printing on personalized therapeutic. AAPS PharmSciTech 24(8):228

Anwar A, Imran M, Iqbal HM (2023) Smart chemistry and applied perceptions of enzyme-coupled nano-engineered assemblies to meet future biocatalytic challenges. Coord Chem Rev 493:215329

Aparicio JD, Raimondo EE, Saez JM, Costa-Gutierrez SB, Alvarez A, Benimeli CS, Polti MA (2022) The current approach to soil remediation: A review of physicochemical and biological technologies, and the potential of their strategic combination. J Environ Chem Eng 10(2):107141

Babu A, Templeton AK, Munshi AM (2016) Engineered nanoparticles: opportunities and avenues for regulatory agencies. Nanotoxicology 10(2):139–152

Bakshi M, Abhilash PC (2020) Nanotechnology for soil remediation: revitalizing the tarnished resource. In: Nano-materials as photocatalysts for degradation of environmental pollutants. Elsevier, pp 345–370

Banerjee A, Banerjee C, Negi S, Chang JS, Shukla P (2018) Improvements in algal lipid production: a systems biology and gene editing approach. Crit Rev Biotechnol 38(3):369–385

Bawa R, Bawa SR, Maebius SB, Flynn T, Wei C (2005) Protecting new ideas and inventions in nanomedicine with patents. Nanomedicine 1(2):150–158

Baxter J, Bian Z, Chen G, Danielson D, Dresselhaus MS, Fedorov AG et al (2009) Nanoscale design to enable the revolution in renewable energy. Energy Environ Sci 2(6):559–588

Begley MR, Gianola DS, & Ray TR (2019). Bridging functional nanocomposites to robust macroscale devices. Science, 364(6447), eaav4299

Behera A, Behera A (2022) Nanomaterials. Advanced materials: an introduction to modern materials science, 77–125

Betz UA, Arora L, Assal RA, Azevedo H, Baldwin J, Becker MS et al (2023) Game changers in science and technology-now and beyond. Technol Forecast Soc Chang 193:122588

Bhagyaraj SM, Oluwafemi OS (2018) Nanotechnology: the science of the invisible. In: Synthesis of inorganic nanomaterials. Woodhead Publishing, pp 1–18

Bharti S (2024) Harnessing the potential of bimetallic nanoparticles: exploring a novel approach to address antimicrobial resistance. World J Microbiol Biotechnol 40(3):89

Bhatia D, Paul S, Acharjee T, Ramachairy SS (2024) Biosensors and their widespread impact on human health. Sens Int 5:100257

Bibri SE, Krogstie J, Kaboli A, Alahi A (2024) Smarter eco-cities and their leading-edge artificial intelligence of things solutions for environmental sustainability: A comprehensive systematic review. Environ Sci Ecotechnol 19:100330

Bilal M, Nguyen TA, Iqbal HM (2020) Multifunctional carbon nanotubes and their derived nanoconstructs for enzyme immobilization–a paradigm shift in biocatalyst design. Coord Chem Rev 422:213475

Bilal M, Qamar SA, Carballares D, Berenguer-Murcia Á, Fernandez-Lafuente R (2023) Proteases immobilized on nanomaterials for biocatalytic, environmental and biomedical applications: advantages and drawbacks. Biotechnol Adv 108304

Biz A, Proulx S, Xu Z, Siddartha K, Indrayanti AM, Mahadevan R (2019) Systems biology based metabolic engineering for non-natural chemicals. Biotechnol Adv 37(6):107379

Borneman AR, Desany BA, Riches D, Affourtit JP, Forgan AH, Pretorius IS, Egholm M (2006) The genome sequence of the wine yeast VIN7 reveals an allotriploid hybrid genome with Saccharomyces cerevisiae and Saccharomyces kudriavzevii origins. FEMS Yeast Res 6(8):1034–1046

Bornscheuer UT, Schwaneberg U (2019) Directed enzyme evolution principles and applications. John Wiley & Sons

Brar KK, Magdouli S, Othmani A, Ghanei J, Narisetty V, Sindhu R et al (2022) Green route for recycling of low-cost waste resources for the biosynthesis of nanoparticles (NPs) and nanomaterials (NMs)-A review. Environ Res 207:112202

Cao J, Huang D, Peppas NA (2020) Advanced engineered nanoparticulate platforms to address key biological barriers for delivering chemotherapeutic agents to target sites. Adv Drug Deliv Rev 167:170–188

Chafiq M, Chaouiki A, Ko YG (2023) Advances in COFs for energy storage devices: harnessing the potential of covalent organic framework materials. Energy Storage Mater 63:103014

Chapman J, Ismail AE, Dinu CZ (2018) Industrial applications of enzymes: recent advances, techniques, and outlooks. Catalysts 8(6):238

Chaudhary S, Sharma P, Chauhan P, Kumar R, Umar A (2019) Functionalized nanomaterials: a new avenue for mitigating environmental problems. Int J Environ Sci Technol 16:5331–5358

Cheah BC, Macdonald AI, Martin C, Streklas AJ, Campbell G, Al-Rawhani MA et al (2016) An integrated circuit for chip-based analysis of enzyme kinetics and metabolite quantification. IEEE Trans Biomed Circuits Syst 10(3):721–730

Chen X, Shen L, Guo X et al (2019) Recent advances in nanoparticle-enzyme conjugates for biomedical applications. Front Bioeng Biotechnol 7:409

Chen S, Gao Y, Liu Y, Liu J, Lin Y, Ma M, Li J (2021) Nanotechnology for enzyme immobilization: principles, methods, and applications. Small 17(25):2006306

Chimento SK, Mohanty DK, Qian Z, Weber JK (2018) Nanopillars for biofilm control and enzyme immobilization. Biotechnol Prog 34(3):656–664

Cirino PC, Arnold FH, Cort JR (2013) Enzyme engineering. Principles and practice. Humana Press

Clark KM, Ray TR (2023) Recent advances in skin-interfaced wearable sweat sensors: opportunities for equitable personalized medicine and Global Health diagnostics. ACS Sens 8(10):3606–3622

Dalziel M, Crispin M, Scanlan CN, Zitzmann N, Dwek RA (2014) Emerging principles for the therapeutic exploitation of glycosylation. Science 343(6166):1235681

Das B, Franco JL, Logan N, Balasubramanian P, Kim MI, Cao C (2021) Nanozymes in point-of-care diagnosis: an emerging futuristic approach for biosensing. Nanomicro Lett 13:1–51

De La Rica R, Aili D, Stevens MM (2012) Enzyme-responsive nanoparticles for drug release and diagnostics. Adv Drug Deliv Rev 64(11):967–978

Dean SN, Turner KB, Medintz IL, Walper SA (2017) Targeting and delivery of therapeutic enzymes. Ther Deliv 8(7):577–595

Del Grosso A, Parlanti G, Mezzena R, Cecchini M (2022) Current treatment options and novel nanotechnology-driven enzyme replacement strategies for lysosomal storage disorders. Adv Drug Deliv Rev 188:114464

Ding J, Zheng Z, Song Z, Ding S, Wen J, Liu K et al (2023) Recent progress in chemiresistive gas sensors with 1D nanostructured sensing materials: insights into the structure-morphology-performance relationship. Mater Des 235:112360

Ditta A (2012) How helpful is nanotechnology in agriculture? Adv Nat Sci Nanosci Nanotechnol 3(3):033002

Dwivedi YK, Jeyaraj A, Hughes L, Davies GH, Ahuja M, Albashrawi MA et al (2024) "Real impact": challenges and opportunities in bridging the gap between research and practice–making a difference in industry, policy, and society. Int J Inf Manag 102750

Elnahal AS, El-Saadony MT, Saad AM, Desoky ESM, El-Tahan AM, Rady MM et al (2022) The use of microbial inoculants for biological control, plant growth promotion, and sustainable agriculture: A review. Eur J Plant Pathol 162(4):759–792

Fayaz M, Rabani MS, Wani SA, Thoker SA (2021) Nano-agriculture: a novel approach in agriculture. Microbiota and biofertilizers: a sustainable continuum for plant and soil health, 99–122

Feynman RP (1959) There's plenty of room at the bottom. Eng Sci 23(5):22–36

Fu S, Zhang T, Jiang H, Xu Y, Chen J, Zhang L, Su X (2021) DNA nanotechnology enhanced single-molecule biosensing and imaging. TrAC Trends Anal Chem 140:116267

Garg A, Basu S, Shetti NP, Reddy KR (2021) 2D materials and its heterostructured photocatalysts: synthesis, properties, functionalization and applications in environmental remediation. J Environ Chem Eng 9(6):106408

Gayathiri E, Prakash P, Kumaravel P, Jayaprakash J, Ragunathan MG, Sankar S et al (2023) Computational approaches for modeling and structural design of biological systems: A comprehensive review. Prog Biophys Mol Biol 185:17

Gericke M, Amaral AJ, Budtova T, De Wever P, Groth T, Heinze T et al (2024) The European polysaccharide network of excellence (EPNOE) research roadmap 2040: advanced strategies for exploiting the vast potential of polysaccharides as renewable bioresources. Carbohydr Polym 326:121633

Gholami-Shabani M, Shams-Ghahfarokhi M, Gholami-Shabani Z, Razzaghi-Abyaneh M (2016) Microbial enzymes: current features and potential applications in nanobiotechnology. Advances and applications through fungal nanobiotechnology, 91–127

Gkantzou E, Chatzikonstantinou AV, Fotiadou R, Giannakopoulou A, Patila M, Stamatis H (2021) Trends in the development of innovative nanobiocatalysts and their application in biocatalytic transformations. Biotechnol Adv 51:107738

Grewal D, Guha A, Satornino CB, Schweiger EB (2021) Artificial intelligence: the light and the darkness. J Bus Res 136:229–236

Gupta O, Roy S (2020) Recent progress in the development of nanocomposite membranes. Nanocomposite membranes for water and gas separation, 29–67

Haleem A, Javaid M, Singh RP, Rab S, Suman R (2023) Applications of nanotechnology in medical field. Glob Health J 7:70

Harish V, Ansari MM, Tewari D, Yadav AB, Sharma N, Bawarig S et al (2023) Cutting-edge advances in tailoring size, shape, and functionality of nanoparticles and nanostructures: a review. J Taiwan Inst Chem Eng 149:105010

Heux S, Meynial-Salles I, O'Donohue MJ, Dumon C (2015) White biotechnology: state of the art strategies for the development of biocatalysts for biorefining. Biotechnol Adv 33(8):1653–1670

Hidangmayum A, Debnath A, Guru A, Singh BN, Upadhyay SK, Dwivedi P (2023) Mechanistic and recent updates in nano-bioremediation for developing green technology to alleviate agricultural contaminants. Int J Environ Sci Technol 20(10):11693–11718

Honarparvar B, Govender T, Maguire GE, Soliman ME, Kruger HG (2014) Integrated approach to structure-based enzymatic drug design: molecular modeling, spectroscopy, and experimental bioactivity. Chem Rev 114(1):493–537

Iñiguez-Moreno M, Calderón-Santoyo M, Ascanio G, Ragazzo-Calderón FZ, Parra-Saldívar R, Ragazzo-Sánchez JA (2023) Harnessing emerging technologies to obtain biopolymer from agro-waste: application into the food industry. Biomass Convers Biorefinery:1–18

Issaka E, Amu-Darko JNO, Adams M, Yakubu S, Gyimah E, Ali N et al (2023a) Zinc imidazolate metal–organic frameworks-8-encapsulated enzymes/nanoenzymes for biocatalytic and biomedical applications. Catal Lett 153(7):2083–2106

Issaka E, Wariboko MA, Mohammed A, Enyan M, Aguree S (2023b) Trends in enzyme mimics for enhanced catalytic cascade systems for bio-sensing of environmental pollutants—a review. Chem Eng J Adv 100510

Jamuna G, Yasodha S, Thamarai P, Vickram AS, Swaminaathan P, Saravanan A, Yaashikaa PR (2023) Design strategies, utilization and applications of nano-engineered biomaterials for the enhancement of bioenergy: A sustainable approach. Biochem Eng J 200:109104

Johannes TW, Zhao H (2006) Directed evolution of enzymes and biosynthetic pathways. Curr Opin Microbiol 9(3):261–267

Ju H (2018) Functional nanomaterials and nanoprobes for amplified biosensing. Appl Mater Today 10:51–71

Kang Y, Song J, Wu X et al (2018) Applications of enzymes in industry and medicine: opportunities and challenges. Biotechnol Adv 36(8):2050–2067

Khan SH (2020) Green nanotechnology for the environment and sustainable development. Green materials for wastewater treatment, 13–46

Khanzada AK, Al-Hazmi HE, Śniatała B, Joseph TM, Majtacz J, Abdulrahman SA et al (2023) Hydrochar-nanoparticle integration for arsenic removal from wastewater: challenges, possible solutions, and future horizon. Environ Res 238:117164

Kharbikar L, Konwarh R, Chakraborty M, Nandanwar S, Marathe A, Yele Y et al (2023) 3Bs of CRISPR-Cas mediated genome editing in plants: exploring the basics, bioinformatics and biosafety landscape. Physiol Mol Biol Plants 1-26:1825

Khoshgoftarmanesh AH, Shariatmadari H, Kariman K (2019) Nanotechnology and nano-nutrition for improvement of the nutrient use efficiency of crop plants. In: Nutrient use efficiency: from basics to advances. Academic Press, pp 439–477

Kong X, Gao P, Wang J, Fang Y, Hwang KC (2023) Advances of medical nanorobots for future cancer treatments. J Hematol Oncol 16(1):74

Kulkarni RP (2007) Nano-bio-genesis: tracing the rise of nanotechnology and nanobiotechnology as' big science'. J Biomed Discov Collab 2(1):1–16

Kumar A, Singh V, Jazayeri SD (2018a) Nanotechnology, an innovative tool for sustainable agriculture: A review. J Clean Prod 197:1435–1444
Kumar V, Shahi SK, Singh S (2018b) Bioremediation: an eco-sustainable approach for restoration of contaminated sites. Microbial bioprospecting for sustainable development, 115–136
Lawson CE, Martí JM, Radivojevic T, Jonnalagadda SVR, Gentz R, Hillson NJ et al (2021) Machine learning for metabolic engineering: A review. Metab Eng 63:34–60
Le Feuvre RA, Scrutton NS, Palmer DJ (2018) Designer enzymes for a greener future: high-throughput screening for biocatalysts. Curr Opin Chem Biol 43:139–148
Li X, Chen Z, Bland GD (2020) Nanotechnology for soil remediation: benefits, challenges, and perspectives. J Environ Sci 87:4–19
Li W, Yigitcanlar T, Browne W, & Nili A (2023). The making of responsible innovation and technology: An overview and framework. Smart Cities, 6(4):1996–2034
Liang J, Liang K (2021) Nano-bio-interface engineering of metal-organic frameworks. Nano Today 40:101256
Lim WA. (2010). Designing customized cell signalling circuits. Nature reviews Molecular cell biology, 11(6):393–403
Lim HR, Kim HS, Qazi R, Kwon YT, Jeong JW, Yeo WH (2020) Advanced soft materials, sensor integrations, and applications of wearable flexible hybrid electronics in healthcare, energy, and environment. Adv Mater 32(15):1901924
Liu DM, Dong C (2020) Recent advances in nano-carrier immobilized enzymes and their applications. Process Biochem 92:464–475
Lu Y, Luo Q, Jia X, Tam JP, Yang H, Shen Y, Li X (2023) Multidisciplinary strategies to enhance therapeutic effects of flavonoids from Epimedii folium: integration of herbal medicine, enzyme engineering, and nanotechnology. J Pharm Anal 13(3):239–254
Magnabosco P, Masi A, Shukla R, Bansal V, Carletti P (2023) Advancing the impact of plant biostimulants to sustainable agriculture through nanotechnologies. Chem Biol Technol Agric 10(1):117
Manikandan S, Krishnan RY, Vickram S, Subbaiya R, Kim W, Govarthanan M, Karmegam N (2023) Emerging nanotechnology in renewable biogas production from biowastes: impact and optimization strategies–a review. Renew Sust Energ Rev 181:113345
Mansoori GA (2002) Advances in atomic & molecular nanotechnology. United Nations Tech Monitor, 53–59
Mansoori GA, Soelaiman TF (2005) Nanotechnology—An introduction for the standards community. ASTM Int 2:13110
Manu G, Firdose N, Jayanthi MK, Raj R, Patil SM, Chandra SJ et al (2021) Current status and perspectives of oral therapeutic protein and peptide formulations: A review. J Pharm Res Int 33(43A):117–137
Mauter MS, Zucker I, Perreault F, Werber JR, Kim JH, Elimelech M (2018) The role of nanotechnology in tackling global water challenges. Nat Sustain 1(4):166–175
Menell PS, Scotchmer S (2007) Intellectual property law. Handbook Law Econ 2:1473–1570
Mishra A, Mal S, Shin K, Kim H, Lee Y (2020) Nanotechnology in biofuels: a rational approach for biodiesel production. Renew Sust Energ Rev 129:109932
Nasrollahzadeh M, Sajadi SM, Sajjadi M, Issaabadi Z (2019) An introduction to nanotechnology. In: Interface science and technology, vol 28. Elsevier, pp 1–27
National Academy of Sciences (2015) Facilitating interdisciplinary research
National Institutes of Health (2020) Ethical considerations in biomedical research
Navya PN, Daima HK (2016) Rational engineering of physicochemical properties of nanomaterials for biomedical applications with nanotoxicological perspectives. Nano Converg 3:1–14
Nel A, Xia T, Mädler L, Li N (2009) Toxic potential of materials at the nanolevel. Science 311(5761):622–627
Nile SH, Baskar V, Selvaraj D, Nile A, Xiao J, Kai G (2020) Nanotechnologies in food science: applications, recent trends, and future perspectives. Nanomicro Lett 12:1–34

Nothling MD, Xiao Z, Bhaskaran A, Blyth MT, Bennett CW, Coote ML, Connal LA (2018) Synthetic catalysts inspired by hydrolytic enzymes. ACS Catal 9(1):168–187

Nultsch K, Germershaus O (2018) Matrix metalloprotease triggered bioresponsive drug delivery systems–design, synthesis and application. Eur J Pharm Biopharm 131:189–202

Oke MA, Ojo SA, Fasiku SA, Adebayo EA (2023) Nanotechnology and enzyme immobilization: a review. Nanotechnology 34:385101

Oliveira VH, Díez S, Dolbeth M, Coelho JP (2024) Restoration of degraded estuarine and marine ecosystems: a systematic review of rehabilitation methods in Europe. J Hazard Mater 469:133863

Otero JM, Panagiotou G, Olsson L (2007) Fueling industrial biotechnology growth with bioethanol. Biofuels:1–40

Othman A, Karimi A, Andreescu S (2016) Functional nanostructures for enzyme based biosensors: properties, fabrication and applications. J Mater Chem B 4(45):7178–7203

Padma S, Dey S, Banerjee M, Modak BK, Mukherjee S (2023) Nanomaterials in bioprocessing and their biomedical applications: an update. In: Nanomaterials for bioreactors and bioprocessing applications. Elsevier, pp 365–401

Pal T, Ruidas B, Das Mukhopadhyay C (2021) Modern advancement of nanotechnology over conventional drug therapies. In: Advances in bioprocess engineering and technology: select proceedings ICABET 2020. Springer Singapore, pp 493–500

Pandey N, Srivastava G, Goel R (2018) Applications of nanotechnology in biofuels production: challenges and future perspectives. Bioresour Technol 271:464–473

Paramasivam G, Sanmugam A, Palem VV, Sevanan M, Sairam AB, Nachiappan N et al (2024) Nanomaterials for detection of biomolecules and delivering therapeutic agents in theragnosis: A review. Int J Biol Macromol 254:127904

Patel JK, Patel A, Bhatia D (2021) Introduction to nanomaterials and nanotechnology. In: Emerging technologies for nanoparticle manufacturing. Springer International Publishing, Cham, pp 3–23

Pfleger BF, Takors R (2023) Recent progress in the synthesis of advanced biofuel and bioproducts. Curr Opin Biotechnol 80:102913

Polyportis A, Pahos N (2024) Navigating the perils of artificial intelligence: a focused review on ChatGPT and responsible research and innovation. Humanit Soc Sci Commun 11(1):1–10

Porter AE, Gass M, Muller K, Skepper JN (2019) Uptake of non-viral gene delivery nanoparticles: microscopical analysis of the morphology of nanoparticles in peritoneal dialysis effluent after intraperitoneal administration. J Microsc 234(3):263–270

Preethi B, Karmegam N, Manikandan S, Vickram S, Subbaiya R, Rajeshkumar S et al (2024) Nanotechnology-powered innovations for agricultural and food waste valorization: a critical appraisal in the context of circular economy implementation in developing nations, vol 184. Process Saf Environ Protect, p 477

Rafeeq H, Hussain A, Ambreen A, Waqas M, Bilal M, Iqbal HM (2022) Functionalized nanoparticles and their environmental remediation potential: a review. J Nanostructure Chem 12(6):1007–1031

Ramanathan A, Vivanco MD, Deo RP (2019) Nanotechnology in water treatment: applications and perspectives. Crit Rev Environ Sci Technol 49(5):463–499

Rani V, Deep G, Singh RK, Palle K, Yadav UC (2016) Oxidative stress and metabolic disorders: pathogenesis and therapeutic strategies. Life Sci 148:183–193

Rao MA, Scelza R, Acevedo F, Diez MC, Gianfreda L (2014) Enzymes as useful tools for environmental purposes. Chemosphere 107:145–162

Ravichandran R (2010) Nanotechnology applications in food and food processing: innovative green approaches, opportunities and uncertainties for global market. Int J Green Nanotechnol Phys Chem 1(2):P72–P96

Reshmy R, Philip E, Sirohi R, Tarafdar A, Arun KB, Madhavan A et al (2021) Nanobiocatalysts: advancements and applications in enzyme technology. Bioresour Technol 337:125491

Ribeiro LF, Amarelle V, Alves LDF, Viana de Siqueira GM, Lovate GL, Borelli TC, & Guazzaroni ME (2019). Genetically engineered proteins to improve biomass conversion: New advances and challenges for tailoring biocatalysts. Molecules, 24(16):2879

Rinaldi R, Jastrzebski R, Clough MT, Ralph J, Kennema M, Bruijnincx PC, Weckhuysen BM (2016) Paving the way for lignin valorisation: recent advances in bioengineering, biorefining and catalysis. Angew Chem Int Ed 55(29):8164–8215

Roco MC (2011) The long view of nanotechnology development: the National Nanotechnology Initiative at 10 years. J Nanopart Res 13:427–445

Roco MC, Mirkin CA, Hersam MC (2010) Nanotechnology research directions for societal needs in 2020. Retrospective and outlook. National Science Foundation (Sponsor)

Rodríguez-Núñez K, Bernal C, Bolívar JM (2023) The enzyme, the support, and the immobilization strategy: the key findings to a desirable biocatalyst. In: Biocatalyst immobilization. Academic Press, pp 1–16

Sahu MK, Yadav R, Tiwari SP (2023) Recent advances in nanotechnology. Int J Nanomater Nanotechnol Nanomed 9(1):015–023

Salata O (2004) Applications of nanoparticles in biology and medicine. J Nanobiotechnol 2(1):3

Sanchez-Rexach E, Johnston TG, Jehanno C, Sardon H, Nelson A (2020) Sustainable materials and chemical processes for additive manufacturing. Chem Mater 32(17):7105–7119

Sargazi S, Mukhtar M, Rahdar A, Bilal M, Barani M, Díez-Pascual AM et al (2022) Opportunities and challenges of using high-sensitivity nanobiosensors to detect long noncoding RNAs: A preliminary review. Int J Biol Macromol 205:304–315

Sarkar A, Jyothi Lakshmi N, Chandrasekaran N (2019) Nanotechnology in agriculture: scope and current developments. J Sci Ind Res 78(11):783–791

Serag E, El-Maghraby A, El Nemr A (2022) Recent developments in the application of carbon-based nanomaterials in implantable and wearable enzyme-biofuel cells. Carbon Lett 32(2):395–412

Shahcheraghi N, Golchin H, Sadri Z, Tabari Y, Borhanifar F, Makani S (2022) Nano-biotechnology, an applicable approach for sustainable future. 3 Biotech 12(3):65

Shahid M, Khan MS, Singh UB (2023) Pesticide-tolerant microbial consortia: potential candidates for remediation/clean-up of pesticide-contaminated agricultural soil. Environ Res 236:116724

Shahriari M, Zahiri M, Abnous K, Taghdisi SM, Ramezani M, Alibolandi M (2019) Enzyme responsive drug delivery systems in cancer treatment. J Control Release 308:172–189

Sharifi M, Sohrabi MJ, Hosseinali SH, Hasan A, Kani PH, Talaei AJ et al (2020) Enzyme immobilization onto the nanomaterials: application in enzyme stability and prodrug-activated cancer therapy. Int J Biol Macromol 143:665–676

Sharma G, Sharma S, Bhattacharya M et al (2019) Nanotechnology-based approaches for targeting and delivery of drugs and genes. Mol Membr Biol 36(3–4):82–98

Sheikh ZUD, Bajar S, Devi A, Rose PK, Suhag M, Yadav A et al (2023) Nanotechnology based technological development in biofuel production: current status and future prospects. Enzym Microb Technol 171:110304

Singh A, Prajapati P, Vyas S, Gaur VK, Sindhu R, Binod P et al (2023) A comprehensive review of feedstocks as sustainable substrates for next-generation biofuels. Bioenergy Res 16(1):105–122

Somu P, Narayanasamy S, Gomez LA, Rajendran S, Lee YR, Balakrishnan D (2022) Immobilization of enzymes for bioremediation: A future remedial and mitigating strategy. Environ Res 212:113411

Sonvico F, Mornet S, Vasseur S, Dubernet C, Jaillard D, Degrouard J, Couvreur P (2009) Folate-conjugated iron oxide nanoparticles for solid tumor targeting as potential specific magnetic hyperthermia mediators: synthesis, physicochemical characterization, and in vitro experiments. Bioconjug Chem 20(6):1181–1188

Soozanipour A, Ejeian F, Boroumand Y, Rezayat A, Moradi S (2023) Biotechnological advancements towards water, food and medical healthcare: a review. Chemosphere 312:137185

Soppimath KS, Aminabhavi TM, Kulkarni AR, Rudzinski WE (2001) Biodegradable polymeric nanoparticles as drug delivery devices. J Control Release 70(1–2):1–20

Suhag D, Thakur P, Thakur A (2023) Introduction to nanotechnology. In: Integrated nanomaterials and their applications. Springer Nature Singapore, Singapore, pp 1–17

Sujitha P, Shanthi C (2023) Importance of enzyme specificity and stability for the application of proteases in greener industrial processing-a review. J Clean Prod 425:138915

Syed MH, Zahari MAKM, Khan MMR, Beg MDH, Abdullah N (2023) An overview on recent biomedical applications of biopolymers: their role in drug delivery systems and comparison of major systems. J Drug Deliv Sci Technol 80:104121

Tan Z, Cheng H, Chen G, Ju F, Fernández-Lucas J, Zdarta J et al (2023) Designing multifunctional biocatalytic cascade system by multi-enzyme co-immobilization on biopolymers and nanostructured materials. Int J Biol Macromol 227:535–550

Tang KHD, Lock SSM, Yap PS, Cheah KW, Chan YH, Yiin CL et al (2022) Immobilized enzyme/microorganism complexes for degradation of microplastics: A review of recent advances, feasibility and future prospects. Sci Total Environ 832:154868

Tegart G (2004) Nanotechnology: the technology for the twenty-first century. Foresight 6(6):364–370

Terns MP, Terns RM (2017) CRISPR-based technologies: impact of RNA-targeting CRISPR systems. Mol Cell 65(5):677–681

Thanigaivel S, Priya AK, Dutta K, Rajendran S, Sekar K, Jalil AA, Soto-Moscoso M (2022) Role of nanotechnology for the conversion of lignocellulosic biomass into biopotent energy: A biorefinery approach for waste to value-added products. Fuel 322:124236

Thapa S, Li H, OHair J, Bhatti S, Chen FC, Nasr KA et al (2019) Biochemical characteristics of microbial enzymes and their significance from industrial perspectives. Mol Biotechnol 61:579–601

Tripathi S, Siddiqui MH, Kumar A, Vimal A (2023) Nanoparticles: a promising vehicle for the delivery of therapeutic enzymes. Int Nano Lett 13(3):209–221

Trivedi P, Mattupalli C, Eversole K, Leach JE (2021) Enabling sustainable agriculture through understanding and enhancement of microbiomes. New Phytol 230(6):2129–2147

Trivedi D, Chavali M, Vohra S, Salunkhe P, Tripathi S (2023) Advantages of using nanobiotechnology in enhancing the economic status of the country. In: Nanobiotechnology for the livestock industry. Elsevier, pp 369–392

ul Haque S, Nasar A, Duteanu N, Pandey S (2023) Carbon based-nanomaterials used in biofuel cells–A review. Fuel 331:125634

Vasina M, Velecký J, Planas-Iglesias J, Marques SM, Skarupova J, Damborsky J et al (2022) Tools for computational design and high-throughput screening of therapeutic enzymes. Adv Drug Deliv Rev 183:114143

Verma ML, Kumar P, Sharma D, Verma AD, Jana AK (2019) Advances in nanobiotechnology with special reference to plant systems. In: Plant nanobionics: Volume 1, advances in the understanding of nanomaterials research and applications, pp 371–387

Villalba-Rodríguez AM, Martínez-Zamudio LY, Martínez SAH, Rodríguez-Hernández JA, Melchor-Martínez EM, Flores-Contreras EA et al (2023) Nanomaterial constructs for catalytic applications in biomedicine: nanobiocatalysts and nanozymes. Top Catal 66(9–12):707–722

Wang G, Wang H, Mu X (2019) Nanozymes: an emerging field bridging nanotechnology and biology. Sci China Mater 62(9):1239–1260

Wang T, Lai L, Huang Y, Su E (2023a) Nanozyme: an emerging tool for food packaging. Food Control 110104

Wang Z, Shao Y, Zhu Z, Wang J, Gao X, Xie J et al (2023b) Novel gold nanozyme regulation strategies facilitate analytes detection. Coord Chem Rev 495:215369

Wei F, Lillehoj PB, Ho CM (2010) DNA diagnostics: nanotechnology-enhanced electrochemical detection of nucleic acids. Pediatr Res 67(5):458–468

Wei H, Dong C, Zhu J et al (2020) Nanotechnology-enhanced enzymes for biofuel production: recent advances and perspectives. Biotechnol Biofuels 13(1):211. https://doi.org/10.1186/s13068-020-01810-4

World Intellectual Property Organization. (2019). Intellectual property and development: current debates and policy challenges
Xu C, Zhou S, Song H, Hu H, Yang Y, Zhang X et al (2023) Green tea polyphenols-derived hybrid materials in manufacturing, environment, food and healthcare. Nano Today 52:101990
Yadav A, Yadav K, Abd-Elsalam KA (2023) Exploring the potential of nanofertilizers for a sustainable agriculture. Plant Nano Biol 5:100044
Yin PT, Shah S, Chhowalla M, Lee KB (2016) Design, synthesis, and characterization of graphene–nanoparticle hybrid materials for bioapplications. Chem Rev 115(7):2483–2531
Yu Y, Lee SJ, Theerthagiri J, Fonseca S, Pinto LM, Maia G, Choi MY (2022) Reconciling of experimental and theoretical insights on the electroactive behavior of C/Ni nanoparticles with AuPt alloys for hydrogen evolution efficiency and non-enzymatic sensor. Chem Eng J 435:134790
Zdarta J, Kołodziejczak-Radzimska A, Bachosz K, Rybarczyk A, Bilal M, Iqbal HM et al (2023) Nanostructured supports for multienzyme co-immobilization for biotechnological applications: achievements, challenges and prospects. Adv Colloid Interf Sci 315:102889
Zhang C, Pan D, Li J, Hu J, Bains A, Guys N et al (2017) Enzyme-responsive peptide dendrimer-gemcitabine conjugate as a controlled-release drug delivery vehicle with enhanced antitumor efficacy. Acta Biomater 55:153–162
Zhang S, Yang Q, Tư Y, Sun Q, Zhang L, Pan G (2020) Intelligent nanozymes: recent advances and perspectives for healthcare and environmental applications. *Adv Healthc Mater*, 9(16):2000745. https://doi.org/10.1002/adhm.202000745
Zhang JL, Wang YH, Huang K, Huang KJ, Jiang H, Wang XM (2021) Enzyme-based biofuel cells for biosensors and in vivo power supply. Nano Energy 84:105853
Zhang N, Domínguez de María P, Kara S (2024) Biocatalysis for the synthesis of active pharmaceutical ingredients in Deep eutectic solvents: state-of-the-art and prospects. Catalysts 14(1):84
Zheng L, Cao M, Du Y, Liu Q, Emran MY, Kotb A et al (2024) Artificial enzyme innovations in electrochemical devices: advancing wearable and portable sensing technologies. Nanoscale 16(1):44–60
Zhong W, Li H, Wang Y (2023) Design and construction of artificial biological systems for one-carbon utilization. BioDesign Res 5:0021
Zhuang J, Midgley AC, Wei Y, Liu Q, Kong D, Huang X (2023) Machine-Learning-assisted nanozyme design: lessons from materials and engineered enzymes. Adv Mater 36:2210848

In-silico Investigations on Exploring Various Aspects of Biorefinery: A Futuristic Paradigm Through Microbiological Perspectives

Chandrama Chanda, Santosh Kumar, Mingma Thundu Sherpa, Sunil S. Thorat, and Sayak Das

1 Introduction

Producing goods from agricultural commodities, or biomass, is an age-old practice since time immemorial. However, it is comparatively new, to use biomass as an input to create various products using intricate processing techniques - referred to as biorefinery. This method is comparable to what happens in a petroleum refinery when fossil fuels are used as input. Carbohydrates, lignins, proteins, lipids, and, to a lesser extent, a variety of additional substances like vitamins, colours, and tastes make up biomass. A biorefinery's objective is to convert these abundant biological ingredients into valuable products by combining various technologies and procedures. A biorefinery's primary objective is to use a number of unit operations to create high-value low-volume (HVLV) and low-value high-volume (LVHV) products (Fernando et al. 2006). By turning LVHV intermediates into energy, the processes are intended to optimize the valuable extractables while reducing the waste

C. Chanda · S. Das (✉)
Department of Life Science & Bioinformatics, Har Gobind Khurana School of Life Sciences, Assam University, Silchar, Assam, India
e-mail: sayak.das@aus.ac.in

S. Kumar
Department of Microbiology, School of Life Sciences, Sikkim University, Gangtok, Sikkim, India

M. T. Sherpa
Department of Microbiology, Nar Bahadur Bhandari Government College, Tadong, Sikkim, India

S. S. Thorat
Institute of Bioresources and Sustainable Development (IBSD) – Meghalaya, Upper Shillong, Meghalaya, India

K. Agrawal, P. Verma (eds.), *Biotechnological Advances in Biorefinery*, Interdisciplinary Biotechnological Advances,
https://doi.org/10.1007/978-981-97-5544-8_14

streams. While the high-volume fuels assist in meeting the world's energy needs, the high-value goods increase profitability. A biorefinery's electricity output contributes to the overall cost savings. A biorefinery, as opposed to a petroleum refinery, uses renewable resources and generates chemicals and fuels with a lower impact on the environment.

Biorefinery is a concept of conversion of any biomass or residue that has been directly or indirectly obtained from any microorganisms, algae, plant-based materials, etc. to various valuable by-products (biofuel, bioenergy, biochemicals, etc.) which are environmentally friendly and are economically important (Nair et al. 2022; Agrawal and Verma 2020; Kumar et al. 2020; Bhardwaj et al. 2019). The primary objective of biorefinery is to generate by-products from diverse sources and subsequently employ various conventional technologies to recycle and repurpose these by-products. Studies have shown that most valuable by-products have been obtained through sustainable methods from waste biorefinery and are also reused for valorization of the system. Thus, it gives rise to a circular bio-economy through a sustainable approach as it will generate the useful products. Climate change stands out as a global concern, prompting researchers to explore diverse biorefinery processes originating from various sources. Numerous studies have been conducted, focusing on the transformation of acetate into a range of biochemicals, biofuels, and other valuable products (Sakuntala and Kim 2022). Also, methanol (CH_3OH) is produced from lignocellulosic wastes (Haydary et al. 2020), lowering the ammonia (NH_3) content in municipal biowaste for anaerobic (in absence of oxygen) digestion by separating gardening waste from soluble components (Logan and Visvanathan 2019), and using a portion of municipal solid trash that is organic to produce polyhydroxyalkanoates (type of biopolymer and biogas) (Moretto et al. 2020). Another study revealed that a model library is created to maximize the treatment for the recovery of nutrients and energy in order to valorize biowaste from pig dung (Rodríguez-Alegre et al. 2023). A biorefinery in Hong Kong turns bakery residues into chemicals like succinic acid and bioplastics like poly(3-hydroxybutyrate) (PHB) (Karmee and Lin 2014) and manufactures volatile fatty acids from potato peels and kitchen garbage that are then utilized in biorefineries to create plastics, leathers, textiles, and food items (Adetunji et al. 2023). Also, some researchers investigated how to increase the production of omega-3 fatty acids in biorefineries by integrating dark fermentation, an anaerobic digestion process (Ren et al. 2019). The procedure involves isolating and refining pyrolysate obtained from lignin biowaste to produce bisphenol A, which is subsequently used in the manufacturing of plastic products (Shah et al. 2020). Most research on butanol production centers around first-generation sugarcane biorefineries, exploring various butanol ($C_4H_{10}O$) fermentation technologies to produce ethanol, sugar, power, and n-butanol (Mariano et al. 2013). Biorefinery technique has advanced significantly worldwide. Among them, extremophiles are crucial to the biorefinery process because they offer new metabolic pathways and enzymes with high catalytic stability, which enables them to function as biocatalysts on their own in challenging industrial settings. Thermophilic, psychrophilic, alkaliphilic, acidophilic, and halophilic bacteria are some of the common habitat based examples of extremophilic bacteria (Zhu et al.

2020). Other bacteria including lactic acid bacteria (LAB) are used mainly in industrial sector like food fermentation, probiotics, etc. Biomass obtained from diverse plant sources has the potential to be converted into valuable products.

Performing biological experiments through computational means is termed 'in-silico'. In pharmaceutical research, bioinformatics play a crucial role in the creation of new drugs, a method commonly referred to as computer-aided drug design (CADD) or computer-assisted molecular design (CAMD). Numerous studies have utilized CADD, also recognized as in-silico drug designing. This approach encompasses a variety of virtual screening tools capable of forecasting the behaviour of chemical compounds and their interactions with diverse biological targets. Reducing the issues with traditional experiments and increasing their cost and time efficiency are the primary objectives of in-silico drug discovery technology (Shaker et al. 2021).

In-silico research can be used to assess designs and tactics that enhance a bioprocess and to make it easier to comprehend the various intricate physicochemical and biological processes involved in the co-cultures of bacteria and microalgae (Perera et al. 2019). Chemical pesticides find widespread application in modern agricultural practices. In an attempt to meet food demands and sustain agricultural output, this is being done (Nicolopoulou-Stamati et al. 2016). The key element in agriculture revolves around the effectiveness of organic pesticides, commonly referred to as biopesticides. Utilizing the in-silico method is a prominent strategy in this regard. By using molecular docking, this method can be utilized to simulate drug discovery and assess its bioactivity. As a program's initial technical effort, the in-silico approach may help government initiatives to actualize eco-friendly agriculture (Kholila et al. 2021). Many valuable fungal products are obtained through the biorefinery process. In-silico methods facilitate pharmaceuticals by ensuring that any medicine entering the market is first recognized as an effective drug capable of curing diseases while also being free of toxicity and easily absorbed, distributed, excreted, and metabolized within the body. In-silico approach includes many processes like drug-likeness study, ADME-TOX screening, QSAR, simulation, etc. In-silico is a process with cost-efficient and time-saving technique. In order to stay cost-competitive, a biorefinery of the future must target both high- and low-value applications, valorize all components of biomass, and use abundant and inexpensive (waste) streams as raw materials (Asghar et al. 2022).

2 Biorefinery

A biorefinery is a sustainable production process that transforms diverse biomass sources such as wood, algae, straw, coal, ethanol, manure, and fuel into various products. This approach aims to generate eco-friendly products by utilizing renewable resources obtained from different origins. Biorefinery is considered a key route of sustainability. The ability to produce and extract the biochemicals needed at a low cost and effective extraction on an industrial scale are the major obstacles to realizing the potential for algal biomass as a source of sustainable feed for food production, biotechnology, or green biofuel. In this respect, the optimization of biomass production is considered to be a crucial step towards the realization of high-value products (Borowitzka 2013). It requires time to determine the best ratio of light to the availability of nutrients to promote the creation of surplus carbon-rich biomass rather than protein-rich biomass; optimizing biomass production does not automatically translate into optimizing biofuel generation. This is due to the fact that increased lipid production and maximal growth frequently conflict (Scott et al. 2010). While N stress causes lipids to concentrate in algae (Kenny and Flynn 2014), this situation can be used to develop techniques to maximize oil and fatty acid production by modifying cell physiology (Greenwell et al. 2010). However, it is not conducive to rapid biomass growth. In continuous culture systems, ideal conditions can be reached by choosing dilution rates that balance growth and nitrogen limitation. In batch culture systems, this can be done in two steps: first, biomass is allowed to grow to its maximum potential, and then it is starved of nitrogen. The additional challenge is that, if at all possible, a dense N-starved culture is usually too self-shaded to permit enough light penetration to maximize the lipid yield in an acceptable amount of time (Su et al. 2011).

2.1 Classification of Biorefinery System

Biorefinery systems are basically classified into four ways:

1. Platforms—These generally refers to the intermediates formed from various sources which then transformed into any bio-based products or any other intermediates, e.g. biogas, syngas, hydrogen, sucrose, starch, cellulose and hemicellulose, xylose, arabinose, lignin, pyrolysis oil, etc.
2. Products—In the biorefinery system, products are categorized into two groups: energy-driven or energetic products (they are the material-driven) or non-energetic products (they are the main bio-based products obtained directly (chemicals, food, etc.)).
3. Feedstock—These primarily originate from energy crops sourced through agricultural means or aquaculture or the residues obtained from sources such as agriculture, industries, etc.
4. Processes—These are defined as the process in which the conversion of biomass takes place into products. Processes are of four types:

   (a) Physical or mechanical—It includes milling, separation, distillation, pressing, etc. In this process of biorefinery, the chemical structures are maintained, but the shape, size, and amount vary.
   (b) Biochemical—It requires microorganisms or enzymes at a mild temperature and pressure such as fermentation process.
   (c) Chemical—It includes the chemicals for any kind of changes in the conversion such as hydrolysis, esterification, oxidation, reduction, etc.

(d) Thermochemical—It requires high temperature and pressure, such as in processes like gasification and pyrolysis. It may take place with or without catalytic actions.

2.2 *Significance of Exploring Biorefinery through Microbiological Perspectives*

Understanding biorefinery from a microbiological point of view is crucial for sustainable development, producing renewable energy, and effectively using biomass resources. Microorganisms are essential to the biorefinery process at several stages, from the synthesis of valuable chemicals to the breakdown of biomass (Kumar and Verma 2021). Gaining a thorough grasp of these microbial processes yields valuable insights for boosting biorefinery efficiency and optimizing performance (Zhu et al. 2020). Several salient factors clarify the need of investigating biorefinery from a microbiological perspective.

2.2.1 Biological Catalysts

Through their diverse enzymatic capabilities, microorganisms possess the unique ability to break complex sources of biomass into simpler molecules (Yang et al. 2011). This crucial process plays a key role in transforming various resources such as lignocellulosic biomass, agricultural waste, and organic pollutants into valuable products (Kumar et al. 2008). As one of the most abundant polymers in the world, lignocellulose has garnered immense attention for its potential in biorefinery applications (Lu et al. 2022). This is due to its wide distribution and the fact that it is not suitable for human consumption. By harnessing the metabolic power of microorganisms, the utilization of biomass resources can be achieved with minimal energy input (Aristidou and Penttilä 2000), making it an efficient and sustainable process.

2.2.2 Versatile Metabolic Pathways

Microbial metabolism offers a wide array of pathways for the synthesis of biofuels, biochemicals, and biopolymers from renewable feedstocks (Cowan et al. 2023; Adegboye et al. 2021). By deciphering microbial metabolic networks, researchers can engineer microorganisms to produce target compounds with enhanced yields and specific properties. This metabolic engineering approach enables the customization of biorefinery processes to meet diverse industrial demands, ranging from bioethanol and biodiesel production to the synthesis of chemicals and biodegradable plastics (Hasunuma et al. 2013; Mouterde and Allais 2020).

2.2.3 Environmental Sustainability

Microbially driven biorefinery technologies provide a sustainable substitute for traditional petrochemical-based industries, which are prominent for their unfavourable effects on the environment and reliance on restricted fossil fuel supplies (Singh et al. 2022a). Compared to traditional chemical treatments, microbial biorefinery systems use less energy, produce less contaminants, and have a greater resource efficiency (Moreno-Garcia et al. 2017). Microbial biorefineries have a vital role to play in combating climate change and promoting sustainable development by harnessing renewable biomass sources and reducing greenhouse gas emissions, thus helping to preserve biodiversity (Kumar and Verma 2021).

2.2.4 Technological Innovation

Developments in biological technology, bioinformatics, and microbiology have completely changed how biorefinery processes are designed and optimized. Biological refinery innovations may be rapidly developed and scaled up with the help of the addition of synthetic biology tools, advanced screening techniques, and computational modelling (Kondo et al. 2013). Through the incorporation of predictive models and experimental data, scientists may expedite the shift towards an economy based on biotechnology by forecasting microbial activity, refining process parameters, and creating customized biocatalysts for particular uses (Goswami et al. 2022).

2.3 *Algal Biorefinery*

In the twenty-first century, due to an increase in population growth, the production of food is becoming a major challenge sustainably and also the conversion of biomass into a desirable product. Algal biorefinery is a major area in which one can improve global economic conditions and meet global requirements such as nutrition, feed, fuel, and medicine. Algal biomass is an advanced source which provides high productivity in biomass (Ubando et al. 2020). One of the important biorefinery sustainable sources is algae, which are of two major types: microalgae (Chew et al. 2017) and macroalgae (De La Torre et al. 2019) biomass. Macroalgae are seaweeds or a diverse group of marine organisms which are widely known as chlorophyll-containing microorganisms and generally exist as green (Chlorophyta), brown (Phaeophyta), or red algae (Rhodophyta). The nutritional composition of macroalgae varies greatly. The contents of protein and essential amino acids can be low, especially in brown algal species, and indigestible polysaccharides negatively affect the energy value (Øverland et al. 2018). Fucoxanthin is the pigment that gives Phaeophyta species their brown colour; phycobilins give Rhodophyta species their red colour; and various pigments, including chlorophyll a and b, carotenes, and xanthophylls, give Chlorophyta species their green colour. More and more people throughout the world are becoming aware of macroalgae and macroalgal products as possibly sustainable feed components for terrestrial and monogastric livestock (Wells et al. 2016). Formulated feeds can utilize macroalgae as valuable contributors to protein and bioactive compounds, indirectly integrating these elements into the human food chain. Noteworthy instances involve the elevated protein content derived from green algae (*Ulva* sp.) and red algae (*Porphyra* sp.), surpassing that of brown algae, making them promising alternative protein sources for animal feed. Farm animals benefit from the nutritional richness of macroalgae like brown algae (*Laminaria* sp. and *A. nodosum*), which are rich in essential minerals such as iodine, iron, zinc, potassium, magnesium, calcium, and phosphorus. This makes them a valuable dietary supplement for livestock (Øverland et al. 2018). The implementation of biorefinery approaches can enhance nutrient availability by generating microalgal products. These alternative sources, derived from macroalgae, offer a sustainable means of production.

Mathematical models of algae development can be used to create and run virtual systems simulated in-silico, offering theoretical forecasts of potential yields, which can help design better growing techniques. The time and expense required to conduct these computer-based tests are far less than that of their 'real-world' counterparts. The in-silico models used for research (Kenny and Flynn 2014) are an acclimative mechanistic model that can be adjusted to work in a variety of virtual contexts and capture the dynamics of multinutrient interactions. Although there are models available to simulate specific aspects of algal physiology (Xin et al. 2013), the crucial aspect in this case is how biomass production is related to light and nutrients. Between the utter minimalism of simulations generally utilized in previous investigations (Weyer et al. 2010) and systems biology approaches that are still unable to capture growth dynamics, this calls for an intermediate level of description. A thorough portrayal of changeable element stoichiometry caused by variable regimes of light, temperature, and nutrients should be explored by the features of the model that is utilized. Its efficacy is tested against datasets of different species, covering a range of growth-related parameters under different circumstances, such as biomass and photoacclimation (Flynn et al. 2013).

One of the important biorefinery sustainable sources is microalgae (oxygen-releasing photosynthetic organisms). Microalgae contribute a major part in biorefinery; for example, a microalga, *Spirulina*, contains a high percentage of proteins, iron, vitamins, unsaturated fatty acids, and some secondary metabolites such as phytocyanins and carotenoids. The biomass obtained from *Spirulina* is with high economic viability. The production of biomass as well as secondary metabolites is mainly depending on various factors such as level of salinity, temperature, amount of light intensity, nutrient enrichment, particular geographical location, etc. The rate of consumption exceeds production, with studies indicating that around 80 nations use it as a food source, while fewer than 20 countries are engaged in its production. Species of *Spirulina* include *S. arthrospira*, *S. platensis*, and *S. maxima*, which are commonly used as an edible food as they have a high protein content (Costa et al. 2019). Research has been conducted on bio-products derived from algae-based biorefinery in wastewater. The findings indicate that *Chlorella* and *Scenedesmus* exhibit high suitability for growth in wastewater due to their fast growth and robust resistance to pollutants. Nevertheless, wastewater can also facilitate the growth of additional freshwater microalgae such as *Desmodesmus* and *Cyanobacteria*, along with marine microalgae like *Nannochloropsis* and *Tetraselmis* (Abdelfattah et al. 2023). The production of microalgae biomass for bio-products is influenced by various factors when utilizing wastewater.

2.4 Waste Biorefinery

Several case studies have explored bioprocesses utilizing waste to generate biopolymers and bio-lipids. Additionally, there is research on bioprocesses that integrate wastewater treatment. In the context of resolving global concerns, the circular bioeconomy and waste biorefinery approach can assist in addressing greenhouse gas emissions and carbon management. By lowering greenhouse gas emissions, a waste biorefinery-circular bio-economy approach embodies a low-carbon economy and offers excellent opportunities for a sustainable and environmentally friendly future (Leong et al. 2021). Research has uncovered that the orange peel waste (OPW) generally consists of many value-based products such as biogas and bioethanol (also called bioenergy), essential oils, pectin, polyphenols, and antioxidants, which are also considered marketed compounds and have been derived from the outer layer of the orange, i.e. orange peel.

Additional research findings reveal that the orange peel waste (OPW) sample is rich in fibre, essential oils (specifically, the compound D-limonene), polyphenolic compounds (particularly hesperidin), and pectin. The determination of biogas production relies on the biochemical methane potential (BMP) test. Results indicate that removing the essential oil content from OPW enhances biogas production. This improvement is attributed to the fact that the presence of D-limonene (a compound found in essential oil obtained from orange peel) slows down the rate of biogas production (Ortiz-Sanchez et al. 2021; Mariana et al. 2021; De La Torre et al. 2019). Research has indicated that the extracts from the discarded components of tropical fruits, such as mango leaves and peels, guava seeds and leaves, jatoba bark, tamarind seeds, pomegranate seeds, tomato seeds, grape seeds, etc., are rich in valuable compounds (Villacís-Chiriboga et al. 2021). Several studies have explored the potential applications of maize cob waste (MCW). Maize, being the most widely produced cereal globally, results in a significant surplus of MCW. Typically abandoned in crop fields, MCW's limited biodegradability has a minimal impact on soil fertility. Additionally, it serves as an excellent precursor for producing activated carbons (ACs) and can be employed as a substrate to maintain the C/N ratio in anaerobic co-digestion (AcoD) when combined with other biodegradable substrates (Surra et al. 2019). Astaxanthin, a red carotenoid found naturally, is synthesized by fungi, plants, and microorganisms. This xanthophyll, derived from carotenes through oxidation, accumulates in various fish, birds, and crustaceans as they progress through the food chain. Traditionally sourced from fungi and microalgae, astaxanthin may also be derived from waste in the seafood industry for feed and nutraceutical production, as suggested by cost analysis and reported benefits (Stachowiak and Szulc 2021). Several extraction methods have been explored in the quest for astaxanthin extraction, but achieving targeted extraction may prove impractical or economically unviable. Biorefinery strategies are employed to extract various high-value products from diverse sources.

2.5 *Fungal Biorefinery*

Fungi, especially those belonging to the Zygomycetes and Ascomycetes groups, are of greater economic value in the industrial sector as they produce many important human food, antibiotics, organic acids, and enzymes. Moreover, these fungi possess significant potential to transform waste materials into valuable products, playing a crucial role in the environment. As almost all microorganisms occur in a diverse range, they have the potential to produce high-yielding secondary products, especially microorganisms like filamentous fungi. Research indicates that Ascomycetes, a group of fungi, shows promise in transforming various waste materials into environmentally friendly and valuable products (Ferreira et al. 2016; Ghosh et al. 2023). Research has revealed that fungal biomasses derived from the fungi *Aspergillus flavus* and *Rhizopus stolonifer* promote the sustainable production of biodiesel by catalysing the trans-esterification of triglycerides (Elhussiny et al. 2023). The genome-scale metabolic study was done on the fungus *Trichoderma reesei*, and also *Pichia stipitis* is used in the field of biofuel production such as fermentation of xylose (Gu et al. 2019).

2.6 *Lignocellulosic Biorefinery*

Lignocellulosic biomass is widely distributed worldwide, and it is a plentiful source of biomass. This biomass is utilized to produce bio-based products, including biofuels and biochemicals (Yogalakshmi et al. 2023). Cellulose serves as a substantial source of glucose due to its easy accessibility, renewability, and lack of competition with food resources. A new cycle of research and development is being observed in the hydrolysis of cellulose, wherein this reaction is conducted over solid catalysts and combined with other reactions to achieve increased efficiency (Rinaldi and Schüth 2009). Conventional solvents usually do not dissolve cellulose, and it is highly resistant to both chemical and biological changes (Alexandridis et al. 2018). The breakdown of cellulose through hydrolysis plays a crucial role in the biorefinery process, where carbohydrates are utilized to generate biofuels and biochemicals. While over 1 billion tons of lignocellulose is generated worldwide annually, a mere 3% of the total biomass has been employed in the creation of bio-based products (Baruah et al. 2018).

The adoption of enzymatic hydrolysis for biofuel production from lignocellulosic biomass is anticipated to become a mainstream technology by 2020, as indicated in a 2007 report from the European Union. This is envisioned as a significant stride in the advancement of the bio-economy (Ubando et al. 2020). The development of energetically efficient methods to transform cellulose either directly or indirectly into biofuels and chemicals is essential to the long-term viability of the biorefinery concept.

2.7 Utilization of Agro-Industrial Residues in Biorefinery Processes

Utilizing agro-industrial residues in the biorefinery process is a crucial approach for promoting both sustainable resource management and the development of bio-based products. These biorefineries, modeled after traditional petroleum refineries but utilizing biomass as the raw material, are key players in the shift towards a bio-economy (Casau et al. 2022). With ample amounts of organic materials like carbohydrates, lignocellulose, proteins, and lipids, agro-industrial residues, which encompass leftovers and discarded materials from agriculture and food production, represent a vast and underutilized resource (Sharma et al. 2022). Agro-industrial residues not only address environmental issues related to waste disposal but also open up the possibilities for enhancing value and promoting circular economy practices (Yaashikaa et al. 2022). A major benefit of utilizing these residues is their plentiful availability and diverse range. These residues encompass various types of biomasses like crop residues (such as straw, husks, and stalks), fruit and vegetable peels, distillation leftovers, and dairy processing by-products. By making use of these resources, biorefineries can diversify their sources, decrease reliance on new biomasses, and decrease competition with food production (Singh et al. 2022b).

Integrating agro-industrial residues into biorefinery processes requires cutting-edge technologies and carefully constructed optimization strategies (Wagh et al. 2024). By utilizing a variety of conversion methods such as biochemical, thermochemical, and physicochemical routes, these residues can be transformed into a range of bio-based products such as ethanol, biodiesel, biochemicals, biopolymers, and biofertilizers (Blasi et al. 2023). With the implementation of integrated biorefinery platforms, the residues can undergo cascading valorization, where different components are sequentially extracted and transformed into valuable end products (Jin et al. 2018). This approach maximizes the use of resources and minimizes waste production. The agro-industrial residues come with its fair share of challenges, which mainly centre around three key areas: feedstock variability, logistical hurdles, and techno-economic considerations (Scarlat et al. 2015; Bhatia et al. 2020). This is because the composition and characteristics of residues can differ greatly depending on factors like location, farming methods, and processing methods. As a result, specialized pre-processing and conversion techniques must be used to effectively utilize these residues. Furthermore, the sheer bulkiness of the residues poses challenges for their collection, storage, and transport, making it crucial to have a well-organized supply chain and adequate infrastructure in place (Hiloidhari et al. 2023).

Despite the obstacles that lay ahead, the incorporation of agro-industrial remnants into biorefinery operations presents a wealth of opportunity for promoting sustainability and bolstering rural economies (Hiloidhari et al. 2020; Jain et al. 2022). By uniting the efforts of researchers, businesses, and policymakers, we can pioneer ground-breaking strategies to fully tap into the potential of agro-industrial leftovers, paving the path towards a more environmentally friendly and productive tomorrow (Khot et al. 2018).

2.7.1 Maize Cob Waste (MCW) and its Potential Contributions

The by-product of maize cultivation and processing, known as maize cob waste (MCW), holds great potential in numerous fields (Surra et al. 2019). Despite being frequently disregarded, this agricultural residue boasts significant contributions. In particular, it serves as a valuable resource in the realm of renewable energy production. Utilizing techniques such as anaerobic digestion and pyrolysis, MCW can be transformed into biogas, biofuels, and biochar (Surra et al. 2021). This not only decreases reliance on fossil fuels but also aids in curbing greenhouse gas emissions. MCW represents the discarded husks and cobs left after harvesting the kernels. Despite often being considered agricultural waste, MCW possesses inherent characteristics that render it an ideal candidate for biorefinery processes (Surra et al. 2019). MCW's success is largely due to its strong cellulose, hemicellulose, and lignin combination. The production of sustainable biofuels, biochemicals, and biomaterials depends greatly on these essential components (Hiloidhari et al. 2020). These valuable components have the potential to be efficiently processed into a wide variety of useful goods, reducing our reliance on non-renewable resources and providing a remedy for environmental degradation.

MCW is used as a sustainable feedstock in biorefineries to produce bioethanol, a renewable fuel that may be used separately or combined with gasoline to power automobiles (Bakala et al. 2023). The polysaccharides included in MCW may be broken down into fermentable sugars by enzymatic hydrolysis as well as fermentation processes. Microbial organisms then use these sugars to produce ethanol. The bioethanol exhibits potential as a more environmental friendly substitute for fossil fuels, including less emissions of greenhouse gases and improved energy stability (Baeyens et al. 2015). MCW shows promise for the synthesis of biochemicals, including basic chemicals, acidic compounds, and biopolymers. MCW may be valorized into a wide range of valuable chemicals by utilizing modern conversion technologies, which include thermochemical and biochemical methods. This may promote the growth of a bio-based economy. These biochemicals are used in a wide range of sectors, like agriculture, cosmetics, and medicines, boosting sustainable practices and reducing dependency on goods generated from petrochemicals (Yu et al. 2021; Awasthi et al. 2023).

2.7.2 Orange Peel Waste (OPW) as a Valuable Biomass Source

Orange peel waste (OPW) is a significant and essentially untapped source of biomass that has immense potential for use in biorefinery processes (Mariana et al. 2021). The enormous amounts of leftover orange peels following the process of juice extraction and utilization offer a strong chance for resource conservation. Rich in fibres, carbohydrates, and essential oils, OPW is a versatile resource that may be used for a wide range of tasks. OPW is a valuable feedstock in the context of biorefineries, providing pathways for bioconversion into high-value goods via effective and eco-friendly processes (Mariana et al. 2021; Ortiz-Sanchez et al. 2021). Because OPW is mostly composed of cellulose, hemicellulose, and pectin, it may be converted using both biochemical and thermochemical processes (De La Torre et al. 2019; Ortiz-Sanchez et al. 2020). The OPW carbohydrate fraction may be hydrolysed enzymatically to provide fermentable sugars, which will aid in the synthesis of bioethanol, a sustainable biofuel. Moreover, the process of extracting essential oils from orange peels offers prospects for the creation of bio-based compounds and flavours, supporting the environmentally friendly manufacturing of food additives, perfumes, and medications (Mohsin et al. 2022).

OPW shows potential as a precursor to high-value by-products such as dietary fibres, organic acids, and antioxidants (Teigiserova et al. 2019). Using OPW in biorefinery operations helps to build a circular economy model in which waste materials are converted into highly valuable goods while also reducing environmental degradation linked to waste disposal. Additionally, by lowering reliance on fossil fuels and limiting emissions of greenhouse gases, the use of OPW in biorefinery processes is consistent with the principles of sustainability (Mariana et al. 2021).

2.8 Sustainable Approaches to Utilizing Agricultural By-products

Circular economy and resource efficiency concepts are in line with sustainable methods of using agricultural by-products in biorefineries (Igbokwe et al. 2022), which constitute an important part of current industrial and agricultural operations. Maximizing value extraction from resources such as biomass and reducing waste and environmental effects are the goals of biorefinery processes. Agricultural by-products including straw, husk, stalk, bagasse, and residue are used as feedstocks for biorefinery operations in these sustainable methods (Dhanya 2022). These by-products are abundant in organic materials, such as lignin, cellulose, and hemicellulose, which may be processed into biofuels, biochemicals, biopolymers, and bio-based materials, among other useful goods (Hiloidhari et al. 2020).

A key approach in sustainable biorefinery methods is to use integrated processes to optimize the use of biomass components (Budzianowski and Postawa 2016). For instance, fractionation procedures may be used to separate the constituent elements of biomass, which can subsequently be independently processed into various products. This strategy minimizes waste production while maximizing the value extracted from the biomass (Aristizábal-Marulanda and Cardona Alzate 2019). Sustainable biorefinery methods prioritize the use of energy- and ecologically-conscious processes. Various technologies are utilized to transform agricultural by-products into useful bio-based products, including gasification, pyrolysis, fermentation, and enzymatic hydrolysis (Saleem 2022). These procedures are designed to use the least amount of energy, water, and greenhouse gas emissions possible, which lessens the environmental impact of biorefinery operations. The establishment of closed-loop systems, in which waste streams and by-products are recycled and utilized either inside the biorefinery or in other industrial processes, is a crucial component of sustainable biorefinery techniques (Ahmed et al. 2023). The biorefinery's sustainability and overall efficiency may be significantly improved by putting these circular economy concepts into practice.

2.9 In-silico Approaches Towards Biorefinery

Utilizing an in-silico methodology in biorefinery explores the generation of diverse strain characteristics, nutrient supply levels, and more. In contrast to first-generation biorefineries, the second-generation focuses on the production of valuable, environmentally friendly bio-based products through advanced methods. The construction of mechanistic models based on in-silico approaches is a common practice in second-generation biorefineries. Numerous research studies have been undertaken on *E. coli* to harness its innate capabilities as a cellular factory for the synthesis of advantageous bio-based compounds and materials (Kenny and Flynn 2014). Different computational tools are employed to enhance efficiency and precision in experimentation, significantly reducing time requirements. Glycerol, a compound known for its substantial biomass yield, serves as a valuable carbon source in the biodiesel industry (Almeida et al. 2023). *Caldicellulosiruptor saccharolyticus* and *Acidothermus cellulolyticus* play crucial roles in the process of biomass biodegradation. *Zymomonas mobilis* is employed for biofuel production, while *Streptomyces coelicolor* and *Mycobacterium tuberculosis* are recognized as novel species utilized in the study of 'genome-scale' metabolic networks (GSMN). Recent research involved sequencing the genome of the bacterium *Clostridium thermocellum* (Gu et al. 2019). In-silico models can examine the effects of the substrate present, which makes in-vivo experiments easier. Other studies reveal that fungi are considered for the production of a high amount of cellulase enzyme that acts as a biocatalyst and is used to hydrolyse the cellulosic biomass. The substrate molecular binding affinity is to be done by the process of molecular docking which is an in-silico approach (Rabby et al. 2022).

Research has been conducted on utilizing agro-industrial waste from bananas to generate bio-based products, including various bioplastics like PHBs and PLA, as well as nanocellulose fibres (Redondo-Gómez et al. 2020). The in-silico research serves as a valuable tool for assessing strategies and designs aimed at enhancing the efficiency of a bioprocess. It also plays a crucial role in streamlining the intricate physicochemical and biological processes inherent in biorefinery operations. One prominent strategy involves utilizing the in-silico approach. This method is commonly utilized, incorporating in-silico techniques such as molecular docking, multiple sequence alignment, and molecular dynamics simulation. These approaches are increasingly employed in conjunction with the expansion of biological databases and improved computational capabilities (Zhang et al. 2020). By using molecular docking, this method can be utilized to simulate drug discovery and assess its bioactivity. As a program's initial technical effort, the in-silico approach may help government initiatives to actualize eco-friendly agriculture. In-silico approaches include many processes like drug-likeness study, ADME-TOX screening, QSAR, simulation, etc. It is a process with cost-efficient and time-saving technique. In order to stay cost-competitive, a biorefinery of the future must target both high- and low-value applications, valorize all components of biomass, and use abundant and inexpensive waste streams as raw materials. The methods play a vital role in advancing the exploration of catalytic processes. They make substantial contributions to researching novel enzymes; analysing substrates, products, or inhibitors; and developing effective techniques for immobilizing or deactivating food enzymes within the realm of food bioprocessing (Zhang et al. 2020).

3 Conclusions

The contemporary concepts of both in-silico approaches and biorefinery processes are noteworthy. Biorefinery involves the conversion of biomass derived from diverse sources, including plant derivatives, microorganisms, etc. microalgae, particularly *Spirulina*, are pivotal in biorefineries due to their rich content of proteins, iron, vitamins, and unsaturated fatty acids, as well as secondary metabolites such as carotenoids and phytocyanins. Various agro-industrial residues like orange peel waste (OPW), maize cob waste (MCW), banana waste, and fungi such as zygomycetes and ascomycetes contribute significantly to the biorefinery process. In addition to fungi, bacteria like lactic acid bacteria (LAB) and extremophiles, along with glucose extracts like cellulose, play vital roles in biorefinery processes. The formation of elements like nitrogen and carbon from biomass is crucial for the economically feasible production of algal-based products. Microalgae, through in-silico optimization, can generate biomass and biofuel feedstocks. Numerous intermediate products, including carbohydrates, proteins, and fats, are produced through various pathways (biochemical or thermochemical), facilitating their transformation into bio-based products that cater to human needs. While inexpensive agricultural wastes, such as cellulose from substrates, are essential, challenges related to production costs and the development of new microbial strains remain significant obstacles in advancing the biorefinery process.

4 Future Prospects: In-silico Investigations in Biorefinery

The combination of computational models with biorefinery processes marks a transformative era in biomass resource utilization. This innovative approach, merging in-silico methodologies with biorefinery techniques, is set to revolutionize natural resource use in diverse applications, aligning with environmental goals and offering economic benefits. Microorganisms, including extremophiles and microalgae, play key roles in biorefinery, offering unique metabolic pathways and enzymes for industrial settings. The repurposing of agro-industrial residues like orange peel waste and maize cob waste illustrates biorefinery's potential in converting waste into valuable products. In-silico models optimize conversion processes, enhancing the generation of biogas, bioethanol, essential oils, pectin, polyphenols, and antioxidants, promoting the advancement of a circular bio-economy and waste biorefinery strategy to address global challenges. In-silico models prove essential across biorefinery facets, reducing time and experimental costs while increasing accuracy. They aid in designing microbial strains, assessing biochemical processes, and facilitating drug discovery and bioactivity studies. As the biorefinery landscape advances, in silico approaches become imperative for sustainability and cost competitiveness, optimizing both high- and low-value applications and utilizing abundant waste streams. In-silico models play a crucial role in the examination of catalytic processes, enzyme identification, and the development of efficient methods for food bioprocessing.

References

Abdelfattah A, Ali SS, Ramadan H, El-Aswar EI, Eltawab R, Ho S, Elsamahy T, Li S, El-Sheekh MM, Schagerl M, Kornaros M, Sun J (2023) Microalgae-based wastewater treatment: mechanisms, challenges, recent advances, and future prospects. Environ Sci Ecotechnol 13:100205. https://doi.org/10.1016/j.ese.2022.100205

Adegboye MF, Ojuederie OB, Talia PM, Babalola OO (2021) Bioprospecting of microbial strains for biofuel production: metabolic engineering, applications, and challenges. Biotechnol Biofuels 14(1):1–21

Adetunji AI, Oberholster PJ, Erasmus M (2023) From garbage to treasure: a review on biorefinery of organic solid wastes into valuable biobased products. Bioresour Technol Rep 24:101610. https://doi.org/10.1016/j.biteb.2023.101610

Agrawal K, Verma P (2020) Production optimization of yellow laccase from Stropharia sp. ITCC 8422 and enzyme-mediated depolymerization and hydrolysis of lignocellulosic biomass for biorefinery application. Biomass Convers Biorefinery:1–20. https://doi.org/10.1007/s13399-020-00869-w

Ahmed SF, Kabir M, Mehjabin A, Oishi FTZ, Ahmed S, Mannan S et al (2023) Waste biorefinery to produce renewable energy: bioconversion process and circular bioeconomy. Energy Rep 10:3073–3091

Alexandridis P, Ghasemi M, Furlani EP, Tsianou M (2018) Solvent processing of cellulose for effective bioresource utilization. Curr Opin Green Sustain Chem 14:40–52. https://doi.org/10.1016/j.cogsc.2018.05.008

Almeida EL, Olivo JE, Andrade CMG (2023) Production of biofuels from glycerol from the biodiesel production process—a brief review. Fermentation 9(10):869. https://doi.org/10.3390/fermentation9100869

Aristidou A, Penttilä M (2000) Metabolic engineering applications to renewable resource utilization. Curr Opin Biotechnol 11(2):187–198

Aristizábal-Marulanda V, Cardona Alzate CA (2019) Methods for designing and assessing biorefineries. Biofuels, Bioproducts and Biorefining 13(3):789–808

Asghar A, Afzal MT, Haider R, Ahmad MS, Mehmood MA (2022) Scope and characteristics of the biomass sources suitable for biorefinery applications. In: Clean energy production technologies, pp 1–12. https://doi.org/10.1007/978-981-19-4312-6_1

Awasthi MK, Sar T, Gowd SC, Rajendran K, Kumar V, Sarsaiya S et al (2023) A comprehensive review on thermochemical, and biochemical conversion methods of lignocellulosic biomass into valuable end product. Fuel 342:127790

Baeyens J, Kang Q, Appels L, Dewil R, Lv Y, Tan T (2015) Challenges and opportunities in improving the production of bio-ethanol. Prog Energy Combust Sci 47:60–88

Bakala HS, Devi J, Ankita, Sarao LK, Kaur S (2023) Utilization of wheat and maize waste as biofuel source. In: Agroindustrial waste for green fuel application. Springer Nature Singapore, Singapore, pp 27–66

Baruah J, Nath B, Sharma R, Kumar S, Deka RC, Baruah D, Kalita E (2018) Recent trends in the pretreatment of lignocellulosic biomass for value-added products. Front Energy Res 6. https://doi.org/10.3389/fenrg.2018.00141

Bhardwaj N, Kumar B, Agarwal K, Chaturvedi V, Verma P (2019) Purification and characterization of a thermo-acid/alkali stable xylanases from Aspergillus oryzae LC1 and its application in Xylo-oligosaccharides production from lignocellulosic agricultural wastes. Int J Biol Macromol 122:1191–1202. https://doi.org/10.1016/j.ijbiomac.2018.09.070

Bhatia RK, Ramadoss G, Jain AK, Dhiman RK, Bhatia SK, Bhatt AK (2020) Conversion of waste biomass into gaseous fuel: present status and challenges in India. Bioenergy Res 13:1046–1068

Blasi A, Verardi A, Lopresto CG, Siciliano S, Sangiorgio P (2023) Lignocellulosic agricultural waste valorization to obtain valuable products: an overview. Recycling 8(4):61

Borowitzka MA (2013) High-value products from microalgae—their development and commercialisation. J Appl Phycol 25(1):743–756

Budzianowski WM, Postawa K (2016) Total chain integration of sustainable biorefinery systems. Appl Energy 184:1432–1446

Casau M, Dias MF, Matias JC, Nunes LJ (2022) Residual biomass: a comprehensive review on the importance, uses and potential in a circular bioeconomy approach. Resources 11(4):35

Chew KW, Yap JY, Show PL, Suan NH, Juan JC, Ling TC, Lee DJ, Chang JS (2017) Microalgae biorefinery: high value products perspectives. Bioresour Technol 229:53–62

Costa JAV, De Freitas BCB, Da Rosa GM, Moraes L, De Morais MG, Mitchell BG (2019) Operational and economic aspects of spirulina-based biorefinery. Bioresour Technol 292:121946. https://doi.org/10.1016/j.biortech.2019.121946

Cowan AE, Klass SH, Winegar PH, Keasling JD (2023) Microbial production of fuels, commodity chemicals, and materials from sustainable sources of carbon and energy. Curr Opin Syst Biol 100482

De La Torre I, Martín-Domínguez V, Acedos MG, Esteban J, Santos VE, Ladero M (2019) Utilisation/upgrading of orange peel waste from a biological biorefinery perspective. Appl Microbiol Biotechnol 103(15):5975–5991. https://doi.org/10.1007/s00253-019-09929-2

Dhanya MS (2022) Perspectives of agro-waste biorefineries for sustainable biofuels. In: Zero waste biorefinery. Springer Nature Singapore, Singapore, pp 207–232

Elhussiny NI, Mohamed AH, El-Refai HA, Mohamed SS, Shetaia YM, Amin HA, Klöck G (2023) Biocatalysis of triglycerides transesterification using fungal biomass: a biorefinery approach. Fungal Biol Biotechnol 10(1). https://doi.org/10.1186/s40694-023-00160-3

Fernando S, Adhikari S, Chandrapal C, Murali N (2006) Biorefineries: current status, challenges, and future direction. Energy Fuels 20(4):1727–1737. https://doi.org/10.1021/ef060097w

Ferreira JA, Mahboubi A, Lennartsson PR, Taherzadeh MJ (2016) Waste biorefineries using filamentous ascomycetes fungi: present status and future prospects. Bioresour Technol 215:334–345. https://doi.org/10.1016/j.biortech.2016.03.018
Flynn KJ, Mitra A, Greenwell HC, Sui J (2013) Monster potential meets potential monster: pros and cons of deploying genetically modified microalgae for biofuels production. Interface Focus 3(1):20120037. https://doi.org/10.1098/rsfs.2012.0037
Ghosh S, Rusyn IB, Dmytruk OV, Dmytruk KV, Onyeaka H, Gryzenhout M, Gafforov Y (2023) Filamentous fungi for sustainable remediation of pharmaceutical compounds, heavy metal and oil hydrocarbons. Front Bioeng Biotechnol 11. https://doi.org/10.3389/fbioe.2023.1106973
Goswami L, Kayalvizhi R, Dikshit PK, Sherpa KC, Roy S, Kushwaha A et al (2022) A critical review on prospects of bio-refinery products from second and third generation biomasses. Chem Eng J 448:137677
Greenwell HC, Laurens LM, Shields RJ, Lovitt RW, Flynn KJ (2010) Placing microalgae on the biofuels priority list: a review of the technological challenges. J R Soc Interface 7(46):703–726. https://doi.org/10.1098/rsif.2009.0322
Gu C, Kim GB, Kim WJ, Kim HU, Lee SY (2019) Current status and applications of genome-scale metabolic models. Genome Biol 20(1). https://doi.org/10.1186/s13059-019-1730-3
Hasunuma T, Okazaki F, Okai N, Hara KY, Ishii J, Kondo A (2013) A review of enzymes and microbes for lignocellulosic biorefinery and the possibility of their application to consolidated bioprocessing technology. Bioresour Technol 135:513–522
Haydary J, Šuhaj P, Husár J (2020) Waste biomass to methanol–optimisation of gasification agent to feed ratio. Biomass Convers Biorefinery 11(2):419–428. https://doi.org/10.1007/s13399-020-00692-3
Hiloidhari M, Bhuyan N, Gogoi N, Seth D, Garg A, Singh A et al (2020) Agroindustry wastes: biofuels and biomaterials feedstocks for sustainable rural development. In: Refining biomass residues for sustainable energy and bioproducts. Academic Press, pp 357–388
Hiloidhari M, Sharno MA, Baruah DC, Bezbaruah AN (2023) Green and sustainable biomass supply chain for environmental, social and economic benefits. Biomass Bioenergy 175:106893
Igbokwe VC, Ezugworie FN, Onwosi CO, Aliyu GO, Obi CJ (2022) Biochemical biorefinery: a low-cost and non-waste concept for promoting sustainable circular bioeconomy. J Environ Manag 305:114333
Jain A, Sarsaiya S, Awasthi MK, Singh R, Rajput R, Mishra UC et al (2022) Bioenergy and bio-products from bio-waste and its associated modern circular economy: current research trends, challenges, and future outlooks. Fuel 307:121859
Jin Q, Yang L, Poe N, Huang H (2018) Integrated processing of plant-derived waste to produce value-added products based on the biorefinery concept. Trends Food Sci Technol 74:119–131
Karmee SK, Lin CSK (2014) Valorisation of food waste to biofuel: current trends and technological challenges. Sustain Chem Process 2(1). https://doi.org/10.1186/s40508-014-0022-1
Kenny P, Flynn KJ (2014) In silico optimization for production of biomass and biofuel feedstocks from microalgae. J Appl Phycol 27(1):33–48. https://doi.org/10.1007/s10811-014-0342-2
Kholila BN, Soesilo TEB, Herdiansyah H (2021) The potential of in-silico approach to support government programs to achieve eco-friendly agriculture: studies on urban agriculture. IOP Conf Ser. https://doi.org/10.1088/1755-1315/716/1/012007
Khot M, Katre G, Zinjarde S, RaviKumar A (2018) Single cell oils (SCOs) of oleaginous filamentous fungi as a renewable feedstock: a biodiesel biorefinery approach. Fungal Biorefineries:145–183
Kondo A, Ishii J, Hara KY, Hasunuma T, Matsuda F (2013) Development of microbial cell factories for bio-refinery through synthetic bioengineering. J Biotechnol 163(2):204–216
Kumar B, Verma P (2021) Biomass-based biorefineries: an important architype towards a circular economy. Fuel 288:119622
Kumar R, Singh S, Singh OV (2008) Bioconversion of lignocellulosic biomass: biochemical and molecular perspectives. J Ind Microbiol Biotechnol 35(5):377–391

Kumar B, Bhardwaj N, Agrawal K, Chaturvedi V, Verma P (2020) Current perspective on pretreatment technologies using lignocellulosic biomass: an emerging biorefinery concept. Fuel Process Technol 199:106244. https://doi.org/10.1016/j.fuproc.2019.106244

Leong HY, Chang CK, Khoo KS, Chew KW, Chia SR, Lim JW, Chang JS, Show PL (2021) Waste biorefinery towards a sustainable circular bioeconomy: a solution to global issues. Biotechnol Biofuels 14(1). https://doi.org/10.1186/s13068-021-01939-5

Logan M, Visvanathan C (2019) Management strategies for anaerobic digestate of organic fraction of municipal solid waste: current status and future prospects. Waste Manag Res 37(1_suppl):27–39. https://doi.org/10.1177/0734242x18816793

Lu H, Yadav V, Bilal M, Iqbal HM (2022) Bioprospecting microbial hosts to valorize lignocellulose biomass–environmental perspectives and value-added bioproducts. Chemosphere 288:132574

Mariana OS, Camilo STJ, Ariel CAC (2021) A comprehensive approach for biorefineries design based on experimental data, conceptual and optimization methodologies: the orange peel waste case. Bioresour Technol 325:124682

Mariano AP, Dias M, Junqueira TL, Da Cunha MP, Bonomi A, Filho RM (2013) Butanol production in a first-generation Brazilian sugarcane biorefinery: technical aspects and economics of greenfield projects. Bioresour Technol 135:316–323. https://doi.org/10.1016/j.biortech.2012.09.109

Mohsin A, Hussain MH, Zaman WQ, Mohsin MZ, Zhang J, Liu Z et al (2022) Advances in sustainable approaches utilizing orange peel waste to produce highly value-added bioproducts. Crit Rev Biotechnol 42(8):1284–1303

Moreno-Garcia L, Adjallé K, Barnabé S, Raghavan GSV (2017) Microalgae biomass production for a biorefinery system: recent advances and the way towards sustainability. Renew Sust Energ Rev 76:493–506

Moretto G, Russo I, Bolzonella D, Pavan P, Majone M, Valentino F (2020) An urban biorefinery for food waste and biological sludge conversion into polyhydroxyalkanoates and biogas. Water Res 170:115371. https://doi.org/10.1016/j.watres.2019.115371

Mouterde LM, Allais F (2020) Bio-based production of chemicals through metabolic engineering. In: Biobased products and industries. Elsevier, pp 171–202

Nair LG, Agrawal K, Verma P (2022) An insight into the principles of lignocellulosic biomass-based zero-waste biorefineries: a green leap towards imperishable energy-based future. Biotechnol Genet Eng Rev 38(2):288–338. https://doi.org/10.1080/02648725.2022.2082223

Nicolopoulou-Stamati P, Maipas S, Kotampasi C, Stamatis PJ, Hens L (2016) Chemical pesticides and human health: the urgent need for a new concept in agriculture. Front Public Health 4. https://doi.org/10.3389/fpubh.2016.00148

Ortiz-Sanchez M, Solarte-Toro JC, González-Aguirre JA, Peltonen KE, Richard P, Alzate CAC (2020) Pre-feasibility analysis of the production of mucic acid from orange peel waste under the biorefinery concept. Biochem Eng J 161:107680

Mariana OS, Camilo ST, Ariel CA (2021) A comprehensive approach for biorefineries design based on experimental data, conceptual and optimization methodologies: the orange peel waste case. Bioresour Technol 325:124682. https://doi.org/10.1016/j.biortech.2021.124682

Ortiz-Sanchez M, Solarte-Toro JC, Orrego-Alzate CE, Acosta-Medina CD, Cardona-Alzate CA (2021) Integral use of orange peel waste through the biorefinery concept: an experimental, technical, energy, and economic assessment. Biomass Convers Biorefinery 11:645–659

Øverland M, Mydland LT, Skrede A (2018) Marine macroalgae as sources of protein and bioactive compounds in feed for monogastric animals. J Sci Food Agric 99(1):13–24. https://doi.org/10.1002/jsfa.9143

Perera IA, Abinandan S, Subashchandrabose SR, Venkateswarlu K, Naidu R, Megharaj M (2019) Advances in the technologies for studying consortia of bacteria and cyanobacteria/microalgae in wastewaters. Crit Rev Biotechnol 39(5):709–731. https://doi.org/10.1080/07388551.2019.1597828

Rabby MRI, Ahmed ZB, Paul GK, Chowdhury NN, Akter F, Razu MH, Karmaker P, Khan M (2022) A combined study on optimization, in silico modeling, and genetic modifica-

tion of large scale microbial cellulase production. Biochem Res Int 2022:1–14. https://doi.org/10.1155/2022/4598937
Redondo-Gómez C, Quesada MR, Astúa SV, Zamora JPM, Lopretti M, Vega-Baudrit J (2020) Biorefinery of biomass of agro-industrial banana waste to obtain high-value biopolymers. Molecules 25(17):3829. https://doi.org/10.3390/molecules25173829
Ren H, Kong F, Zhao L, Ren N, Ma J, Nan J, Liu B (2019) Enhanced co-production of biohydrogen and algal lipids from agricultural biomass residues in long-term operation. Bioresour Technol 289:121774. https://doi.org/10.1016/j.biortech.2019.121774
Rinaldi R, Schüth F (2009) Acid hydrolysis of cellulose as the entry point into biorefinery schemes. ChemSusChem 2(12):1096–1107. https://doi.org/10.1002/cssc.200900188
Rodríguez-Alegre R, Zapata-Jiménez J, You X, Pérez-Moya M, Sanchis S, García-Montaño J (2023) Nutrient recovery and valorisation from pig slurry liquid fraction with membrane technologies. Sci Total Environ 874:162548. https://doi.org/10.1016/j.scitotenv.2023.162548
Sakuntala M, Kim EJ (2022) Recent advances and challenges in the bioconversion of acetate to value-added chemicals. Bioresour Technol 364:128064. https://doi.org/10.1016/j.biortech.2022.128064
Saleem M (2022) Possibility of utilizing agriculture biomass as a renewable and sustainable future energy source. Heliyon 8:(2)e08905. https://doi.org/10.1016/j.heliyon.2022.e08905
Scarlat N, Dallemand JF, Monforti-Ferrario F, Nita V (2015) The role of biomass and bioenergy in a future bioeconomy: policies and facts. Environ Dev 15:3–34
Scott SA, Davey MP, Dennis JS, Horst I, Howe CJ, Lea-Smith DJ, Smith AG (2010) Biodiesel from algae: challenges and prospects. Curr Opin Biotechnol 21(3):277–286. https://doi.org/10.1016/j.copbio.2010.03.005
Shah PN, Acharige MJT, Kim N, Ryan DK, DeSisto WJ, Lee Y (2020) Green bisphenol A: a high valued building block isolated from lignin biowaste. Waste Biomass Valorization. https://doi.org/10.1007/s12649-020-01032-2
Shaker B, Ahmad S, Lee J, Jung C, Na D (2021) In silico methods and tools for drug discovery. Comput Biol Med 137:104851. https://doi.org/10.1016/j.compbiomed.2021.104851
Sharma V, Tsai ML, Nargotra P, Chen CW, Kuo CH, Sun PP, Dong CD (2022) Agro-industrial food waste as a low-cost substrate for sustainable production of industrial enzymes: a critical review. Catalysts 12(11):1373
Singh N, Singhania RR, Nigam PS, Dong CD, Patel AK, Puri M (2022a) Global status of lignocellulosic biorefinery: challenges and perspectives. Bioresour Technol 344:126415
Singh N, Ogunseitan OA, Wong MH, Tang Y (2022b) Sustainable materials alternative to petrochemical plastics pollution: a review analysis. Sustainable Horizons 2:100016
Stachowiak B, Szulc P (2021) Astaxanthin for the food industry. Molecules 26(9):2666. https://doi.org/10.3390/molecules26092666
Su C-H, Chien L-J, Gomes J, Lin Y-S, Yu Y-K, Liou J-S, Syu R-J (2011) Factors affecting lipid accumulation by *Nannochloropsis oculata* in a two-stage cultivation process. J Appl Phycol 23(1):903–908
Surra E, Bernardo M, Lapa N, Esteves IAAC, Fonseca I, Mota JPB (2019) Biomethane production through anaerobic co-digestion with Maize Cob Waste based on a biorefinery concept: a review. J Environ Manag 249:109351. https://doi.org/10.1016/j.jenvman.2019.109351
Surra E, Esteves IA, Lapa N (2021) Life cycle analysis of a biorefinery for activated carbon and biomethane production. Biomass Bioenergy 149:106080
Teigiserova DA, Hamelin L, Thomsen M (2019) Review of high-value food waste and food residues biorefineries with focus on unavoidable wastes from processing. Resour Conserv Recycl 149:413–426
Ubando AT, Felix CB, Chen WH (2020) Biorefineries in circular bioeconomy: a comprehensive review. Bioresour Technol 299:122585. https://doi.org/10.1016/j.biortech.2019.122585
Villacís-Chiriboga J, Vera E, Van Camp J, Ruales J, Elst K (2021) Valorization of byproducts from tropical fruits: a review, part 2: applications, economic, and environmental aspects of biorefin-

ery via supercritical fluid extraction. Compr Rev Food Sci Food Saf 20(3):2305–2331. https://doi.org/10.1111/1541-4337.12744

Wagh MS, Sowjanya S, Nath PC, Chakraborty A, Amrit R, Mishra B, … Mohanta YK (2024) Valorisation of agro-industrial wastes: circular bioeconomy and biorefinery process–a sustainable symphony. Process Safety and Environmental Protection

Wells ML, Potin P, Craigie JS, Raven JA, Merchant S, Helliwell KE, Smith AG, Camire ME, Brawley SH (2016) Algae as nutritional and functional food sources: revisiting our understanding. J Appl Phycol 29(2):949–982. https://doi.org/10.1007/s10811-016-0974-5

Weyer K, Bush D, Darzins A, Willson B (2010) Theoretical maximum algal oil production. Bioenergy Res 3:204–213

Xin CP, Yang J, Zhu XG (2013) A model of chlorophyll a fluorescence induction kinetics with explicit description of structural constraints of individual photosystem II units. Photosynth Res 117(1–3):339–354. https://doi.org/10.1007/s11120-013-9894-2

Yaashikaa PR, Kumar PS, Varjani S (2022) Valorization of agro-industrial wastes for biorefinery process and circular bioeconomy: a critical review. Bioresour Technol 343:126126

Yang B, Dai Z, Ding SY, Wyman CE (2011) Enzymatic hydrolysis of cellulosic biomass. Biofuels 2(4):421–449

Yogalakshmi KN, Mohamed Usman TM, Kavitha S, Sachdeva S, Thakur S, Adish Kumar S, Rajesh Banu J (2023) Lignocellulosic biorefinery technologies: a perception into recent advances in biomass fractionation, biorefineries, economic hurdles and market outlook. Fermentation 9(3):238. https://doi.org/10.3390/fermentation9030238

Yu IK, Chen H, Abeln F, Auta H, Fan J, Budarin VL et al (2021) Chemicals from lignocellulosic biomass: a critical comparison between biochemical, microwave and thermochemical conversion methods. Crit Rev Environ Sci Technol 51(14):1479–1532

Zhang Y, Aryee ANA, Simpson BK (2020) Current role of in silico approaches for food enzymes. Curr Opin Food Sci 31:63–70. https://doi.org/10.1016/j.cofs.2019.11.003

Zhu D, Adebisi WA, Ahmad F, Sethupathy S, Danso B, Sun J (2020) Recent development of extremophilic bacteria and their application in biorefinery. Front Bioeng Biotechnol 8. https://doi.org/10.3389/fbioe.2020.00483

Conceptual to Technological Reality: Biorefineries Using Techno-Economic and Life Cycle Assessment Models

Shreya Sharma, Pradeep Verma, and Komal Agrawal

1 Introduction

The concept of biorefinery was proposed by Cherubini (2010) as a facility to overcome the concerns of depleting energy resources and increasing emission of greenhouse gases and maximizing the utility of biomass to produce eco-friendly products like lactic acid, citric acid (Laltha et al. 2022), biohydrogen (Woo et al. 2022; Goswami et al. 2021), bioethanol (Valladares-Diestra et al. 2021; Bhardwaj et al. 2020), biodiesel (Degfie et al. 2019; Goswami et al. 2022), and butanol (Tsai et al. 2020). In addition, it is also utilized for the production of enzymes like cellulase (Agrawal et al. 2023; Bhardwaj et al. 2021; Kumar et al. 2018; Pramanik et al. 2021), laccase (Agrawal and Verma 2020; Atilano-Camino et al. 2020; Agrawal et al. 2019), and xylanase (Bakry et al. 2022; Bhardwaj et al. 2019; Bhardwaj et al. 2017) for application in various industries. It is a collective network of thermochemical, biochemical, microbial, and combustion conversion technologies integrated to proficiently manufacture sustainable high-value biobased products from various sources of biomass and waste.

In the case of thermochemical methods, its various types are direct combustion (burning the biomass directly, 1000–2000 °C), torrefaction (heating it without

S. Sharma · K. Agrawal (✉)
Department of Microbiology, School of Bio Engineering and Biosciences, Lovely Professional University, Phagwara, Punjab, India

P. Verma
Bioprocess and Bioenergy Laboratory (BPEL), Department of Microbiology, Central University of Rajasthan, Kishangarh, Ajmer, Rajasthan, India
e-mail: pradeepverma@curaj.ac.in

K. Agrawal, P. Verma (eds.), *Biotechnological Advances in Biorefinery*, Interdisciplinary Biotechnological Advances,
https://doi.org/10.1007/978-981-97-5544-8_15

oxygen, <300 °C), hydrothermal liquefaction (heating it with water and high pressure, 250–500 °C and pressure of 5–35 MPa), pyrolysis (heating without O_2 to convert into gas, liquid, and solid), and gasification (heating it with small amount of O_2 to make syngas, 800–1300 °C). On the other hand, biochemical methods use biological agents (e.g. enzymes, microorganisms) to convert biomass into liquid fuels that can be alternative to fossil fuels (Osman et al. 2021).

The sustainable development goals of 2023 proposed by the United Nations Department of Economic and Social Affairs focuses on various topics from gender equality to climate change with many goals concentrating on promoting sustainable agriculture (goal 2), sustainable energy (goal 7), and industrialization (goal 9). These goals along with revised strategies like the European Commission's bioeconomy strategy (2018) have opened the gates for increasing the numbers of biorefinery projects that focus on valorizing lignin, cellulose, and hemicellulose to produce value-added products. The generation of bioenergy from waste materials can be accomplished by adopting the concept of biorefinery. Wastes are one of the most important parameters for the sustainable bioconversion of biomass to bio-oils, biochar, single-cell protein, and gases because of their excess abundance globally. For a circular economy, this conversion of biomass into value-added products is an excellent example of carbon source utilization (Vuppaladadiyam et al. 2023). Wastes like agriculture, domestic, industrial, and body excreta are the most significant resources for primary energy (55% of this energy is used domestically) (Chan et al. 2019). Food waste is composed of nutrients like proteins, carbohydrates, fats, ions, etc. which function as a great resource to produce green energy (Elgarahy et al. 2023). To enhance the effectiveness and success of the transformation of waste into bioenergy, researchers have conducted waste characterization, as demonstrated by Skaggs et al. (2018). In the previously mentioned work, a detailed database was developed for feedstocks in the United States. This included an analysis of municipal wastewater sludge, food processing waste, animal manure, fats, oils, and grease. The waste-to-energy feedstocks were initially characterized by their general overview and production potential. This was followed by an assessment using hydrothermal liquefaction (assumed conversion techniques for the biomass) to produce biocrude oil, and spatial distribution of feedstock was also carried out.

Lignocellulosic biomass is one of the most abundant feedstocks available and is a great alternative to first-generation food crops that are land-requiring and are essential in times when starvation is still prevailing. Out of the 1.3 billion tons per year of lignocellulosic biomass (LCB) produced, more than 95% remains unused for biorefinery purposes (Baruah et al. 2018).

Algal biomass can be classified into two categories, macroalgae (Dharshini et al. 2020) and microalgae (Goswami et al. 2020), and is a promising third-generation biomass feedstock. Algal biomass offers advantages such as reduced land usage and greater biomass productivity and yield when compared to first- and second-generation biomass. All of these are great carbon sources for the manufacture of biofuels (Goswami et al. 2022), biochemicals, etc. (Table 1). A review by Goswami et al. (2022) focused on microalgae biorefinery where a comparative analysis, conceptual scenario, and integrated mechanism for biofuel production were studied to

Table 1 Insight into different feedstock and their utilization in sustainable biorefineries

Bioproducts	Feedstock	Method	Result	Reference
Chemicals				
2,3-Butanediol (BD)	Bread waste	*Enterobacter ludwigii* Fed-batch fermentation	• Glucose yield: 330–530 g/kg BW • After 96 h, BDO titres: 144.5, 135.4, and 138.8 (g/L), with a yield of 0.47, 0.42, and 0.48 (g/g)	Narisetty et al. (2022)
Biohydrogen	Unpretreated cattle manure	*Caldicellulosiruptor bescii*	• H_2 increased by 47% on using gas sparging. • 142% on using adaptive lab evolution • The feedstock utilization gave a yield of 161.3 ± 1.6 mL H2/g VS_{added} at 15 g VS/L	Tunca et al. (2023)
Hydroxypropionic acid (3 HP)	Lignocellulosic feedstocks	*Rhodosporidium toruloides* Fed-batch fermentation	• 45.4 g/L of 3HP production	Liu et al. (2023)
Hydroxypropionic acid	Corn Stover	*Aspergillus Niger*	• Deacetylated and mechanically refined corn Stover enhanced yield to 0.48 C-Mol resulting in a final titre of 36.0 g/L	Dai et al. (2023)
D-Arabitol	Cheese whey powder	*Metschnikowia pulcherrima* E1	• Yield of 60.12 g/L D-arabitol • Yield of 28.26 g/L galactitol	Zhang et al. (2022a, b)
D-tagatose and bioethanol	Cheese whey powder	*Saccharomyces cerevisiae* NL22	• Yield of 23.5 g/L D tagatose • Yield of 26.9 g/L bioethanol	Zheng et al. (2019)
Butyric acid	Corncob	*Clostridium tyrobutyricum* Fed-batch fermentation	• Butyrate production: 29.8 g/L, with productivity of 0.23 g/L	Wan et al. (2023)
Biogas				
Biogas	Sugarcane bagasse	*Saccharomyces cerevisiae*	• Ethanol: 190.8 L ton $_{SB}{}^{-1}$ • Biogas yield: 27.46 NL_{CH4} $kg_{SB}{}^{-1}$	Bittencourt et al. (2019)
Biogas	Cow manure	Macroalgae *Sargassum plagiophyllum*	• Methane yield: -266.18 mL/gram of volatile solids (VS) • Biogas yield: 371.76 mL/g	Farobie et al. (2023)

(continued)

Table 1 (continued)

Bioproducts	Feedstock	Method	Result	Reference
Biogas/biomethane	Seaweed Thalli	*Ascophyllum nodosum*, *Laminaria digitata*, *Laminaria hyperborea*, *Saccharina latissima*, *Saccorhiza polyschides*	• *L. digitata* gave the highest biomethane yield: 286 L CH_4 kg/VS • *L. Hyperborean* gave the lowest yield: 118 L CH_4 kg/VS	Tabassum et al. (2018)
Biomethane	Sugarcane	Solid-state anaerobic digestion	• On modifying Gompertz's model, for KOH pretreated trash, a biomethane yield of 214.2 L/kg VS • Productivity of 6.9 L/kg VS/day	Ketsub et al. (2022)
Bioplastics				
Polyhydroxyalkanoates (PHA)	Lignocellulosic biomass	Microbial culture	• PHA yield reached 23% (w/w)	Bravo-Porras et al. (2021)
Poly(3-hydroxybutyrate)	Grape pomace waste	*Cupriavidus necator*, *Halomonas halophila* *Halomonas organivorans*	• The mean molecular weight was 512.2 kDa • Polydispersity was 1.2	Kovalcik et al. (2020)
Polyhydroxyalkanoates	Macroalgae Biomass	*Haloferax mediterranei*	• The highest cell dry weight was 3.8 ± 0.2 g·L^{-1} • PHA conc. Was 2.2 ± 0.12 g·L^{-1}	Ghosh et al. (2019)
Polyhydroxybutyrate (PHB)	Vinasse and sugarcane molasses	*Cupriavidus necator*	• Yield of 11.7 g/L PHB with 56% product accumulation	Dalsasso et al. (2019)
Biofuels				
Bioethanol	Orange peel waste	*Saccharomyces cerevisiae strain* BT1	• The yield obtained was 0.95%	Jha et al. (2019)
Bioethanol	Sugarcane leaf, sorghum stalk	*Saccharomyces cerevisiae* TISTR5020	• A yield of 8.374 ± 1.813 g/L (2 days) • A yield of 5.234 ± 0.907 g/L (4 days)	Manmai et al. (2020)
Biobutanol, biohydrogen	Food waste	*Amylase from clostridium* sp. strain BOH3	• Butanol was improved to 16.6 g/L • H_2 increased to 18.2 mmol	Zhang et al. (2020)
Bioethanol, biohydrogen	Rice bran and husk, brewer's spent grain, soybean, and wheat waste	Diluted acid hydrolysis	• 199.61 mil. L/y of bioethanol • 96.14 mil. m^3/y of biohydrogen was estimated to be produced in Brazil	Soares et al. (2022)

maintain a circular bioeconomy. Despite the major benefits of biorefinery on the environment as well as its transformative feasibility, this green approach has not been able to market itself commercially nor at a large-scale industrial level. Though the production of renewable energy is upheld by the policies of the nation, it is still not being implemented on a commercial scale (Ortiz 2020), primarily because of its comparatively reduced profitability and elevated selling prices when compared with petroleum-based products and due to a lack of commercialization. Modification and rectification of the biorefinery process require extensive analysis and research. Small enthusiastic steps towards the elevation and advancement of biorefineries are taking place. Interlinking the prior knowledge about physiochemical techniques and vision to investigate and improve the economic, energy, and environmental effects of biorefineries is a key factor for modern technology. New approaches like techno-economic analysis (TEA) and life cycle assessment (LCA) are implemented in biorefineries for sustainability and practical upscaling. TEA is a tool for the evaluation of capital and operation costs of sustainable biorefineries and a method for designing and improving biomanufacturing from both technological and economic aspects. Environmental impacts are crucial for sustainable technology and process development, but they are not considered in TEA, which is where LCA can be implemented. LCA is the "cradle-to-grave" life cycle evaluation and environmental impact of the production process and products.

Thus, in the current chapter, important analysis techniques recognized by the International Organization for Standardization (ISO), such as TEA and LCA strategies, are comprehensively discussed. The TEA aims to evaluate the technical and economic aspects of manufacturing a product (Kobos et al. 2020). LCA is a certified technology where the shelf life of products, environmental effects like pollution, and overall cost reduction are explored (Liu et al. 2021). Different steps and strategies for the implementation of these technologies, as well as their application in biorefineries for sustainable development, are explored. Further, the chapter sheds light on the collaboration with artificial intelligence and other cyber techniques along with inconsistencies, challenges, and future outlooks as well.

2 Techno-Economic Assessment (TEA)

One of the more notable and oldest papers that includes the TEA started in the 1980s. Few of these papers that used TEA in the earlier days are the studies by Viraraghavan et al. (1983) for the economic analysis of two biogasification processes and an economic comparison of these processes for peat utilization; Luong and Tseng (1984) for the techno-economics of ethanol production by immobilized cells; and Busche and Allen (1989) for the techno-economics of butanol extractive fermentation in a multiphase fluidized bed bioreactor. One such paper that explored the economic aspects found on the Web of Science was by Elliott et al. (1991), where different processes involved in the direct hydrochemical liquefaction of biomass were analysed from a period ranging from 1983 to 1990. It is known to be one of the earlier works in using TEA for economic assessment. Since then, the implementation of TEA as a green technology has increased with a focus on improving the overall social, economic, and environmental impact of biorefinery and its products. It comprises of quantifying the overall costs of biorefinery which includes capital and operational costs. Techno-economic analysis (TEA) is an efficient and valuable method for the assessment of the economic viability of biomass conversion technologies for commercialization. It can estimate the cash flows over a lifetime, revenues, and profits of a hypothetical commercial-scale plant, using available data and assumptions (Kosamia et al. 2022; Restrepo et al. 2022). It can help in identifying the most promising biomass conversion pathways, compare different scenarios and their potential economic feasibility, and guide research and development efforts. There are still huge variations and knowledge methodology gaps for TEA (Thomassen et al. 2017). Whole process simulation is used to evaluate TEA, and because there is not enough data produced by renewable energy, certain hypothetical parameters can be set. TEA for various pretreatment pathway technologies has been carried out for green sustainable product development and for the calculation of the minimum product selling price (MPSP) as well. TEA can be used for analyzing the performance of thermochemical conversion of lignocellulosic biomass (Patel et al. 2016). Overall results are greatly affected by factors like labour costs, tax, raw material costs, land prices, etc. (Mungodla et al. 2019). Understanding different pathways for biomass conversion helps in evaluating the TEA of the end products. García-Velásquez and Cardona (2019) carried out a study that evaluated ethanol and synthetic gas products from biochemical and thermochemical routes. Mexican weeping pine (*Pinus patula*) was used as the substrate for fermentation and gasification, and TEA and energetic and environmental assessment were carried out for both pathways.

The traditional TEA methodology usually includes four major steps. The first step TEA involves the market study which includes extensive research about the industry and its product prices. This enables identifying market potential and its related parameters which helps in gazing and improving commercialization ability. It mainly includes defining the purpose, goal, and boundaries of the TEA, as well as the assumptions, data sources, and methods to be used. In the second step, a process flow diagram (PFD) and mass and energy balance (MEB) were estimated. It also includes explaining and graphically representing the parameters such as the input process, output process, unit operations, and equipment of the system. The third step is the economic evaluation where the economic profit feasibility is calculated based on investment criteria and techno-economic variables. It involves parameters like costs of equipment, installation, labour, materials, utilities, and maintenance. Indicators like total capital investment (TCI) and annual operating cost (AOC) are first determined and measured in this stage followed by MPSP, the net present value (NPV), the internal rate of return (IRR), the payback period, and the break-even point. In the fourth and final step, an evaluation of the economic performance and viability of the process is carried out, which includes an assessment of critical variables that affect economic performance as well as an assessment of overall uncertainties (Thomassen et al. 2019) (Fig. 1). TEA software is selected depending on the

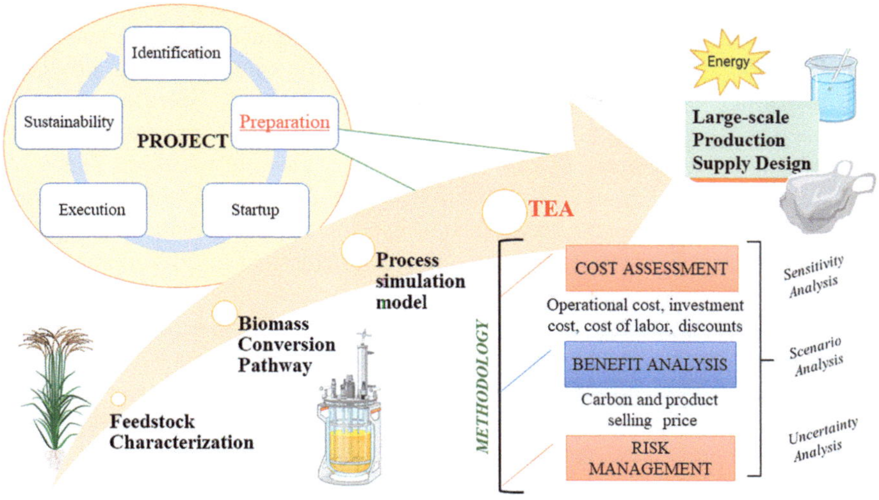

Fig. 1 An overview of TEA approach in the preparation step of a biorefinery project

availability, detail, and complexity required, user-friendliness, and compatibility. Some examples of TEA software are Aspen Plus (Zheng et al. 2023a, b), SuperPro Designer (Lim et al. 2020), EMSO (Elias et al. 2021), BioSTEAM (Stewart et al. 2023), Pro/II (Talero and Kansha 2023), and CHEMCAD (Hos and Herskowitz 2022).

2.1 Aspen Plus

Aspen Plus v9.0 software was used to calculate mass and energy balances as a simulation tool. In the work by García-Velásquez and Cardona (2019), the production of ethanol demonstrated a low production cost of 0.549 USD/L, superior energy efficiency of 45.1%, and higher CO_2 emissions of 0.17 kg per MJ EtOH when compared to gasification which produced electricity with energy efficiency of 29% and CO_2 emissions of 0.11 kg per MJ energy. Fajimi et al. (2022) examined the production of methanol and electricity using bagasse. Four scenarios (S) were contemplated: (S1) one involving gasification and torrefaction combined with methanol and electricity production, (S2) one for electricity production with torrefaction, and (S3–S4) two without torrefaction. Bagasse was processed through gasification, syngas cleaning (Rectisol process), and methanol synthesis (hydrogenation). The torrefaction scenarios produced increased syngas and higher-quality syngas (S1 and S2). The economic analysis was calculated by Aspen Plus using a kinetic-free equilibrium model which showed results that torrefaction was the most financially viable option as it showed higher methanol (0.48 kg MeOH/kg) yield. Syngas production using torrefaction had heightened with lower heating values (LHV) of 9.25 MJ/Nm3 (S1) and 8.71 MJ/Nm^3 (S2) compared to scenarios without torrefaction. Studies like these are important to determine the important contributors to conversion technologies and determine the best options when designing TEA.

Further, Dees et al. (2023) analysed the cost and life cycle of ethanol produced with an oxyfuel boiler and evaluated to capture and store the CO_2 released. TEA was performed to calculate the MPSP, energy balance calculations, and sensitivity stimulation tool, and assessment of CO_2 released was carried out by Aspen Plus V11. 58% reduction in greenhouse gas intensity (CI) (52$/t CO_2e) is seen for the capture and storage of CO_2 emissions. Oxyfuel boiler system integration reduced CI by 75% to 15 gCO_2e/MJ at an MPSP of $2.24/gallon, an increase of $0.31/gallon over the baseline; no intervention case was also observed. Sensitivity analysis showed that the combination of low CI substitutes with an oxyfuel boiler system could produce carbon-neutral or carbon-negative ethanol. Integrated fermentation with an oxyfuel boiler reduced the CI of ethanol by a net 44–50 gCO_2/MJ. If all the sensitivity analysis's suggested interventions were put into practice, CI would drop by a net 79–85 gCO_2/MJ. The present studies mainly focus on methodology designing and implementation in research for applications in biomanufacturing.

2.2 *BioSTEAM*

In the work by Li et al. (2021), TEA and LCA of lignocellulosic lactic acid production were carried out. This was done by the software BioSTEAM (Biorefinery Simulation and Techno-Economic Analysis Modules), a Python-based open-source platform for the design and evaluation of biorefineries in the presence of uncertainties, utilizing its rapid and adaptable framework. BioSTEAM facilitates swift and reliable process design, as well as the calculation of mass and energy balances, environmental sustainability assessments, and TEA. The result showed that the biorefinery could produce lactic acid at a competitive price (between $1.7 and 2.1 per kg) compared to the MPSP (between $1.38 and 1.91 per kg, with a baseline of $1.57 per kg). GWP of 2.79 kg CO_2 eq/kg, fossil energy consumption of 31.7 MJ/kg, and MPSP decreased to $1.09 could be seen on modification of technology, like an increase in fermentation yield. Sensitivity analyses of categories like yield, pH, titre, productivity, feedstock content, and price were performed to see the impact on the environment and its sustainability.

BioSTEAM's effectiveness was illustrated in a study by Cortés-Peña et al. (2022), focusing on two segments: first was the simultaneous production of biodiesel and ethanol using lipid cane, and the second was the production of ethanol from corn stover. The TEA of the production process by BioSTEAM reported that the biofuel production must be done by creating Species (pure-component chemical data), Stream (thermodynamic, transport, and vapour-liquid equilibrium), and Unit System (network of unit operations) objects. The thermo package of the open-source Chemical Engineering Design Library was used for the measurement of thermodynamic properties. The study concluded that BioSTEAM is a user-friendly, transparent, and well-documented tool for uncertainty assessment for efficient decision-making within the biorefinery research community. The future advancements include incorporating equations of state for mixtures, enhancing multicomponent distillation algorithms, and integrating a LCA component. A web-based interface using the Django framework is also in progress to make BioSTEAM more accessible in the study. Singh et al. (2023) evaluated various acidic and basic ion-exchange resins, as well as activated charcoal for their ability to adsorb microbially produced triacetic acid lactone (TAL). Activated charcoal and resin demonstrated similar adsorption capacity of 0.18 ± 0.002 g/g, and ethanol achieved a 72% recovery of TAL absorbed by activated charcoal and 50% recovery of TAL absorbed by resin. A biorefinery design was developed, and TEA under uncertainty for producing a microbial triacetic acid lactone from sucrose was carried out. The results observed the MPSP for current technology to be \$4.27/kg (\$3.71–4.94/kg), and for potential near-term improvements to fermentation, it was \$2.83/kg (\$2.46–3.29/kg). These results were comparatively lower to sorbic acid, making this adsorptive pathway for triacetic acid lactone recovery an upgraded and efficient method.

2.3 Integrated Techno-Economic Assessments

Integrated TEAs are established to increase performance and feasibility by modifying strategies and introducing diverse connections between various technical and economic elements. This integration helps rectify some existing flaws in the system. Adding sub-variable parameters in the traditional strategies enables a more detailed-oriented investigation of technology and cost and helps in better decision-making (Tschulkow et al. 2020). These developments impact process flow diagram (PFD), mass and energy balance (MEB), expenditures, revenues, and net present value (NPV). There are many ways to evaluate new technologies from economic and environmental perspectives, especially as they are still developing. This has led to a growing need for standardized assessments, which various projects on emerging green technologies are working to create (Thomassen et al. 2017). Technology readiness levels (TRLs) of different green development technologies are done to investigate the stage of maturity and practicality. Low TRL indicates that the technology is still just an idea and is unclear (1–3 level). High TRL indicates that the idea is more mature and is more adaptable and ready for industry application (4–9 level). TRL application in biorefinery is important for an optimized implementation of technology. It is important to carry out TRL from the early stages as that includes cost estimation. In the review by Sorunmu et al. (2020), they reviewed the literature on thermochemical pathways and their modification for upgraded pyrolysis of bio-oil via techno-economic analysis (TEA) and LCA methods together with TRL evaluation. In the study conducted by Tschulkow et al. (2020), an integrated TEA strategy was developed to elevate the existing technology for lignin-first biorefinery targeting the valorization of woods. Additional sub-parameters were introduced like assessment of moisture content, feedstock species, and capacity level. Monte Carlo analysis for sensitivity and Break-even point analysis for economic shock resilience of wood was also performed. The results of the study showed that waste wood is the most profitable feedstock, with a NPV of MV 59 and an internal rate of return (IRR) of 26%. The importance of optimal location choice and supply availability was emphasized. It highlighted the implementation of the detailed integrated TEA as an essential tool for economic feasibility, scale selection, and feedstock combinations in biorefineries. In the work by Hytönen and Stuart (2011), to analyse the huge uncertainty of forest biorefinery, TEA was carried out by employing Monte Carlo stimulation. Monte Carlo analysis is a statistical method used to figure out how the properties of a model are changed within specific limits, according to a set statistical distribution. It is a valuable tool in scientific computing because it is straightforward and can be applied to many situations. It is a way to efficiently estimate uncertainty in various scenarios. Its importance lies in its simplicity and its ability to be used in a wide range of situations (Khalid 2022).

Emphasis on the importance of incorporating diverse disciplines gives a viewpoint that is dynamic and detailed. Disciplines like political and social domain are being incorporated with TEA and LCA for better insights of biorefinery reality. Social techno-economic studies for the evaluation of the social impact of the area and the human behaviours of workers involved can be seen in the work by Shahbeig and Nosrati (2020). Harish et al. (2022) attempted to analyse and apply a decentralized energy system for the development of rural microgrids in India. Comparative analysis of parameters and GHG emissions as well as socio-techno-economic studies was carried out. The socio-techno-economic analysis focused on the quota of user willingness to pay as a stakeholder. Kanagaraj et al. (2023) performed a techno-socio-economic sustainable analysis in terms of CO_2 emission and energy demand of Portland cement, geopolymer, and limestone calcined clay cement (LC^3) composite systems. The analysis was done in the categories of politics, economics, social, technical, environmental, and legal (PESTEL). Social equity and community engagement were some of the social aspects studied. This included the involvement of locals in decision-making, indigenous rights and policies, health impacts, biodiversity protection, risk assessment, and other discussions.

The tools and standardized databases help tremendously in the evaluation of indicators and stimulations. Different optimization techniques for advancing TEA are being developed. TEA is being incorporated with simulation models to experience its complete potential. The incorporation of artificial intelligence for the analysis of data is still a work in progress especially in the sector of biorefineries (Sollai et al. 2023; Arias et al. 2023).

In a case study by Ahmad and Zhang (2021), design development was carried out to integrate renewables and perform techno-economic analysis using grid-connected and islanded modes to meet community and commercial load requirements. Three key analytical tasks performed were optimization, simulation, and sensitivity analysis (four sensitivity cases each with four sensitivity loops). TEA parameters included capital costs, replacement costs, operating and maintenance expenses, salvage costs, fuel costs, real-time load data, and climate information. Sensitivity case 2 showed the best results with an islanded mode net present value of $1381080.42 and an annualized cost in sensitivity loop 1 of $106832.62. A net value of $512792.1 and annualized cost of $39,666.71 were seen in the grid-connected utility mode. The results showed the grid-connected mode's cost is half that of the islanded mode, which supports the importance of integrated TEA tools.

Life cycle costing (LCC) is one of the other methodologies quite similar to TEA in terms of economic evaluation. But in LCC, labour and equipment prices are excluded and focus on the cost of products over their total life cycle, which makes coming to conclusions and decision strategies difficult. Contrarily TEA is more feasible as it works based on economic profitability from an investor's perspective and adapts systems from further life cycle stages (Mahmud et al. 2021). In the work by Pena et al. (2022), the life cost analysis of the tomato production cycle with an integrated urban agriculture system was performed. The findings of the work suggested that labour expenses (24.7%), the construction of rooftop greenhouses (15%), external pest control (12.6%), and rainwater consumption (9.5%) are the primary cost factors in tomato production, contributing a total of 61.8% to the overall costs. Consequently, cost-cutting measures were examined for various sensitivity scenarios, which encompass aspects like the design of rooftop greenhouse structures, water sources like tap water and rainwater, and the size of tanks.

3 Life Cycle Assessment

LCA is one of the prevailing tools that are being utilized in research to analyse the whole product life cycle as well as the environmental impact on natural and human resources in biorefineries to produce green value-added products based on the 14,040 series. ISO 14040 provides the general principles and framework for conducting LCA, ISO 14041 focuses on the goal and scope definition of the LCA, ISO 14042 focuses on the methods of LCA, and ISO 14043 focuses on the interpretation phase of LCA. These ISO series collectively form a guideline framework for conducting LCA. This involves collecting data on inputs and outputs from the production system and deriving conclusions based on the research (Ubando et al. 2020). LCA is an internationally recognized solution to identify and reduce the environmental burdens from a process by quantifying energy and materials input. LCA was first integrated with optimization by Azapagic and Clift (1998). Despite LCA implementation in research, manufacturing (Dabbaghi et al. 2021), power generation (Bahlawan et al. 2021), and agriculture (Khanali et al. 2021) as well as in construction (Zhang et al. 2021), further upgradation/modification is still required depending on the type of biorefineries. Thus, to design a biorefinery in a way that is compatible with a circular bioeconomy, it is important to use the LCA method to measure the environmental impacts of the whole life cycle of the products, from "cradle" to "gate". This can help identify areas for improvement and reduce waste. Areas of improvement can be identified by highlighting the hotspots in the product life cycle and the comparison between products and processes to increase the potential of the development system. However, one of the challenges of using the LCA method is the availability of data on the environmental burdens and waste generation of each life cycle stage (Nizami et al. 2017).

There are four main stages of LCA which are defined by ISO 14040 (Principles and Framework) and ISO 14044 (Requirements and Guidelines). The first procedure is the goal and scope. This stage defines the overall objectives, goal, and scope of the system as well as specifies the functional units of the parameters. The functional unit is a measure of the function or performance that the product system delivers, and it should be related to the function rather than the physical product. The system boundaries define the processes and activities that are included or excluded in the LCA (Osman et al. 2021). The decision regarding which system attributes need to be analysed for their environmental effects is inferred in this stage. Decisions regarding all the requirements that may be needed for the life cycle impact assessment (LCIA) are concluded in this stage.

The second stage is the life cycle inventory analysis. As the name suggests, in this stage there is a collection of data related to the life cycle of the product and process. The environmental flows include the inputs from nature and the outputs to air, water, land, and waste streams. The scope and complexity of the inventory analysis depend on how many systems are included in the system boundary. More systems mean more inventory is essential. These indicators are divided into midpoint and endpoint indicators. The most common databases used for collecting the input and output data are SimaPro LCA databases, Ecoinvent, GREET (Greenhouse Gases, Regulated Emissions, and Energy Use in Transportation), etc. SimaPro comes with the Ecoinvent v3 database, covering over 10,000 processes from the Netherlands. For new users of SimaPro, it is recommended to use the LCA wizard to set up your LCA project. This involves answering a series of questions (e.g. Do you want to store product stages created? Or do you want to create a product system?), and based on your answers, a process structure is developed. Also, the store product stages can be chosen to create a product system for further analysis. One of the latest versions of SimaPro available is SimaPro ver. 9.5 (Goedkoop et al. 2016). GaBi was developed in Germany and used in life cycle engineering (Spatari et al. 2001), CLCD was developed in China, and GREET was developed by Argonne laboratory (Wang 2007).

Moretti et al. (2022) used SimaPro 9.1.1 software for the analysis of the inventory for local Dutch biorefining industries (Avantium and Vertoro) for asphalts in the Netherlands. LCA was carried out to investigate the environmental impact of using lignin-based asphalts instead of conventional asphalts. Cortés-Peña et al. (2022) used the GREET 2020 model and the Ecoinvent Life Cycle 2020 to carry out the collection of the input inventory of oilcane-based biorefineries. Some agricultural models like GLOBIOM (Fulvio et al. 2019), STAMINA (Ni et al. 2019), and BEPAM (Chen et al. 2021) are also used to analyse land use and its environmental impact. One of the most common impact categories is global warming potential (GMP) which incorporates greenhouse gases (kg CO_2) equivalent to a time horizon of 20 years or 100 years (Cusenza et al. 2021; Derwent 2023). It is a widespread category that includes emission of gases like CO, CO_2, CH_4, and N_2O, and the combined climate impact of these emissions is measured in GMP. The main contributors to GMP are electricity and fossil energy (Zheng et al. 2023a, b). Other categories are abiotic depletion potential (ADP) to assess the impact of metal and mineral resources like fossil deposits; photochemical oxidation measures in kg the non-methane volatile organic compounds that usually harm health and ecosystem; ecotoxicity potential is evaluated in kg 1,4-dichlorobenzene equivalent or cumulative toxic units; acidification is calculated in kg SO_2; and water depletion analyses the water utilized in the entire production (m^3). Particulate matter formation (PMF) was evaluated as expressed as particulate matter 2.5 equivalent and/or particulate matter 10. RI impact includes PM 2.5/PM10, gas emissions like nitrogen oxides, etc. Eutrophication was evaluated as kg PO_4 to measure the excess amount of nutrients, and human toxicity is measured as 1,4-dichlorobenzene equivalent/cumulative toxic units (carcinogens and noncarcinogens), etc. (Osman et al. 2021).

The third stage is environmental/LCIA. Using different measures, the environmental impact of the collected inventory is evaluated followed by scenario, sensitivity, and uncertainty analyses. It consists of four steps: categorization, classification, characterization, and normalization. The impact assessment methodologies are not that well defined and are still diverse depending on different databases. These tools and models can be modified based on the parameters and preferences of stakeholders. Some of the examples of impact assessment methodology used are Tool for the Reduction and Assessment of Chemical and Other Environmental Impacts (TRACI) (Shi and Guest 2020), ReCiPe (Dekker et al. 2020), Centre of Environmental Science (CML) (problem-oriented classification into midpoint indicators) (Prieler et al. 2019), International Reference Life Cycle Data System (ILCD) (covers 15 impact categories) (Chomkhamsri et al. 2011), IMPACT 2002+ (covers 14 impact categories) (Khounani et al. 2021), etc. Abu-Bakar et al. (2023) performed an LCIA of glucose production from empty and partially filled paddy grain (EPFG) in Malaysia. It was conducted using the ReCiPe methodology. The results concluded that enzyme production was the main source of environmental impact and agricultural phase harmed climate change. The LCA study showed that scenario 2 (where the whole EPFG fractionation was used) had lower environmental impacts than scenario 1 (where only the glucose fraction of the EPFG was used). Sreekumar et al. (2020) performed LCA of ethanol production rice straw in India. The life cycle inventory data was taken from the Ecoinvent database, and LCIA was performed using ILCD 2011. The result showed the GWP was 2.82 kg of CO_2 eq. per litre of ethanol, impact was reduced to 0.07 kg of CO_2 eq. per litre of ethanol using hydroelectricity, and impact was 3.35 kg of CO_2 without allocation and implementing system expansion. Monte Carlo analysis can be used to measure uncertainty and for different scenarios. Sensitivity analysis analyses the minute environmental footprints. It is important to perform scenario, sensitivity, and uncertainty analyses simultaneously for the optimized LCA.

The final stage is the life cycle interpretation, where the measured and analysed parameters are inferred and a conclusion is reached. It involves deciding and

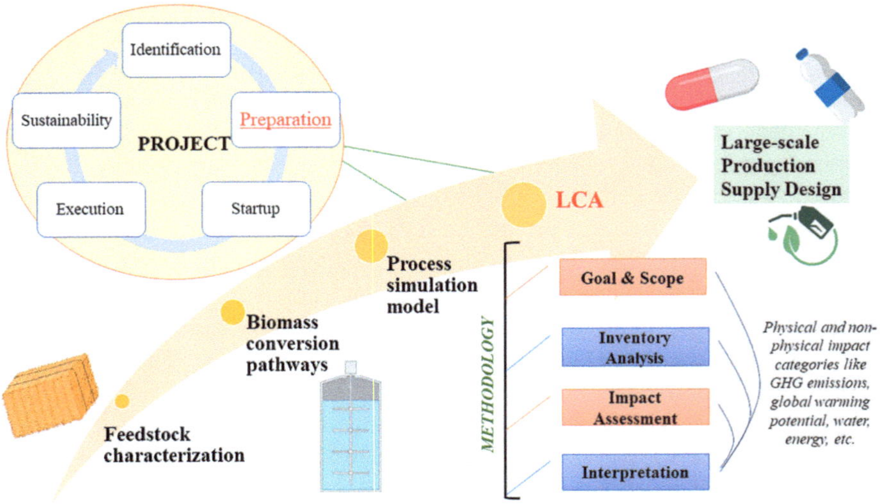

Fig. 2 An overview LCA approach in the preparation step of a biorefinery project

identifying the points in the process that are required to be modified for further improvement (Mulya et al. 2022). Sometimes these steps involve communicating with the stakeholders and presenting the results (Fig. 2).

Contrary to initial predictions, there is growing apprehension regarding the potential adverse social consequences arising from the prevailing conditions in feedstock production. This study employs the social life cycle assessment (SLCA) methodology to evaluate these impacts. SLCA was introduced by the United Nations Environmental Programme/Society of Environmental Toxicology and Chemistry (UNEP/SETAC) and is more challenging than the traditional LCA. Costa and Oliveira (2022) carried out an SLCA to evaluate the possibility of a negative impact on society during the production of feedstock and biodiesel in Brazil. A subcategory assessment method and R language were utilized in this. The results showed wage differences between the wealthy and poor regions as well as the influence of corporations on workers. Associations between gender, ethnicity, job responsibilities, and income were also established. The results of the study concluded that despite the presence of established policies, social inequalities still prevailed in the masses involved in production. SLCA is still primitive in nature and needs to be developed extensively to be used along with other assessments like LCC, LCA, TEA, and even geographic information systems. An integrated LCA involving LCA, SLCA, and LCC is known as life cycle sustainability assessment and when implemented in biorefineries will be a pathway for efficient circular economy. A review by Larsen et al. (2022) extensively discussed the research gap in this assessment technology as well as factors like the importance of early involvement of stakeholders and reuse/recycling. Some LCA studies can just focus on one aspect of the gap and can be explored concerning real-world practicality. Morris and Hicks (2022) reviewed the LCA of bioplastics with a focus on the role of human behaviour and the environmental impact of bioplastic products caused by real-world usage and disposal.

Unlike LCA, TEA is limited in the "cradle-to-gate" system evaluation of the production as it excludes water utilization and treatment. To acquire an optimized sustainable process development, it is wise to integrate TEA and LCA. This integration provides a systematic and inclusive analysis of the relationships between economic, technological, and environmental aspects. Zheng et al. (2023a, b) conducted a widespread comparison between two processes—hydrogenation upgrading (FP-HU) and supercritical ethanol upgrading (FP-SU)—for biofuel (1 MJ) production derived from fast pyrolysis (FP) of cornstalk (2000 t/day). TEA analysis to calculate the minimum fuel selling prices (MFSP) and mass and energy flows was performed by Aspen Plus software, and LCA analysis was evaluated by eBalance software V4.7 and CLCD. MFSP for FP-HU was measured as \$0.0417 per MJ and FP-SU was \$0.0383 per MJ. The main cost contributors to MFSP for FP-HU were the cornstalk cost (\$0.0084 per MJ) and the ethanol input cost (\$0.012 per MJ), constituting 18.8 and 31.3% of their respective MFSP. LCA results demonstrated that FP-HU had lower ADP, COD, and GWP values compared to FP-SU. The combined environmental impact of FP-HU was 4.5E−12 and FP-SU was 5.2E−12 (eco-points). FP-HU and FP-SU showed reduced ADP, GWP, and respiratory inorganics (RI) in comparison with conservative diesel. It was determined that the electricity consumption parameter had the highest overall impact. The result stated that the cost of FP-SU was 8.9% less than FP-HU, but the environmental impact was 14.2% greater in FP-SU compared to FP-HU.

4 Applications

One of the major compositions of these analysis techniques is a literature review and consistent data collection. There is no standardized methodology for a particular biorefinery application yet, but work is still in progress for designing strategies and introducing default user-friendly guides in assessment software. TEA and LCA are some of the key techniques that minimize risk when taking biomanufacturing of biobased value products to a large-scale supply chain (Table 2). Various biorefinery sectors and the latest utilization of TEA and LCA for the investigation of cost and environmental feasibility are discussed below.

4.1 TEA and LCA in the Biomanufacturing of Biofuels

Fossil fuel production causes massive environmental and social issues every year. These fossil-based productions have a time limit attached to them as they utilize nonrenewable sources which are depleting every minute for energy production. While supercritical circumstances allow for quick reaction times and continuous

Table 2 Application of TEA and LCA for the production process for diverse biobased value products

Product	Feedstock	Software	Result	Reference
Life cycle assessment				
Polyol fuel	Corn Stover	eBalance software with a Chinese life cycle database	• The net fossil energy input for polyol fuel is 0.760 MJ • The net global warming potential of 0.0444 $kgCO_2$	Heng et al. (2018)
Biochar	Cornstalk	Monte Carlo model	• LCA of separate hydrolysis and fermentation, simultaneous saccharification and fermentation, and consolidate bioprocessing process indicate that using biochar resulted in better environmental performance • The main contributor to the environmental impact was electricity consumption • By-products were found to affect life cycle impacts	Wang et al. (2022)
Non-nutritive sweeteners aspartame and neotame		SimaPro 9.3 software and Ecoinvent 3.3 database	• The GWP of 1 kg of aspartame was 29.25 $kgCO_2$-eq/kg, while 1 kg of neotame was 43.42 $kgCO_2$-eq/kg • NNS production showed reduced environmental impact • This study reveals that the environmental impact of the additional resources required to produce neotame from aspartame is more than compensated by the increase in perceived sweetness, which is 200–8000 times higher	Suckling et al. (2023a, b)
Bioethanol	Corn Stover	SimaPro 9.3 software and Ecoinvent 3.3 database	• On using the ILCD Midpoint 2011 methodology, results showed HCL and H_2SO_4 having better environmental performance than nitric and acetic acid • This chemical hydrolysis followed by enzymatic showed the highest relative contribution to environmental impact	Mata et al. (2022)
Techno-economic assessment				
Levulinic acid (LA), formic acid (FA), acetic acid (AA), and furfural	Pine sawdust	Apache OpenOffice Calc software And ASPEN plus software for furfural-rich phase	• The analysis concluded energy to be one of the major parts of production costs • Using 95% of residues improved the internal rate of return (IRR). • Heat integration cuts LA steam consumption by 10%, boosting pellet production capacity	Clauser et al. (2018)

Acrylic acid	Sugarcane A molasses	Aspen plus® software	• The superior levulinic acid intermediate was 11% higher than the actual AA market price and net GHG emissions of 4.4 kg_{CO2}/kg. Also yield of 1 g LA/gram glucose • The second-best 3-HP intermediate of 2425 US t^{-1} and green premium of 30% with a low yield of 0.51 g LA/g glucose • Glycerol intermediate proved to be the lowest yielding, with 0.50 g GLY/g glucose	Brobbey et al. (2023)
Furfural, lignin, and ethanol	Switchgrass	Aspen plus® software	• The analysis resulted in min. Furfural selling price (MFSP) of $625 per metric ton (lowered by 37%) • Sensitivity analysis revealed parameters like temp. And solid loading to have a higher impact on MFSP than some economic factors	Zang et al. (2020)
Biomethanol	Palm oil wastes	Aspen plus® software	• 3 cases of palm kernel shell feedstock showed net profit per methanol production at $201, $132, and $208 million • 3 cases of empty fruit bunches showed net profit per methanol production at $170, $130, and $179 million	AlNouss et al. (2022)
Biomethanol		Aspen HYSYS software	• The economic analysis showed a production cost of $ 286 with an annual profit of 1.2 M$ (1 kg methanol) and MPSP of 0.31 US$/kg • Simulation results for methanol production were estimated at 0.619 kg/kg biomass	Zhu et al. (2022a, b)
TEA and LCA				
3-HP, acrylic acid	Lignocellulosic biomass	BioSTEAM	• Analysis showed yield, titre, and saccharification solids can improve the biorefinery potentially by reducing MPSP to $1.29–1.52 per kg • Global warming potential (GWP) was 3.00 kg CO_2-eq per kg and fossil energy consumption of 39.9 MJ per kg	Bhagwat et al. (2021)

(continued)

Table 2 (continued)

Product	Feedstock	Software	Result	Reference
Ammonia	Corn Stover	Ecoinvent database 3.6	• TEA results showed in reforming pathways: Total capital investment (TCI): $1,254,702,937 MPSP: $0.992/kg Sensitivity and optimal analysis: Low as $0.39/kg GWP: 1.11 kg$CO_2$-eq/kg • TEA results showed in partial oxidation pathway: TCI: $1,273,047,866 MPSP: $1.05/kg GWP: 1.66 kg$CO_2$-eq/kg • LCA (eco-indicator 99 and cumulative energy demand) – Analysis concluded reforming pathway is the better ammonia production	Zheng et al. (2023a, b)
2,3-Butanediol (2,3-BDO), biostimulants			• TEA and LCA of the production of methyl ethyl ketone (MEK) through catalytic dehydration of 2,3-BDO. MPSP was estimated to be $1.90/kg. GWP_{100} was 0.37 kg CO_2 eq/kg, and FEC was found to be 3.1 MJ/kg	Lee et al. (2023)
Methanol and wet ethanol 80	Corn Stover		• The LCA results showed a 113% increase in alcohol fuel energy production/hectare crop and energy split of co-produce of 77%/23% • 14 gCO_2e/MJ of total carbon intensity and a cost of $2.63/gallon of gasoline equivalent were seen	Gainey et al. (2023)

production the deterioration of the environment cannot be overlooked as well. Thus, over the time sustainable fuel production methods have been developed globally. Amongst the biofuels produced bioethanol is has been intensively studied and uses lignocellulosic biomass (e.g. corn stover, switchgrass, and hardwood) as a substrate. They are, usually treated via dilute-acid pretreatment, enzymatic hydrolysis, and microorganism fermentation. These sustainable production systems can be a step towards bioeconomy and commercialized. As most production system starts with feedstock selection, land acquisition, and finalizing all the requirements that might be needed for biomass conversion. It is then followed by the incorporation of different biomass conversion pathways to produce bioethanol, biobutanol, biodiesel, etc. and related techno- and environmental assessment. For an optimized biofuel production, enhanced technology design for techno-economic and environmental analysis is needed. In one such study, TEA was performed by Figueroa-Torres and Theodoropoulos (2023) for a microalgae-based biorefinery for the integrated production of biobutanol, biodiesel, ethanol, acetone, and glycerol. Results were simulated using Aspen Plus v8.8, which showed that on acid hydrolysis followed by solvent extraction, the minimum fuel selling price (MFSP) reached \$2.11 per kilogram of total biofuel. This price was 75% lower than the scenario for producing only biobutanol at \$8.57 per kg and 11% lower than the scenario for producing only biodiesel at \$2.36 per kg. MFSP was reduced significantly by the usage of lipid extraction on feedstock cost, inlet rate, and solvent usage. Variations in biodiesel prices make the MFSP of biobutanol extremely sensitive. Mailaram et al. (2022) conducted a TEA study of the economic viability of producing microbial 1,4-butanediol (BDO) from C5 and C6 sugars obtained from brewers' spent grain (BSG) (fermentation). TEA was performed and optimized for different BDO concentrations and plant sizes using pinch technology by using Aspen Plus software (pinch technology serves as a valuable tool for integrating heat energy and improving the utilization of heat within the process). A 34% reduction in consumption of hot utility (lowest consumption 4.59 MW) and an 18% reduction in cold utility (lowest consumption 10.97 MW) were observed. The production cost of BDO was \$1.84/kg (80 g/L BDO), \$1.76/kg (100 g/L BDO), and \$1.74/kg (120 g/L BDO). The unit production cost of BDO was significantly lower at \$1.07/kg when processing 2000 MT of BSG (100 g/L BDO). For a BDO concentration of 100 g/L, the MPSP of BDO was \$3.63/kg (100MT) and \$2.00/kg (2000 MT) of BSG, respectively, with an 8.5% return on investment and a 5-year payback period. To assess the environmental footprints and emissions and upscaling production to an industrial scale, LCA is performed for biofuel biorefineries. Wahyono et al. (2022) evaluated the environmental impact of multi-feedstock biodiesel simulated production and compared it to palm oil biodiesel production in Indonesia via LCA. A 9.89 tCO_2 GHG emissions per tonne of biodiesel were witnessed on converting scrubland into plantations for crops like soybean, canola, sunflower, and oil palm, whereas transition to oil palm plantation resulted in −3.43 tCO_2. Fertilizer use was observed to be the reason for many environmental issues like ecotoxicity, eutrophication, acidification, and global warming. Planting multiple crops is more environmentally harmful than oil palm alone due to higher fertilizer usage. This study conveyed that the transition of this multi-feedstock biofuel biorefinery to a large scale will cause a negative environmental impact. Waste decomposition produces landfill gas (LFG) as a by-product, which harms the ecosystem. This can be prevented by the conversion of LFG into value products. Choe et al. (2022) studied the utilization of LFG to produce methanol, and its TEA and LCA were performed. TEA results showed that biomethanol production cost was 879.16 $/ton (steam methane reforming), 724.52 $/ton (methane dry reforming), and 1130.74 $/ton (membrane separation). Environmental assessment results showed a climate impact of 2.360 kg CO_2-eq kg/MeOH (steam methane reforming), 0.835 kg CO_2-eq kg/MeOH (methane dry reforming), and 0.605 kg CO_2-eq kg/MeOH (membrane separation). Decision analysis conducted on the study concluded that methane dry reforming and membrane separation have the potential to be applied for methanol production while keeping economic and environmental impact in consideration.

4.2 TEA and LCA in the Biomanufacturing of Biohydrogen

Hydrogen is a clean energy resource which is an essential component of modern sustainability practices. On 4 January 2023, the Central Government of India approved the National Green Hydrogen Mission (NGHM) with an outlay of ₹19,744 crores from 2023 to 2030, with ₹400 crores focusing on R&D development (https://indiaesa.info/events/past-events/india-green-hydrogen-council-ighc). There are numerous applications of biohydrogen, from generating electricity to utilization in fuel cells, etc. Various biochemical methods of H_2 production are dark fermentation (Hosseinzadeh et al. 2022), photo fermentation (Lu et al. 2022), anaerobic digestion (Kovalev et al. 2020), and integrated dark and photo fermentation (Das and Basak 2022). For an improved and optimized biorefinery of hydrogen, analyses like TEA and LCA are implemented. In the study by Kumar et al. (2019), a TEA and sensitivity analysis were carried out, and the calculations were performed using the Aspen Process Icarus Evaluator (APIE), and information from diverse literature sources was collected. It resulted in a total process equipment cost and fixed capital investment of $86.9 and 411.89 million for 2000-tonnes/day biomass plant. Also, the sensitivity analysis indicated that the biohydrogen yield and capital costs have the most substantial impact on its product value, whereas other parameters (with almost +20% variations) have a relatively minor effect on PV. The work has investigated the technical and economic feasibility of producing biohydrogen from biomass through hydrothermal liquefaction process, estimating a cost of $2.84 ± 0.10/kg. Ongoing technology development, with sensitivity analysis revealing the significant influence of biohydrogen yield and capital costs on product value in this research and other parameters, showed a relatively minor impact.

In the study conducted by Akhbari et al. (2021), LCA of biohydrogen production using a pilot-scale up-flow anaerobic sludge blanket fixed-film reactor was carried out in Malaysia. The substrate used was palm oil mill effluent (POME). To measure the ecological impact of this biohydrogen production, LCA was performed using the SimaPro LCA software (version 9) and the ReCiPe 2016 impact assessment method. The results of the study revealed that electricity consumption (50–98% of the total impacts) was accounted as the most significant contributor to environmental impacts. An improvement analysis performed led to a reduction in all impact categories, particularly the global warming impact, showing a 77% decrease, from 818 to 189 kg CO_2-eq per kg of biohydrogen produced. It was predicted that shifting the pilot reactor to Sarawak (a cleaner grid region) could reduce the environmental impact to 142 kg CO_2-eq. Using POME can further decrease the global warming impact up to 54.9 kg CO_2-eq. Integration of LCA in biorefinery is a huge step forward towards sustainability and reduction of overall greenhouse emissions. A study carried out by Gholkar et al. (2021) assessed TEA and LCA to produce biohydrogen or methane by introducing a single-step, tar-free catalytic reactive flash volatilization (RFV) process for microalgae. For TEA a comprehensive Aspen Plus flowsheet model was designed using experimental data acquired from hydrogen and methane production. The analysis focused on a plant in India that processes 12,790 kg/h of microalgae, resulting in the production of 1239 kg/h of H_2 or 3484.96 kg/h of CH_4. For LCA both experimental and simulation data were used. The results revealed a 3.78-year payback period for hydrogen production with a 22% internal rate of return (IRR), and the life cycle climate change impact for hydrogen was 7.56 kg CO_2 eq./kg of H_2 produced, a 36.47% reduction compared to steam reforming of methane. A sensitivity analysis observed that the payback period was most sensitive to changes in feedstock cost, with a 25% increase resulting in a 35.38% longer payback period. The interlinkage of TEA and LCA when used simultaneously has a grand effect in establishing a trade-off system between economics and the environment and elevating the competitive biorefineries (Kosamia et al. 2022). The biorefinery of chemicals not only helps in reducing the environmental burden, but the products from these chemical biorefineries have diverse functions in the food, pharmaceutical, and agriculture industries. Morakile et al. (2022) simulated six different options for making levulinic acid (LA), glutamic acid (GA), and/or xylitol by using the software Aspen Plus v8.8. These options were considered in two settings: one using only first-generation (1G) feedstock from molasses and the other combining first- (1G) and second-generation (2G) feedstock from molasses and sugarcane lignocellulose. The TEA of all options/scenarios showed IRR ranging from 21.3% to 59.8%. GA showed increased market prices and product throughput and decreased complexity of technology. The best economic performance of 59.8% IRR and MPSP of $1687/t was seen by 1G GA producing 11.1 t/h GA. Using mature xylitol technology in a multiproduct facility (1G2G GA-xylitol) can improve economic performance and make it more robust. Other works that analyse the biochemical production are by Del Pozo and Cloete (2022), performed TEA to study the performance of ammonia production using natural gas/electricity. Cash flow analysis of gas switching reforming (blue ammonia) and electrolytic (green ammonia) as energy carriers and LCA of methanol and ammonia (known hydrogen carrier) was performed using GaBi 10 software in the work by Zhu et al. (2022a, b). Also the optimization of ammonia production based on various raw materials to identify the best approach using LCA in China was carried out in the work by Zhang et al. (2022a, b), etc.

4.3 TEA and LCA in the Biomanufacturing of Enzyme

Biological pretreatment of LCB enables the removal of hemicellulose, cellulase, or lignin layer, utilizing enzymes and then hydrolysing the remaining biomass for the production of value-added products. Enzyme-mediated pretreatment and hydrolysis of recalcitrant LCB are one of the major factors in the overall success of sustainable biorefineries (Reis et al. 2023). Enzymes like cellulase, xylanase, amylase, ligase, protease, tannase, pectinase, lipase, and zymase offer an eco-friendly alternative for the production of biofuels, bioplastics, textiles, fertilizers, and biochemicals. The rapid transition of biorefineries from linear to circular bioeconomy has impacted the market demand for enzymes. Due to the increasing market prices, to produce enzymes inexpensive LCB like agricultural waste or food waste is used (Bankeeree et al. 2024). Alternatives like minimizing the enzyme dosage but maintaining effectivity, enzyme recycling, and pretreated biomass washing can also contribute to reducing the use of expensive enzymes and chemicals. Usually, the commercial enzyme used is produced offsite. But offsite enzymes bring with them costs like storage, transportation, and stability maintenance prices. One of the best ways to increase economic profitability is onsite enzyme production because of the reduced cost of flexibility, reliability, and transport (Carpio et al. 2022).

Biorefineries use enzymes that can be optimized by integration with other enzymes; genetic engineering enzyme-producing microbes like bacteria, fungi, and actinomycetes; techniques to improve activity like enzyme immobilization; and environmental, economic, and social sustainability simultaneously. Research on enzyme application in value-added production focuses on analysis technique application on the entire process and bioproducts but rarely on the primary steps of the hotspot, that is, the enzyme production aspect. LCA methodologies to range the environmental impact of systems are important, but LCA results are oversimplified, non-existent, or sometimes not disclosed when accounting for the life or cost flow in TEA (Nogueira et al. 2022). Carpio et al. (2022) conducted a study that performed a comparative TEA of enzymes produced onsite vs offsite for utilization for biogas and ethanol production. The study used process systems engineering tools to assess four configurations of a second-generation ethanol plant attached to an existing independent distillery. The analysis discussed water and electricity usage and economic viability. The requirement for a second fungus strain for onsite cellulase production was identified. The downstream section's impact on cost was also evaluated. Ferreira et al. (2021) performed TEAs on the production process of cellulase and other lignocellulose-degrading enzymes within the last two decades. The production costs of factors like raw materials, capital investments, utilities, labour expenses, and more were compared. The study observed the dominance of the process of submerged culture with *Trichoderma reesei* for cellulase production. The analysis showed both the pros and cons of onsite and offsite production, and the raw material and facility-dependent costs are the main contributors to enzyme cost. Sensitive analysis results showed fermentation substrate, enzyme yield, and production scale to have an important impact on the enzyme cost. Singh et al. (2022) produced cellulase-free xylanase from de-oiled neem cake by using *Trichoderma* species under solid-state fermentation. LCA for enzyme production was performed using the software SimaPro. A significant impact was seen due to electricity consumption (92.82%) and ammonium sulphate salt usage (6.17%). The increase in the release of reducing sugars (mg/g) for pine needle was 41.43 ± 1 mg/g, 137.54 ± 0.96 mg/g for corn cob, 105.17 ± 0.7 mg/g for sugarcane bagasse, and 76.66 ± 1.29 mg/g for PS. Nogueira et al. (2022) evaluated the effect of onsite enzyme production integration with 2G ethanol and electricity biorefinery using sugarcane. The findings indicated that onsite enzyme production can lower the overall environmental impact compared to offsite production, and this pattern holds true for various enzyme dosages and different inventory methods for modeling offsite production. The offsite production literature showed a lack of transparency and detail, leading to significant disparities between impact indicators. In comparison with 1G ethanol, 2G ethanol with onsite enzyme production establishes an approximately 80% reduction in the GWP. Ecotoxicity, mineral scarcity, eutrophication, and GWP reduction showed trade-offs when using cellulosic ethanol, depending on the enzyme inventory used.

4.4 TEA and LCA in the Biomanufacturing of Biobased Products

Plastic offers durability, resistance, transparency, lightweight, and low cost. Recycling plastics can be a method to enable circularity and a wise step towards sustainability, but still, the recycling rate in the United States is 10% (Di et al. 2021). Plastic recycling has been increasing at a fast rate in India; however, there is still a long way to go. India holds the 12th position in the global ranking of countries with significant issues related to mismanaged plastics. However, experts predict that by the year 2025, India is likely to climb to the fifth position in the ranking (Neo et al. 2021). This situation can be attributed to inadequate waste management practices and waste disposal management which are responsible for a substantial 71% of mishandled plastics in the Asian region (Neo et al. 2021). Plastic waste is a worldwide concern, as it is not only a cause of land pollution but is also a major cause of danger to the ecosystem. This has led biorefineries to produce biodegradable plastics as an alternative to nonbiodegradable plastics. LCA analysis for the bioplastic life cycle and TEA for the system cost estimation are carried out in biorefineries to optimize large-scale bioplastic production. Some of the latest research on the techno-economic analysis of bioplastic production using different feedstocks is that of Chalermthai et al. (2023), which performed a techno-economic analysis to evaluate the co-production of edible bioplastics (extrusion method) and *Spirulina* powder food supplements using microalgae. Approximately 1200 MT/y of plastic and approximately 1000 MT/y of *Spirulina* were the production scales of this system. The TEA was carried out by using the software SuperPro Designer v12.3, and the duration of the project was 10 years. For TEA, data was accumulated from different literature and sources. The selected location of the *Spirulina* biorefinery plant was in a mountainous area of Chiang Mai in the northern province of Thailand. This was followed by cultivation, harvesting, treatment, and then TEA analysis. The results of TEA depicted that this co-production showed an attractive venture with a payback time (PBT) of 2.6 years (min.) and an ROI of 38.5% (max.) because of the high revenue of $55.6 million/year despite high operational and capital costs. Sensitivity analysis showed that the profitability exhibited variations in response to changes in crucial parameters. The sensitivity analysis also identified the split ratio of biomass allocated to food supplements versus bioplastic production and the selling price of the bioplastic as the most sensitive factors.

Polyhydroxyalkanoates (PHAs) are a condensation polyester that is a promising bioplastic compound and is being extensively used in bioplastic production. They come in three categories: short-chain length (scl) with 3–5 C atoms, medium-chain length (mcl) with 6–14 C, or long-chain length (lcl) with 15 C or more. Bioplastics can be produced via microbes like fungi, bacteria, and cyanobacteria. Microbial poly(3-hydroxybutyrate) and biobased polyurethane (Wiatrowski et al. 2023) are great alternatives to reduce traditional PHB production costs. Price et al. (2022) studied the current state of PHB (10,000 tonnes capacity) produced using photosynthetic cyanobacteria by utilizing CO_2 as a metabolic carbon substrate. The TEA was performed to explore the financial profitability of cultivating cyanobacteria on open raceway ponds as well as solvent extraction on biomass. The TEA results showed that $193.5 M was the capital cost and $147.3 M the operating cost per year when 10% of the PHB yield was used. A 20-year NPV of -$1.8B and breakeven MPSP of PHB of $18,339 were calculated. Sensitive analysis demonstrated that PHB yield was one of the major factors behind PHB production. Strategies like wastewater treatment, additional revenue for pigment extraction, use of PHB ripening ponds, and others were explored to increase the economic viability of the processing system. Flocculants can reduce operational costs; energy and nutrient prices are reduced by onsite anaerobic digesters and solar power purchase agreements. MPSP of $7704 per tonne was observed if all factors were integrated. Another bioresource that is gaining momentum for bioplastic production is seaweed. Ayala et al. (2023) performed the pilot LCA of *Saccharina latissima* seaweed derived for the production of alginate-based bioplastic. The parameters included in this assessment were the account of seaweed cultivated, alginate extraction, pilot-scale level of bioplastic production, and pathways. ReCiPe 2016 Midpoint (E) was used for LCA, and SimaPro v9.2.0.2 and Brightway2 were used for the calculation of the impacts and for uncertainty analysis. Scenario modeling was utilized for the assessment of alginate extraction, and carbon balance and uncertainty analysis was also performed. The LCA results concluded that the last step of production (i.e., film fabrication) is the major hotspot due to the usage of glycerol. A significant variation was observed depending on the end of the life of the bioplastics produced. The results also pointed out that there was a 30% reduction in the environmental impact of composting in comparison with incineration. A higher protein feedstock can generate higher proteinaceous and higher-quality bioplastics. Algal-based bioplastics can also be used as a replacement for traditional plastics. Beckstrom et al. (2020) carried out a study to investigate the bioplastic feedstock production from algae biomass and fuel. Three growth pathways were examined for economic viability and environmental impact, and three biomass conversion scenarios were examined. The total number of scenarios that were evaluated via TEA and LCA in the study was nine. The findings of the assessment revealed that the minimum selling price of bioplastic feedstock varies as low as $970 USD per ton. Results from the LCA highlighted significant enhancements in the performance of the produced bioplastic feedstock, showing reduced greenhouse gas emissions ranging from 67% to 116% compared to petroleum-based plastic feedstock. These outcomes supported algae-based bioplastic feedstock biorefineries. The TEA results showed the drying pathway to be the most economically profitable bioconversion pathway out of all the others evaluated. The TEA study also delved into sensitivity analysis, alternative co-products, and the potential bioconversion pathway out of all others that were evaluated.

Biobased products that are gaining momentum are non-nutritive sweeteners which are produced to replace added sugar in food. One of the main reasons for this rise is to improve consumer health. The sustainability of natural sweetener production has not been extensively explored. To explore the economic, sustainability, and environmental impact of natural sweeteners like xylitol and sorbitol production, LCA and TEA need to be performed. Suckling et al. (2023a, b) carried out research which was a part of project SWEET in which LCA was performed for examining the production of 60% and 95% pure rebaudioside A (RA60) steviol glycoside mix derived from *Stevia rebaudiana* leaves cultivated in Europe. The findings were evaluated using the ReCiPe 2016 (H) method, and the categories explored were GWP, eutrophication, water, and land usage. RA60 has 200–300× sweetness to sugar, and environmental impact in this study was compared with sugar produced by beets and cane. The findings of this work found GWP to be 20.25 kgCO_2-eq/kg on a mass basis (RA60) and 0.081 kgCO_2-eq/kg on a sweetness equivalence basis (SE). A significant contributor to the environmental impact was field production for stevia leaves and extraction of RA60 for others. GWP reduction by 18.8% was seen on the removal of international transport. In comparison with sugars, RA60 has approximately 5.7–10.2% GWP impact and 5.6–7.2% land use impact and showcases a lower impact in most of the categories. These findings suggested that RA60 should be used to reduce environmental impact and replace sugars. Shaji et al. (2022) studied and proposed a methodology (upstream and downstream) and assessed a xylitol biorefinery from sugarcane bagasse in India. The analysis was calculated for a plant with 4 t/h of dry bagasse quantity. The facility generated 437.4 kg/h of xylitol from the 0.54 g of xylitol per g of xylose produced via fermentation. Upstream data was collected from the literature, and downstream data was measured by Aspen Plus flowsheet stimulation. The TEA results showed that the production cost was US $3.17/kg (230 INR/kg) for xylitol and that the MPSP for a four-year payback period was US $6.2/kg (450 INR/kg). A significant contributor to the production cost was fermentation and pretreatment. The LCA depicted GHG emissions are 2.759 kg CO_2-eq/kg of xylitol. GHG emissions were majorly due to electrical requirements and can be enhanced by reducing fermentation duration. Some other examples of evaluation by TEA and LCA of biobased product development are cellulose nanocrystal production from wood pulp by Rajendran et al. (2023); xylooligosaccharide and 1,3-propanediol production utilizing A molasses and sugarcane bagasse by van Heerden et al. (2023); and production of pectin-derived oligosaccharides (food additive) and bulky terephthalic acid using sugar beet pulp by García-Velásquez and Van der Meer (2022).

5 Challenges

Despite recent studies implementing TEA and LCA in research for transitioning the conceptual idea of biobased sustainable production to large upscale production, there are still some challenges associated with these assessments concerning the biorefinery studies and biomanufacturing of bioenergy, chemicals, biobased products, etc. During assessment, there is a focus only on one method rather than trying to find and apply alternative methods for the same result. Although the lab scale works well, the production cost and efficiency drop sharply on a large scale, and real-time applications face difficulties.

Biofuel production is known to focus on pyrolysis and gasification. Even a single pathway includes numerous parameters that impact techno-economic and environmental feasibilities. The TEA studies can analyse a project with a duration range of more than 5 years but still are a single-point estimate. There is a need to measure the capital and operational price, but despite this, there is a variation and inconsistency in the result. One of the biggest concerns with TEA is the unavailability of concrete and reliable process data which can be rectified by developing result calibration techniques. Many TEA studies assume all their input variables to be completely certain, but the parameters are taken from the current market price. This assumption is often violated by the high variability of many economic factors over time, which lowers the reliability of TEA outcomes (Fu et al. 2023). Often these unrealizabilities and uncertainties require sensitivity analysis where simulation models like Monte Carlo stimulate numerous scenarios focused on varying probability distribution. Sensitivity analysis implementation is a necessity when carrying out TEA for developed and accurate results of biorefinery development which has been lacking in studies.

In LCA, one of the challenges arises due to the lack of multi-facilitated study of multiple feedstock and product analyses. LCA study needs to detail and comprehensively analyse upstream and downstream processes to facilitate integrated multiproduct evaluation and to conduct a wise selection of biomass, separation, and storing waste lignocellulose biomass as it lacks theoretical and practical basis. Due to a lack of data associated with not enough research being carried out for integrated biomass conversion pathways or multiple feedstock utilization, the life cycle inventory is primitive and undersupplied. The fourth step of LCA is crucial for drawing meaningful conclusions and providing valuable insights based on the results obtained from the previous phases. Unfortunately, this interpretation phase is often rudimentary in studies and lacks detailed expansion. When performing LCA and collecting information on methodology, it becomes difficult to understand "how" the results were concluded and its soundness. Work needs to be implemented in the elaboration of the inventory and interpretation step, and if the reason for not elaborating is justified, it needs to be mentioned for a more harmonized and credible LCA (Laurent et al. 2020). LCA is solely based on the existing guidance from previous studies and needs expert supervision for the validity of the work.

Parameters like feedstock price, availability, land use charge, and energy systems are some of the major factors impacting environmental performance but are not stagnant and change with time and space (Hosseinzadeh-Bandbafha et al. 2021). These changes need to be taken into consideration, and to obtain more certain and realistic outcomes of LCA, the traditional and static LCA needs to be replaced with more dynamic and advanced assessment technology. There is also a need to understand the use of inexpensive plant species and develop more low-maintenance biomass. Potentially reducing the environmental impact by integrating energy systems such as solar energy can have an adverse negative effect economically, as the cost might change volatility. Therefore, it is important to carry out a sensitive, energy analysis (focused on ecosystem energy) and exergy analysis (comprehensive thermodynamic evaluation) (Tahir et al. 2021).

As discussed before, SLCA needs to be implemented when it comes to the evaluation of factors like race, ethics, gender, and salary differences which are associated with the plant workers and their scientific decision-making at every step when it comes to sustainability. One of the other challenges of assessment technologies like TEA and LCA is the fact that due to a lack of methodology to validate the results, there is a huge risk of misuse by corporations and other interests. Unethical corporations can manipulate results to sprout biased results for approval of their work for funding, marketing, and large-scale advancement. There are upcoming studies that are focused on developing TEA tools and some for developing LCA tools for life-cycle assessment, but there is an apparent lack of a singular tool that enables the integration and performance of TEA and LCA in a single notch (Fig. 3). For an integrated study managing two different tools and their comprising parameters, datasheets and boundaries make the decision-making difficult. As per our knowledge no evident and practical software are able to perform TEA-LCA for all possible process systems and applications. Some examples of the tools that can perform TEA-LCA only for some applications are the Techno-economic Energy & Carbon Heuristic Tool for Early-State Technologies v1.0 (TECHTEST) which measures energy, carbon, and cost impacts (Carpenter et al. 2023), BioSTEAM (Shi and Guest 2020), Aspen Plus (Borugadda et al. 2020), etc. Studies that focus on developing integrated LCA and TEA frameworks are essential for the optimization and trade-off analysis.

The study of Ferdous et al. (2023) proposed a framework for an integrated TEA-LCA framework that can also process simulation and optimization that can evaluate and optimize the sustainability of pulse protein extraction pathways with an integrated assessment with geographic information systems (GIS) and the analytic

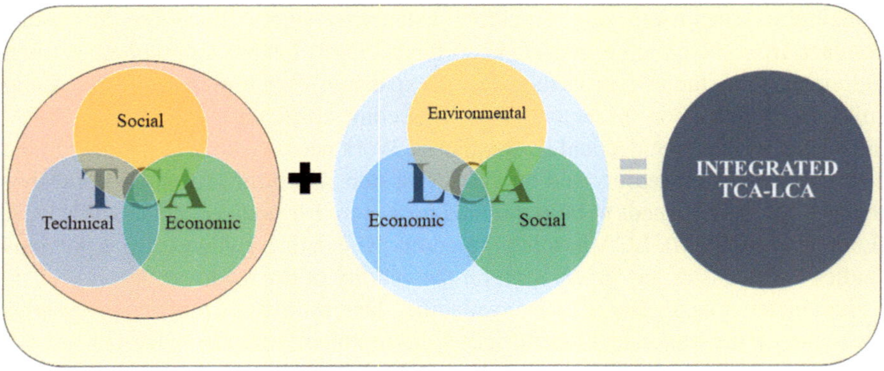

Fig. 3 Venn diagram highlighting the integrated sectors in TEA and life cycle assessment and their collaboration

hierarchy process (AHP). Based on technical, economic, and environmental criteria, the steps summarized for integration of process simulation with LCA and TEA in this chapter are as follows: (1) define system boundary, (2) process simulation, (3) mass and energy balance, (4) build LCI and OPEX/CAPEX inventory, and (5) estimate environmental and economic impacts. Mahmud (2022) worked on developing an integrated TEA-LCA framework, identified methodology, and highlighted challenges of using these at low TRLs. The study demonstrated a framework for two applications—(1) economic and environmental assessment for aerobic treatment of wastewater treatment plant (CUWWTP) and (2) economic and environmental assessment for anaerobic membrane treatment of wastewater treatment plant (CUWWTP). Bacatelo et al. (2023) developed a multi-criteria framework that was used in a case study for the selection of viable CO_2 capture technology in a Portuguese cement plant. Integrated TEA-LCA with an analytic hierarchy process and a weighted sum model showed that out of the three technologies (monoethanolamine, calcium looping post-combustion, and oxyfuel), calcium looping showed the best performance. The integration of TEA and LCA for one single tool will have challenges like lack of availability of data, differences in data formats for both assessments that would be needed to convert into one single format, and deciding on which data would be compatible with each other (Mahmud et al. 2021).

6 Outlook

On reviewing different strategies and applications of TEA and LCA, it can be stated that these tools are promising approaches to investigating the techno-economic and environmental feasibility of biorefineries. Moving forward, work focusing on elaborating the inventory and interpretation steps in both TEA and LCA and research incorporating collaborative TEA-LCA and other integrated assessment technologies as well as SLCA need to be implemented for the evaluation of factors associated with plant workers and their scientific decision-making. Work needs to be done for developing an advanced tool that can perform TEA-LCA in a singular take.

In the future of smart industries, for the advancement of analysis technologies, studies on development strategies have begun incorporating data-driven technologies with TEA and LCA. Chai et al. (2022) carried out a review of TEA and its integration with these cyber technologies. They concluded that the genetic algorithm was the most used technology integrated with TEA. There was a scarce application of technology like machine learning, artificial neural network, and blockchain, and other technologies like cyber-physical systems, the Internet of Things (IoT), cloud computing, big data analytics, digital twin (DT), and metaverse were not seen being incorporated at all. In 2023, there has been more incorporation of artificial intelligence (AI) in LCA as well as TEA. AI helps find the best solution to uncertainty and can be applied to many fields to identify and resolve critical issues. The genetic algorithm is one of the methodologies often used for process optimization and is incorporated with TEA and LCA. Hedayati Moghaddam et al. (2023) worked on biomethanol production through a gasification route, and optimal operational conditions were identified using the genetic algorithm. Kaab et al. (2019) employed artificial neural networks (ANN) and an adaptive neuro fuzzy inference system (ANFIS) model for LCA for the prediction of output energy and environmental impacts of sugarcane production. The findings of ANN models demonstrated high R2 values of 0.923–0.986, whereas the ANFIS model has a higher R2 value of 0.912–0.999 for both environmental impacts and output energy in planted and ratoon farms. These results indicate that ANFIS is a better tool than ANN for prediction. Manoj Kumar and Chopra (2023) designed a grapevoltaic farm where the grape farmer and the solar plant operator share the land, water, and energy. They also used blockchain technology to make sure they follow the rules of sharing and help each other when there is a climate problem. The paper studied 12 places in India where grapes are grown. The research was conducted to help the grape farmer and the solar plant operator make more money, use less water, reduce greenhouse gas emissions, and connect to the government's GrapeNet platform. This study shows a new way of making grape farms and energy networks work together. The paper uses blockchain technology to make sure the sharing rules are followed, and the partners help each other when there is a climate problem. The study was performed using a method called Resilience Performance, Life Cycle Analysis, and Techno-Economic (RePLiCATE). Machine learning (ML) can be applied in various biorefineries. For example, ML can be applied in biofuel production to predict the amount and quality of biofuels from different feedstocks and conditions. They can be incorporated with TEA for optimized measurements (Jeon et al. 2023). Cyber-physical systems (CPS) are one of the background infrastructures with great impact. Cortes-Cornax et al. (2023) studied how to design smart systems that have less environmental impact by considering different aspects and choices by discussing system configuration, data management, and the goal and functionalities of the cyber-physical system (CPS). LCA was used to measure the environmental impact of the devices and data management in the CPS and help analyse CPS's trade-offs. Kumar et al. (2021) designed

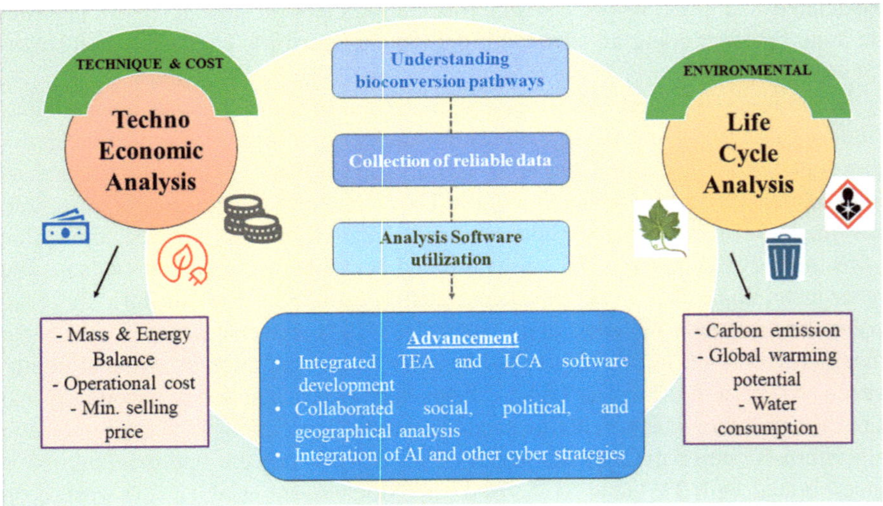

Fig. 4 Graphical depiction of strategies for transforming the primitive TEA and LCA into reliable and advance analysis techniques

a CPS framework for live LCA to evaluate the environmental impact of 3D printed products produced by different process methodologies and parameters. Research for developing TEA and LCA software which has a graphical user interface and other interfaces like question-and-answer user interface, form-based user interface, and command line interface should be investigated. Ezzat Nousir et al. (2023) worked on developing TEA software based on a graphical user interface to calculate the array and battery size of a photovoltaic (PV) system. The new software uses C sharp language and was developed into software design off-grid, on-grid, and hybrid systems. The new software features observed were constant load, daily load, and night load, selection of the best budget options, time of energy unavailability, economic analysis of cost transactions from the bank to know monthly instalments and total cost, parameter editing, and user-friendliness. Reduced system cost by lowering the reliability slightly (e.g. from 100% to 99.7% or 99.3%) was seen. It is important to implement reliable data inventory, method standardization, development software, and artificial intelligence techniques for transitioning TEA and LCA into advanced efficient practical assessment technology (Fig. 4).

7 Conclusion

TEA and LCA are strong techniques for assessing environmental consequences and cost feasibility of the production system of value-added biobased biorefinery products. These strategies are a key factor for a sustainable circular economy and serve as a footing for decision-making. Despite the evident lack of data and standardized methods for these techniques, research focused on methodology designing, reliable data collection via analysis software, and utilizing it in diverse application sectors can be a stepping stone for the development of bio-based industries. The influence of biomass conversion pathways and different feedstocks on these techniques along with the integration of can positively impact the entire process. Thus, the TEA and LCA are one of the most important analysis strategies that in future can play a very critical role for the commercialization of green biorefinery plants and contribute towards the goal of sustainable development.

References

Abu-Bakar NA, Roslan AM, Hassan MA et al (2023) Environmental impact assessment of rice mill waste valorisation to glucose through biorefinery platform. Sci Rep 13:14767. https://doi.org/10.1038/s41598-023-28487-2

Agrawal K, Verma P (2020) Production optimization of yellow laccase from Stropharia sp. ITCC 8422 and enzyme-mediated depolymerization and hydrolysis of lignocellulosic biomass for biorefinery application. Biomass Convers Biorefinery 12:1–20. https://doi.org/10.1007/s13399-020-00869-w

Agrawal K, Bhardwaj N, Kumar B, Chaturvedi V, Verma P (2019) Process optimization, purification and characterization of alkaline stable white laccase from *Myrothecium verrucaria* ITCC-8447 and its application in delignification of agroresidues. Int J Biol Macromol 125:1042–1055. https://doi.org/10.1016/j.ijbiomac.2018.12.108

Agrawal K, Nair LG, Chaturvedi V, Verma P (2023) Designing microbial cellulases using genetic engineering approach: a promising strategy towards zero-waste cellulosic biorefinery. Biocatal Agric Biotechnol 52:102830. https://doi.org/10.1016/j.bcab.2023.102830

Ahmad T, Zhang D (2021) Renewable energy integration/techno-economic feasibility analysis, cost/benefit impact on islanded and grid-connected operations: a case study. Renew Energy 180:83–108. https://doi.org/10.1016/j.renene.2021.08.041

Akhbari A, Onn CC, Ibrahim S (2021) Analysis of biohydrogen production from palm oil mill effluent using a pilot-scale up-flow anaerobic sludge blanket fixed-film reactor in life cycle perspective. Int J Hydrog Energy 46(68):34059–34072. https://doi.org/10.1016/j.ijhydene.2021.07.186

AlNouss A, Shahbaz M, Mckay G, Al-Ansari T (2022) Bio-methanol production from palm wastes steam gasification with application of CaO for CO_2 capture: techno-economic-environmental analysis. J Clean Prod 341:130849. https://doi.org/10.1016/j.jclepro.2022.130849

Arias A, Feijoo G, Moreira MT (2023) How could artificial intelligence be used to increase the potential of biorefineries in the near future? A review. Environ Technol Innov 32:103277. https://doi.org/10.1016/j.eti.2023.103277

Atilano-Camino MM, Álvarez-Valencia LH, García-González A, García-Reyes RB (2020) Improving laccase production from Trametes versicolor using lignocellulosic residues as cosubstrates and evaluation of enzymes for blue wastewater biodegradation. J Environ Manag 275:111231. https://doi.org/10.1016/j.jenvman.2020.111231

Ayala M, Thomsen M, Pizzol M (2023) Life cycle assessment of pilot scale production of seaweed-based bioplastic. Algal Res 71:103036. https://doi.org/10.1016/j.algal.2023.103036

Azapagic A, Clift R (1998) Linear programming as a tool in life cycle assessment. Int J Life Cycle Assess 3:305–316. https://doi.org/10.1007/BF02979340

Bacatelo M, Capucha F, Ferrao P, Margarido F (2023) Selection of a CO_2 capture technology for the cement industry: an integrated TEA and LCA methodological framework. J CO2 Util 68:102375. https://doi.org/10.1016/j.jcou.2022.102375

Bahlawan H, Morini M, Pinelli M, Spina PR, Venturini M (2021) Simultaneous optimization of the design and operation of multi-generation energy systems based on life cycle energy and economic assessment. Energy Convers Manag 249:114883. https://doi.org/10.1016/j.enconman.2021.114883

Bakry MM, Salem SS, Atta HM, El-Gamal MS, Fouda A (2022) Xylanase from thermotolerant Bacillus haynesii strain, synthesis, characterization, optimization using box-Behnken design, and biobleaching activity. Biomass Convers Biorefinery 14:1–14. https://doi.org/10.1007/s13399-022-03043-6

Bankeeree W, Abd-Aziz S, Prasongsuk S, Lotrakul P, Ibrahim SN, Punnapayak H (2024) Cellulase as biocatalyst produced from agricultural wastes. In: Chemical substitutes from agricultural and industrial by-products: bioconversion, bioprocessing, and biorefining. Wiley, New York, pp 319–336. https://doi.org/10.1002/9783527841141.ch16

Baruah J, Nath BK, Sharma R, Kumar S, Deka RC, Baruah DC, Kalita E (2018) Recent trends in the pretreatment of lignocellulosic biomass for value-added products. Front Energy Res 6(DEC):1–19. https://doi.org/10.3389/fenrg.2018.00141

Beckstrom BD, Wilson MH, Crocker M, Quinn JC (2020) Bioplastic feedstock production from microalgae with fuel co-products: a techno-economic and life cycle impact assessment. Algal Res 46:101769. https://doi.org/10.1016/j.algal.2019.101769

Bhagwat SS, Li Y, Cortés-Peña YR, Brace EC, Martin TA, Zhao H, Guest JS (2021) Sustainable production of acrylic acid via 3-hydroxypropionic acid from lignocellulosic biomass. ACS Sustain Chem Eng 9(49):16659–16669. https://doi.org/10.1021/acssuschemeng.1c05441

Bhardwaj N, Chanda K, Kumar B, Prasad HK, Sharma GD, Verma P (2017) Statistical optimization of nutritional and physical parameters for xylanase production from newly isolated *Aspergillus oryzae* LC1 and its application in the hydrolysis of lignocellulosic agro-residues. Bioresources 12(4):8519–8538

Bhardwaj N, Kumar B, Agarwal K, Chaturvedi V, Verma P (2019) Purification and characterization of a thermo-acid/alkali stable xylanases from *Aspergillus oryzae* LC1 and its application in Xylo-oligosaccharides production from lignocellulosic agricultural wastes. Int J Biol Macromol 122:1191–1202. https://doi.org/10.1016/j.ijbiomac.2018.09.070

Bhardwaj N, Kumar B, Agrawal K, Verma P (2020) Bioconversion of rice straw by synergistic effect of in-house produced ligno-hemicellulolytic enzymes for enhanced bioethanol production. Bioresou Technol Rep 10:100352. https://doi.org/10.1016/j.biteb.2019.100352

Bhardwaj N, Kumar B, Agrawal K, Verma P (2021) Current perspective on production and applications of microbial cellulases: a review. Bioresour Bioprocess 8:1–34. https://doi.org/10.1186/s40643-021-00447-6

Bittencourt GA, da Silva Barreto E, Brandão RL, Baêta BEL, Gurgel LVA (2019) Fractionation of sugarcane bagasse using hydrothermal and advanced oxidative pretreatments for bioethanol and biogas production in lignocellulose biorefineries. Bioresour Technol 292:121963. https://doi.org/10.1016/j.biortech.2019.121963

Borugadda VB, Kamath G, Dalai AK (2020) Techno-economic and life-cycle assessment of integrated Fischer-Tropsch process in ethanol industry for bio-diesel and bio-gasoline production. Energy 195:116985. https://doi.org/10.1016/j.energy.2020.116985

Bravo-Porras G, Fernández-Güelfo LA, Álvarez-Gallego CJ et al (2021) Influence of the total concentration and the profile of volatile fatty acids on polyhydroxyalkanoates (PHA) production by mixed microbial cultures. Biomass Conv Bioref 14:239. https://doi.org/10.1007/s13399-021-02208-z

Brobbey MS, Louw J, Görgens JF (2023) Biobased acrylic acid production in a sugarcane biorefinery: a techno-economic assessment using lactic acid, 3-hydroxypropionic acid and glycerol as intermediates. Chem Eng Res Des 193:367–382. https://doi.org/10.1016/j.cherd.2023.03.034

Busche RM, Allen BR (1989) Technoeconomics of butanol extractive fermentation in a multiphase fluidized bed bioreactor. Appl Biochem Biotechnol 20:357–374. https://doi.org/10.1007/BF02936496

Carpenter A, Shimizu K, Armstrong K, Reese S, Liddell H, Nimbalkar S (2023) Energy & materials resource flows (no. NREL/PO-6A20–86135). NationalRenewable Energy Labouratory (NREL), Golden, CO

Carpio RR, Secchi SG, Barros RO, Oliveira RA, Queiroz S, Teixeira RSS, Bon EPS, Secchi AR (2022) Techno-economic evaluation of second-generation ethanol from sugarcane bagasse: commercial versus on-site produced enzymes and use of the xylose liquor. J Clean Prod 369:133340. https://doi.org/10.1016/j.jclepro.2022.133340

Chai SYW, Phang FJF, Yeo LS, Ngu LH, How BS (2022) Future era of techno-economic analysis: insights from review. Front Sustain 3:924047. https://doi.org/10.3389/frsus.2022.924047

Chalermthai B, Charoensuppanimit P, Nootong K, Olsen BD, Assabumrungrat S (2023) Techno-economic assessment of co-production of edible bioplastic and food supplements from spirulina. Sci Rep 13(1):1–15. https://doi.org/10.1038/s41598-023-37156-3

Chan YH, Cheah KW, How BS, Loy ACM, Shahbaz M, Singh HKG, Shuhaili AFA, Yusup S, Ghani WAWAK, Rambli J, Kansha Y (2019) An overview of biomass thermochemical conversion technologies in Malaysia. Sci Total Environ 680:105–123. https://doi.org/10.1016/j.scitotenv.2019.04.211

Chen L, Debnath D, Zhong J, Ferin K, VanLoocke A, Khanna M (2021) The economic and environmental costs and benefits of the renewable fuel standard. Environ Res Lett 16(3):034021. https://doi.org/10.1088/1748-9326/abd7af

Cherubini F (2010) The biorefinery concept: using biomass instead of oil for producing energy and chemicals. Energy Convers Manag 51(7):1412–1421. https://doi.org/10.1016/j.enconman.2010.01.015

Choe C, Byun M, Lee H, Lim H (2022) Techno-economic and environmental assessments for sustainable bio-methanol production as landfill gas valorization. Waste Manag 150:90–97. https://doi.org/10.1016/j.wasman.2022.06.040

Chomkhamsri K, Wolf MA, Pant R (2011) International reference life cycle data system (ILCD) handbook: review schemes for life cycle assessment. In: Finkbeiner M (ed) Towards life cycle sustainability management. Springer, Dordrecht. https://doi.org/10.1007/978-94-007-1899-9_11

Clauser NM, Gutiérrez S, Area MC, Felissia FE, Vallejos ME (2018) Techno-economic assessment of carboxylic acids, furfural, and pellet production in a pine sawdust biorefinery. Biofuels Bioprod Biorefin 12(6):997–1012. https://doi.org/10.1002/bbb.1915

Cortes-Cornax M, Lago P, Roncancio C (2023) A case study of environmental considerations and opportunities in cyber-physical systems. IET Softw 17(4):424–434. https://doi.org/10.1049/sfw2.12130

Cortés-Peña YR, Kurambhatti C, Eilts K, Singh V, Guest JS (2022) Economic and environmental sustainability of vegetative oil extraction strategies at integrated Oilcane and oil-sorghum biorefineries. ACS Sustain Chem Eng 10(42):13980–13990. https://doi.org/10.1021/acssuschemeng.2c04204

Costa MW, Oliveira AA (2022) Social life cycle assessment of feedstocks for biodiesel production in Brazil. Renew Sust Energ Rev 159:112166. https://doi.org/10.1016/j.rser.2022.112166

Cusenza MA, Longo S, Guarino F, Cellura M (2021) Energy and environmental assessment of residual bio-wastes management strategies. J Clean Prod 285:124815. https://doi.org/10.1016/j.jclepro.2020.124815

Dabbaghi F, Tanhadoust A, Nehdi ML, Nasrollahpour S, Dehestani M, Yousefpour H (2021) Life cycle assessment multi-objective optimaization and deep belief network model for sustainable lightweight aggregate concrete. J Clean Prod 318:128554. https://doi.org/10.1016/j.jclepro.2021.128554

Dai Z, Pomraning KR, Deng S, Kim J, Campbell KB, Robles AL, Hofstad BA, Munoz N, Gao Y, Lemmon T, Swita MS (2023) Metabolic engineering to improve production of 3-hydroxypropionic acid from corn-Stover hydrolysate in Aspergillus species. Biotechnol Biofuels Bioprod 16(1):53. https://doi.org/10.1186/s13068-023-02288-1

Dalsasso RR, Pavan FA, Bordignon SE, de Aragão GMF, Poletto P (2019) Polyhydroxybutyrate (PHB) production by Cupriavidus necator from sugarcane vinasse and molasses as mixed substrate. Process Biochem 85:12–18. https://doi.org/10.1016/j.procbio.2019.07.007

Das SR, Basak N (2022) Optimization of process parameters for enhanced biohydrogen production using potato waste as substrate by combined dark and photo fermentation. Biomass Convers Biorefinery 14:1–21. https://doi.org/10.1007/s13399-022-02588-w

Dees J, Oke K, Goldstein H, McCoy ST, Sanchez DL, Simon AJ, Li W (2023) Cost and life cycle emissions of ethanol produced with an oxyfuel boiler and carbon capture and storage. Environ Sci Technol 57(13):5391–5403. https://doi.org/10.1021/acs.est.2c04784

Degfie TA, Mamo TT, Mekonnen YS (2019) Optimized biodiesel production from waste cooking oil (WCO) using calcium oxide (CaO) nano-catalyst. Sci Rep 9(1):18982. https://doi.org/10.1038/s41598-019-55403-4

Dekker E, Zijp MC, van de Kamp ME et al (2020) A taste of the new ReCiPe for life cycle assessment: consequences of the updated impact assessment method on food product LCAs. Int J Life Cycle Assess 25:2315–2324. https://doi.org/10.1007/s11367-019-01653-3

Del Pozo CA, Cloete S (2022) Techno-economic assessment of blue and green ammonia as energy carriers in a low-carbon future. Energy Convers Manag 255:115312. https://doi.org/10.1016/j.enconman.2022.115312

Derwent RG (2023) Global warming potential (GWP) for hydrogen: sensitivities, uncertainties and meta-analysis. Int J Hydrog Energy 48(22):8328–8341. https://doi.org/10.1016/j.ijhydene.2022.11.219

Dharshini RS, Fathima AA, Dharani SR, Ramya M (2020) Utilization of alginate from brown macroalgae for ethanol production by clostridium phytofermentans. Appl Biochem Microbiol 56:173–178. https://doi.org/10.1134/S0003683820020040

Di Fulvio F, Forsell N, Korosuo A, Obersteiner M, Hellweg S (2019) Spatially explicit LCA analysis of biodiversity losses due to different bioenergy policies in the European Union. Sci Total Environ 651:1505–1516. https://doi.org/10.1016/j.scitotenv.2018.08.419

Elgarahy AM, Eloffy MG, Alengebawy A, El-Sherif DM, Gaballah MS, Elwakeel KZ, El-Qelish M (2023) Sustainable management of food waste; pre-treatment strategies, techno-economic assessment, bibliometric analysis, and potential utilisations: a systematic review. Environ Res 225:115558. https://doi.org/10.1016/j.envres.2023.115558

Elias AM, Longati AA, de Campos Giordano R, Furlan FF (2021) Retro-techno-economic-environmental analysis improves the operation efficiency of 1G-2G bioethanol and bioelectricity facilities. Appl Energy 282:116133. https://doi.org/10.1016/j.apenergy.2020.116133

Elliott DC, Beckman D, Bridgwater AV, Diebold JP, Gevert SB, Solantausta Y (1991) Developments in direct thermochemical liquefaction of biomass: 1983-1990. Energy Fuel 5(3):399–410. https://doi.org/10.1021/ef00027a008

Ezzat Nousir S, Refaat Anis W, Abouelatta M, Samir Adly G, Abdelaal MA (2023) Developing techno-economic GUI-based software for on-grid, and off-grid photovoltaic systems sizing. Ain Shams Eng J 15:102508. https://doi.org/10.1016/j.asej.2023.102508

Fajimi LI, Chrisostomou J, Oboirien BO (2022) A techno-economic analysis (TEA) of a combined process of torrefaction and gasification of lignocellulose biomass (bagasse) for methanol and electricity production. Biomass Conv Bioref 14:12501. https://doi.org/10.1007/s13399-022-03499-6

Farobie O, Amrullah A, Anis LA, Hartulistiyoso E, Syaftika N, Saefurahman G, Bayu A (2023) Valorization of brown macroalgae Sargassum plagiophyllum for biogas production under different salinity conditions. Bioresour Technol Rep 22:101403. https://doi.org/10.1016/j.biteb.2023.101403

Ferdous J, Bensebaa F, Pelletier N (2023) Integration of LCA, TEA, process simulation and optimization: a systematic review of current practices and scope to propose a framework for pulse processing pathways. J Clean Prod 402:136804. https://doi.org/10.1016/j.jclepro.2023.136804

Ferreira RG, Azzoni AR, Freitas S (2021) On the production cost of lignocellulose-degrading enzymes. Biofuels Bioprod Biorefin 15(1):85–99. https://doi.org/10.1002/bbb.2142

Figueroa-Torres GM, Theodoropoulos C (2023) Techno-economic analysis of a microalgae-based biorefinery network for biofuels and value-added products. Bioresour Technol Rep 23:101524. https://doi.org/10.1016/j.biteb.2023.101524

Fu R, Kang L, Zhang C, Fei Q (2023) Application and progress of techno-economic analysis and life cycle assessment in biomanufacturing of fuels and chemicals. Green Chem Eng 4(2):189–198. https://doi.org/10.1016/j.gce.2022.09.002

Gainey B, Yan Z, Gandolfo J, Lawler B (2023) Methanol and wet ethanol as interchangeable fuels for internal combustion engines: LCA, TEA, and experimental comparison. Fuel 333:126257. https://doi.org/10.1016/j.fuel.2022.126257

García-Velásquez CA, Cardona CA (2019) Comparison of the biochemical and thermochemical routes for bioenergy production: a techno-economic (TEA), energetic and environmental assessment. Energy 172:232–242. https://doi.org/10.1016/j.energy.2019.01.073

García-Velásquez C, Van der Meer Y (2022) Mind the pulp: environmental and economic assessment of a sugar beet pulp biorefinery for biobased chemical production. Waste Manag 155:199–210. https://doi.org/10.1016/j.wasman.2022.10.038

Gholkar P, Shastri Y, Tanksale A (2021) Renewable hydrogen and methane production from microalgae: a techno-economic and life cycle assessment study. J Clean Prod 279:123726. https://doi.org/10.1016/j.jclepro.2020.123726

Ghosh S, Gnaim R, Greiserman S, Fadeev L, Gozin M, Golberg A (2019) Macroalgal biomass subcritical hydrolysates for the production of polyhydroxyalkanoate (PHA) by Haloferax mediterranei. Bioresour Technol 271:166–173. https://doi.org/10.1016/j.biortech.2018.09.108

Goedkoop M, Oele M, Leijting J, Ponsioen T, Meijer E (2016) Introduction to LCA with SimaPro8, 2022. https://pre-sustainability.com/legacy/download/SimaPro8IntroductionToLCA.pdf

Goswami RK, Agrawal K, Mehariya S, Molino A, Musmarra D, Verma P (2020) Microalgae-based biorefinery for utilization of carbon dioxide for production of valuable bioproducts. In: Chemo-biological systems for CO_2 utilization. CRC Press, Boca Raton, pp 203–228

Goswami RK, Mehariya S, Obulisamy PK, Verma P (2021) Advanced microalgae-based renewable biohydrogen production systems: a review. Bioresour Technol 320:124301. https://doi.org/10.1016/j.biortech.2020.124301

Goswami RK, Agrawal K, Upadhyaya HM, Gupta VK, Verma P (2022) Microalgae conversion to alternative energy, operating environment and economic footprint: an influential approach towards energy conversion, and management. Energy Convers Manag 269:116118. https://doi.org/10.1016/j.enconman.2022.116118

Harish V, Anwer N, Kumar A (2022) Applications, planning and socio-techno-economic analysis of distributed energy systems for rural electrification in India and other countries: a review. Sustain Energy Technol Assess 52:102032. https://doi.org/10.1016/j.seta.2022.102032

Hedayati Moghaddam A, Esfandyari M, Jafari D, Sakhaeinia H (2023) Multi-factor optimization of bio-methanol production through gasification process via statistical methodology coupled with genetic algorithm. Result Eng 20:101477. https://doi.org/10.1016/j.rineng.2023.101477

Heng L, Zhang H, Xiao J, Xiao R (2018) Life cycle assessment of polyol fuel from corn Stover via fast pyrolysis and upgrading. ACS Sustain Chem Eng 6(2):2733–2740. https://doi.org/10.1021/acssuschemeng.7b04378

Hos T, Herskowitz M (2022) Techno-economic analysis of biogas conversion to liquid hydrocarbon fuels through production of lean-hydrogen syngas. ACS Engineering Au 2(5):450–460. https://doi.org/10.1021/acsengineeringau.2c00019

Hosseinzadeh A, Zhou JL, Altaee A, Li D (2022) Machine learning modeling and analysis of biohydrogen production from wastewater by dark fermentationprocess. Bioresour Technol 343:126111. https://doi.org/10.1016/j.biortech.2021.126111

Hosseinzadeh-Bandbafha H, Aghbashlo M, Tabatabaei M (2021) Life cycle assessment of bioenergy product systems: a critical review. E-Prime-Adv Electr Eng Electron Energy 1:100015. https://doi.org/10.1016/j.prime.2021.100015

Hytönen E, Stuart P (2011) Techno-economic assessment and risk analysis of biorefinery processes. In: Computer aided chemical engineering, vol 29. Elsevier, pp 1376–1380. https://doi.org/10.1016/B978-0-444-54298-4.50054-4

Jeon PR, Moon J, Ogunsola NO, Lee SH, Ling JLJ, You S, Park Y (2023) Recent advances and future prospects of thermochemical biofuel conversion processes with machine learning. Chem Eng J 471:144503. https://doi.org/10.1016/j.cej.2023.144503

Jha P, Singh S, Raghuram M, Nair G, Jobby R, Gupta A, Desai N (2019) Valorisation of orange peel: supplement in fermentation media for ethanol production and source of limonene. Environ Sustain 2:33–41. https://doi.org/10.1007/s42398-019-00048-2

Kaab A, Sharifi M, Mobli H, Nabavi-Pelesaraei A, Chau K (2019) Combined life cycle assessment and artificial intelligence for prediction of output energy and environmental impacts of sugarcane production. Sci Total Environ 664:1005–1019. https://doi.org/10.1016/j.scitotenv.2019.02.004

Kanagaraj B, Anand N, Samuvel Raj R, Lubloy E (2023) Techno-socio-economic aspects of Portland cement, Geopolymer, and limestone calcined clay cement (LC3) composite systems: a-state-of-art-review. Constr Build Mater 398:132484. https://doi.org/10.1016/j.conbuildmat.2023.132484

Kobos PH, Drennen TE, Outkin AV, Webb EK, Paap SM, Wiryadinata S (2020) Techno-economic analysis: Best practices and assessment tools (No. SAND-2020-13473). Sandia National Lab. (SNL-CA), Livermore, CA (United States); Sandia National Lab.(SNL-NM), Albuquerque, NM (United States); Hobart and William Smith Colleges, Geneva, NY (United States). https://doi.org/10.2172/1738878

Ketsub N, Whatmore P, Abbasabadi M, Doherty WO, Kaparaju P, O'Hara IM, Zhang Z (2022) Effects of pretreatment methods on biomethane production kinetics and microbial community by solid state anaerobic digestion of sugarcane trash. Bioresour Technol 352:127112. https://doi.org/10.1016/j.biortech.2022.127112

Khalid M (2022) Monte Carlo Analysis for Probabilistic Risk Assessment. https://doi.org/10.1016/B978-0-12-824315-2.00109-3

Khanali M, Akram A, Behzadi J, Mostashari-Rad F, Saber Z, Chau KW, Nabavi-Pelesaraei A (2021) Multi-objective optimization of energy use and environmental emissions for walnut production using imperialist competitive algorithm. Appl Energy 284:116342. https://doi.org/10.1016/j.apenergy.2020.116342

Khounani Z, Hosseinzadeh-Bandbafha H, Moustakas K, Talebi AF, Goli SAH, Rajaeifar MA, Khoshnevisan B, Jouzani GS, Peng W, Kim KH, Aghbashlo M (2021) Environmental life cycle assessment of different biorefinery platforms valorizing olive wastes to biofuel, phosphate salts, natural antioxidant, and an oxygenated fuel additive (triacetin). J Clean Prod 278:123916. https://doi.org/10.1016/j.jclepro.2020.123916

Kosamia NM, Samavi M, Piok K, Rakshit SK (2022) Perspectives for scale up of biorefineries using biochemical conversion pathways: technology status, techno-economic, and sustainable approaches. Fuel 324:124532. https://doi.org/10.1016/j.fuel.2022.124532

Kovalcik A, Pernicova I, Obruca S, Szotkowski M, Enev V, Kalina M, Marova I (2020) Grape winery waste as a promising feedstock for the production of polyhydroxyalkanoates and other value-added products. Food Bioprod Process 124:1–10. https://doi.org/10.1016/j.fbp.2020.08.003

Kovalev AA, Kovalev DA, Litti YV, Katraeva IV (2020) Biohydrogen production in the two-stage process of anaerobic bioconversion of organic matter of liquid organic waste with recirculation of digister effluent. Int J Hydrog Energy 45(51):26831–26839. https://doi.org/10.1016/j.ijhydene.2020.07.124

Kumar B, Bhardwaj N, Alam A, Agrawal K, Prasad H, Verma P (2018) Production, purification and characterization of an acid/alkali and thermo tolerant cellulase from Schizophyllum commune NAIMCC-F-03379 and its application in hydrolysis of lignocellulosic wastes. AMB Express 8:1–16. https://doi.org/10.1186/s13568-018-0696-y

Kumar M, Oyedun AO, Kumar A (2019) Biohydrogen production from bio-oil via hydrothermal liquefaction. In: Biofuels: alternative feedstocks and conversion processes for the production of liquid and gaseous biofuels. Academic Press, New York, pp 715–732. https://doi.org/10.1016/B978-0-12-816856-1.00029-4

Kumar R, Padma Vilochani PG, Kahnthinisha S, Patil O, Cerdas F, Sangwan KS, Herrmann C (2021) Live life cycle assessment implementation using cyber-physical production system framework for 3D printed products. Proc CIRP 105:284–289. https://doi.org/10.1016/j.procir.2022.02.047

Laltha M, Sewsynker-Sukai Y, Kana EG (2022) Simultaneous saccharification and citric acid production from paper wastewater pretreated banana pseudostem: optimization of fermentation medium formulation and kinetic assessment. Bioresour Technol 361:127700. https://doi.org/10.1016/j.biortech.2022.127700

Larsen VG, Tollin N, Sattrup PA, Birkved M, Holmboe T (2022) What are the challenges in assessing circular economy for the built environment? A literature review on integrating LCA, LCC and S-LCA in life cycle sustainability assessment, LCSA. J Building Eng 50:104203. https://doi.org/10.1016/j.jobe.2022.104203

Laurent A, Weidema BP, Bare J, Liao X, Maia de Souza D, Pizzol M, Sala S, Schreiber H, Thonemann N, Verones F (2020) Methodological review and detailed guidance for the life cycle interpretation phase. J Ind Ecol 24(5):986–1003. https://doi.org/10.1111/jiec.13012

Lee JW, Bhagwat SS, Kuanyshev N, Cho YB, Sun L, Lee Y-G, Cortés-Peña YR et al (2023) Rewiring yeast metabolism for producing 2, 3-butanediol and two downstream applications: techno-economic analysis and life cycle assessment of methyl ethyl ketone (MEK) and agricultural biostimulant production. Chem Eng J 451:138886. https://doi.org/10.1016/j.cej.2022.138886

Li Y, Bhagwat SS, Cortés-Peña YR, Ki D, Rao CV, Jin YS, Guest JS (2021) Sustainable lactic acid production from lignocellulosic biomass. ACS Sustain Chem Eng 9(3):1341–1351. https://doi.org/10.1021/acssuschemeng.0c08055

Lim SM, Lau MSL, Tiong EIJ et al (2020) Process design and economic studies of two-step fermentation for production of ascorbic acid. SN Appl Sci 2:816. https://doi.org/10.1007/s42452-020-2604-8

Liu Y, Lyu Y, Tian J, Zhao J, Ye N, Zhang Y, Chen L (2021) Review of waste biorefinery development towards a circular economy: from the perspective of a life cycle assessment. Renew Sust Energ Rev 139:110716. https://doi.org/10.1016/j.rser.2021.110716

Liu D, Hwang HJ, Otoupal PB, Geiselman GM, Kim J, Pomraning KR, Kim YM, Munoz N, Nicora CD, Gao Y, Burnum-Johnson KE (2023) Engineering Rhodosporidium toruloides for production of 3-hydroxypropionic acid from lignocellulosic hydrolysate. Metab Eng 78:72–83. https://doi.org/10.1016/j.ymben.2023.05.001

Lu C, Jiang D, Jing Y, Zhang Z, Liang X, Yue J, Li Y, Zhang H, Zhang Y, Wang K, Zhang N (2022) Enhancing photo-fermentation biohydrogen production from corn stalk by iron ion. Bioresour Technol 345:126457. https://doi.org/10.1016/j.biortech.2021.126457

Luong JHT, Tseng MC (1984) Process and technoeconomics of ethanol production by immobilized cells. Appl Microbiol Biotechnol 19:207–216. https://doi.org/10.1007/BF00251838

Mahmud R (2022) Integration of techno-economic analysis (TEA) and life cycle assessment (LCA) for sustainable process design. https://tigerprints.clemson.edu/all_dissertations/3139

Mahmud R, Moni SM, High K, Carbajales-Dale M (2021) Integration of techno-economic analysis and life cycle assessment for sustainable process design–A review. J Clean Prod 317:128247. https://doi.org/10.1016/j.jclepro.2021.128247

Mailaram S, Narisetty V, Ranade VV, Kumar V, Maity SK (2022) Techno-economic analysis for the production of 2, 3-butanediol from brewers' spent grain using pinch technology. Ind Eng Chem Res 61(5):2195–2205. https://doi.org/10.1021/acs.iecr.1c04410

Manmai N, Unpaprom Y, Ponnusamy VK, Ramaraj R (2020) Bioethanol production from the comparison between optimization of sorghum stalk and sugarcane leaf for sugar production by chemical pretreatment and enzymatic degradation. Fuel 278:118262. https://doi.org/10.1016/j.fuel.2020.118262

Manoj Kumar N, Chopra SS (2023) Integrated techno-economic and life cycle assessment of shared circular business model based blockchain-enabled dynamic grapevoltaic farm for major grape growing states in India. Renew Energy 209:365–381. https://doi.org/10.1016/j.renene.2023.03.064

Mata TM, Rodrigues S, Caetano NS, Martins AA (2022) Life cycle assessment of bioethanol from corn Stover from soil phytoremediation. Energy Rep 8:468–474. https://doi.org/10.1016/j.egyr.2022.01.059

Morakile T, Mandegari M, Farzad S, Görgens JF (2022) Comparative techno-economic assessment of sugarcane biorefineries producing glutamic acid, levulinic acid and xylitol from sugarcane. Ind Crop Prod 184:115053. https://doi.org/10.1016/j.indcrop.2022.115053

Moretti C, Hoefnagels R, van Veen M, Corona B, Obydenkova S, Russell S, Jongerius A, Vural-Gürsel I, Junginger M (2022) Using lignin from local biorefineries for asphalts: LCA case study for The Netherlands. J Clean Prod 343:131063. https://doi.org/10.1016/j.jclepro.2022.131063

Morris MIR, Hicks AL (2022) A human-centered review of life cycle assessments of bioplastics. Int J Life Cycle Assess 27:157–172. https://doi.org/10.1007/s11367-021-02001-0

Mulya KS, Zhou J, Phuang ZX, Laner D, Woon KS (2022) A systematic review of life cycle assessment of solid waste management: methodological trends and prospects. Sci Total Environ 831:154903. https://doi.org/10.1016/j.scitotenv.2022.154903

Mungodla SG, Linganiso LZ, Mlambo S, Motaung T (2019) Economic and technical feasibility studies: technologies for second generation biofuels. J Eng Design Technol 17(4):670–704. https://doi.org/10.1108/JEDT-07-2018-0111

Narisetty V, Zhang L, Zhang J, Sze Ki Lin C, Wah Tong Y, Loke Show P, Kant Bhatia S, Misra A, Kumar V (2022) Fermentative production of 2,3-Butanediol using bread waste—A green approach for sustainable management of food waste. Bioresour Technol 358:127381. https://doi.org/10.1016/j.biortech.2022.127381

Neo ERK, Soo GCY, Tan DZL, Cady K, Tong KT, Low JSC (2021) Life cycle assessment of plastic waste endof-life for India and Indonesia. Resour Conserv Recycl 174:105774. https://doi.org/10.1016/j.resconrec.2021.105774

Ni Y, Mwabonje ON, Richter GM, Qi A, Yeung K, Patel M, Woods J (2019) Assessing availability and greenhouse gas emissions of lignocellulosic biomass feedstock supply–case study for a catchment in England. Biofuels Bioprod Biorefin 13(3):568–581. https://doi.org/10.1002/bbb.1966

Nizami AS, Rehan M, Waqas M, Naqvi M, Ouda OK, Shahzad K, Miandad R, Khan MZ, Syamsiro M, Ismail IM, Pant D (2017) Waste biorefineries: enabling circular economies in developing countries. Bioresour Technol 241:1101–1117. https://doi.org/10.1016/j.biortech.2017.05.097

Nogueira GP, Capaz RS, Franco TT, Dias MOS, Cavaliero CKN (2022) Enzymes as an environmental bottleneck in cellulosic ethanol production: does on-site production solve it? J Clean Prod 369:133314. https://doi.org/10.1016/j.jclepro.2022.133314

Ortiz FJG (2020) Techno-economic assessment of supercritical processes for biofuel production. J Supercrit Fluids 160:104788. https://doi.org/10.1016/j.supflu.2020.104788

Osman AI, Mehta N, Elgarahy AM, Al-Hinai A, Al-Muhtaseb AAH, Rooney DW (2021) Conversion of biomass to biofuels and life cycle assessment: a review. Environ Chem Lett 19:4075–4118. https://doi.org/10.1007/s10311-021-01273-0

Patel M, Zhang X, Kumar A (2016) Techno-economic and life cycle assessment on lignocellulosic biomass thermochemical conversion technologies: a review. Renew Sust Energ Rev 53:1486–1499. https://doi.org/10.1016/j.rser.2015.09.070

Pena A, Rovira-Val MR, Mendoza JMF (2022) Life cycle cost analysis of tomato production in innovative urban agriculture systems. J Clean Prod 367:133037. https://doi.org/10.1016/j.jclepro.2022.133037

Pramanik SK, Mahmud S, Paul GK, Jabin T, Naher K, Uddin MS, Zaman S, Saleh MA (2021) Fermentation optimization of cellulase production from sugarcane bagasse by bacillus pseudomycoides and molecular modeling study of cellulase. Curr Res Microb Sci 2:100013. https://doi.org/10.1016/j.crmicr.2020.100013

Price S, Kuzhiumparambil U, Pernice M, Ralph P (2022) Techno-economic analysis of cyanobacterial PHB bioplastic production. J Environ Chem Eng 10(3):107502. https://doi.org/10.1016/j.jece.2022.107502

Prieler M, Lindorfer J, Steinmüller H (2019) Life-cycle assessment of green biorefinery process options. Biofuels Bioprod Biorefin 13(6):1391–1401. https://doi.org/10.1002/bbb.2022

Rajendran N, Runge T, Bergman RD, Nepal P, Houtman C (2023) Techno-economic analysis and life cycle assessment of cellulose nanocrystals production from wood pulp. Bioresour Technol 377:128955. https://doi.org/10.1016/j.biortech.2023.128955

Reis CER, Junior NL, Bento HB, de Carvalho AKF, de Souza Vandenberghe LP, Soccol CR, Aminabhavi TM, Chandel AK (2023) Process strategies to reduce cellulase enzyme loading for renewable sugar production in biorefineries. Chem Eng J 451:138690. https://doi.org/10.1016/j.cej.2022.138690

Restrepo JC, Izidoro DL, Násner AML, Venturini OJ, Lora EES (2022) Techno-economical evaluation of renewable hydrogen production through concentrated solar energy. Energy Convers Manag 258:115372. https://doi.org/10.1016/j.enconman.2022.115372

Shahbeig H, Nosrati M (2020) Pyrolysis of municipal sewage sludge for bioenergy production: Thermo-kinetic studies, evolved gas analysis, and techno-socio-economic assessment. Renew Sust Energ Rev 119:109567. https://doi.org/10.1016/j.rser.2019.109567

Shaji A, Shastri Y, Kumar V, Ranade VV, Hindle N (2022) Sugarcane bagasse valorization to xylitol: techno-economic and life cycle assessment. Biofuels Bioprod Biorefin 16(5):1214–1226. https://doi.org/10.1002/bbb.2368

Shi R, Guest JS (2020) BioSTEAM-LCA: an integrated modeling framework for agile life cycle assessment of biorefineries under uncertainty. ACS Sustain Chem Eng 8(51):18903–18914. https://doi.org/10.1021/acssuschemeng.0c05998

Singh G, Samuchiwal S, Hariprasad P, Sharma S (2022) Melioration of Paddy straw to produce cellulase-free xylanase and bioactives under solid state fermentation and deciphering its impact by life cycle assessment. Bioresour Technol 360:127493. https://doi.org/10.1016/j.biortech.2022.127493

Singh R, Bhagwat SS, Viswanathan MB, Cortés-Peña YR, Eilts KK, McDonough G, Cao M, Guest JS, Zhao H, Singh V (2023) Adsorptive separation and recovery of triacetic acid lactone from fermentation broth. Biofuels Bioprod Biorefin 17(1):109–120. https://doi.org/10.1002/bbb.2427

Skaggs RL, Coleman AM, Seiple TE, Milbrandt AR (2018) Waste-to-energy biofuel production potential for selected feedstocks in the conterminous United States. Renew Sust Energ Rev 82(July 2017):2640–2651. https://doi.org/10.1016/j.rser.2017.09.107

Spatari S, Betz M, Florin H, Baitz M, Faltenbacher M (2001) Using GaBi 3 to perform life cycle assessment and life cycle engineering. Int J Life Cycle Assess 6:81–84. https://doi.org/10.1007/BF02977842

Soares JF, Confortin TC, Todero I, Luft L, Ugalde GA, Tovar LP, Mayer FD, Mazutti MA (2022) Estimation of bioethanol, biohydrogen, and chemicals production from biomass wastes in Brazil. Clean Soil Air Water 50(8):2200155. https://doi.org/10.1002/clen.202200155

Sollai S, Porcu A, Tola V, Ferrara F, Pettinau A (2023) Renewable methanol production from green hydrogen and captured CO_2: A techno-economic assessment. J CO2 Util 68:102345. https://doi.org/10.1016/j.jcou.2022.102345

Sorunmu Y, Billen P, Spatari S (2020) A review of thermochemical upgrading of pyrolysis bio-oil: techno-economic analysis, life cycle assessment, and technology readiness. GCB Bioenergy 12(1):4–18. https://doi.org/10.1111/gcbb.12658

Sreekumar A, Shastri Y, Wadekar P et al (2020) Life cycle assessment of ethanol production in a rice-straw-based biorefinery in India. Clean Techn Environ Policy 22:409–422. https://doi.org/10.1007/s10098-019-01791-0

Stewart DW, Cortés-Peña YR, Li Y, Stillwell AS, Khanna M, Guest JS (2023) Implications of biorefinery policy incentives and location-specific economic parameters for the financial viability of biofuels. Environ Sci Technol 57(6):2262–2271. https://doi.org/10.1021/acs.est.2c07936

Suckling J, Morse S, Murphy R et al (2023a) Environmental life cycle assessment of production of the high intensity sweetener steviol glycosides from *Stevia rebaudiana* leaf grown in

Europe: the SWEET project. Int J Life Cycle Assess 28:221–233. https://doi.org/10.1007/s11367-022-02127-9

Suckling J, Morse S, Murphy R, Raats M, Astley S, Halford JCG, Harrold JA, Le-Bail A, Koukouna E, Musinovic H, Raben A (2023b) Environmental life cycle assessment of production of the non-nutritive sweeteners aspartame (E951) and neotame (E961) from chemical processes: the SWEET project. J Clean Prod 424:138854. https://doi.org/10.1016/j.jclepro.2023.138854

Tabassum MR, Xia A, Murphy JD (2018) Biomethane production from various segments of brown seaweed. Energy Convers Manag 174:855–862. https://doi.org/10.1016/j.enconman.2018.08.084

Tahir MF, Haoyong C, Guangze H (2021) A comprehensive review of 4E analysis of thermal power plants, intermittent renewable energy and integrated energy systems. Energy Rep 7:3517–3534. https://doi.org/10.1016/j.egyr.2021.06.006

Talero G, Kansha Y (2023) Atom economy or product yield to determine optimal gasification conditions in biomass-to-olefins biorefinery. Chem Eng Res Des 199:689–699. https://doi.org/10.1016/j.cherd.2023.10.024

Thomassen G, Van Dael M, Lemmens B, Van Passel S (2017) A review of the sustainability of algal-based biorefineries: towards an integrated assessment framework. Renew Sust Energ Rev 68:876–887. https://doi.org/10.1016/j.rser.2016.02.015

Thomassen G, Van Dael M, Van Passel S, You F (2019) How to assess the potential of emerging green technologies? Towards a prospective environmental and techno-economic assessment framework. Green Chem 21(18):4868–4886. https://doi.org/10.1039/C9GC02223F

Tsai TY, Lo YC, Dong CD, Nagarajan D, Chang JS, Lee DJ (2020) Biobutanol production from lignocellulosic biomass using immobilized clostridium acetobutylicum. Appl Energy 277:115531. https://doi.org/10.1016/j.apenergy.2020.115531

Tschulkow M, Compernolle T, Van den Bosch S, Van Aelst J, Storms I, Van Dael M, Van den Bossche G, Sels B, Van Passel S (2020) Integrated techno-economic assessment of a biorefinery process: the high-end valorization of the lignocellulosic fraction in wood streams. J Clean Prod 266:122022. https://doi.org/10.1016/j.jclepro.2020.122022

Tunca B, Kutlar FE, Kas A, Yilmazel YD (2023) Enhanced biohydrogen production from high loads of unpretreated cattle manure by cellulolytic bacterium Caldicellulosiruptor bescii at 75 °C. Waste Manag 171:401–410. https://doi.org/10.1016/j.wasman.2023.09.028

Ubando AT, Felix CB, Chen WH (2020) Biorefineries in circular bioeconomy: a comprehensive review. Bioresour Technol 299:122585. https://doi.org/10.1016/j.biortech.2019.122585

Valladares-Diestra KK, de Souza Vandenberghe LP, Torres LAZ, Nishida VS, Zandoná Filho A, Wojciechowski AL, Soccol CR (2021) Imidazole green solvent pre-treatment as a strategy for second-generation bioethanol production from sugarcane bagasse. Chem Eng J 420:127708. https://doi.org/10.1016/j.cej.2020.127708

van Heerden C, Farzad S, Görgens JF (2023) Techno-economic and environmental assessments of 1, 3-Propanediol and Xylooligosaccharide production annexed to a typical sugar mill. ACS Sustain Chem Eng 11:16453. https://doi.org/10.1021/acssuschemeng.3c03338

Viraraghavan T, Cocci A, Landine R, Steeves A (1983) A techno-economic analysis of biogasification of peat. Biomass 5(2):137–159. https://doi.org/10.1016/0144-4565(84)90053-2

Vuppaladadiyam AK, Vuppaladadiyam SSV, Sahoo A, Murugavelh S, Anthony E, Bhaskar T, Zheng Y, Zhao M, Duan H, Zhao Y, Antunes E, Sarmah AK, Leu SY (2023) Bio-oil and biochar from the pyrolytic conversion of biomass: a current and future perspective on the trade-off between economic, environmental, and technical indicators. Sci Total Environ 857(Pt 1):159155. https://doi.org/10.1016/j.scitotenv.2022.159155. Epub 2022, 159155

Wahyono Y, Hadiyanto H, Gheewala SH, Budihardjo MA, Adiansyah JS (2022) Evaluating the environmental impacts of the multi-feedstock biodiesel production process in Indonesia using life cycle assessment (LCA). Energy Convers Manag 266:115832. https://doi.org/10.1016/j.enconman.2022.115832

Wan L, Qin S, Wei H, Luo L, Liu S, Suo Y (2023) Enhanced tolerance of clostridium tyrobutyricum to furan derivatives for efficient butyric acid production by overexpressing native transporter genes. Ind Crop Prod 206:117604

Wang M (2007) Overview of GREET model development at Argonne. Center for Transportation Research, Argonne National Labouratory, GREET User Workshop, Argonne, IL

Wang ZH, Li LQ, Zhao L, Chen C, Yang SS, Ren NQ (2022) Comparative life cycle assessment of biochar-based lignocellulosic biohydrogen production: sustainability analysis and strategy optimization. Bioresour Technol 344:126261. https://doi.org/10.1016/j.biortech.2021.126261

Wiatrowski M, Tan EC, Davis R (2023) TEA and LCA of bio-based polyurethanes. In: Rethinking polyester polyurethanes. Elsevier, Amsterdam, pp 153–176. https://doi.org/10.1016/B978-0-323-99982-3.00004-3

Woo WX, Koh HS, Tan JP, Yeap SK, Abdul PM, Luthfi AAI, Manaf SFA (2022) An overview on cell and enzyme immobilization for enhanced biohydrogen production from lignocellulosic biomass. Int J Hydrog Energy 47(96):40714–40730. https://doi.org/10.1016/j.ijhydene.2022.08.164

Zang G, Shah A, Wan C (2020) Techno-economic analysis of an integrated biorefinery strategy based on one-pot biomass fractionation and furfural production. J Clean Prod 260:120837. https://doi.org/10.1016/j.jclepro.2020.120837

Zhang C, Li T, Su G, He J (2020) Enhanced direct fermentation from food waste to butanol and hydrogen by an amylolytic clostridium. Renew Energy 153:522–529. https://doi.org/10.1016/j.renene.2020.01.151

Zhang H, Hewage K, Prabatha T, Sadiq R (2021) Life cycle thinking-based energy retrofits evaluation framework for Canadian residences: a pareto optimization approach. Build Environ 204:108115. https://doi.org/10.1016/j.buildenv.2021.108115

Zhang G, Zabed HM, An Y, Yun J, Huang J, Zhang Y, Li X, Wang J, Ravikumar Y, Qi X (2022a) Biocatalytic conversion of a lactose-rich dairy waste into D-tagatose, D-arabitol and galactitol using sequential whole cell and fermentation technologies. Bioresour Technol 358:127422. https://doi.org/10.1016/j.biortech.2022.127422

Zhang Y, Liu H, Li J et al (2022b) Life cycle assessment of ammonia synthesis in China. Int J Life Cycle Assess 27:50–61. https://doi.org/10.1007/s11367-021-02010-z

Zheng Z, Xie J, Liu P, Li X, Ouyang J (2019) Elegant and efficient biotransformation for dual production of D-tagatose and bioethanol from cheese whey powder. J Agric Food Chem 67(3):829–835. https://doi.org/10.1021/acs.jafc.8b05150

Zheng JL, Zhu YH, Dong YY, Chen Y, Zhu MQ (2023a) Techno-economic analysis and life cycle assessment of industrial production of ammonia via bio-oil conversion. Energy 280:128223. https://doi.org/10.1016/j.energy.2023.128223

Zheng X, Zhong Z, Zhang B, Du H, Wang W, Li Q, Yang Y, Qi R, Li Z (2023b) Techno-economic analysis and life cycle assessment of hydrogenation upgrading and supercritical ethanol upgrading processes based on fast pyrolysis of cornstalk for biofuel. Biomass Convers Biorefinery:1–17. https://doi.org/10.1007/s13399-023-04096-x

Zhu L, Wang Y, Wang Y et al (2022a) The design of new process, parametric analysis, technical and economic analysis of methanol production from biogas. Multiscale Multidiscip Model Exp and Des 5:351–364. https://doi.org/10.1007/s41939-022-00121-0

Zhu R, Wang Z, He Y, Zhu Y, Cen K (2022b) LCA comparison analysis for two types of H2 carriers: methanol and ammonia. Int J Energy Res 46(9):11818–11833. https://doi.org/10.1002/er.7947

Strategies to Develop an Efficient Biorefinery System Using Techno-Economic Evaluation and Life Cycle Assessment

Nirza Moktan, Anindita Banerjee, and Mohan Bir Subba

1 Introduction

Without a doubt, the industrial revolution influenced the foundations of our modern civilization in both positive and negative aspects. This revolution was initially driven primarily by coal as its energy source, and over the ensuing decades, it transitioned to an economy centered around crude oil. The United Nations Climate Report unequivocally states that significant and prompt measures are necessary to keep the rise in global temperature within the still-manageable range of +1.5 °C. There has been a substantial increase in public support for reducing greenhouse gas (GHG) emission and shifting to more socially, economically, and environmentally sustainable development and energy production. Thus, a colossal change involving highly efficient and cost-effective biorefinery system for the production of energy and chemical compounds will have to be adopted. Biorefinery thus can be thought of as an efficient sustainable system that effectively produces high-value bioproducts like biofuels, biochemicals, bioenergy, and other high-value bioproducts by integrating different conversion technologies like thermochemical, biochemical, combustion, and microorganism growth platforms (Kumar et al. 2020; Kumar and Samadder 2020). Recently, the idea of biorefinery has been developed

N. Moktan
Department of Microbiology, Dhruba Chand Halder College, Dakshin Barasat, WB, India

UG, PG & Research Department of Microbiology, St. Xavier's College (autonomous), Kolkata, WB, India

A. Banerjee (✉)
UG, PG & Research Department of Microbiology, St. Xavier's College (autonomous), Kolkata, WB, India

M. B. Subba
Department of Mathematics, K. K. Das College, Kolkata, WB, India

K. Agrawal, P. Verma (eds.), *Biotechnological Advances in Biorefinery*, Interdisciplinary Biotechnological Advances,
https://doi.org/10.1007/978-981-97-5544-8_16

and applied to process several biomass feedstocks, including lignocelluloses (Bhardwaj et al. 2019; Kumar et al. 2018; Agrawal et al. 2019), algal biomass (Goswami et al. 2021a, b), and different kinds of food and microbial treated wastes to produce high-value bioproducts (Cherubini 2010; Ferreira et al. 2018). Thus, the biorefinery concept aims to maximize biomass utilization and simultaneously minimize wastes and emissions.

General crops such as maize and rapeseeds are utilized as feedstocks in first-generation biorefineries (Cherubini 2010). However, there are ethical concerns regarding the use of general crops as feedstocks. Growing of crops for energy than for food would be more profitable for the farmers. Moreover, monocrop farming would be more preferred where the crops shall be more prone to diseases and plaques, thereby jeopardizing the food supply chain and also the livelihood of farmers. As a consequence of the flaws of first-generation biorefineries, second-generation biorefineries that utilize residual and waste streams such as lignocellulosic waste, farm waste, kitchen waste, and industry waste were developed, and thus they are also sometimes called biowaste biorefineries (Cherubini 2010; Mohr and Raman 2013). Though biorefineries are mostly linked with the production of biofuels, the production of platform chemicals and its subsequent conversion into other products such as plastics and fibers through interconversion technologies are considered to be a more sustainable process. A variant of the products can be obtained from biorefineries, depending upon the feedstock and technique employed (Cherubini 2010).

By seeking ways to reuse the byproducts and integrating different conversion platforms, the biorefinery concept seeks to increase the economic potential of bioproducts production (Sy et al. 2018). Biorefineries transform bioresources into usable compounds, much how refineries transform fossil fuels into energy and chemical compounds (Naik et al. 2010); Mohan et al. 2016). Circular bioeconomy or sustainable bioeconomy refers to the sustainable valorization of biomass while also opening possibilities for reuse, recycling, and remanufacture of wastes and byproducts through integrated technologies. Thus, a sustainable approach towards circular bioeconomy is biowaste biorefineries (Vaskan et al. 2017). The basic conditions for successful valorization of biomass in biorefineries from a techno-economic point of view are a high conversion rate and a low cost of derivative product separation (Liu et al. 2019). Thus, techno-economic evaluation aims at analyzing the technical and economic criteria in product manufacture. Techno-economic assessment (TEA) is an evaluation technique that enables academicians and business owners to analyze the expenses and earnings of a particular biorefinery or plant, through cost analysis, to evaluate the expenses associated with raw materials, equipment, manpower, utilities, etc. Techno-economic assessment models serve as a framework for manufacturing processes and set boundaries for performance and cost. TEA thus helps measure the economic feasibility of biorefinery platform. Techno-economic evaluation has been employed in several biorefineries such as corn stover, lignocellulosic feedstocks, sugarcane, and food residue (Lauer 2008; Shen et al. 2019).

Life cycle assessment (LCA) is a "cradle to grave" approach for identifying and measuring the effect of a good service or process on the environment during the course of its life cycle. As per the iterative LCA principles, environmental life cycle assessment is conducted using a standardized approach in compliance with International Organization for Standardization (ISO) 14,040 and 14,044 standards. This framework comprises (a) goal and scope definition which help specify functional units, impact categories, and system boundaries, (b) inventory analysis which helps track the system flow, (c) impact assessment which helps translate the inventory information into effect categories, and (d) data interpretation (Chan et al. 2018).

Initially economic valuation and environmental evaluation through LCA (life cycle assessment) were utilized to define or measure biorefinery performance. However, in order to analyze the sustainability assessment of biorefineries, a triple bottom line framework that comprises economic, environmental, and social aspects has been proposed (Tuazon et al. 2013; Tuazon and Gnansounou 2017). The LCA and TEA results offer valuable information for decision-making at the commercial level concerning investment opportunities and alternative methods of bioenergy production. Findings may also be useful to policy makers aiming for expansion of the bioenergy and bioproducts sectors.

2 Biorefinery Conversion Platforms

Biorefinery platform represents a conversion technology where the biomass is segregated into a number of important intermediate products that are further processed for purity. The platforms serve as a connecting bridge between the raw materials and the final finished goods. Three major biorefinery platforms have been recognized depending upon the biomass conversion technology, viz., biotechnological, thermochemical, and hybrid platforms (Benjamin et al. 2015). Considering the diversity of biomass and feedstock, the same biorefinery platform can process different biomasses, or different platforms can be implemented by the same biorefinery.

Biotechnological platforms utilize microorganisms, fermentation, and enzymatic processes (Bhardwaj et al. 2021; Kumar and Verma 2020). The biomass is converted into hydrolysate stream through enzymatic hydrolysis and then fermented using industrial (genetically modified) strains. They are the perfect platforms for producing highly specialized products as they have high product selectivity. The desired products are eventually separated from byproducts/fermentation broth by downstream processing methods. The use of biomass feedstock, most specifically lignocellulosic biomass, has been found to exhibit certain obstacles in the otherwise simple biological conversion platform. The enzyme cellulase brings about the degradation of the cellulosic material in the lignocellulose; however, it is unable to hydrolyze the lignin fraction that remains in the hydrolysate as a solid mass. Thus, a major drawback of this method is its inability to catalyze the hydrolysis of a significant fraction of the lignocellulosic biomass, resulting in substantial carbon source loss (Chen et al. 2018; Michailos et al. 2019). Additionally, the lignin fraction has also been found to exhibit an inhibition on cellulose and hemicellulose conversion, while its prior removal is a costly process. High-value chemicals are mostly produced through this platform. However, because there is a dearth of experimental data and mechanistic understanding of the action of enzymes and microorganisms, the biological conversion platform is currently quite restricted.

Thermochemical platform utilizes general chemical processes where the entire biomass feedstock is decomposed and converted to the desired product through reactors or catalytic synthesis under high pressure and temperature either under oxygenic or anoxygenic environment. There is a more efficient use of feedstock as compared to biotechnological platform (Michailos et al. 2019; Trung et al. 2019). The primary technologies applied in this platform include the following:

(a) Combustion (Mandø 2013): It is the oxidation of biomass to produce energy. Because it is cost-effective and highly reliable, this method is widely used. For instance, the paper and pulp industry utilizes its byproduct for energy production by combustion.
(b) Carbonization (Amer and Elwardany 2020): It is the enhancement of the biomass carbon content under reduced-oxygen and high-temperature environment. It is utilized for the production of charcoal-like bio-based products.
(c) Gasification (Heidenreich et al. 2016): It is the production of light gases such as hydrogen, carbon monoxide, carbon dioxide, and methane under very high temperature and in the presence of pure oxygen in stoichiometric amount. The gas produced can be directly used as a fuel or can be converted to other desired products.
(d) Pyrolysis (Madár et al. 2018): It is the conversion of biomass into bioproducts such as biogas, bio-oil, and biochar under external heating and inert atmosphere.
(e) Hydrothermal processing (Grande et al. 2021; Bhagea et al. 2019): It is the conversion of biomass into biocrude oil under subcritical water conditions. A major drawback of this platform is its relatively low yields and low product selectivity.

A hybrid platform utilizes a combination of the positive aspects of both biotechnological and thermochemical platforms and is often comprised of a biological conversion and a thermochemical pretreatment step (Michailos et al. 2019). This thus leads to an optimum utilization of biomass, high carbon efficiency, and optimum

Table 1 Biorefinery conversion platforms

Biotechnological	Thermochemical	Hybrid
1. Enzymatic conversion 2. Microbial fermentation 3. Aerobic & anaerobic transformation	1. Combustion 2. Carbonization 3. Gasification 4. Pyrolysis 5. Hydrothermal processing	1. Enzymatic conversion 2. Microbial fermentation 3. Combustion 4. Carbonization 5. Gasification 6. Pyrolysis 7. Hydrothermal processing

production of targeted products. The main drawback of this platform is the limited availability of models (Table 1).

3 Viability Assessment of Biorefinery System

The sustainability and economic viability of the supply of the required feedstock are essential to predict the economic feasibility of any biorefinery design. It is therefore of prime importance that, while designing a biorefinery, the suppliers be persuaded to produce the biomass required by the biorefinery, assuring them that it would be most beneficial. Thus, modelling of biorefinery supply chain networks has been gaining interest over time now. The viability of biorefinery performance though was previously evaluated through economic valuation and environmental evaluation, where the combinatorial effect of different criteria was often neglected (Tuazon et al. 2013; Li et al. 2018). The sustainability assessment of biorefineries has now been analyzed through a triple bottom framework, namely, environmental, economic, and social aspects (Tuazon et al. 2013; Chan et al. 2018). In order to improve the biorefinery process and enhance its overall efficiency, a cascading approach to treat the liquid and solid components of biowaste has been proposed (Probst et al. 2015). Waste valorization is a sustainable biowaste utilization practice to produce usable products, and the development areas for biowaste biorefinery can be summarized as follows: (a) biowaste collection techniques from various platforms, (b) expansion of biowaste biorefinery to include a host of biowaste feedstocks, and (c) valorization process improvement (Fava et al. 2015).

Integration and optimization of conversion platforms can maximize the biorefinery performance potential and at the same time minimize process disruption (Cherubini 2010). The idea of a "circular economy" has also had a major impact on waste management, shifting the focus from traditional recycling and disposal methods to very effective utilization strategies that ultimately aim to create a zero-waste economy (Shen et al. 2019).

Life cycle assessment (LCA) which is based on the ISO 14040 series is a reliable technique for tracking and assessing the environmental effect of providing a service or generating a product. LCA is based on four important criteria: (1) the objective and purview, (2) the life cycle inventory, (3) the life cycle footprint evaluation, and (4) the explanation of the results. LCA studies have been conducted to report the environmental impact of biorefinery systems. For instance, there has been LCA analysis of lignocelluloses, algae, and agricultural stream biorefineries (Ubando et al. 2021). Life cycle assessment was utilized as a tool to compare the environmental impact of the biorefinery platform against the traditional waste management platform. The global warming impact of biorefinery was estimated to be around −15 kg CO_2 eq/ton of biowaste in ethanol production (Papadaskalopoulou et al. 2019). From the purview of circular bioeconomy, it is essential that a comprehensive design of biorefinery incorporate the use of the life cycle assessment (LCA) technique with a cradle-to-grave system boundary in order to identify areas for

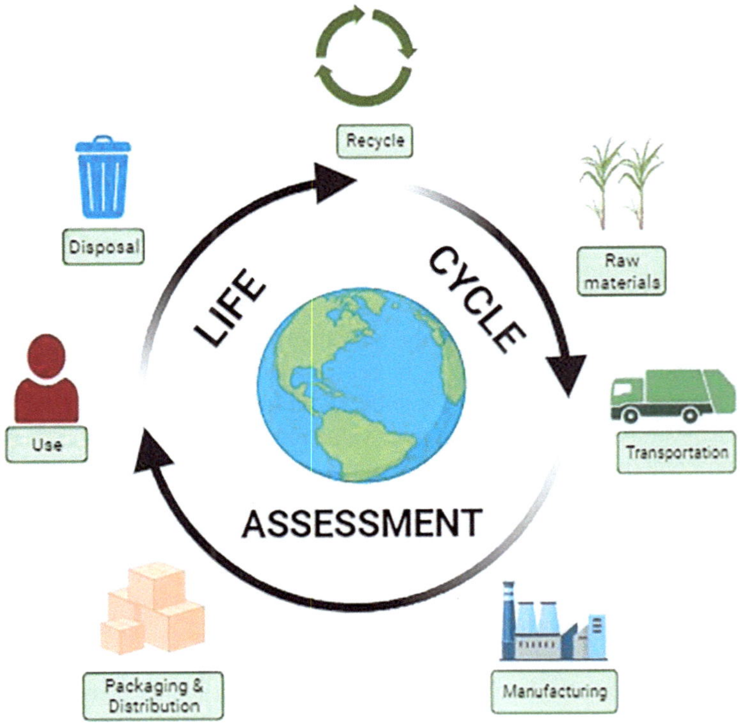

Fig. 1 Life cycle assessment strategy

improvement. With LCA, the environmental costs of a product's life cycle, including waste measurement, may be determined (Nizami et al. 2017) (Fig. 1).

The evaluation of technical and economic aspects of product manufacturing is evaluated through techno-economic assessment (TEA) (Lauer 2008). TEA involves quantifying the capital and operating expenses while taking into account the different technologies used in the biorefinery. TEA has been employed for the assessment of numerous biorefinery systems such as sugarcane, lignocellulose, food residues, and feedstocks.

There have been few studies on the socio-economic assessment (SEA) of biorefinery system and evaluation of the social impact of biorefinery system in circular bioeconomy (Cherubini 2010). SEA is extended to the farms where the biomass is cultivated and is not limited to only production in biorefineries. Process integration (PI) is a framework for integrating the varied conversion platforms by adopting cleaner, sustainable production principles. PI aims to meet production demands while simultaneously minimizing waste and emission, reducing energy and material usage (Özdenkçi et al. 2017).

4 Biorefinery in Action

Biorefinery is a platform that effectively produces high-value bioproducts like biofuels, biochemicals, and energy from a variety of sustainable bio-based product streams by integrating different conversion platforms like thermochemical, biochemical, combustion, and microbial growth platforms (Cherubini 2010; Ferreira et al. 2018).

4.1 Lignocellulosic Biorefinery

Considering the high availability and wide range of plant varieties, lignocellulosic biomass forms an important biorefinery feedstock (Baruah et al. 2018). Essentially barley, straw, rice, sugarcane bagasse, wood, etc. are the predominant sources of lignocellulosic biomass. They are a rich source of lignin, cellulose, and hemicellulose that may be fractionated sustainably to a variety of bioproducts (Grande et al. 2021) through conversion and integration technologies. The use of lignocellulosic biomass by enzymatic hydrolysis for the manufacture of biofuels is anticipated to be a standard technology by 2020 as part of the growth of the bioeconomy, according to a 2007 European Union assessment.

4.2 Algal Biorefinery

When compared to lignocellulosic biomass, algal biomass is thought of as an improved biomass feedstock that offers advantages including a lower need for land and better biomass production and output. Algal biomass falls into two primary categories: macroalgae biomass (Torres et al. 2019; Hasenheit et al. 2016) and microalgae biomass (Chew et al. 2017). Microalgae are photosynthetic microorganisms that effectively use solar energy to produce biomass that can be efficiently and sustainably converted to various microalgal-based products through biorefinery platforms (Agrawal et al. 2023; Goswami et al. 2022, 2021a, b). Different reactor systems may be used to cultivate microalgae, such as photobioreactors, which can be built vertically and need less area than traditional farming methods. Seaweeds, sometimes referred to as macroalgae, are marine microorganisms that are mostly cultivated offshore and found in large quantities along coastal shorelines. They are a rich source of different nutritional components and elements that can be sustainably processed into high-value bioproducts like biofuels, biochemicals, etc. Utilizing integration technology, wastewater treatment is fused with microalgae cultivation to generate water for irrigation and value-added products and has garnered a lot of interest because of its possible advantages for the environment and the economy (Vaskan et al. 2017).

4.3 *Waste Biorefinery*

Immense waste is generated by the continuous economic activities and urbanization. Waste biorefinery route studies, utilizing various kinds of wastes (municipal solid waste, nonedible food waste, lignocellulosic waste), have been carried out in the prospect of circular bioeconomy. Biowaste biorefinery thus has emerged as a sustainable and viable substitute for the production of various bioproducts such as biofuels and biopolymers. The biorefinery platforms are highly efficient enough to bring about the reuse, recycle, and manufacture from biowastes. Waste biorefinery enables a comprehensive approach in addressing various environmental, social, and economic problems (Minelgaitė and Liobikienė 2019) (Fig. 2).

5 Efficient Biorefinery Design with Techno-Economic and Life Cycle Assessment

The excessive reliance on nonrenewable fossil resources has resulted in serious challenges including global climatic changes, rising carbon footprints, global warming, and dwindling supply of nonrenewable fossil resources, which endangers not just the humans but the entire mother earth. It has thus become imperative to replace fossil-based products with sustainable, environmentally friendly bio-based products from renewable resources to address these worldwide concerns (Sheldon and Norton 2020). There are several manufacturing methods for the conversion of biomass into

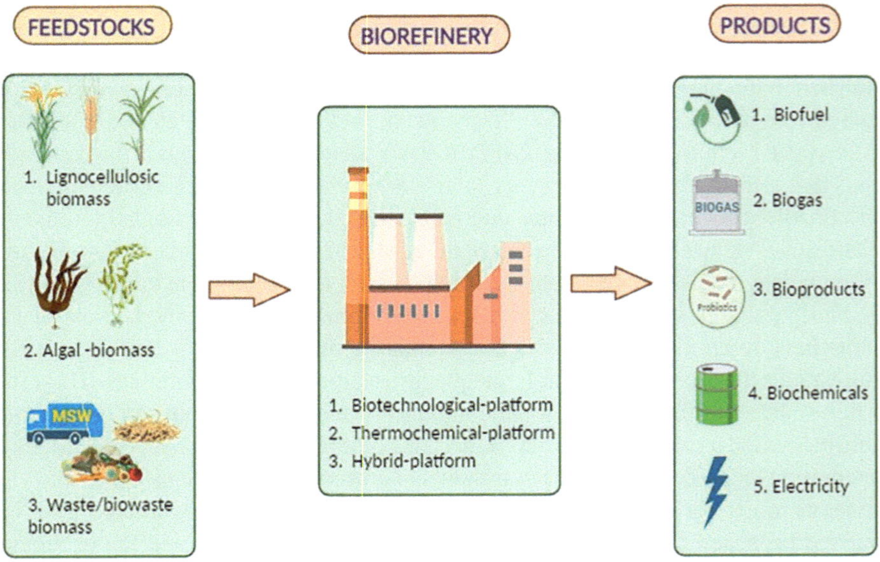

Fig. 2 Biorefinery in action

various products. It is imperative that an optimal methodology or model be chosen for optimum product production in a biorefinery. The three methodologies for designing biorefineries are (1) conceptual design, (2) optimization design, and (3) early-stage method design.

5.1 Conceptual Design

Conceptual design is based on a holistic approach as it takes into consideration a number of new difficulties/factors such as developing local and regional locations and incorporating a variety of feedstocks. Conceptual design of a biorefinery platform is based on three analyses, viz., technical, economic, and environmental. The criteria used for selecting feedstocks and products include cost, availability, and market demands. Uses and applications however do come with some certain limitations. For instance, antioxidants which are value-added products are prioritized over energy goods like biofuels (Sy et al. 2018; Khan et al. 2018). Conceptual design is based on three concepts, i.e., hierarchy, sequencing, and integration, which are related to one another, and in combination they provide a process synthesis method for creating biorefinery models.

The hierarchy idea takes into account how feedstocks, products, and technology break down hierarchically inside a biorefinery. The first analysis is centered around feedstocks and products. Here, the feedstock composition and its potential breakdown to produce a set of goods in the biorefinery platform are examined. The second analysis is centered around technology and its key points. The sequencing approach establishes a logical order for linking the various products to transformation paths and thus helps determine the most promising order. It takes into account biorefinery platforms, technological advancements, and limitations on product quality. Making decisions involves optimization (Temmes and Peck 2019). The integration concept focuses on the optimal utilization of feedstocks in a biorefinery. It defines the potential of integration between feedstocks, processing pathways, and products. When it comes to feedstocks, waste streams may be minimized by using the leftover streams from one operation in another.

5.2 Optimization Design

Bioprocessing can benefit from optimization when evaluating the economics of product portfolios and transformation paths. The global supply chain optimization process maximizes stakeholder value (SC). This technique makes it possible to evaluate several optimization goals, including environmental effects, waste/fixed cost reduction, cost reduction, and yield/profit maximization. Optimization methodology is based on three concepts, i.e., defining scope and problem complexity, modelling and experimental data integration, and biorefinery optimization.

For defining scope and problem complexity, the process system engineering (PSE) approach is used. Even with reduced energy and feedstock use, the PSE may provide significant economic and social benefits. PSE is based on mathematical optimization and process integration that ensure minimum environmental impact, minimum energy and feedstock use, maximum value in product production, and maximum social and economic benefit, considering all possible transformation paths. For integrating modelling and experimental data, the simulations of processes created with commercial software packages are validated by means of experimental data (Sy et al. 2018). This leads to the creation of a comprehensive library made up of yield databases and simulation models of transformation pathways, taking into consideration minimum environmental impact, minimum use of energy, maximum product production, and maximum resource utilization. For optimization framework solutions that maximize economic performance are explored. If a solution is found to meet this goal, the environmental performance is evaluated. An ideal process scheme is said to have been found if a solution satisfies the environmental requirements. Constraints are loosened until satisfactory environmental outcomes are obtained if the environmental solution is negative.

5.3 Early-Stage Sustainability

The “early stage” sustainability evaluation approach evaluates and modifies chemical processes in the laboratory scale using a multi-objective decision framework to ensure sustainability, integrating economic, environmental, safety, and health variables. This methodology is adjusted for bio-based operations, taking into account variables including co-product quantities, hazards, biomass pretreatment, and comparison with comparable petrochemical alternatives. The objective of the early-stage sustainability evaluation approach used in biorefineries is to identify the biorefinery products with the greatest potential from a sustainability standpoint at an early design stage. This approach makes it possible to assess the sustainability of chemical and catalytic processes and determine their benefits and drawbacks (Vaskan et al. 2017; Yang and Xiao 2013). This approach is more akin to the conceptual design approach as it provides the initial data with almost identical tools as conceptual design approach. The “early stage” sustainability evaluation approach is based on method modification, data collection, case study definition, method application, and result analysis (Table 2).

Table 2: Efficient biorefinery design

Methodology	Key aspects	Case study
Conceptual design		
Techno-economic and environmental assessment taking into account various integration levels (such as mass and energy integration, nonconventional technologies) in order to ascertain the financial viability and environmental implications	Utilizing life cycle assessment (LCA) to complete the environmental assessment	Utilizing fresh fruit as the raw material, this technique compares the synthesis of biodiesel, ethanol, and butanol utilizing stand-alone and biorefinery methods as prospective scenarios for bioenergy production Sugarcane biorefinery uses this process to produce sugar, fuel ethanol, PHB, anthocyanins, and power
Optimization		
With a multi-objective optimization that takes into account both environmental (like carbon efficiency) and economic (like net present value) factors, this technique incorporates a variety of products and processing routes	The process offers a practical and efficient way to choose a transformation path	Utilizing lignocellulosic biomass, this technique considers the processing pathways for the manufacture of bioalcohols
Early-stage method		
A multi-criteria approach-based methodology that takes into account five indications as a single value. The indicators cover areas such as dangers related to the environment, economy, and life cycle based on green chemistry principles and techno-economic evaluations	The approach is used for both anticipated future and laboratory procedures	Catalytic conversion of bio-based syngas, biochemical conversion of carbohydrates, and biochemical and catalytic conversion of glycerol

6 Biorefineries Assessed from Techno-Economic and Environmental Perspective

The integration of several processes aimed at generating a range of valuable products (bioenergy, green chemicals, biofuels, and biomaterial compounds) from biomass and waste is known as the biorefinery strategy. This technique enables the final circuit valorization of these substrates. Biorefineries are thus an exceptional way to valorize waste streams and ought to be used increasingly often. Furthermore, employing residual waste as a raw material, the integrated biorefinery concept ideally strives for "zero waste" production in accordance with a circular economy framework (Shah et al. 2022). In the near future, biorefineries should be used to convert biowaste into high-value products and energy. This is a very desired approach that should be put into practice not just for reasons of environmental preservation but also for economic reasons (Ragauskas et al. 2006).

6.1 Economic and Environmental Assessment of Biorefinery Using Winery Waste Streams

In 2018, wine production reached a worldwide total of 29.2 million m^3 (OIV 2019), and waste streams from wineries are produced in substantial volumes. Since wine production is a seasonal operation, it is best to handle the leftover material swiftly. As winery wastes are rich in phenolic and flavonoid compounds and have a high organic load, disposing of such a significant volume of residue can pollute and have a negative impact on the environment. The only long-term sustainable approach for generating a range of bio-based products would be through the management of winery waste (Chowdhary et al. 2021; You et al. 2012). Traditional wineries can be reorganized into sustainable biorefineries through the integration of bio-based chemical production. It is possible to obtain a variety of bio-based products using the integrated biorefinery system, including biochemical and bioenergy products like heat, power, and biofuels. Process integration may have a big impact on profitability, and firms can manage a low demand crisis by diversifying their products appropriately.

Winery waste streams such as grape pomace (which includes skins and seeds), grape stalks, and wine lees (generated after the winemaking process) exhibit significant potential for application in sustainable biorefinery for the production of bio-based products such as polyphenols (high value, low volume products), succinic acid (low value, high volume products), various value-added co-products (such as grape seed oil, crude tannin), food additives, dyes, cosmetics, fertilizer, etc. Numerous important bioactive phytochemicals such as flavonoids and phenolic acids are present in considerable quantities in the residues produced by the wine industry. These phytochemicals are known to have a plethora of beneficial properties such as anticancer, antioxidant, antimicrobial, etc. (Walmsley et al. 2018). According to the biorefinery idea, when the biocompounds are recovered, the residual materials can be utilized to produce energy through thermal processes (pyrolysis and gasification combustion), bioprocesses (biogas, bioethanol, and bio-oil), or bio-fertilizers (compost).

Organic winery waste breakdown uncontrollably can have a variety of adverse consequences on the environment. It is imperative to manage organic winery waste, as even a single metric tonne of decaying organic solid waste may release 50–110 m^3 of carbon dioxide and 90–140 m^3 of methane into the atmosphere. Any waste has an inherent net positive energy that may be captured and recycled through a series of closed-loop bioprocesses to create bio-based goods and biofuels. This process transfers the trend towards a low-carbon, circular bioeconomy. The most effective method for evaluating the environmental impact of extraction to disposal may be the life cycle assessment or LCA. LCA is a tool for gauging how sustainable a process or final product is (Neto et al. 2013). Most wineries dispose of their organic waste in the trash, while some produce compost, animal feed, and biofertilizers. Generating compost from winery waste has been demonstrated to be a practical, sustainable, and an inexpensive way to make use of the massive volume of organic waste produced. The compost can be utilized for enhancing nutrient content, organic matter content, and microbial biomass load (Paradelo et al. 2010).

Biogas constitutes about 74% methane. Bagasse, stems, and dregs are the main winery waste stream. Winery waste stream can be utilized as a suitable substrate for biogas production. Anaerobic digestion's life cycle evaluation was examined in relation to indices of resource impact, ecosystem quality, climate change, and human health. Regarding the environment, it was determined to be far more advantageous than the other options (Boulamanti et al. 2013). A useful source for the manufacturing of second-generation (2G) bioethanol would be residual sugars found in grape skins. From one ton of grape skins, approximately 11 kg of oleanolic acid and 250 L of bioethanol may be produced. Since second-generation bioethanol doesn't compete with food security, it has a higher chance of producing sustainable goods rather than first-generation bioethanol (Mendes et al. 2013). Integrated biorefinery strategy can be used to produce a range of products, including high-value organic compounds, fuel, food, feed, clean energy, and chemical compounds with added value from mixed and multiple winery wastes (Fig. 3).

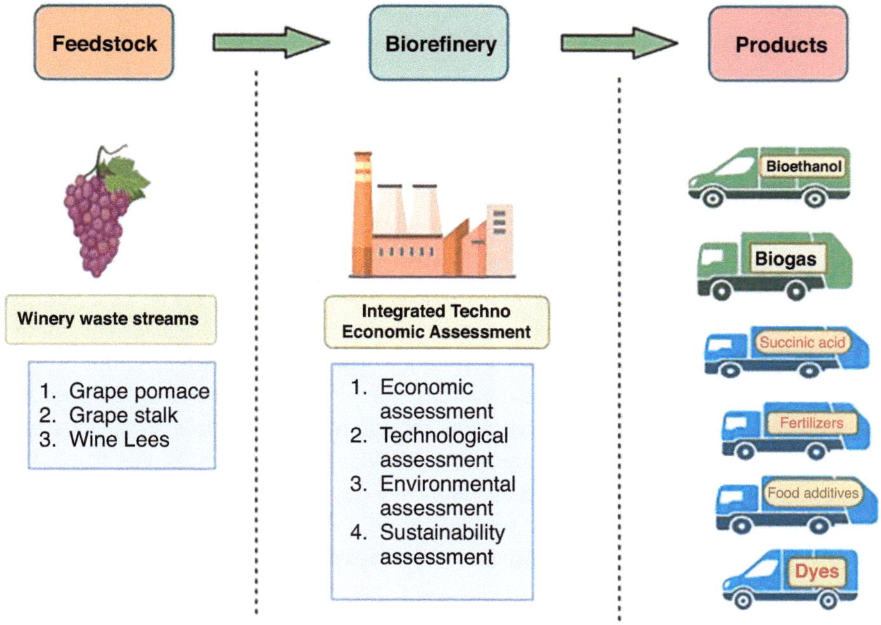

Fig. 3 Process design and TEA of an integrated winery waste biorefinery

6.2 Economic and Environmental Assessment of Biorefinery Using Microalgal Waste Streams

Microalgae-based biorefineries enable the concurrent synthesis of high-value and low-value products and microalgae biomass enriched in a specific macromolecule, provided that the microalgae species and growing conditions are suitable for the intended use. Microalgae are a very valuable feedstock for biofuels, food and feed supplements, and high-value products. Despite all of these benefits, it has been acknowledged that a number of obstacles still need to be removed in order to make their large-scale production economically viable. These obstacles include low productivity, the possibility of contamination in outdoor cultures, and high costs associated with harvesting and downstream processing. It is significant to note that the circular economy idea is adhered to by both the integrated systems for treating wastewater using microalgae and the microalgae-based biorefineries for treating and utilizing wastewater.

Compared to lignocellulosic materials, the use of carbohydrate-rich microalgal biomass for bioethanol production has some advantages because the microalgal carbohydrates are primarily in the form of starch and cellulose (without lignin), which makes them easier to convert to monosaccharides in a biorefinery process (Bhagea et al. 2019). In addition, microalgae biomass has a lot of potential to meet the future demand for clean and sustainable energy due to its use of CO_2 as a carbon source for cultivation, accelerated growth, high yields achieved, and lack of competition for a large surface area of land with agricultural production. Although the production of bioethanol from microalgae is thought to be a very promising approach to address some of the most significant environmental issues, such as climate change and the production of biofuel, there are still obstacles in the way of its large-scale production and commercialization. The most difficult tasks for microalgae to become a competitive feedstock for the manufacture of bioethanol are the selection of species with high biomass productivity and high carbohydrate content, their harvesting, pretreatment, and effective fermentation process (Khan et al. 2018). Exopolysaccharide (EPS)-producing cyanobacteria have been viewed as a promising source of biopolymers of applied interest due to the unique characteristics of the polysaccharides these microorganisms produce. Their exploitation to produce macromolecules appropriate for particular industrial applications has been proposed in numerous studies. Cyanobacterial polysaccharide reportedly has a broad spectrum of activity such as antiviral, antibacterial, antioxidant, and antitumor properties. The goal of microalgae-based biorefineries is thus to harvest biomass from various valued species and use it to produce biofuels, high-value goods like polysaccharides and pigments, and/or low-value products like biostimulants and fertilizers. Future studies should, however, focus on exciting and cutting-edge developments that have the potential to increase the biorefineries' economic benefit-to-operating cost ratio. Using pretreated wastewater, anaerobic digestates, or crude wastewater to supply the majority of the nutrients needed by the microalgae is a promising trend in the field of microalgae-based biorefineries. Thus, the biorefinery's operation cost-to-economic benefit ratio greatly improves if it is designed with a multi-product strategy in mind.

6.3 Economic and Environmental Assessment of Biorefinery Using Sugarcane Molasses and Agave Juice

There are three different types of bioethanol: first, second, and third generation. While the second and third generations of bioethanol are derived from lignocellulosic materials and algae, respectively, the first generation of bioethanol is made from sugar and starchy feedstocks like molasses and maize. Some possible feedstocks that may be employed to manufacture bioethanol and lessen the effects of climate change are agave and sugarcane. Furthermore, agave grows in semiarid soils where food crops cannot be grown. Thus, there is no land competition, and food security issues do not arise (Liu et al. 2019; Zamboni et al. 2009).

Physical pretreatment (such as crushing or chipping), hydrolysis (which is only necessary when both lignocellulosic and algal materials are used as feedstock), fermentation, and distillation are the steps involved in the manufacture of bioethanol. The life cycle assessment (LCA) conducted for the bioethanol production system examined the whole production chain, from agave and sugarcane cultivation to the bioethanol manufacturing process using agave juice and sugarcane molasses. Thus, LCA was based on the cradle-to-gate approach considering four stages, viz., cultivation, extraction of juice, fermentation, and distillation. The functional unit (FU) was chosen to be the generation of bioethanol from molasses and agave juice to provide one MJ of energy. The LCA for ethanol generation using various byproducts from sugar extraction was carried out. It was demonstrated that the largest reduction in nonrenewable energy and global warming potential (GWP) was achieved by sugarcane juice. Furthermore, an assessment of the environmental effects of a sugarcane biorefinery demonstrated that they might be mitigated by including waste valorization.

7 Achievements, Challenges, and Future Opportunities

Unlike petrochemical industry, the amount of essential building blocks and essential materials from biomass in biorefinery platforms is limited. Moreover, costly scale-up and downstream processing and poor performance of biorefinery platforms as compared to petrochemical industry have made our economy mostly dependent on fossil fuels. It has been noted that wastes are the driving force behind the advancement of biorefinery towards circular economy. Sustainability assessment of biorefineries through triple bottom line framework can easily be met through the utilization of wastes (Shen et al. 2019). A consistent and dependable supply of raw materials is expected to be available all year round for biorefinery platforms designed to utilize wastes. Apart from promptly addressing the environmental pollution concern, the readily available feedstock minimizes production costs. Additionally, because the public already has a good understanding of the waste management issue, it is more socially acceptable (Özdenkçi et al. 2017). Therefore, research could be focused more on developing techniques for valorization and refining existing procedures for gathering, storing, sorting, and processing the feedstocks before it is used in biorefinery platforms.

Instead of immediately converting byproducts from primary biorefinery operations into fuel or compost, which are regarded to be the lowest-value products, consideration should be given to further processing of these byproducts to generate less value-added goods in order to maximize economic possibilities. Since feedstocks can share a plethora of valorization technologies, therefore, for a more sustainable approach and to shorten the time that biorefineries are inactive (due to seasonality of feedstocks), platforms should be able to handle more than one kind of feedstock (Sy et al. 2018).

A potential substitute for petroleum-based fuel which has been gaining attention over the past few years is biodiesel/bioethanol. The United States is the top producer of bioethanol followed by Brazil. Despite its massive production, to maintain the long-term sustainability of biodiesel, certain important factors need to be considered. For instance, sugarcane and corn are the major feedstocks used for its production and thus are in conflict with food security. Therefore, procedures for producing biofuels from waste biomass or energy crops that are not competitive with agriculture were developed, resulting in the introduction of a second-generation idea for biofuels (Vaskan et al. 2017).

The following are the primary difficulties with a biowaste biorefinery: (a) transporting and managing biowaste from different sectors, (b) broad adaptability of biorefineries to handle different biowaste compositions, and (c) continuous research and setting up of effective downstream valorization process. In the future, improvements in integration technologies in a biowaste biorefinery and rationalization of premium products could be done to bridge the gap between technology and market. Reportedly there is a significant discrepancy between the performance of biorefineries and the anticipated level of circularity. The assessment of technical efficiency and economic viability are the potent issues related to commercial exploitation of biorefinery (Papadaskalopoulou et al. 2019). Thus, additional research on real-world applications is strongly advised in order to fully grasp the potential of biorefinery platforms for circular bioeconomy and witness its encouraging advantages in real-world settings. The creation and implementation of process models to forecast the economic output of the biorefinery platform can help influence decision-makers and investors to increase their capital expenditures in biorefinery facilities. The precise responsibilities of the many stakeholders, including the academic community, commercial businesses, and national and international government agencies, still remain undefined, and there are still unanswered questions. These questions call for methodical and thorough research (Temmes and Peck 2019). Finally, for fairly analyzing genuine biorefinery installations, sustainability evaluation models need to make an effort to use real data and reduce the number of uncertainties or assumptions.

8 Optimization Methodology

The lack of a reliable source of biomass feedstock throughout the entire year and plant lifetime is one of the primary obstacles impeding the development of biorefineries, leading to economic viability and sustainability. Akgul et al. (Akgul and Papageorgiou 2012a) have created an optimization model for the strategic design of a hybrid first- and second-generation ethanol supply chain. This model tackles difficulties related to sustainability, including the utilization of food crops, the need for second-generation crops to occupy land, and the competition between other sectors for biomass. In another study (Shabani and Sowlati 2013), they suggested a multi-objective optimization model for hybrid first- and second-generation biofuel supply chains in order to examine the trade-offs between environmental and economic goals and the effects of the carbon tax on the supply chain's environmental and economic performance. Several research examined the design of dynamic models throughout several time periods. Shabani and Sowlati (Shabani and Sowlati 2013; Zhang et al. 2008) establish a dynamic, nonlinear mixed integer programming model whose goal is to maximize the total value of the forest biomass supply chain. Multi-objective optimization has also been utilized in various studies to optimize for competing objectives within biofuel supply chain design and management (Trung et al. 2019; You et al. 2012; Zamboni et al. 2009). Typically, this kind of optimization produces a group of Pareto optimal solutions, also known as non-dominated solutions, in which achieving one goal always has an adverse effect on another. Finding a solution that is desirable to the decision-maker is the aim of multi-objective optimization.

8.1 Model Formulation

Maximization of profits, minimization of expenses, and energy consumption are the three main goals of the mathematical optimization model. The profit is the amount that is left over after deducting the purchasing costs of raw materials and energy streams, as well as the fixed costs of each process unit that is used, from the overall costs created by the biorefinery. Let us consider a set of compounds $n \in S$ and such that S^F, S^I and S^P denote the set of feedstocks, intermediates, and final products, respectively. Then it satisfies the following equalities or inequalities:

$$\alpha_n \neq 0, \forall n \in S^F, S^P$$

$$\alpha_n = 0, \forall n \in S^I$$

$$\beta_n \neq 0, \forall n \in S^F$$

$$\beta_n = 0, \forall n \in S^I, S^P$$

$$\gamma_n \neq 0, \forall n \in S^P$$

$$\gamma_n = 0, \forall n \in S^F, S^I$$

where α denotes the purchase/selling price Rs/Year, β denotes feedstock availability Kg/Year, and γ denotes the final product demand Kg/Year. Let $m \in T$ be the set of technologies m, such that T^{in} and T^{out} are the two subsets of the set T because each technology is both a producer and a consumer of compounds. The unit technology cost μ_m, Rs/Kg main input in m, unit energy cost ω_m, Wh/Kg main input in m and mass ration λ_{nm}, *Kg* of output stream of *n*/*kg* of main input in m. For any compunds $n \in S$, the mass balance constrain is given by

$$P_n - \sum_{m \in T^{\text{in}}} \lambda_{nm} R_m = Q_n - \sum_{m \in T^{\text{out}}} \lambda_{nm} R_m, \ \forall n \in S \tag{1}$$

where P_n is the amount of feedstock purchased, Q_n is the amount of product sold, and R_m is the amount of raw material consumed in technology m. In addition to this, we have the following constraints:

$$P_n \leq \beta_n, \ \forall n \in S^F \tag{2}$$

$$Q_n \geq \gamma_n, \ \forall n \in S^P \tag{3}$$

$$P_n = 0, \ \forall n \in S^I \tag{4}$$

$$Q_n = 0, \ \forall n \in S^F, S^I \tag{5}$$

assuming that only final goods may be sold and only feedstock can be bought.

Considering different objective functions in this mathematical model that can be optimized, the total production cost (TC) Eq. (6), profit (PRO) Eq. (7) and energy consumption (EC) Eq. (8):

$$\min \text{TC} = \sum_{n \in S^F} \alpha_n P_n + \sum_m R_m \left(\mu_m + \omega_m \xi_m\right) \tag{6}$$

$$\max \text{PRO} = \sum_{n \in S^P} \alpha_n Q_n - \sum_{n \in S^F} \alpha_n P_n - \sum_m R_m \left(\mu_m + \omega_m \xi_m\right) \tag{7}$$

$$\min \text{EC} = \sum_m R_m \omega_m \tag{8}$$

where ξ_m is the energy purchased cost in Rs/Wh.

9 Conclusion

With the world's consumption of energy and organic compounds escalating, there has also been an enormous increase in dependency on nonrenewable fossil fuels (coal, petroleum, natural gas) to meet basic energy, fuel, and organic polymer needs. As much as 80% of energy needs and 90% of organic chemical needs are derived from fossil fuels, the current need of the hour is to search for substitutes for nonrenewable resources. Factors such as the increased price of crude oil, greenhouse gas emission and thereby environmental deterioration, and government policies and legislations have all contributed to a shift towards renewable, clean, and sustainable resources.

Biomass resources thus provide a sustainable opportunity, as they can be transformed into beneficial bioproducts and bioenergy such as biofuels. Building a sustainable foundation for industrial development, boosting competitiveness, and ensuring social and economic advancement are the principal objectives. Biorefineries act as a key component in this process as they are important platforms where biomass sources are converted into value-added products through integration and valorization techniques. Biorefineries are thus promising modern processing platforms for the sustainable conversion of biomass into a wide spectrum of bioproducts such as bioenergy, biofuels, and biochemicals.

Depending upon their conversion platform, biorefineries are subdivided into three broad categories, viz., thermochemical, biological, and hybrid. If the modelling efforts for biological processes are to be strengthened, then a greater acquisition of experimental data and mechanistic understanding is crucial. Furthermore, developing and broadening the hybrid conversion idea would help the short- and long-term economic exploitation of sustainable biorefinery processes.

Depending upon their origin, biomass can be similarly classified as agricultural feedstock, forestry feedstock, industrial feedstock, or municipal feedstock. Biorefinery network modelling towards using biowaste streams instead of single food crops as feedstock has gained momentum over the years. The feedstocks differ significantly in their individual characteristics and qualities which makes it extremely challenging to design a sustainable integrated biorefinery platform.

The most important objective of biorefineries is to reduce carbon footprints and thereby create industrial processes that are carbon neutral. To realize this objective, there must be communication among many stakeholders as well as a strong push for system integration with other industries and pertinent technology. Biorefineries have advanced to the point where they can begin offering trustworthy substitutes for the majority of the goods we use on a daily basis. However, they have not reached the point where they can completely replace current fossil fuel refinery technologies, and thus the feasible goal of biorefinery system would be to integrate with current technologies and infrastructure and not completely displace the petrochemical platforms.

The implementation of the life cycle assessment tool in conjunction with promising/novel research lines and the circular economy strategy presents a variety of opportunities and challenges for overcoming constraints, particularly with regard to process scaling and obtaining a convenient cost-benefit ratio.

References

Agrawal K, Bhardwaj N, Kumar B, Chaturvedi V, Verma P (2019) Process optimization, purification and characterization of alkaline stable white laccase from Myrothecium verrucaria ITCC-8447 and its application in delignification of agroresidues. Int J Biol Macromol 125:1042–1055

Agrawal K, Ruhil T, Verma P (2023) Microalgae: a way toward sustainable development of a society. In: Basic research advancement for algal biofuels production. Springer Nature Singapore, Singapore, pp 259–277

Akgul N, Papageorgiou LG (2012a) Economic optimisation of a UK advanced biofuel supply chain. Biomass Bioenergy 41:57–72

Akgul N, Papageorgiou LG (2012b) An optimisation framework for a hybrid first/second generation bioethanol supply chain. Comput Chem Eng 42:101–114

Amer M, Elwardany A (2020) Biomass carbonization, in Renewable energy-resources, challenges and applications 90480

Baruah J, Nath BK, Sharma R et al (2018) Recent trends in the pretreatment of lignocellulosic biomass for value-added products. Front Energy Res 6:1–19

Benjamin MFD, Ubando AT, Razon LF et al (2015) Analyzing the disruption resilience of bioenergy parks using dynamic inoperability input–output modeling. Environ Syst Decis 35:351–362

Bhagea R, Bhoyroo V, Puchooa D (2019) Microalgae: the next best alternative to fossil fuels after biomass. A review. Microbiol Res 10:12–24

Bhardwaj N, Kumar B, Agarwal K, Chaturvedi V, Verma P (2019) Purification and characterization of a thermo-acid/alkali stable xylanases from Aspergillus oryzae LC1 and its application in Xylo-oligosaccharides production from lignocellulosic agricultural wastes. Int J Biol Macromol 122:1191–1202

Bhardwaj N, Agrawal K, Kumar B, Verma P (2021) Role of enzymes in deconstruction of waste biomass for sustainable generation of value-added products. Bioprospecting of enzymes in industry, healthcare and sustainable environment, pp. 219–250

Boulamanti AK, Donida Maglio S, Giuntoli J et al (2013) Influence of different practices on biogas sustainability. Biomass Bioenergy 53:149–161

Chan HY, Tan RR, Yusup AT, et al. (2018) Process Integr Optim Sustainability 2, 343

Chen Y, Wu Y, Zhu B et al (2018) Co fermentation of cellobiose and xylose by mixed culture of recombinant Saccharomyces cerevisiae and kinetic modeling. PLoS One 13:e199104

Cherubini F (2010) The biorefinery concept: using biomass instead of oil for producing energy and chemicals. Energy Convers Manag 51(7):1412–1421

Chew KW, Yap JY, Show PL et al (2017) Microalgae biorefinery: high value products perspectives. Bioresour Technol 229:53–62

Chowdhary P, Gupta A, Gnansounou E et al (2021) Current trends and possibilities for exploitation of grape pomace as a potential source for value addition. Environ Pollut 278:116796

Fava F, Totaro G, Diels L et al (2015) Biowaste biorefinery in Europe: opportunities and research & development needs. New Biotechnol 32(1):100–108

Ferreira JA, Brancoli P, Agnihotri S et al (2018) A review of integration strategies of lignocelluloses and other wastes in 1st generation bioethanol processes. Process Biochem 75:173–186

Goswami RK, Mehariya S, Obulisamy PK, Verma P (2021a) Advanced microalgae-based renewable biohydrogen production systems: a review. Bioresour Technol 320:124301

Goswami RK, Agrawal K, Verma P (2021b) Microalgae-based biofuel-integrated biorefinery approach as sustainable feedstock for resolving energy crisis. Bioenergy research: commercial opportunities & challenges, pp.267–293

Goswami RK, Agrawal K, Upadhyaya HM, Gupta VK, Verma P (2022) Microalgae conversion to alternative energy, operating environment and economic footprint: an influential approach towards energy conversion, and management. Energy Convers Manag 269:116118

Grande L, Pedroarena I, Korili SA et al (2021) Hydrothermal liquefaction of biomass as one of the most promising alternatives for the synthesis of advanced liquid biofuels: a review. Materials 14(18):5286

Hasenheit M, Gerdes H, Kiresiewa Z, et al. (2016) Summary report on the social, economic and environmental impacts of the bioeconomy Promoting stakeholder engagement and public awareness for a participative governance of the European bioeconomy 6

Heidenreich S, Müller M, Foscolo PU (2016) Advanced biomass gasification: new concepts for efficiency increase and product flexibility. Academic Press, Cambridge, MA, pp 4–10

Khan MI, Shin JH, Kim JD (2018) The promising future of microalgae: current status, challenges, and optimization of a sustainable and renewable industry for biofuels, feed, and other products. Microb Cell Factories 17:36

Kumar A, Samadder SR (2020) Performance evaluation of anaerobic digestion technology for energy recovery from organic fraction of municipal solid waste: a review. Energy 197:117253

Kumar B, Verma P (2020) Enzyme mediated multi-product process: a concept of bio-based refinery. Ind Crop Prod 154:112607

Kumar B, Bhardwaj N, Alam A, Agrawal K, Prasad H, Verma P (2018) Production, purification and characterization of an acid/alkali and thermo tolerant cellulase from Schizophyllum commune NAIMCC-F-03379 and its application in hydrolysis of lignocellulosic wastes. AMB Express 8:1–16

Kumar B, Bhardwaj N, Agrawal K, Verma P (2020) Bioethanol production: generation-based comparative status measurements. Biofuel production technologies: critical analysis for sustainability, pp.155–201

Lauer M. Methodology guideline on techno economic assessment (TEA). 2008. Thermal Net WP3B Economics 25, 25

Li J, Zhang M, Li J et al (2018) Corn Stover pretreatment by metal oxides for improving lignin removal and reducing sugar degradation and water usage. Bioresour Technol 263:232–241

Liu ZH, Le RK, Kosa M et al (2019) Identifying and creating pathways to improve biological lignin valorization. Renew Sust Energ Rev 105:349–362

Madár V, Bacskai I, Dhaundiyal A et al (2018) Development of biomass-based pyrolysis CHP (R+D). HungAgricEng 33:17–23

Mandø M (2013) Direct combustion of biomass, in biomass combustion science. technology and engineering 23:61–83

Mendes JAS, Xavier AMRB, Evtuguin DV et al (2013) Integrated utilization of grape skins from white grape pomaces. Ind Crop Prod 49:286–291

Michailos S, Parke D, Webb C (2019) Design, sustainability analysis and multi objective optimisation of ethanol production via syngas fermentation. Waste Biomass Valori 10:865–876

Minelgaitė A, Liobikienė G (2019) Waste problem in European Union and its influence on waste management behaviours. Sci Total Environ 667:86–93

Mohan SV, Butti SK, Amulya K et al (2016) Waste biorefinery: a new paradigm for a sustainable bioelectro economy. Trends Biotechnol 34(11):852–855

Mohr A, Raman S (2013) Lessons from first generation biofuels and mplications for the sustainability appraisal of second-generation biofuels. Energy Policy 63:114–122

Naik S, Goud VV, Rout PK et al (2010) Production of first and second-generation biofuels: a comprehensive review. Renew Sust Energ Rev 14:578–597

Neto B, Dias AC, Machado MJ (2013) Life cycle assessment of the supply chain of a portuguese wine: from viticulture to distribution. Int J Life Cycle Assess 18(3):590–602

Nizami AS, Rehan M, Waqas M et al (2017) Waste biorefineries: enabling circular economies in developing countries. Bioresour Technol 241:1101–1117

OIV (2019) Statistical Report on World Viti viniculture. International Organisation of Vine and Wine Intergovernmental Organisation

Özdenkçi K, De Blasio C, Muddassar HR et al (2017) A novel biorefinery integration concept for lignocellulosic biomass. Energy Convers Manag 149:974–987
Papadaskalopoulou C, Sotiropoulos A, Novacovic J et al (2019) Comparative life cycle assessment of a waste to ethanol biorefinery system versus conventional waste management methods. Resour Conserv Recycl 149:130–139
Paradelo R, Moldes AB, Barral MT (2010) Utilization of a factorial design to study the composting of hydrolyzed grape Marc and Vinification lees. J Agric Food Chem 58(5):3085–3092
Probst M, Walde J, Pumpel T et al (2015) Lactic acid fermentation within a cascading approach for biowaste treatment. Appl Microbiol Biotechnol 99:3029–3040
Ragauskas AJ, Williams CK, Davison BH et al (2006) The path forward for biofuels and biomaterials. Science 311:484–489
Shabani N, Sowlati N (2013) A mixed integer non-linear programming model for tactical value chain optimization of a wood biomass power plant. Appl Energy 104:353–361
Shah AV, Singh A, Mohanty SS et al (2022) Organic solid waste: biorefinery approach as a sustainable strategy in circular bioeconomy. Bioresour Technol 349:126835
Sheldon RA, Norton M (2020) Green chemistry and the plastic pollution challenge: towards a circular economy. Green Chem 22:6310–6322
Shen R, Tao L, Yang B (2019) Techno-economic analysis of jet-fuel production from biorefinery waste lignin. Biofuels Bioprod Biorefin 13(3):486–501
Sy CL, Ubando AT, Aviso KB et al (2018) Multi-objective target oriented robust optimization for the design of an integrated biorefinery. J Clean Prod 170:496–509
Temmes A, Peck P (2019) Do forest biorefineries fit with working principles of a circular bioeconomy? A case of Finnish and Swedish initiatives. Forest Policy Econ 101896:1–12
Torres MD, Kraan S, Domínguez H (2019) Seaweed biorefinery. Rev Environ Sci Biotechnol 18:335–388
Trung H, Julien G, Pandya D et al (2019) High-resolution multi-objective optimization of feedstock landscape design for hybrid first and second generation biorefineries. Appl Energy 238:1484–1496
Tuazon D, Gnansounou E (2017) Towards an integrated sustainability assessment of biorefineries. Elsevier B.V
Tuazon D, Corder G, McLellan B (2013) Sustainable development: a review of theoretical contributions. Int J Sustain Future Human Secur 1(1):40–48
Ubando AT, Africa ADM, Maniquiz-Redillas MC et al (2021) Reduction of particulate matter and volatile organic compounds in biorefineries: a state-of-the-art review. J Hazard Mater 403:123955
Vaskan P, Pachón ER, Gnansounou E (2017) Life cycle assessment of sugar crops and starch-based integrated biorefineries. Life-Cycle Assessment of Biorefineries, pp 97–146
Walmsley TG, Varbanov PS, Su R et al (2018) Frontiers in process development, integration and intensification for circular life cycles and reduced emissions. J Clean Prod 201:178–191
Yang J, Xiao YY (2013) Grape phytochemicals and associated health benefits. Criti Rev Food Sci Nutr 53:1202–1225
You F, Tao L, Graziano DJ et al (2012) Optimal design of sustainable cellulosic biofuel supply chains: multi objective optimization coupled with life cycle assessment and input–output analysis. AICHE J 58:1157–1180
Zamboni A, Bezzo F, Shah N (2009) Spatially explicit static model for the strategic design of future bioethanol production systems. Multi-objective environmental optimization. Energy Fuels 23:5134–5143
Zhang YHP (2008) Reviving the carbohydrate economy via multi-product lignocellulose biorefineries. J Ind Microbiol Biotechnol 35(5):367–375

The Prospective Impact of Technological Advances in Value-Sensitive Design (VSD) Used in the Biorefinery Design Selection Processes

Santosh Kumar, Subir Kumar Shill, Chirantan Saha, Nagendra Thakur, and Sayak Das

1 Introduction

The increasing worldwide demand for sustainable energy and materials has gained impetus because of growing environmental consciousness and the imperative need to implement eco-friendly practices (Chen et al. 2023). Increased industrialization, urbanization, and population expansion are among the factors driving this demand, which is further fuelled by a growing awareness of human activity's effects on the environment (York et al. 2003). One of the primary drivers of the anticipated population explosion, which is expected to total around nine billion people by 2050 (Franklin and Andrews 2012), is the surge in energy demand for a variety of uses. Since fossil fuels, in particular, transmit greenhouse gases into the atmosphere, they show a negative influence on the environment and exacerbate climate change. This emphasizes how urgently we need to make the switch to sustainable alternatives (Yoro and Daramola 2020).

As our cities and industries continue to expand, the demand for sustainable materials and energy becomes increasingly urgent. This growth and development puts a strain on our resources and contributes to the deterioration of our environment, disrupting the delicate balance of our ecosystems (Kumar et al. 2022). The construction sector has a major burden for the depletion of resources and the development of waste, underscoring the pressing need to employ environmentally friendly

S. Kumar · C. Saha · N. Thakur
Department of Microbiology, School of Life Sciences, Sikkim University, Gangtok, Sikkim, India

S. K. Shill · S. Das (✉)
Department of Life Science & Bioinformatics, Har Gobind Khurana School of Life Sciences, Assam University, Silchar, Assam, India
e-mail: sayak.das@aus.ac.in

K. Agrawal, P. Verma (eds.), *Biotechnological Advances in Biorefinery*, Interdisciplinary Biotechnological Advances,
https://doi.org/10.1007/978-981-97-5544-8_17

construction supplies and methods. Growing awareness of the detrimental effects of climate change is influencing the demand for sustainable energy and materials. Governments, corporations, and people all understand how critical it is to implement policies that lessen carbon footprints and encourage environmental stewardship (Babiak and Trendafilova 2011). International policies such as the Paris Agreement emphasize the need for cooperative efforts to combat climate change, prompting nations to enact laws supporting environmental sustainability (Blau 2017).

The increasing need for sustainable materials and energy, particularly for renewable energy sources like hydropower, solar power, and wind, requires technological innovation. Developments in these fields have made them much more accessible and affordable, which has led to their widespread use on a global scale. Similarly, developments in material engineering have made it possible to create sustainable alternatives, which have a negative impact on the environment in a number of ways (Ashraf et al. 2021). In contrast to conventional refineries that rely on fossil fuels, biorefineries use sustainable biomass resources to create energy, supplies, substances, and biofuels. By incorporating ethical and societal values into the design process, biorefinery design adheres to the principles of value-sensitive design (VSD). VSD promotes an accountable and sustainable design approach by helping to recognize and address ethical issues (Parada et al. 2022; Palmeros Parada et al. 2020).

Value-sensitive design (VSD) is a conceptual strategy for the development of technology that considers value systems holistically and meticulously throughout the entire process of designing (Friedman et al. 2013). When expressed strictly, "value" refers simply to the financial worth of an item. Technical, empirical, and conceptual research are all combined in a sequential process that forms the foundation of VSD (Friedman and Kahn Jr. 2003). Conceptual investigations: These rely on those that are both directly and indirectly influenced by the layout in consideration (Friedman and Kahn Jr. 2003). The design values are taken into consideration, taking into account the values of data that should be developed, implemented, and run. And it also discusses about the moral principles (the power to confidentiality) and non-moral values (aesthetically pleasing preferences). VSD answers these queries in the category of conceptual explorations. Empirical investigations: The scope of conceptual research is constrained. Depending on the topics at hand, a number of analyses will necessitate empirical investigation on the human context around the technical creation (Friedman and Kahn Jr. 2003). Furthermore, empirical research is usually needed to evaluate the efficacy of a design. Any human behavior that is observed, measured, and documented can be the focus of an empirical study. Because of this, any social science research method—qualitative or quantitative—can be employed in this context. These techniques include user behavior and human physiology measurements, observations, interviews, surveys, experimental manipulations, and document gathering (Lipinski and Britz 2000). Technical investigations: VSD takes a stance that technologies in general and computer and information technologies specifically offer value suitabilities that stem from the technology's inherent characteristics (Friedman and Kahn Jr. 2003). That is to say, a certain technology makes it easier to achieve certain activities and values while making other actions and values less feasible (Sawyer and Rosenbaum 2000). Technical inquiries take one form that focuses on how the features and underlying mechanisms of current technology help or hinder human values. Technical studies in the second kind entail the proactive design of systems to uphold values determined by the conceptual study (Lipinski and Britz 2000). However, they differ greatly with respect to their unit of evaluation. Technical studies focus on the relevant technologies. Empirical research focuses on the individuals, groups, or larger social structures that establish, use, or are affected by the technology.

VSD stands out for being strategic and having the power to influence technology design all the way from the start to the finish (Friedman et al. 2013). It expands the environment in which values are formed to include everyday life, online networks, transport, and manufacturing in addition to the fundamental resources (Sawyer and Rosenbaum 2000). It must integrate and iteratively apply conceptual, empirical, and technical research (Friedman et al. 2013). It should also encompass the spectrum of human values that are morally meaningful. Standardization of protocols is also taken into consideration. It must distinguish between usefulness and morally significant human values. It is necessary to consider all of its beneficiaries, both direct and indirect. It is a societal viewpoint that helps individuals create technologies that more readily uphold specific principles. The goals of those utilizing the technology dictate how it is actually used.

2 The Role of Biorefineries in the Bio-economy

The present human generation is being forced to transition from fossil fuels to a sustainable bio-economy due to the growing demand for energy and resources (Martinez-Hernandez et al. 2013; Banu et al. 2021). The bio-economy requires renewable feedstocks with the capacity to produce a variety of bio-based products (Xu et al. 2020; Mongkhonsiri et al. 2018). This requires a refinery that involves many scientific, management, and engineering disciplines (Dahiya et al. 2018). A sustainable waste-based bio-economy can be developed through the use of biogenic waste as a potential feedstock in this context (Hoang Nguyen Tran et al. 2020). This can be further explored by using biogenic waste in an integrated biorefinery mode to produce bio-based products (Oliva et al. 2020). Strategic waste conversion to commercially advantageous product recovery can be achieved by integrating different supporting bio-processes, such as acidogenesis, fermentation, methanogenesis, solventogenesis, photosynthesis, oleaginous process, bio-electrogenesis, etc. (Fava et al. 2015; Sekoai et al. 2021, 2023). With improved waste treatment efficiency, the biorefinery approach can produce a variety of products, including biofuels (H_2, CH_4, biohythane, biodiesel), platform chemicals (sugars, carboxylic acids, bioethanol, biobutanol), biomass, bioelectricity, biomaterials (biopolymers), biofertilizers, animal feed, etc. (Fig. 1) (Paul et al. 2022; Fal et al. 2023; Chopra et al. 2022; Kawa-Rygielska et al. 2022; Chandra et al. 2019). Additionally, the fermentation results in the co-generation of high-value compounds that can be used as substrates for the

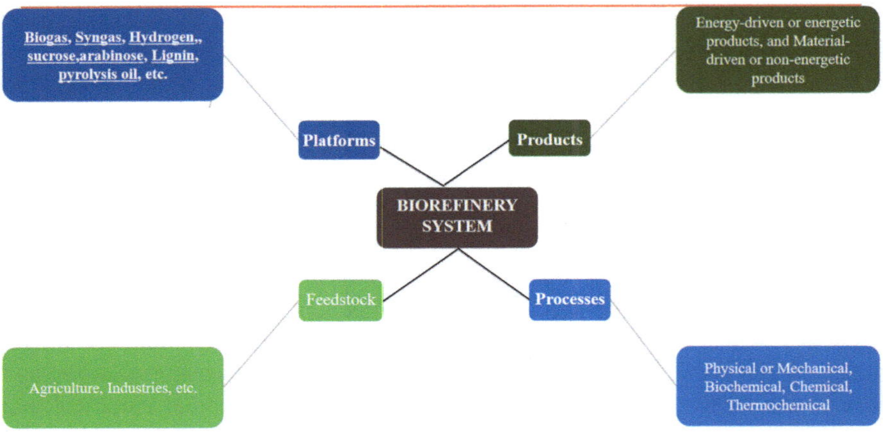

Fig. 1 Schematic representation of the various platforms, products, feedstocks, and processes involved in a biorefinery system

integration strategy of processes, thereby overcoming the constraints of the separate processes and providing an additional benefit of product recovery (Gómez et al. 2022; Elmekawy et al. 2013). The Food-waste (FW) biorefinery can somewhat complement fossil fuel-based refineries by producing a variety of bio-based products, such as biofuels, commodity chemicals, and biofertilizers. It can also address the main drivers of the bio-economy, which are resource security, ecosystem services, and climate change (Venkata Mohan et al. 2016; Venkata Mohan and Sarkar 2017). In the upcoming future, the biorefinery concept may be able to attain a sustainable green route with minimal impact on the environment.

Biorefineries are a significant contributor to the expanding bio-economy and an indicator of the shift towards environmentally friendly methods (Amulya et al. 2014, 2015; Sadhukhan et al. 2018). This sector utilizes renewable resources like biomass to create biofuels, biochemicals, and biomaterials (Nair et al. 2022a, b; Agrawal et al. 2022; Bhardwaj et al. 2020). The industry's ability to successfully produce biofuels, biochemicals, and biomaterials depends in large part on the utilization of renewable resources like biomass (Sadhukhan et al. 2018). Biorefineries are essential for advancing economic development, energy security, and environmental sustainability since they effectively convert different forms of biomass (Gavrilescu 2014). The capacity of biorefineries to use a multiplicity of feedstocks and break free from reliance on finite fossil fuels sets them beyond conventional refineries. This makes it possible for them to digest various types of biomass, which eventually boosts overall resource efficiency, lessens dependency on scarce resources, and encourages sustainable land use (Ghatak 2011). Biorefineries are essential to the process of producing biofuels, offering valuable substitutes for traditional fuels and helping to reduce greenhouse gas emissions (Mossberg et al. 2021).

Furthermore, by removing useful components from biomass, reducing waste, and pushing for a closed-loop system, biorefineries support a circular economy

(Paladino and Neviani 2024). It is impossible to overlook the vital responsibilities that agriculture and the agroindustry play in supplying essential ingredients for biorefineries. They therefore contribute significantly to upgrading the bioeconomy, sustainable agricultural methods, and rural communities (Wagh et al. 2024). This cooperative relationship demonstrates how both industries benefit from one another and make significant contributions to the sustainability of the economy and the environment. This is consistent with the ideas of waste minimization and being environmentally conscious. Employing constant advancement, biorefinery methods are expanding the limits of efficiency, enabling the usage of a wider range of feedstocks and producing derivatives with increased value added (Awasthi et al. 2020). Value-added goods that are produced consequently, such as biofuels and bioplastics, are crucial in moving the market in the direction of a more environmentally friendly and circular economy (Visco et al. 2022). Policy, legislation, and market forces that strongly support bio-based business operations and give preference to environmentally friendly alternatives possess a noteworthy impact on the advancement of the bio-economy. This cooperative strategy opens the door for sustainable growth by fostering a dynamic and competitive environment (Singh et al. 2021; Rodríguez et al. 2019). Within the framework of the bio-economy, the advancement of biorefineries spurs innovation in green chemistry and biotechnology, enhancing enzymatic and microbial processes to increase productivity and financial sustainability (Manzanares 2020).

3 An Overview of Biorefineries and their Conversion of Resources for Renewable Biomass

Biorefineries are crucial in encouraging the sustainable use of sustainable biomass resources, marking a substantial alteration in the global energy and chemical industries (Ubando et al. 2020). These innovative facilities efficiently transform organic resources into a wide range of valuable goods, including organic waste, forestry by-products, and agricultural leftovers. Biorefineries use biological processes to extract energy, fuels, and chemicals using biomass, in contrast to typical refineries that rely mostly on fossil fuels (Kumar and Verma 2021). The primary aim is to optimize resource utilization and mitigate ecological footprint, consequently facilitating the development of a bio-based economy.

Physically separating the biomass holding the precursors is the initial stage in this procedure (Brito et al. 2019; Yakaboylu et al. 2021). Thereafter, further chemical or microbiological processing is applied to the primary and intermediate products. After that, these end products can be sent to conventional refineries for further processing (Kamm and Kamm 2004, 2007; Fitz Patrick et al. 2010). In this context, the term "biorefinery" has particular significance as it captures the essence of both the organic character of the starting materials and the increasing application of biological processes in pre-treatment and processing (Alibardi et al. 2020). Currently, three primary biorefinery models exist that are being researched and developed. The first is referred to as a whole-crop biorefinery and employs raw materials like wheat or maize (Kamm and Kamm 2004; Clark and Deswarte 2015; Thongchul et al. 2022). The second is a green biorefinery, which breaks down biomass that is inherently wet, such as young cereal crops, lucerne, clover, and green grass (Sauer et al. 2014; Santamaría-Fernández and Lübeck 2020). In addition to this, the lignocellulose feedstock (LCF) biological refinery concentrates on naturally dry substances such as waste products and biomass that is rich in cellulose (Uihlein and Schebek 2009; Cheng and Wang 2013; Kumar et al. 2020; Bhardwaj et al. 2019, 2020; Kumar et al. 2018; Agrawal et al. 2019; Agrawal and Verma 2020).

To extract biofuels, bio-based chemicals, and other useful products, biorefinery procedures often involve biochemical or thermochemical conversion approaches such as gasification, pyrolysis, anaerobic digestion, and fermentation (Awasthi et al. 2020; Ubando et al. 2020). Within the biofuels industry, biorefineries are essential for creating substitutes for traditional fossil fuels. Advanced biofuels made from biomass, such as bioethanol and biodiesel, offer a sustainable and environmentally friendly option for industrial and transportation purposes (Demirbas 2007). By lessening the harm to the environment that traditional petrochemical production causes, the manufacturing of bio-based chemicals using biorefinery processes is essential to advancing the growth of circular economy (Saini et al. 2019).

Utilizing a variety of biomass feedstocks, biorefineries may adjust to local availability, promote agricultural and diversification of the economy, and function with an adaptable and integrated approach (Manzanares 2020). It is common practice to combine advanced technologies like biotechnology, nano-technology, and process optimization to boost production and overall efficiency. Strategically positioned biorefineries have the power to greatly influence the global chemical and energy industries, paving the mode for a more resilient and sustainable future. This transformation is propelled by continuous research and development, utilizing an extensive range of resources and expertise, as depicted in Fig. 2. An urgent need exists for fossil resource-dependent sectors to find alternatives (Hunt et al. 2010). These finite resources yield a range of chemical substances as well as electricity. The best option for completing these applications turns out to be biomass (Ghatak 2011; Ragauskas et al. 2006). Recent research focus on using Value Sensitive Design (VSD) to specify design space boundaries in the early stages of biorefinery development, given the complexity of these projects. The fact that biorefineries need large financial investments and the active engagement of several stakeholders demonstrates the importance of values in the design process.

Developing renewable resources is crucial in reducing the environmental consequences of relying on non-renewable energy sources. In particular, biomass has emerged as a ground-breaking solution with the capability to revolutionize numerous industrial uses (Julio et al. 2017). There is a tremendous potential and ability to unleash sustainable biopower and biomaterials by fusing agro-energy crops with cutting-edge biorefinery techniques, opening the door for a significant revolution in manufacturing (Ragauskas et al. 2006). Bioethanol stands out as a major player in the world of renewable energy derived from biomass, showcasing both innovation

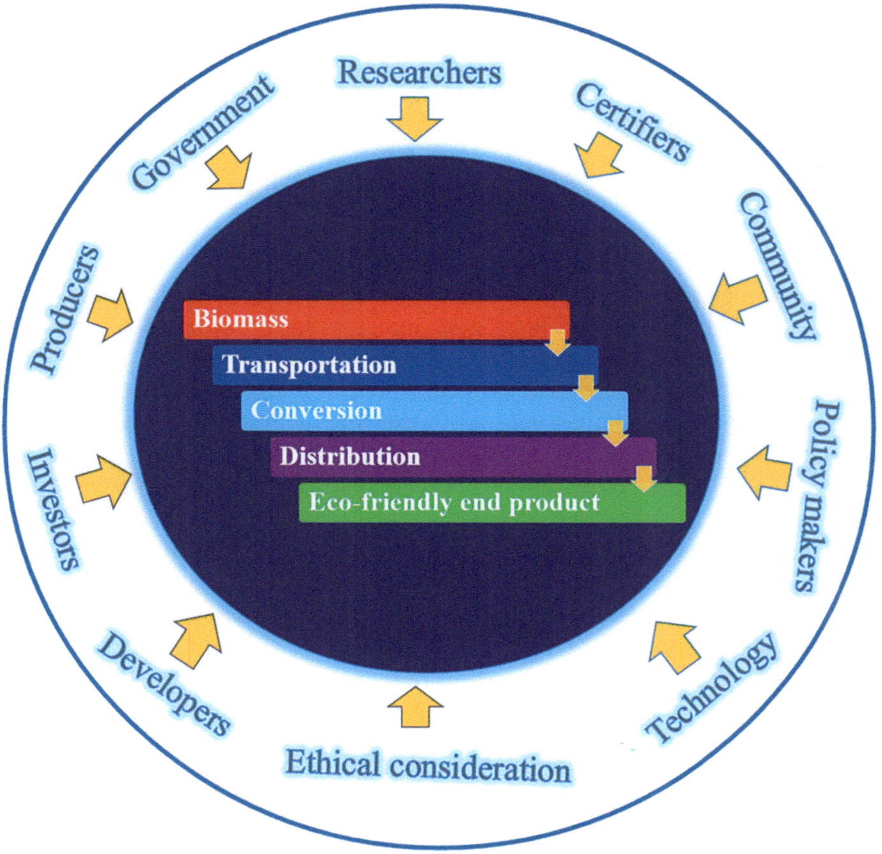

Fig. 2 Outline of the various approaches that are utilized for the efficient conversion of biomass to value-added products of significance in biorefinery

and sustainability in the transportation industry (Majidian et al. 2018; Spagnuolo et al. 2019; Zhang et al. 2019; Ko et al. 2020). Its remarkable environmental advantages make it an ideal, serving as a solution to the pressing issues associated with conventional fossil fuels (Berg 2001). Meanwhile, biodiesel has also emerged as a significant contender in the realm of alternative transportation fuels, leading to a new wave of environmental awareness worldwide (Brunschwig et al. 2012). As we embark on a journey towards a more sustainable future free from reliance on fossil fuels, an exciting opportunity emerges with the joining of agro-energy crops, biorefinery technologies, and biofuels. This convergence has the potential to completely revolutionize our global landscape. It signifies a ground-breaking transformation where innovation meets environmental consciousness, forging a new path towards a greener tomorrow. The composition of biofuel is heavily influenced by the specific raw materials employed. There are three phases to biorefineries based on their input capacity, processing capabilities, and product output (Fernando et al. 2006). In the

USA, Canada, and Brazil, first-generation biorefineries hold a dominant position. However, concern has been leveled at their impact on land utilization and potential competitiveness with food production (Van Dyne et al. 1999). Instead, the USA, China, and Canada have adopted second-generation biorefinery technology, with growing interest from Northern-European countries. The current research and development focus is on third-generation biorefineries, which utilize waste materials and have already gained momentum in Europe (Kaur et al. 2022). Growing biorefineries using locally produced raw materials is a viable strategy to lessen environmental footprints and boost rural economies (Lopolito et al. 2011). In Northern Europe, the idea of a "green biorefinery" is increasing in popularity. This involves separating green biomass into press cake and press juice that contains important components. According to Kamm and Kamm (2004), green juice can produce chemicals, high-quality feed, and cosmetic proteins. It also acts as a substrate for the creation of biogas. Another benefit of a green biorefinery is its effectiveness in using a variety of green biomasses (Kromus et al. 2004).

According to Fitz Patrick et al. (2010), there are four foremost stages in the process of converting natural biomass into bio-based products that may be sold commercially. These steps include pre-treatment, anatomizing (hydrolysis), fermentation, and product recovery. Though it requires a large amount of energy, preprocessing is necessary to convert the strong lignocellulosic structure into responsive cellulosic intermediates (Galbe et al. 2007).

4 Principles of Value-Sensitive Design (VSD)

According to Iversen and Cawthorne (2023), the principles of VSD prioritize the active incorporation and consideration of a wide-range of human values throughout technology development. This involves recognizing the intricate and varied nature of these values, along with their dependence on context and individual and cultural differences. As a result, designers must thoroughly examine the values that are relevant to their target users, both explicit and implicit, to understand how these values may influence user experiences and interactions with the technology. Another crucial aspect of VSD is connecting stakeholders in the design process, as emphasized by Sadek et al. (2023) and Palmeros et al. (2018). By doing so, designers can ensure that the technology reflects the needs and perspectives of various stakeholders, leading to a more inclusive and effective end product. In acknowledging the effects of technology on both individuals and communities, VSD promotes its essential role in its design. By actively involving end users throughout the process, this inclusive approach empowers them to shape technology that reflects their values and meets their needs. Ultimately, this democratization of the design process helps minimize any possible negative consequences and emphasizes the importance of prioritizing the welfare of both individuals and society (Palmeros Parada et al. 2020).

In the world of design, ethics takes center stage in VSD, compelling designers to confidently prioritize the well-being of all (Dennis 2021). This requires a steadfast dedication to recognizing and addressing ethical complexities throughout the entire design process, promoting transparency, accountability, and a proactive approach to ethical decision-making. VSD advocates for a comprehensive and forward-thinking outlook, urging designers to consider the far-reaching implications of their creations over time (Friedman et al. 2013). This means being mindful of potential unintended consequences, adapting to shifting societal values, and designing with sustainability in mind. By taking a holistic approach, VSD strives to produce technologies that align with enduring human values and actively contribute to the betterment of society.

5 Ethical, Social, and Technical Integration in VSD: Navigating Biorefinery Challenges

The concept of VSD is a notable approach that recognizes the complex interplay between ethical, social, and technical aspects in the growth of technological systems (Friedman and Hendry 2019). Particularly in the domain of biorefinery design, where the manufacture of bio-based goods involves intricate processes and choices, VSD becomes a critical consideration. Its interdisciplinary nature necessitates a thorough investigation that exceeds technical feasibility and delves into the ethical implications, social repercussions, and overall desirability for society (Friedman and Hendry 2019).

The biorefinery processes in transforming biomass into valuable products has a profound impact on ecological sustainability, efficient resource allocation, and community well-being (Fernando et al. 2006; Kromus et al. 2004). However, the development of such technologies involves trade-offs and complexities. VSD, with its focus on ethical considerations and societal viewpoints, provides a framework to navigate these complexities. It ensures that biorefinery design aligns with ethical standards, societal values, and the long-term well-being of human communities and the environment. Thus, VSD plays a crucial role in addressing the intricate challenges of biorefinery design by promoting a comprehensive and responsible approach.

6 VSD in Biorefinery Design

In the dynamic field of biorefinery design, the concept of sustainability is predominantly approached from an engineering perspective. The traditional approach involves defining sustainability through assessment, integration, and optimization methods, relying heavily on metrics that mainly measure global warming and energy efficiency impacts (Palmeros Parada et al. 2020). However, this limited focus on a select few indicators, aligned with engineering principles, fails to fully capture the complex nature of sustainability, as previously discussed. To truly design for sustainability, it is essential to incorporate diverse methodologies and tap into a range of disciplinary expertise (Palmeros Parada et al. 2020). An understanding of sustainability and its application in biorefinery design is essential to fully grasp its implications. It is crucial to have this broader perspective to fully comprehend the contextual implications of bio-based projects and to also recognize the diverse values of stakeholders that guide sustainability assessments (Asveld and Stemerding 2017). To successfully achieve this objective, it is crucial to adopt the practice of VSD, which takes into account the values of stakeholders within the unique technological environment. This offers a promising model for creating more environmentally acceptable biorefinery concepts, as demonstrated by the research of Ubando et al. (2020). Social and ethical considerations must be incorporated into the development of new technologies. The VSD methodology elucidates the seamless integration of human values into digital product design (Deng et al. 2016). The consideration of stakeholder values, both direct and indirect, in the design process is fundamental to this methodology (Cousins et al. 2019).

The process known as VSD employs a tripartite structure that consists of three discrete stages of inquiry: conceptual, empirical, and technological (Sadek et al. 2023; Umbrello and Van de Poel 2021). Stakeholder analysis is employed by VSD at the conceptual level to examine how ideas could affect direct and indirect stakeholders. This is a thorough investigation of the potential effects that distinct conceptions and principles may have on various persons or groups involved in the design process (Sadek et al. 2023; Umbrello and Van de Poel 2021). VSD conducts an extensive investigation that includes the collection of behavioral, contextual, experiential, and cognitive data. The goal of this empirical study is to identify and rank the concerns and values of the parties involved. VSD looks at the concrete aspects of user's experiences and behaviors to align the design process with the highest priorities of all parties concerned (Sadek et al. 2023; Umbrello and Van de Poel 2021). On the technological side, VSD delves deeply into self-reflection to determine whether the technology in use supports or undermines the values of the individuals engaged. This stage comprises a detailed analysis of how these values may be effectively integrated into emerging technologies. It aims to smoothly incorporate these principles, going beyond merely recognizing and understanding them (Sadek et al. 2023; Umbrello and Van de Poel 2021). A range of instruments and methods are used during the VSD process to identify and incorporate stakeholder values at various design phases. This guarantees the thorough and continuous incorporation of user values and ethical issues, enhancing the design process with a more complete grasp of the effects of technology on people and society.

7 Challenges and Prospects in the Integration of VSD-Making in the Biorefinery Design Process

VSD is a key factor in defining how biorefinery design decision-making processes are shaped since it addresses social and ethical issues related to sustainable technological advancement. However, there are several challenges in this field. Navigating the complex balance of differing values throughout the stakeholder community is a major challenge. As intricate systems, biorefineries engage a varied kind of stakeholders with different agendas, including local communities, lawmakers, business people, and environmentalists. It is a difficult task to negotiate and reconcile these opposing ideals to reach an agreement that honors the viewpoints of all parties. Applying VSD to the design of biorefineries presents a recurring challenge: determining the best balance between economic feasibility, environmental effects, and social repercussions.

An additional difficulty is the ever-changing landscape of biorefinery processes and the dynamic nature of technology. Continuous improvements in the biotechnological and engineering fields require that value-sensitive factors be continually re-evaluated and adjusted. New technologies bring about changes in the ethical implications, necessitating ongoing evaluations and revisions to design principles. The multidisciplinary character of biorefinery design adds to the complexity, necessitating efficient coordination and collaboration among specialists in several domains. To close the gap between the ethical concerns embedded in VSD and the technological aspects of biorefinery design, interdisciplinary collaboration is required.

In the future, mainstream adoption will provide a hurdle to the broad acceptance of VSD in biorefinery decision-making. A culture transformation inside companies and regulatory agencies is necessary to integrate ethical issues into the temperament of decision-making processes. Reducing resistance and inertia within existing systems will require raising knowledge of the benefits of VSD and fostering a dedication to ethical design standards. Standardized frameworks and rules are essential for controlling the use of VSD in biorefinery planning. Adopting a unified strategy will facilitate the incorporation of moral issues, expedite the decision-making process, and maintain uniformity and responsibility throughout the sector.

8 Conclusions

VSD has great potential to direct the moral decisions made during the design of biorefineries. However, there are challenges to overcome in relations of adjusting to new technology, promoting wider applications, and managing complicated stakeholder interests. Researchers, industry leaders, and legislators must work together to guarantee that biorefineries not only exceed technological and financial criteria but also represent and are in line with moral ideals and societal objectives to successfully overcome these obstacles. As the industry develops, addressing these issues is crucial to realizing VSD's full potential in forming an ethically rigorous and sustainable future for the advancement of biorefinery development.

References

Agrawal K, Verma P (2020) Production optimization of yellow laccase from Stropharia sp. ITCC 8422 and enzyme-mediated depolymerization and hydrolysis of lignocellulosic biomass for biorefinery application. Biomass Convers Biorefinery 12:1–20

Agrawal K, Bhardwaj N, Kumar B, Chaturvedi V, Verma P (2019) Process optimization, purification and characterization of alkaline stable white laccase from Myrothecium verrucaria ITCC-8447 and its application in delignification of agroresidues. Int J Biol Macromol 125:1042–1055

Agrawal K, Vaishnavi S, Verma P (2022) Laccase mediated green composite synthesis: a name synonymous with each other. In: Fungal biopolymers and biocomposites: prospects and avenues. Springer Nature Singapore, Singapore, pp 295–312

Alibardi L, Astrup TF, Asunis F, Clarke WP, De Gioannis G, Dessì P et al (2020) Organic waste biorefineries: looking towards implementation. Waste Manag 114:274–286

Amulya K, Reddy MV, Mohan SV (2014) Acidogenic spent wash valorization through polyhydroxyalkanoate (PHA) synthesis coupled with fermentative biohydrogen production. Bioresour Technol 158:336–342. https://doi.org/10.1016/j.biortech.2014.02.026

Amulya K, Dahiya S, Venkata Mohan S (2015) In: Prasad MNV (ed) Bioremediation and bioeconomy, edition: first, chapter: building a bio-based economy through waste remediation: innovation towards sustainable future (chapter 19). Elsevier, pp 497–521

Ashraf SA, Siddiqui AJ, Abd Elmoneim OE, Khan MI, Patel M, Alreshidi M et al (2021) Innovations in nanoscience for the sustainable development of food and agriculture with implications on health and environment. Sci Total Environ 768:144990

Asveld L, Stemerding D (2017) Social learning in the bioeconomy: the Ecover case. In: New perspectives on technology in society. Routledge, pp 103–124

Awasthi MK, Sarsaiya S, Patel A, Juneja A, Singh RP, Yan B et al (2020) Refining biomass residues for sustainable energy and bio-products: an assessment of technology, its importance, and strategic applications in circular bio-economy. Renew Sust Energ Rev 127:109876

Babiak K, Trendafilova S (2011) CSR and environmental responsibility: motives and pressures to adopt green management practices. Corp Soc Responsib Environ Manag 18(1):11–24

Banu J, Preethi Kavitha S, Tyagi VK et al (2021) LCB based biorefinery: a successful platform towards circular bioeconomy. *Fuel* 302(May):121086. https://doi.org/10.1016/j.fuel.2021.121086

Berg C (2001) World bioethanol production, the distillery and bioethanol network. *línea:* http://www.distill.com/worldethanolproduction.htm (Mayo 2012)

Bhardwaj N, Kumar B, Agarwal K, Chaturvedi V, Verma P (2019) Purification and characterization of a thermo-acid/alkali stable xylanases from aspergillus oryzae LC1 and its application in Xylo-oligosaccharides production from lignocellulosic agricultural wastes. Int J Biol Macromol 122:1191–1202

Bhardwaj N, Kumar B, Agrawal K, Verma P (2020) Bioconversion of rice straw by synergistic effect of in-house produced ligno-hemicellulolytic enzymes for enhanced bioethanol production. Bioresour Technol Rep 10:100352

Blau J (2017) The Paris agreement: climate change, solidarity, and human rights. Springer

Brito GM, Cipriano DF, Schettino MA Jr, Cunha AG, Coelho ERC, Freitas JCC (2019) One-step methodology for preparing physically activated biocarbons from agricultural biomass waste. J Environ Chem Eng 7(3):103113

Brunschwig C, Moussavou W, Blin J (2012) Use of bioethanol for biodiesel production. Prog Energy Combust Sci 38(2):283–301

Chandra R, Iqbal HMN, Vishal G, Lee HS, Nagra S (2019) Algal biorefinery: A sustainable approach to valorize algal-based biomass towards multiple product recovery. Bioresour Technol 278:346–359. https://doi.org/10.1016/j.biortech.2019.01.104

Chen L, Hu Y, Wang R, Li X, Chen Z, Hua J et al (2023) Green building practices to integrate renewable energy in the construction sector: a review. Environ Chem Lett:1–34

Cheng H, Wang L (2013) Lignocelluloses feedstock biorefinery as petrorefinery substitutes. Biomass now–sustainable growth and use, 347–388

Chopra J, Rangarajan V, Sen R (2022) Recent developments in oleaginous yeast feedstock based biorefinery for production and life cycle assessment of biofuels and value-added products. Sustain Energy Technol Assess 53:102621

Clark J, Deswarte F (2015) The biorefinery concept: an integrated approach. Introduction to chemicals from biomass, 1–29

Cousins K, Subramanian H, Esmaeilzadeh P (2019) A value-sensitive design perspective of cryptocurrencies: a research agenda. Commun Assoc Inf Syst 45(1):27

Dahiya S, Kumar AN, Shanthi Sravan J, Chatterjee S, Sarkar O, Mohan SV (2018) Food waste biorefinery: sustainable strategy for circular bioeconomy. Bioresour Technol 248(Pt A):2–12. https://doi.org/10.1016/j.biortech.2017.07.176

Demirbas A (2007) Progress and recent trends in biofuels. Prog Energy Combust Sci 33(1):1–18

Deng X, Joshi KD, Galliers RD (2016) The duality of empowerment and marginalization in microtask crowdsourcing. MIS Q 40(2):279–302

Dennis MJ (2021) Digital Well-being under pandemic conditions: catalysing a theory of online flourishing. Ethics Inf Technol 23(3):435–445

Elmekawy A, Diels L, De Wever H, Pant D (2013) Valorization of cereal based biorefinery byproducts: reality and expectations. Environ Sci Technol 47(16):9014–9027. https://doi.org/10.1021/es402395g

Fal S, Smouni A, Arroussi HE (2023) Integrated microalgae-based biorefinery for wastewater treatment, industrial CO_2 sequestration and microalgal biomass valorization: A circular bioeconomy approach. Environ Adv 12:100365

Fava F, Totaro G, Diels L, Reis M, Duarte J, Carioca OB, Poggi-Varaldo HM, Ferreira BS (2015) Biowaste biorefinery in Europe: opportunities and research & development needs. New Biotechnol 32(1):100–108. https://doi.org/10.1016/j.nbt.2013.11.003

Fernando S, Adhikari S, Chandrapal C, Murali N (2006) Biorefineries: current status, challenges, and future direction. Energy Fuel 20(4):1727–1737

Fitz Patrick M, Champagne P, Cunningham MF, Whitney RA (2010) A biorefinery processing perspective: treatment of lignocellulosic materials for the production of value-added products. Bioresour Technol 101(23):8915–8922

Franklin D, Andrews J (2012) The world in 2050. The Economist, London, UK

Friedman B, Hendry DG (2019) Value sensitive design: shaping technology with moral imagination. Mit Press

Friedman B, Kahn PH Jr (2003) Human values, ethics, and design. In: Jacko JA, Sears A (eds) The human-computer interaction handbook. Erlbaum, Mahwah, pp 1177–1201

Friedman B, Kahn PH, Borning A, Huldtgren A (2013) Value sensitive design and information systems. In: Doorn N, Schuurbiers D, van de Poel I, Gorman M (eds) Early engagement and new technologies: opening up the laboratory. Philosophy of engineering and technology, vol 16. Springer, Dordrecht, pp 55–95. https://doi.org/10.1007/978-94-007-7844-3_4

Galbe M, Sassner P, Wingren A, Zacchi G (2007) Process engineering economics of bioethanol production. Biofuels:303–327

Gavrilescu M (2014) Biomass potential for sustainable environment, biorefinery products and energy. In: Sustainable energy in the built environment-steps towards nZEB: proceedings of the conference for sustainable energy (CSE) 2014. Springer International Publishing, pp 169–194

Ghatak HR (2011) Biorefineries from the perspective of sustainability: feedstocks, products, and processes. Renew Sust Energ Rev 15(8):4042–4052

Gómez JA, Matallana LG, Sánchez ÓJ (2022) Towards a biorefinery processing waste from plantain agro-industry: process design and techno-economic assessment of single-cell protein, natural fibers, and biomethane production through process simulation. Fermentation 8:582

Hoang Nguyen Tran P, Ko JK, Gong G, Um Y, Lee SM (2020) Improved simultaneous co-fermentation of glucose and xylose by *Saccharomyces cerevisiae* for efficient lignocellulosic biorefinery. Biotechnol Biofuels 13:12. https://doi.org/10.1186/s13068-019-1641-2

Hunt AJ, Sin EH, Marriott R, Clark JH (2010) Generation, capture, and utilization of industrial carbon dioxide. ChemSusChem 3(3):306–322

Iversen N, Cawthorne D (2023) Ethics in action: envisioning human values in the early stages of drone development. Social Robots in Social Institutions: Proceedings of Robophilosophy 2022, 366, 445

Julio R, Albet J, Vialle C, Vaca-Garcia C, Sablayrolles C (2017) Sustainable design of biorefinery processes: existing practices and new methodology. Biofuels Bioprod Biorefin 11(2):373–395

Kamm B, Kamm MJAM (2004) Principles of biorefineries. Appl Microbiol Biotechnol 64:137–145

Kamm B, Kamm M (2007) Biorefineries–multi product processes. White Biotechnology, 175–204

Kaur R, Krishna BB, Kumar A, Biswas B, Bhaskar T (2022) Strategy and Design of Innovation Policy Road Mapping for waste biorefineries. In: Handbook of waste biorefinery: circular economy of renewable energy. Springer International Publishing, Cham, pp 833–858

Kawa-Rygielska J, Pietrzak W, Lennartsson PR (2022) High-efficiency conversion of bread residues to ethanol and edible biomass using filamentous fungi at high solids loading: A biorefinery approach. Appl Sci 12:6405

Ko JK, Lee JH, Jung JH et al (2020) Recent advances and future directions in plant and yeast engineering to improve lignocellulosic biofuel production. Renew Sust Energ Rev 134:110390. https://doi.org/10.1016/j.rser.2020.110390

Kromus S, Wachter B, Koschuh W, Mandl M, Krotscheck C, Narodoslawsky M (2004) The green biorefinery Austria-development of an integrated system for green biomass utilization. Chem Biochem Eng Q 18(1):8–12

Kumar B, Verma P (2021) Biomass-based biorefineries: an important architype towards a circular economy. Fuel 288:119622

Kumar B, Bhardwaj N, Alam A, Agrawal K, Prasad H, Verma P (2018) Production, purification and characterization of an acid/alkali and thermo tolerant cellulase from Schizophyllum commune NAIMCC-F-03379 and its application in hydrolysis of lignocellulosic wastes. AMB Express 8:1–16

Kumar B, Bhardwaj N, Agrawal K, Chaturvedi V, Verma P (2020) Current perspective on pretreatment technologies using lignocellulosic biomass: an emerging biorefinery concept. Fuel Process Technol 199:106244

Kumar S, Thakur N, Singh AK, Gudade BA, Ghimire D, Das S (2022) Microbes-assisted phytoremediation of contaminated environment: global status, progress, challenges, and future prospects. Phytoremediation technology for the removal of heavy metals and other contaminants from soil and water, 555–570

Lipinski TA, Britz JJ (2000) Rethinking the ownership of information in the 21st century: ethical implications. Ethics Inf Technol 2(1):49–71

Lopolito A, Prosperi M, Sisto R, Pazienza P (2011) Economic and social implications of the industrial use of renewable raw materials. Renewable raw materials: new feedstocks for the chemical industry, 143–168

Majidian P, Tabatabaei M, Zeinolabedini M et al (2018) Metabolic engineering of microorganisms for biofuel production. Renew Sust Energ Rev 82:3863–3885. https://doi.org/10.1016/j.rser.2017.10.085

Manzanares P (2020) The role of biorefinering research in the development of a modern bioeconomy. Acta Innov 37:47–56

Martinez-Hernandez E, Campbell G, Sadhukhan J (2013) Economic value and environmental impact (EVEI) analysis of biorefinery systems. Chem Eng Res Des 91(8):1418–1426. https://doi.org/10.1016/j.cherd.2013.02.025

Mongkhonsiri G, Gani R, Malakul P et al (2018) Integration of the biorefinery concept for the development of sus-tainable processes for pulp and paper industry. Comput Chem Eng 119:70–84. https://doi.org/10.1016/j.compchemeng.2018.07.019

Mossberg J, Söderholm P, Frishammar J (2021) Challenges of sustainable industrial transformation: S wedish biorefinery development and incumbents in the emerging biofuels industry. Biofuels Bioprod Biorefin 15(5):1264–1280

Nair LG, Agrawal K, Verma P (2022a) An insight into the principles of lignocellulosic biomass-based zero-waste biorefineries: a green leap towards imperishable energy-based future. Biotechnol Genet Eng Rev 38(2):288–338

Nair LG, Agrawal K, Verma P (2022b) An overview of sustainable approaches for bioenergy production from agro-industrial wastes. Energy Nexus 6:100086

Oliva JM, Negro MJ, Álvarez C et al (2020) Fermentation strategies for the efficient use of olive tree pruning biomass from a flexible biorefinery approach. Fuel 277:118171. https://doi.org/10.1016/j.fuel.2020.118171

Paladino O, Neviani M (2024) Interchangeable modular design and operation of photo-bioreactors for Chlorella vulgaris cultivation towards a zero-waste biorefinery. Enzym Microb Technol 173:110371

Palmeros Parada M, Asveld L, Osseweijer P, Posada JA (2020) Integrating value considerations in the decision making for the design of biorefineries. Sci Eng Ethics 26(6):2927–2955

Parada MP, Kehrein P, Xevgenos D, Asveld L, Osseweijer P (2022) Societal values, tensions and uncertainties in resource recovery from wastewaters. J Environ Manag 319:115759

Paul T, Sinharoy A, Baskaran D, Pakshirajan K, Pugazhenth G, Lens PLL (2022) Bio-oil production from oleaginous microorganisms using hydrothermal liquefaction: A biorefinery approach. Crit Rev Environ Sci Technol 52:356–394

Ragauskas AJ, Williams CK, Davison BH, Britovsek G, Cairney J, Eckert CA et al (2006) The path forward for biofuels and biomaterials. Science 311(5760):484–489

Rodríguez AG, Rodrigues MDS, Sotomayor Echenique O (2019) Towards a sustainable bioeconomy in Latin America and the Caribbean: elements for a regional vision

Sadek M, Calvo RA, Mougenot C (2023) Designing value-sensitive AI: a critical review and recommendations for socio-technical design processes. AI and Ethics, 1–19

Sadhukhan J, Martinez-Hernandez E, Murphy RJ, Ng DK, Hassim MH, Ng KS et al (2018) Role of bioenergy, biorefinery and bioeconomy in sustainable development: strategic pathways for Malaysia. Renew Sust Energ Rev 81:1966–1987

Saini JK, Gupta R, Hemansi V, Gaur PA, Saini R, ... Kuhad RC (2019) Integrated lignocellulosic biorefinery for sustainable bio-based economy. Sustainable Approaches for Biofuels Production technologies: from current status to practical implementation, 25–46

Santamaría-Fernández M, Lübeck M (2020) Production of leaf protein concentrates in green biorefineries as alternative feed for monogastric animals. Anim Feed Sci Technol 268:114605

Sauer M, Steiger M, Mattanovich D, Marx H (2014) Biorefineries—concepts for sustainability. Bioprocessing of renewable resources to commodity bioproducts, 1–27

Sawyer S, Rosenbaum H (2000) Social informatics in the information sciences: current activities and emerging direction. Inf Sci 3(2):89–95

Sekoai PT, Ghimire A, Ezeokoli OT, Rao S, Ngan WY, Habimana O, Yao Y, Yang P, Fung AHY, Yoro KO et al (2021) Valorization of volatile fatty acids from the dark fermentation waste Streams-A promising pathway for a biorefinery concept. Renew Sust Energ Rev 143:110971

Sekoai PT, Chunilall V, Ezeokoli O (2023) Creating value from Acidogenic biohydrogen fermentation effluents: an innovative approach for a circular bioeconomy that is acquired via a microbial biorefinery-based framework. Fermentation 9:602

Singh A, Christensen T, Panoutsou C (2021) Policy review for biomass value chains in the European bioeconomy. Glob Transit 3:13–42

Spagnuolo M, Yaguchi A, Blenner M (2019) Oleaginous yeast for biofuel and oleochemical production. Curr Opin Biotechnol 57:73–81. https://doi.org/10.1016/j.copbio.2019.02.011

Thongchul N, Songserm P, Ngaosuwan K (2022) Whole crop feedstocks in biorefinery: A common classification. In: AZ of biorefinery. Elsevier, pp 35–77

Ubando AT, Felix CB, Chen WH (2020) Biorefineries in circular bioeconomy: A comprehensive review. Bioresour Technol 299:122585

Uihlein A, Schebek L (2009) Environmental impacts of a lignocellulose feedstock biorefinery system: an assessment. Biomass Bioenergy 33(5):793–802

Umbrello S, Van de Poel I (2021) Mapping value sensitive design onto AI for social good principles. AI and Ethics 1(3):283–296

Van Dyne DL, Blase MG, Clements LD (1999) A strategy for returning agriculture and rural America to long-term full employment using biomass refineries. In: Perspectives on new crops and new uses. ASHS Press, Alexandria, Va, pp 114–123

Venkata Mohan S, Sarkar O (2017) Waste to biohydrogen: addressing sustainability with biorefinery. In: Raghavan K, Ghosh P (eds) Energy engineering. Springer, Singapore

Venkata Mohan S, Nikhil GN, Chiranjeevi P, Nagendranatha Reddy C, Rohit MV, Kumar AN, Sarkar O (2016) Waste biorefinery models towards sustainable circular bioeconomy: critical review and future perspectives. Bioresour Technol 215:2–12. https://doi.org/10.1016/j.biortech.2016.03.130

Visco A, Scolaro C, Facchin M, Brahimi S, Belhamdi H, Gatto V, Beghetto V (2022) Agri-food wastes for bioplastics: European prospective on possible applications in their second life for a circular economy. Polymers 14(13):2752

Wagh MS, Sowjanya S, Nath PC, Chakraborty A, Amrit R, Mishra B et al (2024) Valorisation of agro-industrial wastes: circular bioeconomy and biorefinery process–a sustainable symphony, vol 183. Process Saf Environ Prot, p 708

Xu Y, Wang P, Xue S, Kong F, Ren H, Zhai H (2020) Green biorefinery—the ultra-high hydrolysis rate and behavior of *Populus tomentosa* hemicellulose autohydrolysis under moderate subcritical water conditions. RSC Adv 10(32):18908–18917. https://doi.org/10.1039/d0ra02350g

Yakaboylu GA, Jiang C, Yumak T, Zondlo JW, Wang J, Sabolsky EM (2021) Engineered hierarchical porous carbons for supercapacitor applications through chemical pretreatment and activation of biomass precursors. Renew Energy 163:276–287

York R, Rosa EA, Dietz T (2003) STIRPAT, IPAT and ImPACT: analytic tools for unpacking the driving forces of environmental impacts. Ecol Econ 46(3):351–365

Yoro KO, Daramola MO (2020) CO2 emission sources, greenhouse gases, and the global warming effect. In: Advances in carbon capture. Woodhead Publishing, pp 3–28

Zhang MM, Xiong L, Tang YJ, Mehmood MA, Zhao ZK, Bai FW, Zhao XQ (2019) Enhanced acetic acid stress tolerance and ethanol production in *Saccharomyces cerevisiae* by modulating expression of the de novo purine biosynthesis genes. Biotechnol Biofuels 12:116. https://doi.org/10.1186/s13068-019-1456-1

Correction to: An Overview of Fungal Enzymes in the Generation of Value-Added Product of Industrial Significance

Deepshikha, Pradeep Verma, and Komal Agrawal

Correction to:
Chapter 4 in: K. Agrawal & P. Verma (eds.), *Biotechnological Advances in Biorefinery*, https://doi.org/10.1007/978-981-97-5544-8_4

The first author's name "Shreya Sharma" in this chapter was incorrect in the initially published version. It has been corrected to "Deepshikha," which is the correct name of the first author.

The updated version of this chapter can be found at
https://doi.org/10.1007/978-981-97-5544-8_4

K. Agrawal, P. Verma (eds.), *Biotechnological Advances in Biorefinery*, Interdisciplinary Biotechnological Advances,
https://doi.org/10.1007/978-981-97-5544-8_18

Index

K. Agrawal, P. Verma (eds.), *Biotechnological Advances in Biorefinery*, Interdisciplinary Biotechnological Advances,
https://doi.org/10.1007/978-981-97-5544-8

MIX
Papier aus verantwortungsvollen Quellen
Paper from responsible sources
FSC® C105338

If you have any concerns about our products,
you can contact us on
ProductSafety@springernature.com

In case Publisher is established outside the EU,
the EU authorized representative is:
Springer Nature Customer Service Center GmbH
Europaplatz 3, 69115 Heidelberg, Germany

Printed by Libri Plureos GmbH
in Hamburg, Germany